CATALYSIS OF ORGANIC REACTIONS

CHEMICAL INDUSTRIES

A Series of Reference Books and Textbooks

Consulting Editor

HEINZ HEINEMANN

ADDITIONAL VOLUMES IN PREPARATION

CATALYSIS OF ORGANIC REACTIONS

edited by

Dennis G. Morrell

Dupont Company
Wilmington, Delaware, U.S.A

CRC Press
Taylor & Francis Group
Boca Raton London New York

CRC Press is an imprint of the
Taylor & Francis Group, an **informa** business

First published 2003 by Marcel Dekker, Inc.

Published 2019 by CRC Press
Taylor & Francis Group
6000 Broken Sound Parkway NW, Suite 300
Boca Raton, FL 33487-2742

First issued in paperback 2019

No claim to original U.S. Government works

ISBN 13: 978-0-367-44694-9 (pbk)
ISBN 13: 978-0-8247-4132-7 (hbk)

This book contains information obtained from authentic and highly regarded sources. Reasonable efforts have been made to publish reliable data and information, but the author and publisher cannot assume responsibility for the validity of all materials or the consequences of their use. The authors and publishers have attempted to trace the copyright holders of all material reproduced in this publication and apologize to copyright holders if permission to publish in this form has not been obtained. If any copyright material has not been acknowledged please write and let us know so we may rectify in any future reprint.

Visit the Taylor & Francis Web site at
http://www.taylorandfrancis.com

and the CRC Press Web site at
http://www.crcpress.com

Library of Congress Cataloging-in-Publication Data
A catalog record for this book is available from the Library of Congress.

Preface

Catalysis of Organic Reactions is a compendium of sixty technical papers and poster summaries presented at the 19th Conference on Catalysis of Organic Reactions in San Antonio, Texas, in April 2002. The conference is sponsored biennially by the Organic Reaction Catalysis Society (ORCS), an affiliate of the North America Catalysis Society. Since the first conference held in 1966, ORCS has provided a technical forum for scientists and engineers from academia and the chemical, pharmaceutical, and catalyst manufacturing industries to present results of recent developments in applications of catalysis to organic synthesis. In addition to contributing scientists and engineers, graduate students and postdoctoral fellows had the opportunity to present current work in catalysis.

These proceedings from the 19th Conference represent the work of over 150 scientists from fourteen countries. The proceedings are organized into the following topics: immobilized and supported catalysts, practical aspects of catalysis, solid acid and base catalysis, heterogeneous catalysis and commercial processes, Raney-type and base metal catalysis, stereoselective catalysis, and hydrogenation and amination.

ORCS and the 19th Conference recognize the financial support of the following industrial sponsors: Crompton Corporation, CRI Catalyst; Degussa Corporation; Engelhard Corporation; Johnson Matthey; Merck Corporation; Nova Molecular; Precious Metals/Activated Metals Corporation; Süd-Chemie; Upjohn, Aventis; and W. R. Grace. The Parr Instrument Company has been a continuing, significant supporter of ORCS. ORCS is also grateful for financial support from the North American Catalyst Society that increased our student and postdoctoral fellow travel scholarships. The American Chemical Society Petroleum Research Fund provided travel grants for academicians from foreign countries.

Plaudits go to all contributors for their efforts in the preparation and presentation of the technical papers and posters and to the session chairpersons for their introductions. Over fifty scientists and engineers refereed the submitted manuscripts, and members of the Editorial Board worked diligently and used their expertise to assure a high quality of technical contributions. The Editorial Board members for the 19th Conference are: Mike Ford, Brian James, Tom Johnson, Sanjay Malhotra, Mike McGuire, Yongkui Sun, Frank Herkes, and John Sowa. Members of the ORCS Executive Committee, including Norman Colby, Robert DeVries, Mike Ford, Frank Herkes, Jeffrey Herring, Steve Jacobson, Brian James, Vic Mylroie, and John Sowa, Jr., provided encouragement and helped direct the course of this conference.

I would especially like to acknowledge the continued support for my ORCS activities by Barbara K. Morrell. Her encouragement and final editing of the manuscripts proved invaluable.

Dennis G. Morrell
Hockessin, Delaware
www.ORCS.org

Acknowledgement

The Editorial Board wishes to acknowledge and thank the following individuals for their generous support in refereeing the manuscripts for the oral and poster presentations.

Michael Amiridis, University of Southern California

Juan Arhancet, Monsanto

Bob Augustine, Seton Hall University

Michael Baird, Queen's University

David G. Bergbrieter, Texas A & M

Clemens Brechtelsbauer, Glaxo SmithKline

Richard Carr, Air Products and Chemicals

Jianping Chen, Engelhard

William Clark, Glaxo SmithKline

George Cooper, Air Products and Chemicals

Cathleen Crudden, Univ. of New Brunswick

Ende David, Pfizer

Mark Davis, California Institute of Technology

Johannes De Vries, DSM, the Netherlands

Phillip Dell'Orco, Glaxo SmithKline

Robert DeVries, Dow

Deryn Fogg, University of Ottawa

Pierre Gallezot, CNRS France

Richard Goodman, Glaxo SmithKline

Kathy Hayes, Air Products and Chemicals

J. Richard Heys, Glaxo SmithKline

Johannes Hieble, Glaxo SmithKline

Wolfgang Holderich, Univ. of Tech., RWTH Aachen

Steve Jacobson, DuPont

Peter Johnson, Johnson Matthey

Ray Jowett, Dow

Anthony King, Merck

John Knifton, Shell

Manoj Koranne, W. R. Grace

Kevin Lassila, Air Products and Chemicals

Simon Leung, Bristol-Myers Squibb

Tamas Mallat, Swiss Federal Inst. of Technol.

Hayao Matsuhashi, Glaxo SmithKline

Steve McDonald, Solutia

Wilford Mendelson, Glaxo SmithKline

Dave Morgenstern, Monsanto

Luis Oro, University of Zaragosa

Walter Partenheimer, DuPont

Lou Pignolet, University of Minnesota

Terry Ranken, Huntsman

Steve Schmidt, W. R. Grace

Scott Schultz, Merck

Roger Sheldon, University of Delft

Art Shu, Glaxo SmithKline

Utpal Singh, Merck
Ambarish Singh, Bristol-Myers Squibb
P. Grant Spoors, Glaxo SmithKline
Jeff Stryker, University of Alberta
William Tamblyn, Johnson Matthey
Setrak Tanielyan, Seton Hall University
David G. Tellers, Merck
Piet Van Leeuwan, University of Amsterdam

Francis Waller, Air Products and Chemicals
Jian Wang, Merck
David G. Ward, W. R. Grace
Hengxu Wei, Glaxo SmithKline
James White, Engelhard
Frederick Wilhem, Air Products and Chemicals
Hua Zhao, New Jersey Institute of Technology

Editorial Board

Contents

Contributors

Mohamed Abid   Université Louis Pasteur, Strasbourg, France

Stephen Anderson   Center for Applied Catalysis, Seton Hall University, South Orange, New Jersey, U.S.A.

Arthur T. Andrews   Merck Research Laboratories, Rahway, New Jersey, U.S.A.

Robert Appell   The Dow Chemical Company, Midland, Michigan, U.S.A.

Isabel W. C. E. Arends   Delft University of Technology, Delft, the Netherlands

J. N. Armor   Air Products and Chemicals, Allentown, Pennsylvania, U.S.A.

Bruce A. Arndtsen   McGill University, Montreal, Quebec, Canada

Venu Arunajatesan   University of Kansas, Lawrence, Kansas, U.S.A.

Robert L. Augustine   Center for Applied Catalysis, Seton Hall University, South Orange, New Jersey, U.S.A.

J. Aumo   Process Chemistry Group, Åbo Akademi University, Åbo/Turku, Finland

James E. Babin   The Dow Chemical Company, South Charleston, West Virginia, U.S.A.

Henrik Backman   Åbo Akademi University, Åbo/Turku, Finland

V. A. Barkhash   Novosibirsk Institute of Organic Chemistry, Novosibirsk, Russia

C. F. J. Barnard   Johnson Matthey, Reading, Berkshire, United Kingdom

Bruce A. Barner   The Dow Chemical Company, South Charleston, West Virginia, U.S.A.

S. Bennett   Johnson Matthey Technology Centre, Reading, Berkshire, United Kingdom

Ursula Bentrup   Institut für Angewandte Chemie Berlin-Adlershof e.V., Berlin, Germany

David E. Bergbreiter   Texas A&M University, College Station, Texas, U.S.A.

I. Bertóti   Research Laboratory of Materials and Environmental Chemistry, Hungarian Academy of Sciences, Budapest, Hungary

Monika Berweiler   Catalysts and Initiators Division, Degussa AG, Hanau, Germany

Virginie Bette   University of Lille, Villeneuve d'Ascq, France

I. Bitter   Budapest University of Technology and Economics, Budapest, Hungary

M. Bolognini Dipartimento di Chimica Industriale e dei Materiali, Università di Bologna, Bologna, Italy

I. Borbáth Chemical Research Center, Institute of Chemistry, Hungarian Academy of Sciences, Budapest, Hungary

Abdillah O. Bouh University of Ottawa, Ottawa, Ontario, Canada

Jodie Brice Department of Chemistry and Biochemistry, Seton Hall University, South Orange, New Jersey, U.S.A.

John R. Briggs The Dow Chemical Company, South Charleston, West Virginia, U.S.A.

Robert Brinkman* Department of Chemistry and Biochemistry, Seton Hall University, South Orange, New Jersey, U.S.A.

M. Campanati Dipartimento di Chimica Industriale e dei Materiali, INSTM, Udr. di Bologna, Bologna, Italy

Jean-François Carpentier† University of Lille, Villeneuve d'Ascq, France

Jason Carson Degussa Corporation, Calvert City, Kentucky, U.S.A.

F. Cavani Dipartimento di Chimica Industriale e dei Materiali, Università di Bologna, Bologna, Italy

Sandrine Chatel Delft University of Technology, Delft, the Netherlands

Baoshu Chen Degussa Corporation, Calvert City, Kentucky, U.S.A.

Bei Chen The University of Akron, Akron, Ohio, U.S.A.

J. P. Chen Chemical Catalysts Research and Development Center, Engelhard Corporation, Beachwood, Ohio, U.S.A.

Yawu Chi The University of Akron, Akron, Ohio, U.S.A.

Steven S. C. Chuang The University of Akron, Akron, Ohio, U.S.A.

C. L. Clayton Department of Chemical Engineering, University of Cambridge, Cambridge, United Kingdom

Christopher J. Cobley Chirotech Technology Ltd., Cambridge, United Kingdom

S. Collard Johnson Matthey, Royston, Hertforshire, United Kingdom

R. J. Cross Department of Chemistry, The University of Glasgow, Glasgow, Scotland

Cathleen M. Crudden University of New Brunswick, Fredericton, New Brunswick, Canada

Pei-Shing E. Dai Texaco Global Products, The Woodlands, Texas, U.S.A.

Rania D. Dghaym McGill University, Montreal, Quebec, Canada

Rajiv Dhawan McGill University, Montreal, Quebec, Canada

M. P. Dolgikh Novosibirsk Institute of Organic Chemistry, Novosibirsk, Russia

Mark Duda CREAVIS Gesellschaft für Technologie und Innovation mbH, Marl, Germany

Jerry R. Ebner Retired from Monsanto Company, St. Louis, Missouri, U.S.A.

S. H. Elgafi Johnson Matthey Technology Centre, Reading, Berkshire, United Kingdom

---

* *Current affiliation*: Albany Molecular Research, Inc., Albany, New York, U.S.A.
† *Current affiliation*: University of Rennes, Rennes, France.

Kari Eränen Åbo Akademi University, Process Chemistry Group, Åbo/Turku, Finland

Meredith Fairgrieve University of New Brunswick, Fredericton, New Brunswick, Canada

C. Felloni Dipartimento di Chimica Industriale e dei Materiali, Università di Bologna, Bologna, Italy

C. Flego EniTecnologie SpA, S. Donato Milanese (MI), Italy

V. V. Fomenko Novosibirsk Institute of Organic Chemistry, Novosibirsk, Russia

M. E. Ford Air Products and Chemicals, Allentown, Pennsylvania, U.S.A.

S. Franceschini Dipartimento di Chimica Industriale e dei Materiali, INSTM, Udr. di Bologna, Bologna, Italy

Yujing Gao Center for Applied Catalysis, Seton Hall University, South Orange, New Jersey, U.S.A.

Andreas Geisselmann Degussa AG, Hanau, Germany

Jaime Gibson Degussa Corporation, Calvert City, Kentucky, U.S.A.

L. F. Gladden Department of Chemical Engineering, University of Cambridge, Cambridge, United Kingdom

S. Göbölös Chemical Research Center, Institute of Chemistry, Hungarian Academy of Sciences, Budapest, Hungary

Pradeep Goel Center for Applied Catalysis, Seton Hall University, South Orange, New Jersey, U.S.A.

Christian Goralski The Dow Chemical Company, Midland, Michigan, U.S.A.

Frank P. Gortsema Merck Research Laboratories, Rahway, New Jersey, U.S.A.

K. G. Griffin Johnson Matthey plc, Royston, Hertfordshire, United Kingdom

Steven M. Hannick Abbott Laboratories, Abbott Park, Illinois, U.S.A.

Azfar Hassan University of Ottawa, Ottawa, Ontario, Canada

Ralf Hausmann Degussa AG, Hanau, Germany

S. Hawker Johnson Matthey, Royston, Hertfordshire, United Kingdom

David Hawn The Dow Chemical Company, Midland, Michigan, U.S.A.

Stuart Hayden* Merck & Co., Inc., Rahway, New Jersey, U.S.A.

M. Hegedüs Chemical Research Center, Institute of Chemistry, Hungarian Academy of Sciences, Budapest, Hungary

R. G. Heidenreich Technische Universität München, Garching, Germany

G. R. Henderson Johnson Matthey Technology Centre, Reading, Berkshire, United Kingdom

Frank E. Herkes The DuPont Company, Wilmington, Delaware, U.S.A.

Matti Hotokka Process Chemistry Group, Åbo Akademi University, Åbo/Turku, Finland

Keith W. Hutchenson The DuPont Company, Wilmington, Delaware, U.S.A.

S. D. Jackson Department of Chemistry, The University of Glasgow, Glasgow, Scotland; R&T Group, Synetix, Billingham, Cleveland, United Kingdom

---

* Current affiliation: The R. W. Johnson Pharmaceutical Research Institute, Raritan, New Jersey, U.S.A.

Brian R. James Department of Chemistry, University of British Columbia, Vancouver, British Columbia, Canada

James E. Jerome Chattem Chemicals, Inc., Chattanooga, Tennessee, U.S.A.

P. Johnston Johnson Matthey plc, Royston, Hertfordshire, United Kingdom

J. R. Jones Department of Chemistry, University of Surrey, Guildford, United Kingdom

Carsten Jost CREAVIS Gesellschaft für Technologie und Innovation mbH, Marl, Germany

S. Kaliq Johnson Matthey, Reading, Berkshire, United Kingdom

Mark A. Keane University of Kentucky, Lexington, Kentucky, U.S.A.

Fredrik Klingstedt Åbo Akademi University, Åbo/Turku, Finland

John F. Knifton Shell Global Solutions, Houston, Texas, U.S.A.

Jason Kodah* Department of Chemistry and Biochemistry, Seton Hall University, South Orange, New Jersey, U.S.A.

K. Köhler Technische Universität München, Garching, Germany

D. V. Korchagina Novosibirsk Institute of Organic Chemistry, Novosibirsk, Russia

J. G. E. Krauter Catalysts and Initiators Division, Degussa AG, Hanau, Germany

Adolf Kühnle CREAVIS Gesellschaft für Technologie und Innovation mbH, Marl, Germany

Narendra Kumar Åbo Akademi University, Åbo/Turku, Finland

J. Kuusisto Process Chemistry Group, Åbo Akademi University, Åbo/Turku, Finland

Matthew D. Le Page Department of Chemistry, University of British Columbia, Vancouver, British Columbia, Canada

Carl LeBlond Merck Research Laboratories, Rahway, New Jersey, U.S.A.

Juny Lee New Jersey Institute of Technology, Newark, New Jersey, U.S.A.

Ian C. Lennon Chirotech Technology Ltd., Cambridge, United Kingdom

Chunmei Li Texas A&M University, College Station, Texas, U.S.A.

Minna Lindroos Åbo Akademi University, Åbo/Turku, Finland

Bernhard Lücke Institut für Angewandte Chemie Berlin-Adlershof e.V., Berlin, Germany

Liyan Ma School of Chemical Engineering and Industrial Chemistry, The University of New South Wales, Sydney, Australia

Nagendranath Mahata Chemical Research Center, Institute of Chemistry, Hungarian Academy of Sciences, Budapest, Hungary; Center for Applied Catalysis, Seton Hall University, South Orange, New Jersey, U.S.A.

Päivi Mäki-Arvela Process Chemistry Group, Åbo Akademi University, Åbo/Turku, Finland

Sanjay V. Malhotra Department of Chemical and Environmental Science, New Jersey Institute of Technology, Newark, New Jersey, U.S.A.

Current affiliation: The R. W. Johnson Pharmaceutical Research Institue, Raritan, New Jersey, U.S.A.

Russell E. Malz, Jr.   Uniroyal Chemical Division, Crompton Corporation, Naugatuck, Connecticut, U.S.A.

J. L. Margitfalvi   Chemical Research Center, Institute of Chemistry, Hungarian Academy of Sciences, Budapest, Hungary

Andreas Martin   Institut für Angewandte Chemie Berlin-Adlershof e.V., Berlin, Germany

Raymond McCague   Chirotech Technology Ltd., Cambridge, United Kingdom

Robert J. McNair   Johnson Matthey, West Deptford, New Jersey, U.S.A.

R. Mezzogori   Dipartimento di Chimica Industriale e dei Materiali, Università di Bologna, Bologna, Italy

J. P. Mikkola   Process Chemistry Group, Åbo Akademi University, Åbo/Turku, Finland

V. Morawsky   Institute of Technical Chemistry, Technical University of Braunschweig, Braunschweig, Germany

Howard E. Morton   Abbott Laboratories, Abbott Park, Illinois, U.S.A.

André Mortreux   University of Lille, Villeneuve d'Ascq, France

Dmitry Yu. Murzin   Process Chemistry Group, Åbo Akademi University, Åbo/Turku, Finland

Nabin K. Nag   Engelhard Corporation, Beachwood, Ohio, U.S.A.

Jayesh J. Nair   Center for Applied Catalysis, Seton Hall University, South Orange, New Jersey, U.S.A.

Ahmad Kalantar Neyestanaki   Åbo Akademi University, Åbo/Turku, Finland

Flora T. T. Ng   Department of Chemical Engineering, University of Waterloo, Waterloo, Ontario, Canada

Ville Nieminen   Process Chemistry Group, Åbo Akademi University, Åbo/Turku, Finland

Tapio Ollonqvist   University of Turku, Turku, Finland

Daniel J. Ostgard   Catalysts and Initiators Division, Degussa AG, Hanau, Germany

Juha Päivärinta   Process Chemistry Group, Åbo Akademi University, Åbo/Turku, Finland

M.-L. Palacios-Alcolado   Johnson Matthey, Royston, Hertfordshire, United Kingdom

Mahesh Pallerla   New Jersey Institute of Technology, Newark, New Jersey, U.S.A.

Peter Panster   Degussa AG, Hanau, Germany

Charles R. Penquite   Chemical Catalysts Research and Development Center, Engelhard Corporation, Beachwood, Ohio, U.S.A.

C. Perego   EniTecnologie SpA, S. Donato Milanese (MI), Italy

O. Piccolo   Chemi SpA, Cinisello Balsamo (MI), Italy

J. Pietsch   Catalysts and Initiators Division, Degussa AG, Hanau, Germany

Daniel J. Plata   Abbott Laboratories, Abbott Park, Illinois, U.S.A.

David Poon   Department of Chemistry, University of British Columbia, Vancouver, British Columbia, Canada

Indra Prakash   The NutraSweet Company, Mount Prospect, Illinois, U.S.A.

U. Prüße Institute of Technical Chemistry, Technical University of Braunschweig, Braunschweig, Germany

Rinaldo Psaro Centro CNR CSSCMTBSO and Dipartimento di Chimica I.M.A., Università di Milano, Milano, Italy

James A. Ramsden Chirotech Technology Ltd., Cambridge, United Kingdom

Tiina-Kaisa Rantakylä Process Chemistry Group, Åbo Akademi University, Åbo/Turku, Finland

Nicoletta Ravasio Centro CNR CSSCMTBSO and Dipartimento di Chimica I.M.A., Università di Milano, Milano, Italy

Sandro Recchia Facoltà di Scienze CCFFMM, Università dell'Insubria, Como, Italy

Janek Reinik Process Chemistry Group, Åbo Akademi University, Åbo/Turku, Finland

Martin Reisinger Degussa AG, Hanau, Germany

Garry L. Rempel Department of Chemical Engineering, University of Waterloo, Waterloo, Ontario, Canada

Roman Renneke Degussa Corporation, Calvert City, Kentucky, U.S.A.

Clementine Reyes Center for Applied Catalysis, Seton Hall University, South Orange, New Jersey, U.S.A.

Stefan Röder Catalysts and Initiators Division, Degussa AG, Hanau, Germany

N. F. Salakhutdinov Novosibirsk Institute of Organic Chemistry, Novosibirsk, Russia

Tapio Salmi Process Chemistry Group, Åbo Akademi University, Åbo/Turku, Finland

Manickam Sasidharan Delft University of Technology, Delft, the Netherlands

D. Scagliarini Dipartimento di Chimica Industriale e dei Materiali, Università di Bologna, Bologna, Italy

Stephen R. Schmidt W.R. Grace & Co., Columbia, Maryland, U.S.A.

Susannah L. Scott University of Ottawa, Ottawa, Ontario, Canada

Louis S. Seif Abbott Laboratories, Abbott Park, Illinois, U.S.A.

Luwan Semere, Merck Research Laboratories, Rahway, New Jersey, U.S.A.

Rodney H. Sergent Chattem Chemicals, Inc., Chattanooga, Tennessee, U.S.A.

Valentina Serra-Holm Process Chemistry Group, Åbo Akademi University, Åbo/Turku, Finland

Padam N. Sharma Abbott Laboratories, Abbott Park, Illinois, U.S.A.

A. P. Sharratt Ineos Fluor Ltd, Cheshire, United Kingdom

Roger A. Sheldon Delft University of Technology, Delft, the Netherlands

Joseph P. Simeone Merck & Co., Inc., Rahway, New Jersey, U.S.A.

É. Sípos Budapest University of Technology and Economics, Budapest, Hungary

Young-Chan Son Chemistry Department, University of Connecticut, Storrs, Connecticut, U.S.A.

I. V. Sorokina Novosibirsk Institute of Organic Chemistry, Novosibirsk, Russia

John R. Sowa, Jr. Department of Chemistry and Biochemistry, Seton Hall University, South Orange, New Jersey, U.S.A.

Guido Stochniol Degussa AG, Hanau, Germany

Bala Subramaniam University of Kansas, Lawrence, Kansas, U.S.A.

Steven L. Suib University of Connecticut, Storrs, Connecticut, U.S.A.
Yuhan Sun Institute of Coal Chemistry, Chinese Academy of Sciences, Taiyuan, P.R. China
Yongkui Sun Merck Research Laboratories, Rahway, New Jersey, U.S.A.
M. Sundell Smoptech Ltd., Turku, Finland
G. Sweeney Johnson Matthey Technology Centre, Reading, Berkshire, United Kingdom
Thomas Tacke Degussa Corporation, Calvert City, Kentucky, U.S.A.
Akira Tai Professor Emeritus, Himeji Institute of Technology, Hyoga, Japan
E. Tálas Chemical Research Center, Institute of Chemistry, Hungarian Academy of Sciences, Budapest, Hungary
Setrak K. Tanielyan Center for Applied Catalysis, Seton Hall University, South Orange, New Jersey, U.S.A.
Robert J. Taylor Texaco Global Products, The Woodlands, Texas, U.S.A.
E. Tfirst Chemical Research Center, Institute of Chemistry, Hungarian Academy of Sciences, Budapest, Hungary
Deepak S. Thakur Chemical Catalysts Research and Development Center, Engelhard Corporation, Beachwood, Ohio, U.S.A.
T. G. Tolstikova Novosibirsk Institute of Organic Chemistry, Novosibirsk, Russia
Pisanu Toochinda The University of Akron, Akron, Ohio, U.S.A.
L. Tóth Research Institute for Technical Physics and Materials Science, Hungarian Academy of Sciences, Budapest, Hungary
Esa Toukoniitty Process Chemistry Group, Åbo Akademi University, Åbo/Turku, Finland
Raymonde Touroude Université Louis Pasteur, Strasbourg, France
A. Tungler Budapest University of Technology and Economics, Budapest, Hungary
A. Vaccari Dipartimento di Chimica Industriale e dei Materiali, INSTM, Udr. di Bologna, Bologna, Italy
Juhani Väyrynen University of Turku, Turku, Finland
G. A. Vedage Air Products and Chemicals, Allentown, Pennsylvania, U.S.A.
Amanda Villoresi Department of Chemistry and Biochemistry, Seton Hall University, South Orange, New Jersey, U.S.A.
K.-D. Vorlop Institute of Technical Chemistry, Technical University of Braunschweig, Braunschweig, Germany
Mark S. Wainwright School of Chemical Engineering and Industrial Chemistry, The University of New South Wales, Sydney, Australia
Francis J. Waller Air Products and Chemicals, Inc., Allentown, Pennsylvania, U.S.A.
Mouhua Wang Institute of Coal Chemistry, Chinese Academic of Sciences, Taiyuan, P.R. China
Dingjun Wang Johnson Matthey, West Deptford, New Jersey, U.S.A.

e-mail: vacange@ms.fci.unibo.it

Johan Wärnå   Process Chemistry Group, Åbo Akademi University, Åbo/Turku, Finland

G. Webb   Department of Chemistry, The University of Glasgow, Glasgow, Scotland

Tong Wei   Institute of Coal Chemistry, Chinese Academy of Sciences, Taiyuan, P.R. China

Wei Wei   Institute of Coal Chemistry, Chinese Academy of Sciences, Taiyuan, P.R. China

Geoffrey T. White   Chemical Catalysts Research and Development Center, Engelhard Corporation, Beachwood, Ohio, U.S.A.

Gregory T. Whiteker   The Dow Chemical Company, South Charleston, West Virginia, U.S.A.

Alan F. Wiese   Chemical Catalysts Research and Development Center, Engelhard Corporation, Beachwood, Ohio, U.S.A.

Gert-Ulrich Wolf   Institut für Angewandte Chemie Berlin-Adlershof e.V., Berlin, Germany

L. Yakhyaeva   Chemical Research Center, Institute of Chemistry, Hungarian Academy of Sciences, Budapest, Hungary

Federica Zaccheria   Centro CNR CSSCMTBSO and Dipartimento di Chimica I.M.A., Università di Milano, Milano, Italy

Antonio Zanotti-Gerosa   Chirotech Technology Ltd., Cambridge, United Kingdom

Hua Zhao   Department of Chemical and Environmental Science, New Jersey Institute of Technology, Newark, New Jersey, U.S.A.

Yuxiang Zheng   Department of Chemical Engineering, University of Waterloo, Waterloo, Ontario, Canada

Bing Zhong   Institute of Coal Chemistry, Chinese Academy of Sciences, Taiyuan, P.R. China

Agnes Zsigmond   Center for Applied Catalysis, Seton Hall University, South Orange, New Jersey, U.S.A.

CATALYSIS OF ORGANIC REACTIONS

1

2001 Paul Rylander Plenary Lecture

Immobilizing Homogeneous Catalysts for Organic Reactions

Francis J. Waller
Air Products and Chemicals, Inc., Allentown, Pennsylvania, U.S.A.

ABSTRACT

There have been many approaches published to "immobilize" a homogeneous catalyst. These heterogeneous catalysts must have better activity, selectivity, lifetime or some other property to warrant their use in an industrial chemical process. This talk will draw upon the author's experiences at DuPont and Air Products with Nafion®, Nafion® on carbon, anionic attached $[Rh(CO)_2I_2]^-$ to Reillex-425, carbons as catalysts for the hydrolysis of esters, heterogeneous catalyst to activate dimethyl ether as a methyl transfer reagent, and monoliths as hydrogenation catalysts. These individual experiences will show the complexities of a simple problem – "immobilization".

INTRODUCTION

There have been many approaches published to "heterogenize," "immobilize" or "anchor" a homogeneous catalyst. Table 1 compares the advantages and disadvantages of the homogeneous and heterogeneous analog with respect to several important industrial issues such as activity, selectivity, lifetime and recyclability.

Table 1 Comparison of Homogeneous and Heterogeneous Catalyst

Issues	Homogeneous	Heterogeneous
activity	high	variable
selectivity	high	variable
lifetime	variable	long
recyclability	expensive	not important
reaction conditions	mild	mild to severe
diffusion	none	usually important
tunability of properties	possible	limited
mechanistic understanding	very developed	developed

Table 2 summarizes some of the methodologies that have been published from the late 1960s to "immobilize" a homogeneous catalyst. Each selected technique depends upon the reaction chemistry and if the reaction chemistry involves a liquid phase, gas-liquid phase, or two immiscible liquid phases with the solid catalyst.

Table 2 "Immobilization" Techniques

Technique	Comment
biphasic catalysts	aqueous (Phase Transfer Catalyst), non-aqueous
supported liquid-phase catalysts (SLPC)	nonvolatile organic liquid in porous support
supported aqueous-phase catalysts (SAPC)	water soluble catalyst in thin aqueous film on hydrophilic high surface area support
catalyst anchored to support via physisorption or entrapment	strict control of porosity of support
catalyst anchored to support via ionic bonding	anionic, cationic
catalyst anchored to support via covalent bonding	Polymeric organic matrix, inorganic matrices

Therefore, in order for the "immobilized" version of the homogeneous catalyst to be successful in an industrial chemical process it must not negatively impact the economics of the process because of its cost, activity, selectivity and lifetime.

Similar issues also exist for heterogeneous catalysts not defined by the "immobilization" techniques listed in Table 2. The more general heterogeneous catalysts have been used in many organic industrial reactions such as hydrogenation, oxidation, disproportionation and isomerization.his work will focus on Nafion®, Nafion® on carbon, anionic attached $[Rh(CO)_2I_2]^-$ to Reillex-425, carbons as catalysts for the hydrolysis of esters, heterogeneous catalyst to activate dimethyl ether as a methyl transfer reagent, and monoliths as hydrogenation catalysts.

EXPERIMENTAL

Each experimental procedure can be found in the references cited throughout this paper and in the reference section.

RESULTS AND DISCUSSION

Nafion® Blends and Nafion® Liquid Compositions

Nafion® is a perfluorinated ion-exchange polymer. It is an ionomer with a chemically resistant polymeric backbone and highly acidic sulfonic acid group (1).

$$\left[-(CF_2CF_2)_n-CFCF_2-\right]_x$$
$$(OCF_2CF)_mOCF_2CF_2SO_3H$$
$$CF_3$$

Nafion® in the bulk phase separates into hydrophobic and hydrophilic SO_3H regions. The ionic domains or clusters are inverted micelles surrounded by a fluorocarbon matrix, and the ionic domains are connected by short channels. Because of the inverted micelle within the fluorocarbon matrix, availability of the acid sites for catalysis is greatly diminished. Two approaches were initially used to increase the catalytic activity of bulk Nafion®: polymer blends (2) and coating a liquid composition of Nafion® on a hydrophobic support (3).

The catalyst blends were prepared by coextruding the thermoplastic form of the polymers. Catalyst A is Nafion® in the sulfonic acid form and catalyst B is a blend of Nafion® in the sulfonic acid form and a perfluorinated polymer containing CO_2H groups. Catalyst C is a blend of Nafion® in the sulfonic acid form and Teflon® FEP and Catalyst D is a blend of Nafion® in the sulfonic acid form and Teflon®. The oligomerization of isobutylene in toluene at 110°C was used to measure the activity of Catalysts A-D. Table 3 summarizes the results.

Table 3 Isobutylene Oligomerization Catalyzed by Nafion® Blends

Catalyst (gram)	% SO_3H	Activity[1]
A (0.25)	100	19
B (0.12)	79	58
C (0.12)	66	68
D (0.12)	62	120

[1] mmole isobutylene reacted per equiv. of total acid sites in one gram of polymer per minute

These initial experiments suggested that the accessibility of the SO_3H groups increased when the Nafion® was blended with another perfluorinated matrix. Isobutylene, like other hydrocarbons, did not swell Nafion® or the Nafion® blends.

In an attempt to explain the activity of Nafion® it was hypothesized that the SO_3H groups were not in the inverted micelle, but rather were exposed at the surface of the micelle. This observation suggested selecting a different hydrophobic material. The next approach was to utilize a liquid composition of Nafion® and coat a hydrophobic support such as carbon. The carbon selected was calcined shot coke, a material with very large pores. The mean pore diameter was about 1000Å.

The liquid composition of Nafion® is commercially available or can be made by a published procedure (4). In this procedure n-propanol, methanol, water and Nafion® are heated to 220°C for 3 hours. Two liquid phases are produced by this method. The upper phase contains ethers; the lower phase is the liquid composition of Nafion®.

Nafion® on Carbon

The liquid composition of Nafion®, approximately 5 wt% solution, was used to coat the calcined shot coke (10-20 mesh, 0.42m²/gram). Loadings were varied up to 3 wt%. The oligomerization of isobutylene in toluene at 110°C was used to measure the activity of the sulfonic acid catalysts. Table 4 compares the activity of several supported and non-supported sulfonic acid polymers. The supported Nafion® on a carbon support was very active when compared to bulk Naflon® and a similar fluorocarbon sulfonic acid polymer on α-alumina available from Dow Chemical. The high polymer loading, 14 wt%, for the DOW catalyst probably was required because α-alumina is more hydrophilic than the carbon. The alumina first was coated with the fluorocarbon sulfonic acid polymer from solution, and the polar sulfonic acid groups interfaced between the alumina and the polymer matrix. Once the surface was coated, the external surface on the alumina was a fluorocarbon. The fluorocarbon sulfonic acid polymer from solution continued to coat the Teflon-like wrapped alumina. Now, however, the fluorocarbon polymer formed the interface between the Teflon-like wrapped alumina and the external surface of sulfonic acid groups.

Table 4 Isobutylene Oligomerization Catalyzed by Heterogeneous Sulfonic Acids

Catalyst	Loading (wt%)	Activity[1]$\times 10^{-3}$
Nafion® on CSC[2]	0.57	164
Nafion® on CSC	0.92	158
Amberlyst 15	-	0.71
Nafion® (10-35 mesh)	-	1.86
XU-40036.01[3]	14	19.7

[1] Moles of isobutylene reacted per equiv. of total acid sites in one gram of polymer per hr
[2] CSC is calcined shot coke
[3] DOW fluorocarbon sulfonic acid polymer on α-alumina.

A similar Nafion® on CSC (0.5 wt%) also hydrolyzed dimethyl adipate to adipic acid (5). Table 5 illustrates the effect of reaction time and temperature on the hydrolysis reaction. Dimethyl ether was also produced during the Nafion® on CSC catalyzed hydrolysis of dimethyl adipate.

A more unexpected result occurred when Catalyst B from Table 2 (blend of perfluorinated polymer (79%) with SO_3H groups and perfluorinated polymer (21%) with CO_2H groups) could be dissolved and a stable liquid composition prepared as a coating solution (6). This coating solution when added to CSC (10-20 mesh), dried, activated with dilute nitric acid and dried again produced a catalyst with a low surface area of $0.12 m^2$/gram. The polymer loading was varied from 0.3 to 0.8 wt%.

Table 6 compares activity of three catalysts: Nafion® on CSC with 100% SO_3H group, Catalyst B on CSC with 79% SO_3H groups and bulk Nafion®. Activity was again measured by isobutylene oligomerization.

Table 5 Hydrolysis of Dimethyl Adipate[1,2]

Temp(°C)	Rxn. Time (hr)	% Product Composition[3] by Weight		
		DMA	MMA	AA
160	1	1.0	31.3	66.0
160	2	0.4	11.6	87.2
160	4	0.1	0.2	99.1
180	2	0.2	0.4	99.1

[1] 0.5% Nafion® on CSC
[2] Reaction carried out in autoclave at a stirring rate of 300 rpm and charged with 100ml H_2O, 12g DMA and 20g catalyst. Additional water (100ml) was pumped into autoclave while water and methanol were removed.
[3] Dimethyl adipate (DMA); monomethyl adipate (MMA); adipic acid (AA)

Table 6 Isobutylene Oligomerization Catalyzed by Nafion®

Catalyst	% SO$_3$H	Loading (wt%)	Activity[1]x10^{-3}
Nafion® CSC[2]	100	0.57	164
Catalyst B on CSC	79	0.56	342
Nafion® (10-35 mesh)	100	-	1.9

[1] moles of isobutylene reacted per equiv. of total acid sites in one gram of polymer per hr
[2] CSC is calcined shot coke

The Nafion® polymers supported on carbon had very high activity based on the weight of active polymer. In general, these supported perfluorinated catalysts were 86-180 times as active as bulk Nafion® and 234-488 times more active than Amberlyst 15 (see Table 4).

Anionic Attached [Rh(CO)$_2$I$_2$]$^-$ to Reillex™ 425

Besides the "immobilized" CF$_3$SO$_3$H, another homogeneous catalyst is anionic [Rh(CO)$_2$I$_2$]$^-$. This was the first active rhodium catalyst for the carbonylation of methanol to acetic acid. Recently, Chiyoda and UOP introduced the Acetica™ process, a novel technology based on an "immobilized" [Rh(CO)$_2$I$_2$]$^-$ on a polyvinyl pyridine resin. Compared with the existing homogeneous process, immobilization increases catalyst concentration in the reaction mixture.

The same homogenous anionic rhodium catalyst is also active for the carbonylation of methyl acetate to acetic anhydride. The "immobilized" form on Reillex™, a polyvinyl pyridine resin, has been reported for the carbonylation of methyl acetate to acetic anhydride. Literature procedures readily allow for the preparation of 2 to 5% by weight of rhodium on Reillex™ (7). These catalysts are prepared by quarternization of the polymeric resin with CH_3I followed by $[Rh(CO)_2I_2]^-$. The polymeric resin has two different sites, rhodium in the +1 oxidation state and quarternized nitrogen.

Another reaction catalyzed by the same anionic $[Rh(CO)_2I_2]^-$ is the hydrogenation of acetic anhydride to ethylidene diacetate (8). Figure I is a plot of

$$2(CH_3CO)_2O + H_2 \rightarrow CH_3CO_2H + CH_3CH(O_2CCH_3)_2$$

ethylidene diacetate concentration (M) verses a series of thirty minute reaction runs for two different syngas compositions. The reaction was carried out at 180°C and 1500 psig. In both examples there is a gradual decay in activity. From the Rh analysis of all the individual samples the average leaching rate is 0.7 mg Rh per hr. The recovered catalyst by ^{13}C NMR did not show any structural differences except for the increase in quarternized methyl sites (9).

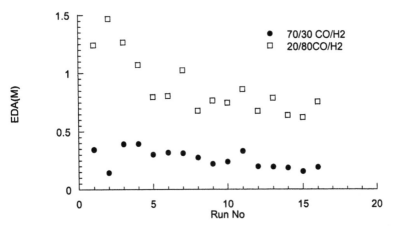

Figure 1 Hydrogenation of acetic anhydride using Reillex™ with 1.1 wt% Rh as $[Rh(CO)_2I_2]^-$.

Carbons as Catalysts for Hydrolysis of Esters

Normally, one does not think of carbons as catalysts. However, carbons are sufficiently acidic or can be treated with nitric acid to be made acidic. These acidic carbons catalyze the hydrolysis of methyl esters. The carbons are more thermally robust than acidic sulfonic ion-exchange resins and can routinely operate at temperatures greater than 100°C. The nature of the acidic groups on carbon responsible for the catalytic activity for ester hydrolysis do not catalyze the dehydration of methanol to dimethyl ether at temperatures up to 200°C. The hydrolysis of methyl acetate was selected to study.

In Table 7, the effect of nitric acid pretreatment at room temperature is illustrated for an OL carbon from Calgon. A 30% converted methyl acetate feed solution was selected for the hydrolysis study. When not treated with nitric acid, the carbon required a contact time between 26 to 52 minutes to reach 40% conversion. When the same carbon is treated with nitric acid, a 42% conversion was reached at a contact time of 12 minutes or less. Dimethyl ether was not observed in either experiment (10,11). The surface acidity was determined by using the Boehm method. The consumed $NaHCO_3$ determined the amount of carboxylic groups on the surface of the carbon. The nitric acid treated carbon possessed 6.3 times more carboxylic groups.

Table 7 Methyl Acetate Hydrolysis with Untreated and Treated Carbon

Catalyst	Contact Time (min)[1]	% Conv.[2]	Acidity[3]
OL Carbon[4,5]	26	38	0.03
	52	40	
OL Carbon[4,6]	12	42	0.19
	24	42	
	47	42	

[1] Reaction temperature is 150°C at 200 psig.
[2] Conversion is based on acetic acid; DME was not detected by GC.
[3] mmoles of acid groups per gram of carbon determined with 0.05M $NaHCO_3$
[4] 30% converted methyl acetate feed solution; 20-35 mesh; 4 grams of catalyst.
[5] Untreated carbon.
[6] Method C treatment.

The Boehme method was also used to determine the increase in carboxylic groups on a Barnaby Sutcliffe activated carbon. Table 8 summarizes this determination.

Table 8 Surface Acidity of Barnaby Sutcliffe Carbon by Boehme Method

Treatment	Acidity[1]
none	0.05
HNO_3	0.46

[1] mmoles of acid groups per gram of carbon determined with 0.05M $NaHCO_3$; average of two determinations at equilibrium times of 70 and 144 hrs.

Heterogeneous Catalyst to Activate Dimethyl Ether for Transfer

A more typical heterogeneous catalyst was encountered during attempts to couple a methyl group to a carbon adjacent to an electron withdrawing group (EWG) using dimethyl ether (12).

$$R(CH_2)_nCH_2EWG + CH_3OCH_3 \rightarrow R(CH_2)_nCH(CH_3)EWG + CH_3OH$$

Now, if an oxidant is added, the methyl group can further be dehydrogenated to an α,β-unsaturated group within the molecule. Gamma-Al_2O_3 was one of several acid catalysts that possessed catalytic activity for this reaction. The results in Table 9 are typical but not optimized. The combined methylation selectivity is 33.7 mole %. If the propionic acid is recycled to extinction, the combined methylation selectivity increases to 61.5 mole %.

Table 9 Alkylation of Methyl Propionate with Dimethyl Ether

	Selectivity[2] (mol%)					
Conversion[1] (%)	MIB	MMA	IBA	MAA	PA	other
DME: 40.7						
	6.5	22.7	1.7	2.8	42.6	17.9
MP: 30.7						

[1]   Temperature (350°C), GHSV 920 hr^{-1}, mole fraction
DME:MP:O_2 is 0.422:0.106:0.021, DME/MP of 3.98, DME/O_2 of 20.6
[2]   methyl isobutyrate (MIB); methyl methacrylate (MMA); isobutyric acid (IBA); methacrylic acid (MAA); propionic acid (PA)

Monoliths as Hydrogenation Catalysts

Presently, the ongoing research is focused on monoliths and monolithic catalysts especially with respect to hydrogenation reactions. Industrial hydrogenation is often performed by using slurry catalysts in stirred-tank reactors. These reaction systems are inherently problematic in chemical process safety, operability and productivity. Finely divided powder catalysts are often pyrophoric and require extensive operator handling during reactor charging and filtration. By the nature of their heat cycles for start-up and shut-down, slurry systems promote co-product formation which can shorten catalysts' life and lower yield. There are alternatives to slurry reactors. These include packed-bed and monolith reactors.

A 400 cpi cordierite monolith is coated with a polymer made from the polymerization of furfuryl alcohol. The coated cordierite is then heat treated (calcination), and after an activation step the active metal, Pd, is impregnated followed by another heat treatment. One characteristic of the monolith catalyst made this way is its low surface area (13). The hydrogenation of nitrobenzene is used to probe the activity of the monolith catalyst. Table 10 summarizes these results for a series of monolith catalysts.

Table 10 Hydrogenation of 40 wt% Nitrobenzene in Isopropanol[1]

Monolith	Layer/L	Surface Area[3]	Rate[2] Init.[4]	Rate[2] Final[5]	Sel.[6]
2% Pd/cordierite[7]	none	<1	33	16	98
1.5% Pd/L/cordierite[7]	polymer	<1	92	91	97
3.1% Pd/L/cordierite[7]	polymer	12	61	74	97
1.7% Pd on carbon composite[8]	carbon composite	466	20	13	98
4.6% Pd on carbon composite[9]	carbon composite	372	36	23	93

[1] 120°C, 200 psig, 1500 rpm in lab-scale monolith autoclave testing unit
[2] moles H_2/m^3 catalyst/second
[3] m^2 per gram
[4] second experimental run
[5] eighth experimental run
[6] aniline(mol%)
[7] 400 cpi
[8] 200 cpi
[9] 250 cpi

We have found that washcoats producing a low surface area monolith catalyst have very high hydrogenation rates for nitrobenzene. These high hydrogenation rates are due to the high hydrogen mass transfer rates within the Taylor flow bubble region throughout the monolith channels.

CONCLUSION

The research described here is a selection of related projects dealing with "immobilizing" a homogeneous catalyst or modifying properties of materials in order to design a catalyst with the correct catalytic function. These functions are activity, selectivity and lifetime. However, as with many heterogeneous catalysts used in gas-liquid environments, gas mass transfer, adsorption of the reactant, desorption of the product and heat management all become important in maximizing activity, selectivity and lifetime. Therefore, as we learn more about the application of monoliths to gas-liquid reaction chemistry problems, a new frontier will emerge for catalysis.

ACKNOWLEDGEMENTS

I want to thank Mrs. T. Hoppe for word processing the manuscript and Air Products and Chemicals, Inc. for permission to publish these reflections. I also want to personally thank all my co-workers whose names appear on the cited references. These researchers have made a difference in challenging the concepts put forth in our publications.

REFERENCES

1. F.J. Waller and R.W. Van Scoyoc, *ChemTech,* 438 (1987).
2. F.J. Waller, US Patent 5,105,047 and 5,124,299 to E.I. DuPont de Nemours and Company (DuPont) (1992).
3. M.H.D. Butt and F.J. Waller, US Patent 5,094,995 to DuPont (1992).
4. W.G. Grot, US Patent 4,433,082 to DuPont (1984).
5. M.H.D. Butt and F.J. Waller, US Patent 5,315,033 to DuPont (1994).
6. F.J. Waller, WO 91/01805 to DuPont (1991).
7. D. Ramprasad and F.J. Waller, US Patent 5,892,110 to Air Products and Chemicals, Inc. (1999).
8. D. Ramprasad and F.J. Waller, US Patent 5,767,307 to Air Products and Chemicals, Inc. (1998).
9. D. Ramprasad and F.J. Waller, in *ACS Preprints,* Division of Petroleum Chemistry, **44 (1)**, 49 (1999).
10. J.B. Appleby, F.J. Waller and S.C. Webb, US Patent 5,872,289 to Air Products and Chemicals, Inc. (1999).
11. F.J. Waller, J.B. Appleby and S.C. Webb, Chem. Ind. (Marcel Dekker), **82**, (Catal. Org. React.), 169 (2000).
12. G.E. Parris and F.J. Waller, US Patent application (allowed) to Air Products and Chemicals, Inc. (2001).
13. A.F. Nordquist, F.C. Wilhelm, F.J. Waller and R.M. Machado, US Patent application (allowed) to Air Products and Chemicals, Inc. (2001).

2

Polymer-Supported Catalyst for the Enantioselective Addition of Diethylzinc to Aldehydes

Sanjay V. Malhotra, Juny Lee, and Mahesh Pallerla
New Jersey Institute of Technology, Newark, New Jersey, U.S.A.

ABSTRACT

Chiral auxiliaries, n-methylephedrine, **3**, and L-tartaric acids **4**, were grafted to Merrifield resin **1** and aminomethylated Merrifield's resin **2**, respectively, to afford polymeric ligands **3A** and **4A**. These polymer-supported ligands were used as auxiliaries in the catalytic asymmetric addition of diethylzinc to various aromatic aldehydes. In the heterogeneous reactions, the products with highest ee of 97% were obtained using 4A, while low ee was seen when 3A was used. Considering the enantioselectivity and reusability, the best results were obtained with 4A. After recycling of the polymer-supported tartaric acid, the catalytic reaction resulted in an ee of 96%. Thus, a new polymer-bound catalytic system has been developed for C-C bond forming reaction.

INTRODUCTION

Synthesis of enantiomerically pure organic compounds of biological importance has always been a challenge and an intense area of research. This goal has lead to a rapid growth in research on catalytic asymmetric process for C-C bond forming reactions. Although the methodologies employing homogenous catalysts have been developed for a long time, there are several disadvantages in terms of separation and selectivity. Therefore, with the exception of only a few, most such methods have not been used for the commercial scale production of chemicals. On the other hand, the reactions with heterogeneous catalysts are

13

becoming important. This is mainly due to the advantages of easy recovery and reuse of the expensive catalyst. The use of polymers as supports has found many applications in catalysis (1-5). The strategy of attaching a chiral ligand onto a polymer support offers the advantages of a heterogeneous system. The polymer-based chiral catalysts also make it possible to carry out the asymmetric reaction in a flow system for continuous production. Site isolation could possibly enhance the catalyst activity and lifetime because it prevents the aggregation of the catalytically active species.

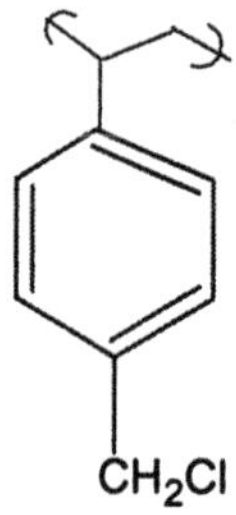

Merrifields resin, 1

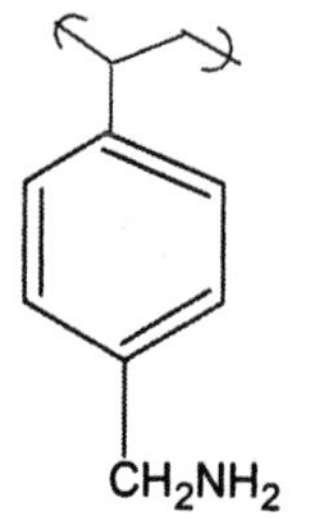

Aminomethylated
resin, 2

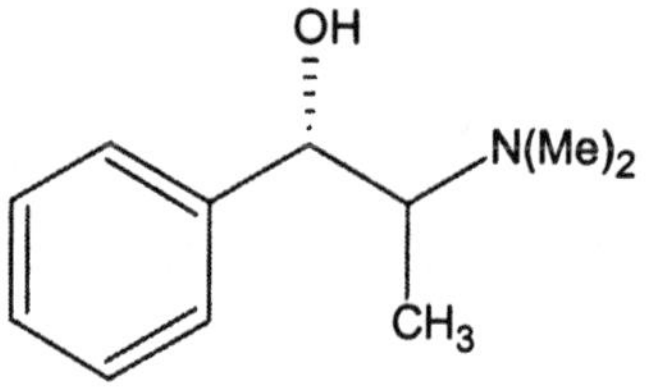

(-)-N-methylephedrine,
3

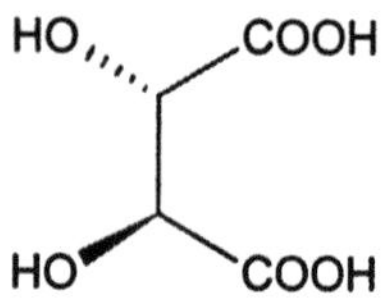

L - Tartaric acid, 4 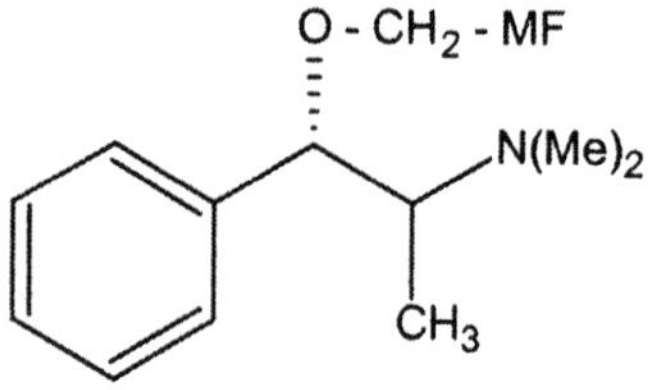   3A 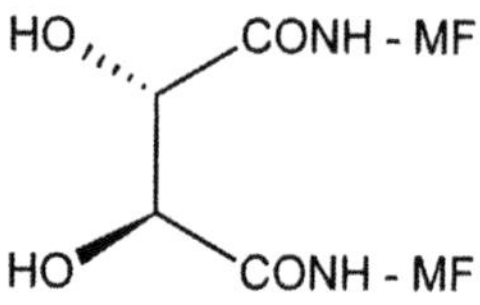   4A

Figure 1

Most of the natural products, pharmaceuticals, and medicinally important compounds have chiral centers. Therefore, it has always been a challenge for the synthetic chemists to come-up with simpler and viable methods to achieve the synthesis of chiral target molecules. Methodologies using chiral ligands, reagents, or auxiliaries are in significant progress. One such target has been the addition of diethylzinc to aldehyde enantioselectively. Various compounds such as chiral amino alcohols and amines have been used as chiral auxiliaries for the addition of organozinc reagents to aldehydes (6) and hydrogenation reactions (7) respectively, and also other organic transformations (8,9).

Based on our earlier success in developing heterogeneous catalysts for allylic alkylation (10), we took up the task of developing a polymer-based heterogeneous catalytic system for the enantioselective addition of diethylzinc to an aldehyde. We herein report the results from our investigation in this area.

EXPERIMENTAL

Materials:　The Merrifield resins, aldehydes and diethylzinc were purchased from Aldrich Chemical Co. and used as such. The solvents hexane and toluene were also from Aldrich Chemical Co., and used after distillation and drying.

Methods:

(1) General experimental Procedures: Sodium hydride (1.1 equivalents) is taken in dry DMF, under nitrogen atmosphere, and cooled to 0°C. To this, chiral catalyst (1 equivalent) is added and stirred for twenty minutes. The polymer (0.9 equivalents) is added, flushed with nitrogen, the reaction allowed to warm to room temperature and stirred for 48 hrs. The reaction mixture is neutralized with 1.0 N hydrochloric acid, and the polymer is filtered, and washed with dichloromethane. When the silyl protection group is to be removed, the reaction mixture is stirred for another 12 hrs with methanol/potassium carbonate. The polymer is dried in vacuum at 80 – 90° C for 48 hrs. IR of the functionalized polymer was recorded by preparing KBr pellet. The decrease in the peak of CH_2Cl – 1265 cm^{-1} and appearance of additional peaks when compared to the starting material (polymer), confirmed the product polymer.

(2) General procedure for addition of diethyzinc to benzaldehyde: The polymer (2-15 mol %) is stirred in dry toluene to swell the polymer under nitrogen atmosphere for 90 minutes. Benzaldehyde (1.0 equivalent) is added, stirred for 20 minutes, cooled to 0° C, and diethyl zinc (2-4 equivalents) is added. The reaction mixture is allowed to warm to room temperature gradually over a period of 1 hr and stirred at RT for 48-72 hrs. This was followed by quenching the reaction mixture with 2.0 N hydrochloric acid solution. The polymer was removed by filtration, and the aqueous layer extracted with diethylether. The crude product was purified over a column of silica gel, eluting with light petrol: ethylacetate (99:1 to 98:2). The purity of the pure product was recorded over GC, and the specific rotation recorded on a polarimeter. The enantiomeric excess is calculated with reference to the literature values.

RESULTS AND DISCUSSION

Reaction of diethylzinc with various aldehydes was used to investigate the use of polymer supported chiral ligands. As an initial study, we explored the homogenous system with various cinchona alkaloids. The reaction with catalytic amount of N-methylephedrine gave the product on the addition of diethylzinc to benaldehayde with 82 % ee. Encouraged with this result, we proceeded to study this reaction by anchoring chiral molecules to a polymer support. Considering both the chemical structure and the easy availability of the Merrifield's resin **1**, and the aminomethylated Merrifield's resin **2**, these were used as support to obtain the polymer-supported heterogeneous systems **3A** and **4A**. Representative reaction procedure (11) for the preparation of these polymer-supported catalysts is shown in Scheme 1 below.

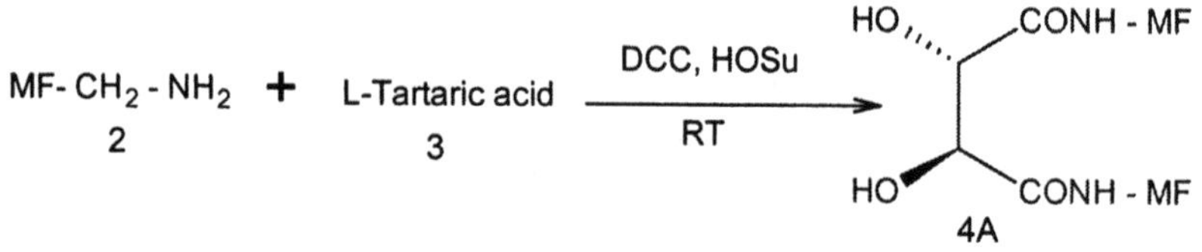

Scheme 1

The formation of polymer bound products was confirmed by infrared spectroscopic analysis. These catalysts were employed for an enantioselctive addition of diethylzinc to aldehyde (Scheme 2). In order to obtain a good reaction medium the reaction was initially studied with varying amounts of catalyst **3A** in hexane, THF and toluene as solvents. Moderate results (ee 20-32%; yield 70-81%) were obtained in hexane and THF. However, the best results were obtained with toluene. Therefore, toluene became the solvent of choice for the rest of the investigation.

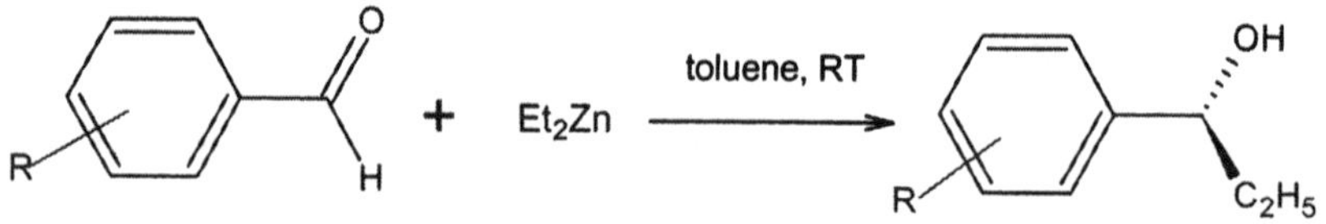

Scheme 2

The best results were obtained using 10 mole % of the catalyst in toluene. The results of work with N-methylephederine are summarized in Table 1. As the table shows, the reaction with unattached N-methylephedrine gave up to 82 % ee of the product, while lower ee were obtained when it was tethered to polymer and used for the same reactions. Clearly, this indicates that it does not provide the necessary cavity for the selective addition of diethylzinc to aldehyde.

Table 1 Addition of diethylzinc catalyzed by 3A to benzaldehyde.

Catalyst / mole %	Et_2Zn, moles	e.e (%)
3 / 15.0	2	82
no catalyst	2	0
3A / 2.0	2	9
3A / 5.0	2	21
3A / 10.0	2	29

There are reports in the literature where diols have been used for binding the metals and carrying out organic transformation with appreciable selectivity. Therefore, we shifted our investigation of this reaction using L-tartaric acid to see if reaction the product can be obtained with high stereoselectivity. Reaction with 3% of L-Tartaric acid under the homogeneous condition gave an ee of 97%. Therefore, L-tartaric acid was attached to aminomethylated Merrifield's resin through the reaction procedure shown in Scheme 1. On using 5 % of this tethered ligand **4A**, our initial investigation of the addition of diethylzinc to benzaldehyde gave a high ee of 96 %. The next three repeated attempts gave the product with 96% ee. Encouraged with these results we proceeded to study other aldehyde systems. The results are summarized in Table 2.

In order to obtain a better understanding of the catalytic system, the reaction was also studied through molecular modeling using 'Materials Studio', a program provided by Accelrys Inc. for calculating the physical data. These calculations support the experimental results.

Table 2 Addition of diethylzinc catalyzed by **4A** to various aldehydes.

Aldehyde	time (day/s)	temp (°C)	yield(%)	ee (%)
Benzaldehyde	1	25	84	97
t-cinnamaldehyde	1	25	99	96
O-salicylicaldehyde	2.5	25	37	93
O-anisaldehyde	1.5	25	86	89
Pyrrol,2-aldehyde	3	25	91	-
O-tolualdehyde	2.5	25	87	71
O-chlorobenzaldehyde	2.5	25	79	97
Piperonal	2.5	25	83	94

The adduct of diethylzinc to diols is known to exist as a dimer. This reaction, therefore, proceeded through a mechanism proposed in Scheme 3. The results in Table 2 show that in the addition of diethylzinc to benzaldehyde the polymer-supported tartaric acid creates a chiral pocket that facilitates a stereoselective addition. The data also indicate that the presence of electronegative groups such as hydroxyl and chloro decrease the binding ability of the aldehyde to zinc and thereby slow the rate of the reaction. Therefore, relatively lower yields of the product are obtained. On the other hand, a methyl group lowers the selectivity through steric hindrance.

CONCLUSION

An investigation of polymer-supported chiral ligands as catalysts has been used for the enantioselective addition diethylzinc to various aldehydes. L-Tartaric acid tethered to amino-methylated Merrifield's resin gives very high yields and enantioselectivity, similar to that seen in case of homogeneous system. This polymer-supported ligand is easily reusable and gives the product with negligible loss in the enantioselectivity.

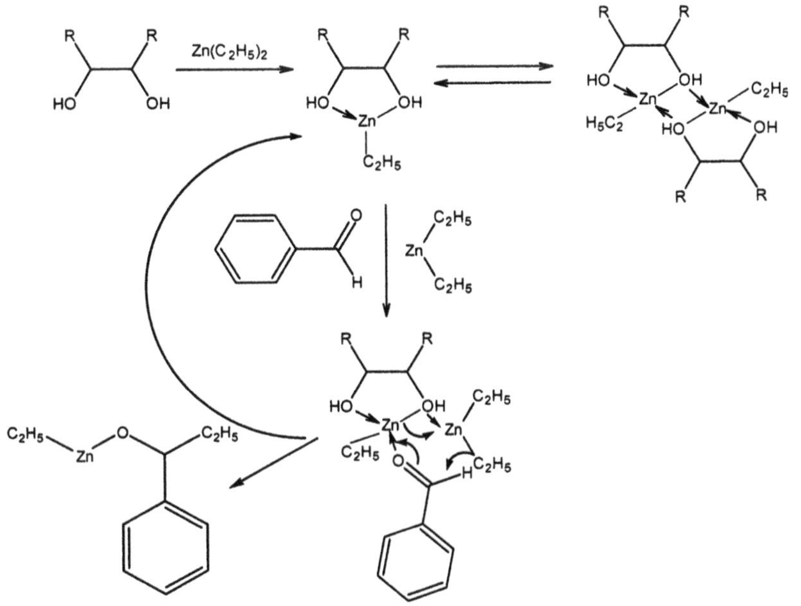

Scheme 3 Mechanism of the addition of diethylzinc to aldehyde

REFERENCES

1.   PHU Herkend, HCJ Ottenheijm, DC Rees. *Tetrahedron*, **53**, 5643, 1997.
2.   PHH Herkens, HCJ Ottenheijm, DC Rees. *Tetrahedron*, **52**, 4527, 1996.
3.   S Kobayashi. *Chem Soc Rev*, **28**, 1, 1999.
4.   T Li, KD Janda, JA Ashley, RA Lerner. *Science* , **264**, 1289, 1994.
5.   A Vidal-Ferran, N Bampos, A Moyano, MA Pericas, A Riere, JKM Sandera. *J Org Chem,*. **63**, 6309, 1998.
6.   S. Bastin, F. A. Niedercorn, J. Brocard, L. Pelinski. *Tetrhedron Asym,* **12**, 2399-2408, 2001.
7.   VF Anton, B Nick. *J Org Chem,* **63**, 6309-6318, 1998.
8.   JB Daniel, BT Catherine, ECP Mario. *Tetrahedron Asym.,* **9**, 2015-2018, 1998.
9.   A Rosana, AH Marie, C Christian. *Tetrahedron Lett.,* **40**, 7091-7094, 1999.
10.   S Christine, H Robter, T Francois. *J Org Met Chem.,* **603**, 30-39, 2000.
11.   SV Malhotra, RL Augustine. *Catal Org React.,* **62**, 553, 1995.

3

Increase of the Long-Term Stability of Chirally Modified Platinum Catalysts Used in the Enantioselective Hydrogenation of Ethyl Pyruvate

V. Morawsky, U. Prüße, and K.-D. Vorlop
Institute of Technical Chemistry, Technical University of Braunschweig, Braunschweig, Germany

ABSTRACT

The long-term stability of chirally modified platinum catalysts used in the enantioselective hydrogenation of ethyl pyruvate was investigated. Using chirally modified platinum colloids, the washing out of the modifier from the catalyst's surface could be avoided. For the retention of the quasi-homogeneous catalyst, a novel heterogenization method has been developed and applied to the model reaction. This new heterogenization method is based on the three-dimensional entrapment of the Pt-colloid by electrostatic attraction between a polyelectrolyte and the oppositely charged catalyst. Different polyanions have been found to be suitable for entrapment of the chirally stabilized Pt-colloid. Using alginate for heterogenization, a constant enantioselectivity and a good activity during 25 hydrogenation cycles were obtained if cyclohexane was chosen as solvent in the enantioselective hydrogenation of ethyl pyruvate. In this solvent the long-term stability of the chiral catalyst could be improved considerably since the destructive hydrogenation of the modifier taking place in acetic acid and causing a drop in enantioselectivity could be suppressed.

INTRODUCTION

Chiral modification of platinum catalysts by cinchona alkaloids reported first by Orito [1] is a successful strategy for introducing enantiodifferentiation into the hydrogenation of α-ketoesters to the corresponding α-hydroxyesters. In recent years this concept has received much attention and detailed studies have been made to optimise the catalytic system (substrate, modifier, catalyst, solvent, reaction conditions) and to extend the use of chirally modified catalysts. Several reviews have been published [2]. For technical application, reuse of catalysts is a feature of great interest. However, the long-term stability of chirally modified platinum catalyst used in the enantioselective hydrogenation of ethyl pyruvate is very low. Only by addition of fresh modifier at the beginning of each new hydrogenation cycle [3, 4] or by continuously feeding cinchonidine (CIN) permanently in continuous operation [5] the enantioselectivity can be maintained. The loss in enantioselectivity is due to the partial hydrogenation of the quinoline ring system of CIN [6] and the washing out of the modifier from the platinum surface during continuous use [5], respectively. Recently, it has been shown that in acetic acid the competitive hydrogenation of the aromatic ring system of the modifier did not begin until complete conversion of ethyl pyruvate occurred [6]. The drop in enantioselectivity on reuse could be avoided by stopping the hydrogenation of ethyl pyruvate at a conversion of approximately 70 %. By these changed reaction conditions, the activity and enantioselectivity of the chirally modified catalyst could have been maintained at high values for more than 10 repeated uses without addition of fresh modifier at the beginning of each hydrogenation cycle. But under these reaction conditions it is necessary to separate 30 % of the starting material from the products which is both expensive and partly very difficult if other α-ketoesters are used which tend to decompose during distillation.

The aim of this study was the development of a catalytic system with a high long-term stability used in the enantioselective hydrogenation of ethyl pyruvate as model reaction. For this purpose chirally stabilized platinum colloids were prepared in order to avoid the washing out of the modifier from the platinum surface. Nevertheless, the application of this quasi-homogeneous catalyst in the enantioselective hydrogenation is limited because of the difficult separation and recycling process. Therefore, a novel heterogenization method for chirally stabilized platinum catalysts was developed and applied to the model reaction. The new heterogenization method is based on the three-dimensional entrapment of catalysts by electrostatic attraction between a polyelectrolyte and an opposite charged catalyst. By entrapment of the Pt-colloid in alginate, a catalyst with constant enantioselectivity and a good activity during 25 hydrogenation cycles in cyclohexane as solvent was obtained. It could be shown that the destructive hydrogenation of the chiral modifier observed in acetic acid could be avoided in this solvent.

EXPERIMENTAL

Heterogenization of Chirally Stabilized Platinum Colloids

Chirally stabilized Pt-colloids can be prepared by reduction of an aqueous solution of $PtCl_4$ in the presence of protonated 10,11-dihydrocinchonidine (DH-CIN) [7]. The positively charged alkaloid has not only a stabilizing effect on the platinum particles, it also can induce enantioselectivity in catalytic hydrogenations [7].

For entrapment, an aqueous solution of the chirally stabilized Pt-colloid is mixed with an aqueous solution of a polyanion while stirring (Figure 1). The resulting mixture is dropped onto a suitable surface (e. g. a polyethylene (PE) film) by a syringe equipped with a 1 mm capillary. Afterwards, the droplets were dried by exposing the film to air for at least 24 hours. Very plain lens-shaped particles of 4-5 mm in diameter and 200-400 μm in height with a homogeneous distribution of the Pt-colloid inside the polyelectrolyte matrix were obtained. After complete drying, the entrapped catalyst particles can easily be removed from the PE film and be used in enantioselective hydrogenation of ethyl pyruvate in non-polar solvents like cyclohexane.

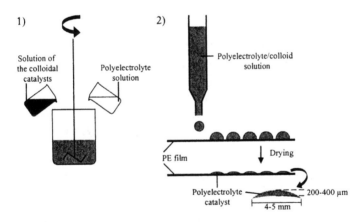

Figure 1 Scheme of the heterogenization procedure for chirally stabilized platinum colloids 1) mixing the solutions, 2) dropping the mixed solutions on a suitable surface e. g. a PE film, and drying the droplets afterwards.

The following polyanions were tested for entrapment of the chirally stabilized Pt-colloid (Figure 2): alginate (PROTANAL LF 20/60, Pronova Biopolymer, Norway), carboxymethylcellulose (Blanose 7MXF, Hercules, Germany), carrageenan (GENUGEL X-0828, Copenhagen Pectin Factory, Denmark), cellulosesulfate (Kelco SCS-LV, Kelco, Norway), pectinate (Pectin Classic AF 707, Herstreith & Fox, Germany), and sulfoethylcellulose (Wolff

Walsrode, Germany). For heterogenization, the amount of Pt-colloid (125 mg) was kept constant. The amount of the carbohydrates was varied so that the molar ratio of Pt to DH-CIN to one monosaccharide unit of the polyelectrolyte was always 1:2:10. The detailed heterogenization procedure has been published previously [8].

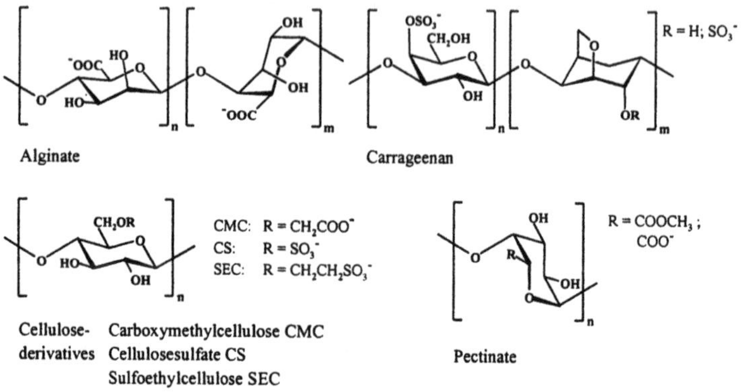

Figure 2 Structure of different carbohydrates used for the heterogenization of chirally stabilized Pt-colloids.

Experimental and Analytical Procedure

Enantioselective hydrogenation of ethyl pyruvate was carried out in a 300 ml stirred autoclave (Parr Instruments, Illinois) at 20 °C and the pressure kept constant at 60 bar with a high pressure H_2 burette. Reaction rates were determined from the drop in H_2 pressure measured in the burette. Conversion and enantiomeric excess were determined by quantitative analysis by a GC equipped with a chiral capillary column for enantiomeric separation (for details see [6]).

The entrapped Pt-colloid was placed in a catalyst basket and used in the hydrogenation of 10 ml ethyl pyruvate (Fluka; used as received) dissolved in 150 ml cyclohexane (Fluka; p. a. grade). No additional modifier was added to the reaction solution. For repeated use of the alginate-entrapped Pt-colloid, the enantioselective hydrogenation of ethyl pyruvate was stopped after complete conversion and the H_2 pressure was released. The catalyst was removed from the reaction solution which was discarded after taking a sample to determine the conversion and the enantiomeric excess. Then the autoclave was refilled with fresh substrate and solvent. The next hydrogenation was started under identical reaction conditions. No fresh modifier was added to the reaction solution at any time. During the whole process, no special care was taken that the catalyst was not exposed to air between the hydrogenation cycles. In all other cases 200 mg conventional Pt/Al$_2$O$_3$ catalyst (E 4759, Engelhard; used as received) modified with 40 mg CIN (Fluka) were used to hydrogenate 10 ml ethyl pyruvate dissolved in 70 ml cyclohexane.

RESULTS AND DISCUSSION

The application of catalysts in industrial processes requires good long-term stability of the catalytic material. However, efficient reuse of the chirally modified platinum catalyst in the enantioselective hydrogenation of ethyl pyruvate has not been achieved. To maintain the enantioselectivity, it is necessary to add fresh modifier at the beginning of each hydrogenation cycle [3, 4].

One of the reasons for the loss in enantioselectivity is the washing out of the modifier from the platinum surface [5]. Using the chiral modifier as stabilizer of Pt-colloids prevents this problem. Therefore, the platinum sol stabilized by the protonated alkaloid DH-CIN prepared according to [7] was chosen as chiral model colloid in the enantioselective hydrogenation of ethyl pyruvate. The alkaloid not only stabilizes the platinum particles, it also induces enantioselectivity in catalytic hydrogenations [7].

Nevertheless, the retention of the quasi-homogeneous catalyst has remained problematical. Therefore, a novel heterogenization method has been developed and applied to the enantioselective hydrogenation of ethyl pyruvate [8]. This new heterogenization method is based on the entrapment of the chirally stabilized Pt-colloid in a three-dimensional network of a polyelectrolyte (e. g. alginate) by electrostatic attraction between the polyelectrolyte and the oppositely charged colloidal catalyst (Figure 3).

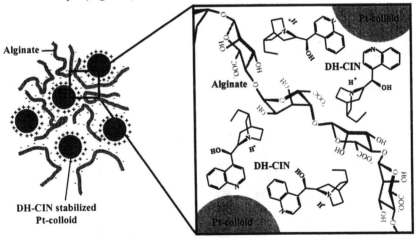

Figure 3 Scheme of the alginate-entrapped platinum colloid stabilized by DH-CIN.

The entrapment can be easily achieved by mixing solutions of the dissolved polymer and the colloidal catalyst (Figure 1) and dropping the resulting mixture onto a suitable surface, e. g. a PE film. After complete drying of these droplets, the retention of the catalyst is so strong that nearly no metal leaching is observed (in general less than 1 %). This novel heterogenization method has also been successfully applied to homogeneous catalysts with ionic-functionalized ligands used in different enantioselective reactions [8]. In contrast to adsorption onto ion-exchange materials, the polymeric network is formed in solution in the presence of the catalytically active metal. Thereby an ideal wrapping of the polymeric matrix around the catalyst and a homogeneous distribution of the catalyst inside the polymer network is allowed. Because of the flexible structure of the polyelectrolyte network, conformational changes of the catalyst causing a decrease or the complete disappearance of enantioselectivity, especially in case of homogeneous catalysts, are assumed to be neglectable in contrast to adsorption on organic or inorganic ion-exchange resins.

Primarily, different carbohydrates (Figure 2) functionalizied by carboxy or sulfate groups, respectively, were tested for the entrapment of the chirally stabilized colloid. Each of these polyanions were suitable to form the polyelectrolyte catalyst which can be used in the hydrogenation of ethyl pyruvate using cyclohexane as solvent (Figure 4). Entrapped in carbohydrates functionalized by carboxy groups, the platinum colloid has been found to be slightly more active than being entrapped in the sulfate functionalized carbohydrates. The type of polyanion used for the entrapment has no significant influence on the enantioselectivity. The pectinate-entrapped platinum colloid has reached the best enantiomeric excess of 29%. The enantioselectivity of the polyelectrolyte catalysts is on the same order as obtained with the non-heterogenized Pt-colloid used in cyclohexane (35 %) [9]. It is very important that the polyelectrolyte catalyst is dried completely during the heterogenization procedure before using in non-polar solvents like cyclohexane or toluene. If the catalyst particles contain water or if they are used in polar solvents like ethanol or acetic acid, the polyelectrolyte catalyst will be resolved because of the breakage of

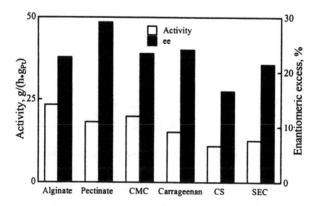

Figure 4 Activity and enantiomeric excess of the chirally stabilized Pt-colloid entrapped in different polyanions and used in the enantioselective hydrogenation of ethyl pyruvate in cyclohexane.

the ionic bonding. Then the platinum colloid could be found in the reaction solution. When using cyclohexane or toluene as solvent, no Pt-leaching was detected. In these solvents no difference in activity and enantioselectivity of the entrapped catalyst was observed.

The decrease in enantioselectivity of chirally modified platinum catalysts on reuse is caused by the partial hydrogenation of the aromatic ring system of the modifier. Recent investigations have shown the chemical changes of CIN vary with the ethyl pyruvate conversion using acetic acid as solvent [6]. Under reaction conditions CIN is hydrogenated sequentially first at the quinuclidine group and then at the quinoline ring (Figure 5). Reactions taken to complete conversion of ethyl pyruvate show multiply-hydrogenated CIN derivatives, the efficiency of which as chiral modifier is reduced because of lowered adsorption strength on the Pt-surface [6].

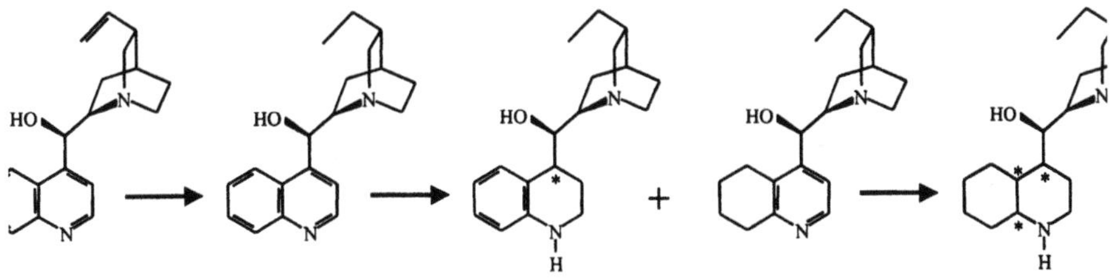

Cinchonidine (CIN)	10, 11-Dihydro-cinchonidine (DH-CIN)	1′, 2′, 3′, 4′, 10, 11-Hexahydrocinchonidine	5′, 6′, 7′, 8′, 10, 11-Hexahydrocinchonidine (Σ HH-CINs)	Dodecahydro-cinchonidine (Σ DDH-CINs)

* stereogenic cent

Figure 5 CIN derivatives formed on Pt/Al$_2$O$_3$ during the enantioselective hydrogenation of ethyl pyruvate.

For investigation of the chemical changes of the chiral modifier during the enantioselective hydrogenation of ethyl pyruvate in cyclohexane as solvent, a conventional Pt/Al$_2$O$_3$ catalyst modified with CIN was used at 60 bar and 20 °C. The temporal course of the conversion of ethyl pyruvate is shown in Figure 6. The concentration of the substrate decreases with time, the products (R)-ethyl lactate (in excess) and (S)-ethyl lactate are formed. After 50 minutes the reaction is finished. In contrast to the enantioselective hydrogenation of ethyl pyruvate performed in acetic acid [6], the enantiomeric excess is not constant during the whole time course; it increases with increasing conversion to a maximum of 40 % ee at the end of the reaction. In acetic acid an enantiomeric excess of nearly 80 % has been obtained under identical reaction conditions [6].

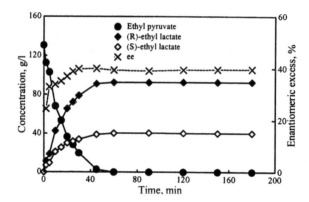

Figure 6 Concentration-time curve of the catalytic hydrogenation of ethyl pyruvate to ethyl lactate on CIN-modified Pt/Al$_2$O$_3$ in cyclohexane at 60 bar and 20 °C.

The concentrations of alkaloid derivatives in the reaction solution formed during this conversion of ethyl pyruvate (Figure 6) by hydrogenation of CIN are shown in Figure 7. In order to compare their amounts in solution, the contents of the alkaloids were normalized. The starting amount of chiral modifier used in all catalytic hydrogenation of ethyl pyruvate (40 mg CIN) corresponds to a relative peak area of 1.

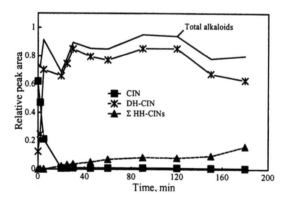

Figure 7 Course of the alkaloid derivatives in reaction solution formed during the catalytic reduction of ethyl pyruvate to ethyl lactate by hydrogenation of CIN on Pt/Al$_2$O$_3$ in cyclohexane at 60 bar and 20°C.

The chiral modifier CIN is transformed mainly to DH-CIN in cyclohexane. But after 180 minutes (more than two hours after complete conversion of the substrate), there are still traces of CIN detectable in the reaction solution. The content of DH-CIN increases rapidly at the beginning of the reduction of ethyl pyruvate, then remains constant between 75 and 85 %. More than one and a half hours after complete conversion of ethyl pyruvate (150 minutes) the concentration of DH-CIN decreases slightly. After 20 minutes different hexahydrogenated CIN derivatives (HH-CINs) are formed in small amounts. At this time more than 70 % of the substrate has been converted. With increasing reaction time, a small enhancement of their content in the reaction solution is observed. The formation of the dodecahydrocinchonidine species (DDH-CINs) could not be found at any point during the enantioselective reduction of ethyl pyruvate in cyclohexane. Figure 7 also shows the total amount of detectable alkaloids in the reaction solution. Only a small quantity of the modifier is adsorbed on the catalyst's surface, as more than 80 % of the total amount of the alkaloids are detectable in the reaction solution. The efficient modifier, DH-CIN, is found to be the major constituent of the alkaloids in solution. The small amount of adsorbed modifier could be the reason for the low enantioselectivity of 40 % reached in cyclohexane.

Figure 8 shows the comparison of the content of CIN derivatives in reaction solution at different times in the enantioselective hydrogenation of ethyl pyruvate formed on conventional Pt/Al_2O_3 at 60 bar and 20 °C using acetic acid [6] or cyclohexane as solvent. In both solvents the modifier CIN is hydrogenated rapidly to DH-CIN which is also an efficient modifier in the enantioselective hydrogenation of ethyl pyruvate. At a conversion of approximately 70 %, DH-CIN is the main alkaloid in reaction solution. Only traces of CIN were detected and a small amount of the different HH-CINs were found both in acetic acid and in cyclohexane at 70 % conversion.

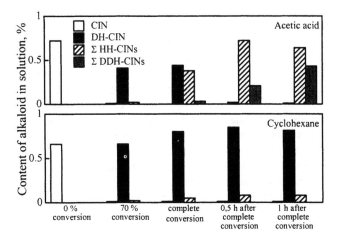

Figure 8 Comparison of the content of alkaloid derivatives in reaction solution at different times in the enantioselective hydrogenation of ethyl pyruvate on Pt/Al_2O_3 using acetic acid or cyclohexane as solvent.

After complete conversion of ethyl pyruvate the concentration of DH-CIN in the reaction solution decreases dramatically using acetic acid as solvent. In contrast, in cyclohexane the amount of DH-CIN in solution remains constant even if the enantioselective hydrogenation of ethyl pyruvate is not stopped at complete conversion. In addition, traces of CIN could be found in the reaction solution more than one hour after complete conversion using cyclohexane as solvent. In acetic acid the formation of the different HH-CINs starts before complete conversion of the substrate is reached. After complete conversion their concentration increases rapidly and half an hour after complete conversion the HH-CINs are the main alkaloid in the reaction solution. By contrast, only a small amount of HH-CINs has been formed after complete conversion in cyclohexane. Their content increases very slightly with increasing reaction time. The formation of the completely hydrogenated DDH-CINs only have been observed in acetic acid. In this solvent, the DDH-CINs have been found at 100 % conversion of ethyl pyruvate and their concentration in reaction solution increased dramatically with increasing reaction time (Figure 8).

These results show that the undesired destruction of the chiral modifier by hydrogenation of the aromatic ring system taking place in acetic acid after complete conversion of the substrate [6] could be suppressed by choosing cyclohexane as solvent. Using the alginate entrapped platinum colloid repeatedly in the enantioselective hydrogenation of ethyl pyruvate in cyclohexane, the catalyst shows a very high long-term stability even without adding fresh modifier at the beginning of each new hydrogenation cycle (Figure 9). An increase in the catalytic activity is observed during the first 13 hydrogenation cycles, after which the activity remains nearly constant at 60 to 70 g/(h·g$_{Pt}$). This might be due to an activation of the entrapped platinum particles which takes place over several hydrogenation cycles since the colloid was exposed to air before and during the heterogenization procedure. Previously it has been shown that a thermal pre-treatment of conventional Pt/Al$_2$O$_3$ catalyst with hydrogen can improve the performance of the catalyst [10]. However, in the case of the entrapped catalytic system a thermal activation before using the catalyst is not possible as the carbohydrates would be destroyed by this procedure. In addition, a rearrangement of the structure of the polyelectrolyte-encapsulated colloid leading to a facilitated diffusion might also contribute to the observed increase in activity. Although no fresh modifier was added and contact with air during individual hydrogenation cycles had not been prevented, the enantiomeric excess remains constant during more than 25 runs. It is assumed that the polymeric matrix prevents an oxidation of the platinum particles after entrapment. After the first hydrogenation cycle a small amount of platinum colloid was found in the reaction solution, probably because of incomplete drying. In all further runs, no leaching of the active metal was observed.

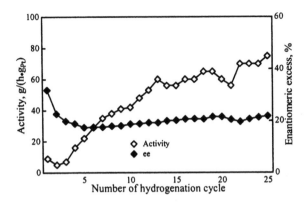

Figure 9 Activity and enantioselectivity of the alginate encapsulated Pt-colloid stabilized by DH-CIN used repeatedly in the enantioselective hydrogenation of ethyl pyruvate in cyclohexane without adding fresh modifier before each hydrogenation cycle.

The long-term stability of chirally modified platinum catalysts could be considerably increased not only by using the chiral modifier as stabilizer of Pt-particles but also by choosing cyclohexane as solvent. The strong solvent effect on the stability of the modifier is responsible for the different long-term performance of the chirally modified catalysts. In acetic acid the modifier is destroyed by hydrogenation causing the drop in activity and enantioselectivity. Therefore it is necessary to stop the enantioselective hydrogenation of ethyl pyruvate at a conversion of 70 % or to add fresh modifier at the beginning of each new hydrogenation cycle to maintain the enantioselectivity. In cyclohexane the hydrogenation of the chiral modifier is much slower. Therefore the long-term stability of the chirally modified catalyst used in this solvent is very high.

One possible explanation for the different long-term behavior of the chiral modifier could be electronic effects on the platinum particles caused by the different solvents. These effects could be the reason both for the different adsorptions strength of CIN on the catalyst and for a faster hydrogenation of the modifier in acetic acid. On the other hand, the protonation of CIN at the aromatic group (pK$_a$ of quinoline 4.90) in acetic acid (pK$_a$ 4.76) could also cause the different long-term stability of the chiral modifier. By this protonation at the quinoline-N-atom, the conformation of the planar aromatic part of CIN is changed (transition from sp^2 to sp^3 hybridization). Eventually, this could be the cause for a faster hydrogenation of the modifier at the aromatic ring system in acetic acid which leads to a weaker adsorption on the platinum surface and consequently to a decrease in enantioselectivity. In contrast, in cyclohexane a protonation of CIN is not possible and the long-term stability of the modifier is much greater.

CONCLUSIONS

The long-term stability of a platinum catalyst used in the enantioselective hydrogenation of ethyl pyruvate can be considerably improved by using chirally stabilized colloids to prevent the washing out of the modifier from the catalyst's surface. For the retention of this quasi-homogeneous catalyst, a novel heterogenization method has been developed and applied to the model reaction. An easy preparation procedure, a simple separation and good retention of the active metal in the polymeric material are regarded as the significant features of this new heterogenization method. Different polyanions have been found to be suitable for entrapment. By entrapment of the Pt-colloid in alginate, a constant enantioselectivity and a good activity during 25 hydrogenation cycles were observed using the catalyst in cyclohexane as solvent. The destructive hydrogenation of the chiral modifier which takes place in acetic acid is avoided in this solvent.

ACKNOWLEDGEMENTS

The authors thank the Ministry of Education, Research, Science and Technology, German Government, for financial support (Grant 03C0272A 4) and L. Witte (TU Braunschweig) for GC/MS determination.

REFERENCES

[1]   Y. Orito, S. Imai, S. Niwa; J. Chem. Soc. Jpn. (1979) 1118; Y. Orito, S. Imai, S. Niwa; J. Chem. Soc. Jpn. (1980) 670; Y. Orito, S. Imai, S. Niwa; J. Chem. Soc. Jpn. (1982) 137.
[2]   H. U. Blaser, H. P. Jalett, M. Müller, M. Studer; Catal. Today 37 (1997) 441; A. Baiker, H. U. Blaser; in: G. Ertl, H. Knözinger, J. Weitkamp (Eds.), Handbook of Heterogeneous Catalysis, Vol. 5. WILEY-VCH, Weinheim, 1997, p.2422; A. Baiker; J. Mol. Catal. A: Chemical 115 (1997) 473; P. B. Wells, A. G. Wilkinson; Topics in Catalysis 5 (1998) 39; A. Pfaltz, T. Heinz; Topics in Catalysis 4 (1997) 229.
[3]   J. T. Wehrli, A. Baiker, D. M. Monti, H. U. Blaser, H. P. Jalett; J. Mol. Catal. 57 (1989) 245.
[4]   U. Böhmer, F. Franke, K. Morgenschweis, T. Bieber, W. Reschetilowski; Catal. Today 167 (2000) 167.
[5]   N. Künzle, R. Hess, T. Mallat, A. Baiker; J. Catal. 57 (1999) 239.
[6]   V. Morawsky, U. Prüße, L. Witte, K.-D. Vorlop; Catal. Commun. 1 (2000) 15.
[7]   H. Bönnemann, G. A. Braun; Angew. Chem. Int. Ed. Engl. 35 (1996) 1992.
[8]   A. Köckritz, S. Bischoff, V. Morawsky, U. Prüße, K.-D. Vorlop; J. Mol. Catal. A: Chem. 180 (2002) 231.
[9]   V. Morawsky; unpublished results.
[10] H.U. Blaser, H. P. Jalett, D. M. Monti, J. T. Wehrli; Appl. Catal. 52 (1989) 19.

4

Application of New Metal Impregnated Knitted Silica-Fiber on Catalytic Oxidation and Hydrogenation Processes

Ahmad Kalantar Neyestanaki, Päivi Mäki-Arvela, Esa Toukoniitty, Henrik Backman, Fredrik Klingstedt, Tapio Salmi, and Dmitry Yu. Murzin
Åbo Akademi University, Åbo/Turku, Finland

INTRODUCTION

Heterogeneous catalysis is increasingly applied in chemical industries to decrease the raw material consumption, pollutants emission and to improve the product selectivity. Catalytically active metals and metal oxides are usually deposited on a carrier or support. The role of the support is to stabilize the active component in highly dispersed small particles and hereby increase their exposed surface area. The activity and selectivity of the catalyst can be significantly altered by the particle size of the active metal/metal oxide and the pore size distribution of the support material.

Depending on the application, different kinds of support geometries such as pellets, monolithic honeycomb structures and fibers are used. Pressure drop induction becomes of importance when incorporating a catalyst to the reactor system. Pellets can contribute to relatively higher pressure drops. The monolithic honeycomb structures are technologically advanced and can offer low pressure drop induction. They are predominantly applied for environmental catalysis applications and hydrogen peroxide production. In order to lower the pressure drop in honeycomb monoliths, the channel geometry, channel diameters and wall thickness should be optimized. However, heat and mass transfer limitations are usually associated with monolithic honeycomb structures [1, 2]. Different fibers such as ceramic fibers, fiber glass and alumina/zirconia fibers have also been used [3, 4]. Fibrous substrates are of interest since they do not induce pressure drop and can provide additional mixing of the reactants, hence, minimizing mass transfer limitations. Another advantage of fibers is that they can be constructed in vide variety of geometries suitable for different applications. The disadvantages of fibers can be pointed to their lower mechanical strength and their tendency to sinter under steam.

In the present study a novel knitted silica-fiber was developed and employed as a catalyst carrier. Different silica-fiber supported catalysts were prepared and studied for their performance in removal of volatile organic compounds (VOC) emissions, gas-phase hydrogenation of o-xylene, liquid-phase hydrogenation of citral and liquid-phase enantioselective hydrogenation of 1-phenyl-1,2-propanedione.

EXPERIMENTAL

Support Preparation and Characterization

The raw material for the support preparation was a hybrid (organic-inorganic) fiber consisting of cellulose (ca. 67%), polysilicic acid (ca. 30%) and sodium aluminate (ca. 3%). These fibers (VISIL fibers) were produced via a modified viscous process in which the cellulosic component was regenerated simultaneously with the polymerization of polysilicic acid and subsequently impregnated with sodium aluminate. Heat treatment of these fibers up to temperatures of 873 - 1223 K results in burning of the cellulosic component and formation of a coherent filament of silica/silica aluminate (Figure 1a) exhibiting an amorphous character [5]. At temperatures above 1373 K, crystallization of silica starts.

Yarns from these Visil fibers were double knitted to obtain a textile weighing 530 g/m^2. The obtained textile was then washed with alcohol and rinsed with hot distilled water in order to take away the impurities from the spinning bath. The hybrid textile was burned at T > 873 K, at a heating rate of 10 K/min, to produce a knitted silica/silica aluminate fiber structure (Figure 1) having a thickness of 1.5 mm which was used as a catalyst support. The XRD pattern of the catalyst support indicated an amorphous structure. The chemical composition of the support, measured by direct current plasma technique, was: Si = 95.26 wt.%, Al = 3.15 wt.% and Na = 1.59 wt.% corresponding to the chemical formula (computed by stoichiometry) of $Na_{0.59}Al_1Si_{29.05}O_{59.92}$.

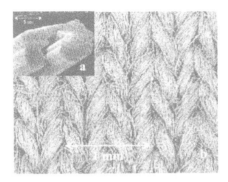

Figure 1 The SEM image of the single fiber (a) and the catalyst support (b).

The specific surface area and pore volume of the catalyst support was determined by nitrogen adsorption using a Sorptomatic 1900 (Carlo Erba Instruments). Increase in the calcination temperature from 973 to 1173 K resulted in a decreased surface area from 184 to 81 m^2/g (Figure 2). The pore volume was also decreased from 0.41 to 0.18 m^3/g. The support sintering was studied by subjecting the support prepared at 1173 K to steam (12 vol% water in nitrogen) at 1073 K for 8 h. The N_2-physisorption data indicated that the water vapor treatment resulted in a decrease of the BET specific surface area of ca. 9% as well as in closure of the pores smaller than 5 nm in diameter. The alkali content of the silica is probably the main contributor to the loss of surface during the hydrothermal treatment [6].

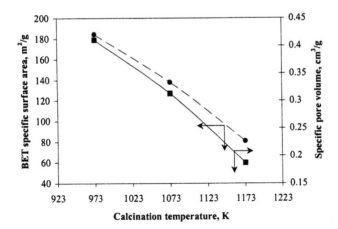

Figure 2 The specific surface area and pore volume of the knitted silica-fiber support as a function of calcination temperature.

Catalyst Preparation

Different knitted silica-fiber supported Pd, Pt and Ni catalysts were prepared by impregnation of the support with the solutions of metal salts. The catalysts for the given application were prepared as:

Catalytic removal of VOC emissions. Pt, Pd and Pd-Pt catalysts were prepared by impregnation of the support precalcined at 1173 K with solutions of H_2PtCl_6 and $PdCl_2(NH_3)_4$. The catalysts were dried and calcined at 823 K for 3 h.

Gas-phase hydrogenation of o-xylene. Supported platinum catalysts were prepared by impregnation of the support, precalcined at different temperatures, with solutions of $Pt(NH_3)_4Cl_2$. The catalysts were dried and stored for the use.

Liquid-phase hydrogenation of citral. The catalysts were prepared by impregnation of the support, precalcined at different temperature, with solutions of nickel nitrate, dried and stored for the use.

Liquid-phase enantioselective hydrogenation of 1-phenyl-1,2-propanedione. Pt-catalysts were prepared by impregnation of the support, precalcined at different temperature, with solutions of H_2PtCl_6 The catalysts were dried and stored for the use.

The catalysts in text will be referred to as the values in parenthesis represent the metal content in weight percentages and the last digits represent the calcinations temperature of the support (*e.g.* the (0.3, 2.1)Pt-Pd/SF-1173 represent a catalyst containing 0.3 wt.% Pt and 2.1 wt.% Pd where the support was calcined at 1173 K prior to metal loading).

Catalyst Characterization

Direct current plasma technique (DCP) was used to determine the metal content of the catalysts. The metal dispersions was measured by hydrogen adsorption at 298 K (363 K for Pd-catalyst). Extrapolation of the adsorption isotherms to zero pressure was applied for the determination of adsorbed hydrogen. The amount of reversibly adsorbed hydrogen was determined by back-sorption method. Dissociative adsorption of hydrogen was considered and the metal particle sizes were calculated assuming a spherical geometry. The mean metallic particle sizes were also investigated by transmission electron microscopy (TEM) and X-ray diffraction (XRD) techniques.

Catalyst Testing

Catalytic removal of VOC emissions. The activity of the catalysts was examined in conversion of a gas mixture containing 200 ppm CH_4, 2500 ppm CO,

50 ppm C_2H_4, 50 ppm naphthalene (model polyaromatic compound), 10 vol.% O_2, 12 vol.% CO_2, 12 vol.% H_2O, and balanced by nitrogen at a GHSV of 20000 h^{-1}. The catalysts were tested for their light-off behavior in a continuos flow tube reactor at temperature ranges of 413 – 1073 K with a heating rate of 3 K/min. The unburned hydrocarbons were analyzed by a HP-5890 GC equpped with a FI detector and a 30 m DB1 column. Carbon monoxide was analyzed using a NDIR instrument.

Gas-phase hydrogenation of o-xylene. Gas-phase hydrogenation of o-xylene was investigated in the temperature range of 360 – 460 K and a GHSV of 11000 h^{-1}. The partial pressure of o-xylene and hydrogen were varied between 0.04 - 0.13 and 0.37 - 0.74 bar; respectively. Argon was used as the make-up gas. Prior to the reaction, the catalyst was reduced in situ in hydrogen flow at 673 K for 2 h. The reaction products were analyzed by a Varian 1400 GC equipped with 60 m HP-1 column (cross linked methyl siloxane) and FI-detector. Calibrated standards were used for peak areas. The products were further confirmed by GC-MS (HP 6890-5973 Instrument).

Liquid-Phase Hydrogenation of Citral. Liquid-phase hydrogenation of citral (34 mol% cis- and 66 mol% trans-citral) was carried out at 343 K in a tubular glass reactor (V = 67 cm^3, d = 18mm, h = 300 mm). Prior to the reaction, the catalyst was reduced in situ at 673 K for 2 h in flowing hydrogen. The solvent (ethanol) and citral (0.1 mol dm^{-3}) were heated and saturated in a different vessel prior to pumping to the reactor where the catalyst was under hydrogen. Hydrogen and the liquid phase were flowing concurrently. Hydrogen passed trough the reactor only once and the liquid phase was recirculated. The citral-to-nickel ratio was kept at 18 in the experiments. The hydrogenation products were analyzed by a GC equipped with FID and capillary column (HP-Innowax) and was further confirmed with GC-MS (HP 6890-5973 Instrument).

Liquid-phase enantioselective hydrogenation of 1-phenyl-1,2-propanedione. Enantioselective hydrogenation was carried out in the batch and fixed bed reactors. The enantioselective hydrogenation was performed at 298 K and 5 bar. In all experiments the concentrations of the reactant and the modifier (cinchonidine) were 0.05 and 0.7×10^{-3} mol dm^{-3}, respectively, and the reactant-to-Pt ratio was maintained at 50.

RESULTS AND DISCUSSION

Catalytic Removal of VOC Emissions

Catalytic removal of emissions from automotive and stationary combustion system has become of great importance in the last two decades. A suitable catalyst for emission control must exhibit low-temperature oxidation activities and high thermal and hydrothermal resistance. Since the operation temperature can readily approach 1073 – 1123 K and large quantities of water (up to 12 vol.%) is present in the flue gases, one should optimize the catalyst's properties to prevent the support sintering. Therefore, for this application a support being calcined at 1173 K for 3 h was employed.

Complete oxidation of model pollutants was achieved under the temperature range investigated. The light-off temperatures (defined as the temperature of 50% conversion, $T_{50\%}$) are given in Table 1. The catalysts were highly active in low temperature oxidation of CO, naphthalene and ethylene. Methane, being the most difficult gas to combust catalytically, required higher oxidation temperatures. Catalyst deactivation under the reactants stream was observed. The combination of Pt and Pd resulted in improved low-temperature activities and decreased the deactivation. The deactivation was found to be the consequence of metal sintering. The sintering of platinum takes place under the oxidizing environment. Whereas, the sintering of palladium is due to decomposition of the PdO (the active state of palladium) at temperature above 923 K to metallic palladium which further agglomerates into larger particles. The H_2-chemisorption data indicated an increase of Pd mean diameter from 6 nm to 8.3 nm during the catalyst ageing.

The temperature programmed decomposition of the fresh and used Pd/SF-1173 catalyst indicated the presence of two PdO species on the catalyst surface. A single O_2-evolution peak with maximum at 903 K took place from the fresh catalyst, whereas, on the aged catalyst two O_2-evolution peaks with maxims at 948 K and 1053 K were observed. Similar twin O_2-evolution peaks are also observed from alumina supported Pd-catalysts [7]. The O_2-evolution peak at lower temperatures is originating from the decomposition of crystalline palladium oxide, while the higher temperature peak is due to the decomposition of the amorphous PdO. A comparison of the performances of the knitted silica-fiber catalysts with those over

Table 1 The light-off temperatures in complete oxidation of model pollutants.

Catalyst	Cycle	Light-off temperature ($T_{50\%}$, K) in:			
		CO	$C_{10}H_8$	C_2H_4	CH_4
(0.3)Pt/SF-1173	1*	470	482	479	923
	2**	523	524	526	977
(2.1)Pd/SF-1173	1	506	515	511	954
	2	546	546	759	994
(0.3,2.3)Pt-Pd/SF-1173	1	488	477	492	946
	2	515	518	521	985

*- Fresh catalyst, **- after being aged under the reactants flow at 1073 K for 6 h.

granules, cordierite monolith and a commercial quartz fiber, in terms of TON, is given in Table 2. The knitted silica-fiber catalysts exhibited good performances as compared to the other geometries. To summarize, the knitted silica-fiber exhibited high thermal and hydrothermal resistance and provided a suitable support for such high temperature applications. Complete removal of the model pollutants was achieved under the sever conditions investigated.

Table 2 Comparison of the TON (at 673 K) for fibrous, granules and monolithic catalysts.

Catalyst	TON x 1000, s^{-1} in oxidation of:		
	CO	$C_{10}H_8$	$CH_4^{\dagger}$
(0.3)Pt/SF-1173	183.09	3.67	2.09
(2.1)Pd/SF-1173	25.31	0.53	0.15
(0.3, 2.3)Pt-Pd/SF-1173	21.40	0.43	0.48
(2.0)Pd/Al$_2$O$_3$ [a]	9.94	0.12	0.79
(6.0, 0.8)Pd-Ce/Al$_2$O$_3$-monolith [b]	29.40	0.59	2.35
(7.0)Pt/quartz fiber [c]	30.87	0.57	2.31

$\dagger$- at 873 K; a- 250-500 micron granules; b- 400 cpsi cordierite (112 g/ft^3 active components); c) commercial fibrous catalyst (Heraeus)

Gas-Phase Hydrogenation of O-Xylene

Hydrogenation of aromatic compounds over Group VIII metals has attracted considerable interest of research, from both theoretical and applied viewpoint [8], as the new environmental regulations strictly limit the aromatic contents of the fuels. In the present study the gas-phase hydrogenation of o-xylene was investigated over Pt/silica-fiber catalysts prepared from the support precalcined at different temperatures. The o-xylene hydrogenation was studied at temperature range of 360 – 460 K and a GHSV of 11000 h^{-1}. The partial pressure of o-xylene and hydrogen were varied between 0.04 - 0.13 and 0.37 - 0.74 bar; respectively.

The H$_2$-chemisorption data indicated that the average platinum particle size was not affected by the metal uptake/loading and all three catalysts had a platinum particle size of ca. 1.5 nm in diameter. The catalyst exhibited high activities in o-xylene hydrogenation. *Cis* and *trans* 1, 2-dimethylcyclohexane were the only reaction products. The hydrogenation rate was found to increase by increased hydrogen partial pressure. Complete hydrogenation of o-xylene could be achieved at 460 K at H$_2$ to o-xylene molar ratio of 12.

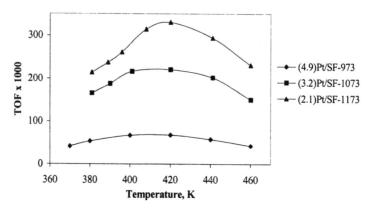

Figure 3 TOF as a function of temperature over the Pt/silica-fiber catalysts (p_{H2} = 0.36. p_{o-xyl} = 0.06 bar).

The temperature dependency of the TOF is given in Figure 3. Since all the catalysts had the same platinum particle size (ca. 1.5 nm) the TOF appears to depend on the support morphology. The catalyst in which the fiber support was calcined at 1173 K exhibited the highest activity. This due to the lack of internal diffusional limitations. The higher calcinations temperature results in closure of smaller pores and decreased pore volume (Figure 2), hence, less internal mass-transfer limitation is taking place.

Cis 1, 2-DMCH hexane was found to be the kinetically favored product whereas the selectivity to the formation of thermodynamically preferred *trans*-isomer increased by increasing the operation temperature (Figure 4). The selectivity to the formation of *trans*-isomer also increases by the enhanced conversion levels. In Table 3 the activity of the fiber supported catalysts in xylene hydrogenation is compared to that of a (4.2)Pt/Al$_2$O$_3$ catalyst having approximately the same mean platinum particle diameter. The (2.1)Pt/SF-1173 catalyst which has the lowest pore volume (Figure 2) is exhibiting similar activity to that of the Pt/alumina catalyst.

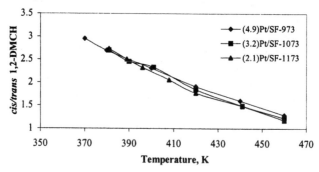

Figure 4 The *cis/trans* 1,2-DMCH dependency on the operation temperature.

Table 3 Comparison of TOF of the fibrous catalysts with that of a Pt/Al$_2$O$_3$.

Catalyst	TOF x 1000, s^{-1}
(2.1)Pt/SF-1173	231
(3.2)Pt/SF-1073	150
(4.9)Pt/SF-973	42
(4.2)Pt/Al$_2$O$_3$ [a]	235

a) 125-15 micron granules (Pt particle diameter = 1.6 nm); p$_{H2}$ = 0.36. p$_{o-xyl}$ = 0.06 bar

Liquid-Phase Hydrogenation of Citral

In citral hydrogenation the aim is to maximize the yield of the unsaturated/desired alcohol, citronellol, which is used in the perfumery industry. The reaction scheme of citral hydrogenation is displayed in Scheme 1. The product distribution in citral hydrogenation is steered by the selection of the catalyst: high selectivities to the desired product, citronellol, can be obtained over nickel catalysts [9].

Liquid-phase hydrogenation of citral (34 mol% *cis*- and 66 mol% *trans*-citral) was carried out at 343 K in a tubular glass reactor over Ni/knitted silica-fiber catalyst containing 5, 10, 15 and 20 wt.% nickel [9]. Citronellal and citronellol were the primary and secondary products. Small amounts of nerol and geraniol were also detected.

The yield of citronellol is given in Figure 5 over catalysts with different metal loading. The hydrogenation rate of citral and the selectivity to citronellol increased with increasing support surface area and metal dispersion (Table 4), characteristic for a structure insensitive reaction. The highest initial hydrogenation rate and 92% selectivity to citronellol (at maximum yield) was achieved over the (5)Ni/SF-973 catalyst. The reaction was highly selective (> 90% yield of the desired product) and exhibited high activity, comparable to a commercial catalyst [9].

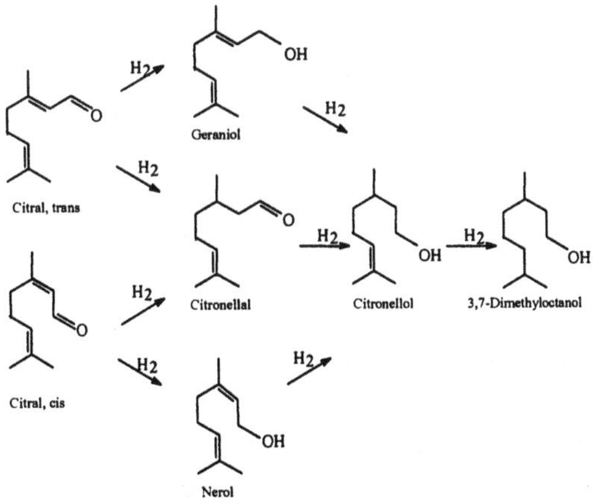

Scheme 1 The reaction scheme in citral hydrogenation over Ni/SF catalyst.

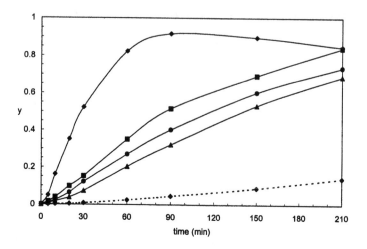

Figure 5 The yield of citronellol in citral hydrogenation over Ni/SF-973 catalysts. (♦)5, (■) 10, (●) 15 and (▲) 20 wt.% Ni. Dashed line: yield of 3, 7-dimethyl octanol over (5)Ni/SF-973

Table 4 The effect of support calcination temperature and catalyst metal loading on the initial reaction rate and the selectivity to citronellol over Ni/silica fibers.

Catalyst	Characteristics		Initial rate, mmol/min.g_{Ni}	$S^{\ddagger}_{citronellol}$, %
	BET^{*}, m^2/g	Ni particle size†, nm		
(5)Ni/SF-973	184	1.3	22	92
(10)Ni/SF-973	184	3.6	14	84
(15)Ni/SF-973	184	5.6	11	74
(15)Ni/SF-1073	138	-	8.9	63
(15)Ni/SF-1173	81	-	8.3	58
(20)Ni/SF-973	184	10.3	6	69

*- BET surface area of the support, †- determined from XRD, ‡- selectivity to citronellol.

Liquid-Phase Enantioselective Hydrogenation of 1-phenyl-1,2-propanedione

Enantioselective hydrogenation in a batch reactor. The liquid-phase enantioselective hydrogenation of 1-phenyl-1,2-propanedione was tested in a batch reactor over Pt/silica-fiber [10]. The hydrogenation route is given in Scheme 2.

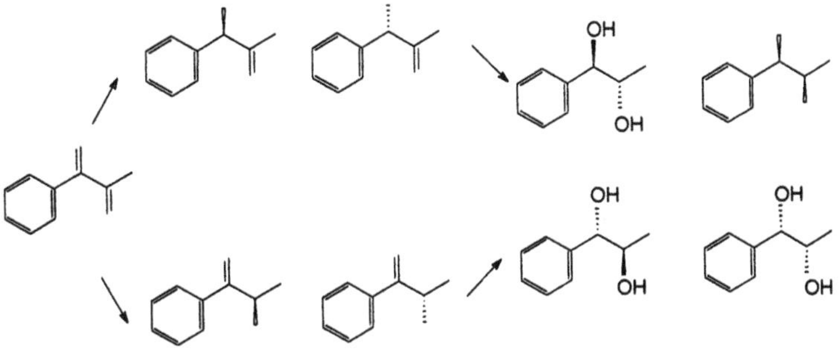

Scheme 2 Reaction scheme in the hydrogenation of 1-phenyl-1,2-propanedione. Symbols: **A**: 1-phenyl-1,2-propanedione, **B**: (R)-1-Hydroxy-1-phenylpropanone, **C**: (S)-1-Hydroxy-1-phenylpropanone, **D**: (R)-2-Hydroxy-1-phenylpropanone, **E**: (S)-2-Hydroxy-1-phenylpropanone, **F**: (1R,2S)-1-Phenyl-1,2-propanediol, **G**: (1R,2R)-1-Phenyl-1,2-propanediol, **H**: (1S,2R)-1-Phenyl-1,2-propanediol and **I**: (1S,2S)-1-Phenyl-1,2-propanediol.

The prochiral substrate, 1-phenyl-1,2-propanedione (**A** in Scheme 2), can be hydrogenated enantioselectiviely by adding small amount of cinchonidine, catalyst modifier, into the system. The catalyst modifier adsorbs on the catalyst surface and steers the adsorption of the reactant in such a way that enantiodifferentiation is enabled and an excess of (R)-1-hydroxy-1-phenylpropanone (**B** in Scheme 2) over the (S)-1-hydroxy-1-phenylpropanone (**C** in Scheme 2) is obtained. The main product **B** is an important intermediate in the synthesis of L-ephedrine, norephedrine, adrenaline and amphetamine [11]. The catalyst properties were optimized for the enantioselective hydrogenation of **A** using crushed silica fiber support material in a batch reactor. The catalysts activities and the enantioselectivities are compiled in Table 5 and Figure 6. The hydrogenation rate increased with decreasing specific surface area and pore volume of the support. The best catalyst, prepared from the support calcined at 1173 K, had the lowest dispersion (27%) and largest mean platinum particle size (3.8 nm, chemisorption data).

These results could be explained by the optimized dispersion and Pt particle size of the catalyst as well as by the suitable catalyst morphology obtained for this particular application. A series of experiments carried out with the best support material calcined at 1173 K containing 2.5, 5, 10 and 15 wt.% of Pt revealed that the hydrogenation activity and enantiomeric excess (defined as : ee = ([B] - [C])/([B] + [C]) x 100%) of **B** strongly depend on the metal loading (Table 5). The best results, *i.e.* the highest activity and ee was obtained for the catalyst containing 5 wt.% Pt (Figure 6). Based on transmission electron microscopy, the best catalyst had relatively broad particle size distribution; 50% of the Pt particles were smaller than 6 nm, with a clear maximum between 2 - 4 nm. Both X-ray diffraction and hydrogen chemisorption data gave a similar trend, *i.e.* increasing the mean Pt particle size with increasing metal content.

Table 5 The effect of support calcinations temperature and Pt-loading on the initial reaction rate and the enantiomeric excess (ee) of (*R*)-1-hydroxy-1-phenylpropanone in the hydrogenation of 1-phenyl-1,2-propanedione.

Catalyst	Characteristics		Initial rate, mmol/min g_{Pt}	ee*, %
	BET, m^2/g	Pt particle size, nm		
(5)Pt/SF-973	184	2.8	10.8	36
(5)Pt/SF-1073	138	3.3	14.0	47
(2.5)Pt/SF-1173	81	2.2	12.0	55
(5)Pt/SF-1173	81	3.8	19.6	54
(10)Pt/SF-1173	81	6.0	2.7	44
(15)Pt/SF-1173	81	6.7	0.7	11

- enantiomeric excess (ee) defined as : ee = ([B] - [C])/([B] + [C]) x 100, nomenclature as in Scheme 2.

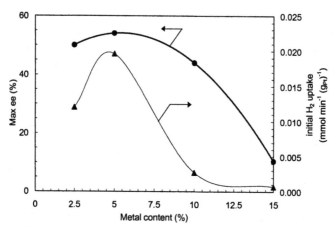

Figure 6 The maximum enantiomeric excess (●) of (*R*)-1-hydroxy-1-phenylpropanone and initial hydrogen uptake (▲) as a function Pt-loading.

It can be concluded that for the hydrogenation of **A** there exists an optimum metal dispersion (ca. 30%) and a mean Pt particle size (ca. 4 nm) for enantiodifferentiation and deviations from this optimum decreased both the reaction rate and enantioselectivity. Large pores are needed to allow facile diffusion of the bulky reactant and the modifier.

Hydrogenation in a fixed bed reactor. After the catalyst optimization in the batch reactor over the catalysts prepared from the crushed support material, a knitted fibrous catalyst containing the optimum amount of metal (5 wt.% Pt over support calcined at 1173 K) was prepared and investigated in a fixed bed reactor.

The results were encouraging, the conversion of **A** (Scheme 2) in the beginning of the operation was high, about 100%, and decreased to 90% during 50 min of reaction, which indicated some catalyst deactivation. The enantiomeric excess of **B** (Scheme 2) increased with increasing time-on-stream from 0% up to the the steady state value of 55%, showing the development of the enantioselectivity to the same level as in the batch reactor (Figure 7). The low initial enantiomeric excess (0%), was caused by the lack of the modifier on the catalyst surface. The modifier feed was switched on simultaneously with the reactant and with increasing time-on-stream, more and more of the modifier was adsorbed on the catalyst resulting in a gradual increase of the enantiomeric excess until the steady state level of 55% was attained. It can be concluded that the continuous operation, using knitted fibrous catalyst proved to be successful in enantioselective hydrogenation of **A,** and after optimization of the operating conditions, the continuous fixed bed operation might offer a competitive alternative for the batch operation.

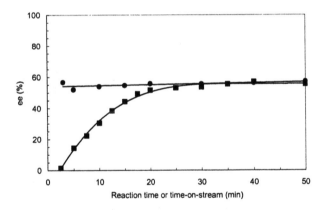

Figure 7 A comparison of the enantiomeric excesses (ee) of (*R*)-1-hydroxy-1-phenylpropanone in a continuous fixed bed reactor (■) and in a batch reactor (●).

CONCLUSIONS

A knitted silica-fiber was produced and employed as a catalyst support. Different Pt, Pd and Ni/knitted silica-fiber catalysts were prepared. The activity of the catalysts was investigated in oxidation of volatile organic compounds, gas-phase hydrogenation of o-xylene, liquid-phase hydrogenation of citral and liquid phase enantioselective hydrogenation of 1-phenyl- 1,2-propanedione.

The support used was found to be very versatile and by varying its calcinations temperature one can readily optimize its surface area and pore size distribution. This in turn enables one flexibly to control the metal loading and mean metallic particle sizes of the final catalysts. After any given application, the novel supported materials excellently kept their integrity. The lifetime of the catalyst, although promising, is still under investigation.

High hydrothermal resistance and metallic dispersions for VOC removal application was obtained by precalcination of the support at higher temperatures. Consequently complete catalytical removal of the model pollutants was achieved.

Efficient gas-phase hydrogenation of xylene requires a highly dispersed catalysts with large pores since higher porosity induces internal diffusion limitations. The goal was achieved by lowering the pore volume and choice of the metal precursor.

In citral hydrogenation, small particles on high surface area support gave rise to the highest activity and selectivity to citronellol. Whereas, since in the enantioselective hydrogenation of 1-phenyl 1,2-propanedione relatively large molecules are involved a larger metal particle will provide the higher rate and enantiodifferentiation. The optimum catalyst was found to consist of relatively large (4 nm in diameter) platinum particles, stabilized on a low surface area support.

ACKNOWLEDGEMENTS

The financial support from the Academy of Finland is gratefully acknowledged. This work is a part of the activities at the Åbo Akademi Process Chemistry Group within the Finnish Center of Excellence Program (2000-2005) by the Academy of Finland.

REFERENCES

1.   A Cybulski, J A Moulijn, Monoliths in Heterogeneous Catalysis, Catal. Rev.-Sci. Eng., 36: 179 - 270, 1994.
2.   J H B J Hoebnik and G B Marin, in Structured catalysts and reactors, ed. A. Cybulski, J. A. Moulijn, New York., Marcel Dekker, 209 - 237, 1998.
3.   V Höller, K Radevik, I Yuranov, L Kiwi-Minsker, A Renken, Reduction of nitrate-ions in water óver Pd-supported on structured fibrous materials, Appl. Catal. B, 32(3): 143 - 150, 2001.
4.   V Höller, I Yuranov, L Kiwi-Minsker, A Renken. Catal. Today, Structured multiphase reactors based on fibrous catalysts: nitrate hydrogenation as a case study 69: 175 -181, 2001.
5.   A Kalantar Neyestanaki, L-E Lindfors, Catalytic Combustion over transition metal oxides and platinum metal oxides supported on knitted silica fibre, Combust. Sci. and Tech., 97:121-136, 1994.
6.   R K Iler, The Chemistry of Silica: Solubility, Polymerization, Colloid and Surface Properties and Biochemistry of Silica Wiley, New York, p.544, 1979.

7.   F Klingsredt, A Kalantar Neyestanaki, R Byggningsbacka, L-E Lindfors, M Lunden, M Petersson, P Tengström, T Ollonqvist, J Väyrynen, Palladium based catalysts for exhaust aftertreatment of natural gas powered vehicles and biofuels combustion, Appl. Catal. A: General, 209: 301 - 316, 2001.

8.   S Smeds, T Salmi, D Murzin, Gas-phase hydrogenation of ethyl benzene over Ni. Comparison of different laboratory fixed bed reactors, Appl. Catal A:General, 201: 55 – 59, 2000.

9.   T Salmi, P Mäki-Arvela, E Toukoniitty, A Kalantar Neyestanaki, L-P Tiainen, L-E Lindfors, R Sjöholm, E Laine. Liquid phase hydrogenation of citral over an immobile silica fibre catalyst. Applied Catalysis A: General 196: 93 - 102, 2000.

10.   E Toukoniitty, P Mäki-Arvela, A Kalantar Neyestanaki, T Salmi, R Sjöholm, R Leino, E Laine, P. J Kooyman, T Ollonqvist, J Väyrynen. Batchwise and continuous enantioselective hydrogenation of 1-phenyl-1,2-propanedione catalysed by new Pt/SiO$_2$ fibres, Applied Catalysis A: General 216: 73 - 83, 2001.

11.   V B Shukala, P R Kulkarni, L-Phenylacetylcarbinol (L-PAC): Biosynthesis and Industrial Applications. World Journal of Microbiology and Biotechnology 16: 499 - 506, 2000.

5

Functionalised Polymer Fibres as Supports for Platinum Group Metal Catalysts

S. Collard
Johnson Matthey, Royston, Hertforshire, United Kingdom

C. F. J. Barnard, S. Bennett, S. H. Elgafi, G. R. Henderson, and G. Sweeney
Johnson Matthey Technology Centre, Reading, Berkshire, United Kingdom

M. Sundell
Smoptech Ltd., Turku, Finland

ABSTRACT

Insoluble polymer fibres are being used to anchor homogeneous catalysts for an increasingly wide range of chemical reactions. Fibres offer significant advantages over polymer beads, being physically robust and simple to recover and recycle. The emphasis of this paper is to compare the use of these catalysts with their more well-known homogeneous equivalents in cross coupling, hydrogenation and oxidation chemistry.

INTRODUCTION

Homogeneous catalysis is now widely used across the organic chemicals industry. It offers a number of advantages over traditional carbon or metal oxide supported catalysts. For example, the design of ligands with particular steric or electronic properties around the metal centre can facilitate both the activity and selectivity of a homogeneous catalyst. The main drawback in the use of these catalysts has often been the need for a clean separation of the catalyst from the product, either on financial or contamination grounds. Heterogeneous catalysts can simply be filtered from the reaction medium, whereas homogeneous catalysts require more sophisticated separation strategies. One such strategy involves anchoring the homogeneous catalyst to an insoluble support. Polymer beads have been used, but not without the penalty of poor physical handling and mass transport limitations across the bead itself. In order to surmount these difficulties, Johnson Matthey and Smoptech have invented a new range of fibrous grafted copolymers, called FibreCat ™. By linking the ligand of a homogeneous catalyst to the fibrous polymer, we have been able to demonstrate the advantage of facile removal from the reaction mixture and potential reuse of the catalyst with minimal loss of catalyst metal into solution. These aspects of FibreCat ™ use have been documented in our earlier paper (1), where we compared FibreCat ™ palladium catalysts with palladium loaded Merrifield beads in a simple Suzuki reaction. The subject of this paper is an extension of the study to cover more challenging systems such as the coupling of chloroaromatics, and also to examine the use of other Precious Metals in FibreCat ™ systems.

FORMATION OF FIBRECAT™ CATALYSTS

Polyethylene fibres (l=250 μm, d=10 μm) are irradiated under a nitrogen atmosphere using an Electrocurtain electron accelerator operating at 175 keV to receive a dose of 200 kGy. Immediately after the irradiation, the fibre is immersed in a nitrogen purged solution of styrene and a phosphine precursor (e.g. styryldiphenylphosphine). The grafting reaction is allowed to continue for 24 h to ensure complete conversion. The poly(ethylene-g-(styrene-co-styryldiphenylphosphine)) fibres are then extracted with dichloromethane to remove any unreacted monomer or homopolymer, and dried under vacuum. The degree of grafting can be determined gravimetrically and by elemental analysis of phosphine. The treated fibre is now ready for reaction with a precious metal salt in order to form the active FibreCat™. (Hartley (2) has also reported a grafting technique using gamma-radiation for the formation of polymer supported metal catalysts).

PALLADIUM CATALYSED CROSS COUPLING REACTIONS

The use of palladium attached to FibreCat™ catalysts in simple Heck and Suzuki reactions has already been described (1). This work has now been extended to cover a wider variety of FibreCat™ systems (Figure 1) in the more difficult Suzuki coupling of aryl chlorides (3, 4). (NB: FibreCat™ 1000-D1, D2 and D3 are experimental samples, whereas 1000-D7, D8 and 1001 are all commercially

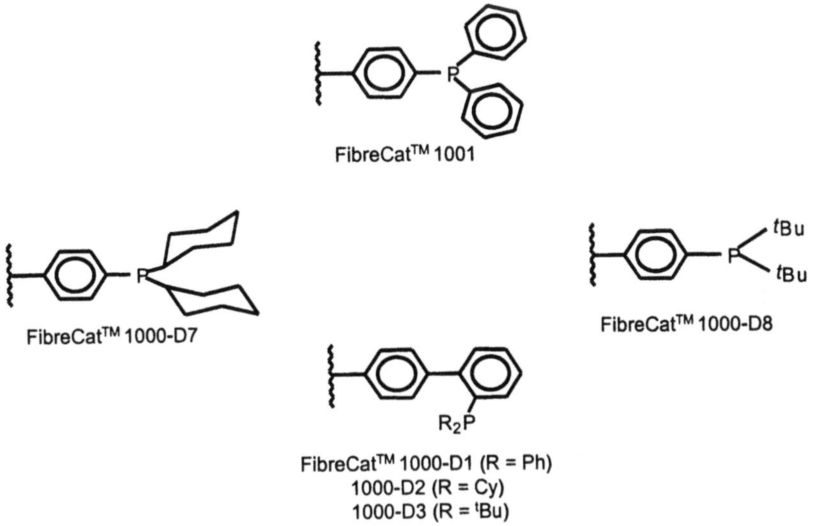

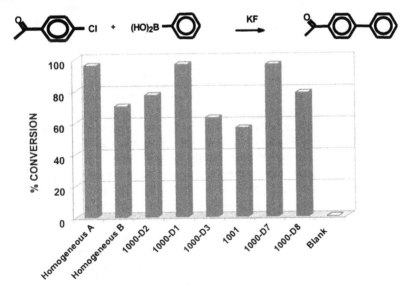

Figure 1 FibreCat™ 1000 series catalysts.

available). 4-Chloro acetophenone and phenylboronic acid were coupled in the presence of a series of FibreCat™ palladium catalysts using potassium fluoride as base (Figure 2). These were compared with a control (no catalyst present) and also two

Figure 2 FibreCat™ in Suzuki coupling of chloro-aromatics.

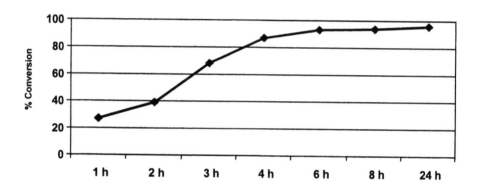

Figure 3 Chloro-aryl Suzuki reaction rate using FibreCat™ 1000-D7.

homogeneous catalysts formed in situ by adding palladium acetate and Buchwald's ligands: (A) 2-(dicyclohexylphosphino) biphenyl or (B) 2-(di-*t*-butylphosphino) biphenyl. Over the course of 24 hours, some of the FibreCat™ samples showed very high conversion to the coupled product: notably FibreCat™ 1000-D1 and 1000-D7. These FibreCat™ samples behave in the same way as the homogeneous equivalents, although the activity is somewhat lower: the latter can give complete conversion within an hour, whereas the 1000-D7, for example, does not exceed 90% conversion until 6 hours have elapsed (Figure 3). Unlike bromo-aryl coupling, where the FibreCat™ catalysts can be repeatedly used in further cycles, we have so far been unable to retain significant activity for recycle of a FibreCat™ from a chloro-aryl coupling after the reaction has gone to completion. As a precaution, these reactions were carried out in an argon atmosphere, although we have no evidence for any serious catalyst oxidation. Moreover, the FibreCat™ catalysts are stored in air without any detriment to their activity. The effect of solvent on FibreCat™ performance has been studied in the Suzuki coupling of 4-bromoanisole to phenyl boronic acid (Figure 4). In this case the catalyst was FibreCat™ 1000-D7 and the base potassium carbonate. Methanol and ethanol are good solvents to use if lower temperatures are necessary. If higher temperatures are allowed, the reaction rates are increased, and those solvents which showed sluggish reaction rates at lower temperatures (e.g. toluene) can become suitable. THF on the other hand is a poor solvent for this reaction. (NB: the 3:1 ethanol water mixture was not included in the 60^0 C screen). FibreCat™ catalysts have shown activity for a broad selection of other substrates. These include activated (electron withdrawing, e.g. 4-chloroacetophenone), mildly deactivated (e.g. 4-chlorotoluene) and more strongly deactivated (electron donating, e.g. 4-chloroanisole) aryl halide bonds.

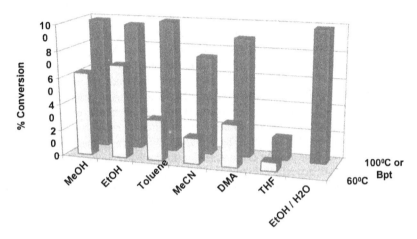

Figure 4 Suzuki coupling of 4-bromoanisole using FibreCat™ 1000-D7: solvent compatibility.

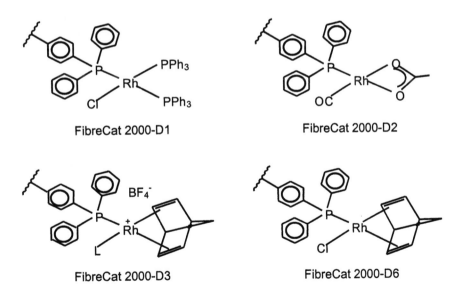

Figure 5 FibreCat™ 2000 series.

SELECTIVE HYDROGENATION USING SUPPORTED RHODIUM CATALYSTS

A number of rhodium catalysts supported on FibreCat™ (the 2000 series, see Figure 5) have been developed for the catalysis of selective hydrogenation reactions (5). FibreCat™ 2000-D1 is the direct analogue of Wilkinson's catalyst. FibreCat™ 2000-D3 and D6 have diene ligands which are removed during hydrogenation to give catalysts with good activity. The hydrogenation of the least hindered carbon-carbon double bond in carvone is illustrated in our first study (Figure 6). The data shows that the supported rhodium catalysts behave in the same way as homogeneous Wilkinson's catalyst, giving dihydrocarvone as the major product. As expected, the palladium on carbon catalyst gives different behaviour, with full hydrogenation to tetrahydrocarvone. The apparent difference in the performance of the FibreCats and Wilkinson's catalyst is due to the latter being allowed to react for longer than was necessary to complete the initial hydrogenation, hence reducing its selectivity below the optimum achievable. We have evidence that these rhodium FibreCat™ samples are air-sensitive. Samples stored in air for approximately three months showed a significant decrease in activity compared with samples stored under argon.

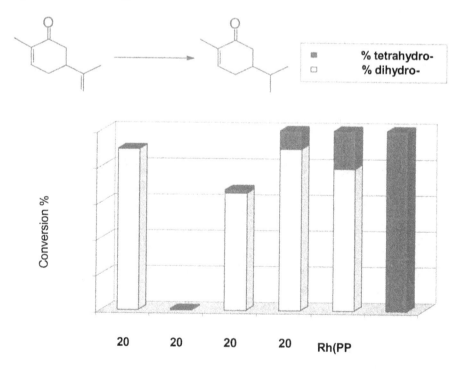

Figure 6 Selective hydrogenation of carvone.

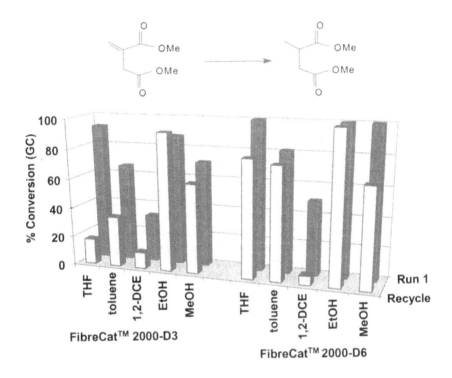

Figure 7 Solvent compatibility in the hydrogenation of dimethylitaconate.

In the second example (Figure 7), the activity of FibreCat[TM] 2000-D3 and D6 were compared in a range of solvents for the hydrogenation of dimethylitaconate to dimethyl methylsuccinate. In each case the FibreCat[TM] catalysts were filtered at the end of the reaction, washed and reused without exposure to air. To some degree catalysis was seen in all the solvents, although in dichloroethane the FibreCat[TM] samples gave poor performance, both in terms of activity and leaching. Both catalysts showed excellent activity and recycle with ethanol, which would therefore be the solvent of choice for this reaction. FibreCat[TM] 2000-D6 was also very active and recyclable in methanol, THF and toluene. After removal of the FibreCat[TM], each of the reaction solutions was wet ashed and analysed for rhodium by ICP-AES. Rhodium levels were found to be below the detection limit (i.e. <0.2 ppm, which is equivalent to <0.02% of the input rhodium). Consequently, rhodium FibreCat[TM] catalysts show very encouraging behaviour for selective hydrogenation.

SELECTIVE OXIDATION USING RUTHENIUM AND OSMIUM ON FIBRECAT™

An increasing awareness of the environment has intensified the search for a benign catalytic route for the oxidation of alcohols to aldehydes and ketones. Selective oxidation using homogeneous ruthenium catalysts has been reported (6), where air or oxygen is used as the oxidant. Despite the advantages of these catalysts however, they have found little use industrially because of the problem of removing the metal complexes from the product after the reaction. This can be overcome by anchoring the catalyst to a polymer. Similarly osmium compounds, such as osmium (VIII) tetroxide, have been known for some time as excellent catalysts for cis-dihydroxylation of carbon-carbon double bonds (7). But again, their large-scale use has been hampered by the problem of preventing the toxic osmium catalyst contaminating either the product or the environment. In a recent paper, Jacobs et al. (8) have demonstrated that osmium catalysts can be anchored successfully to functionalised silica supports. Figure 8 gives some examples of ruthenium and osmium catalysts attached to FibreCat™.

The oxidation of benzyl alcohol to benzaldehyde is reported in Figure 9. Four standard heterogeneous catalysts and two Merrifield bead resins loaded with ruthenium are compared with nine FibreCat™ samples. These were prepared by loading the ruthenate (VI) or perruthenate (VII) salts by ion exchange onto the triethylamine based fibres (as illustrated in Figure 8). In each case 10 mol% catalyst with respect to the benzyl alcohol was added to the reaction. Most of the

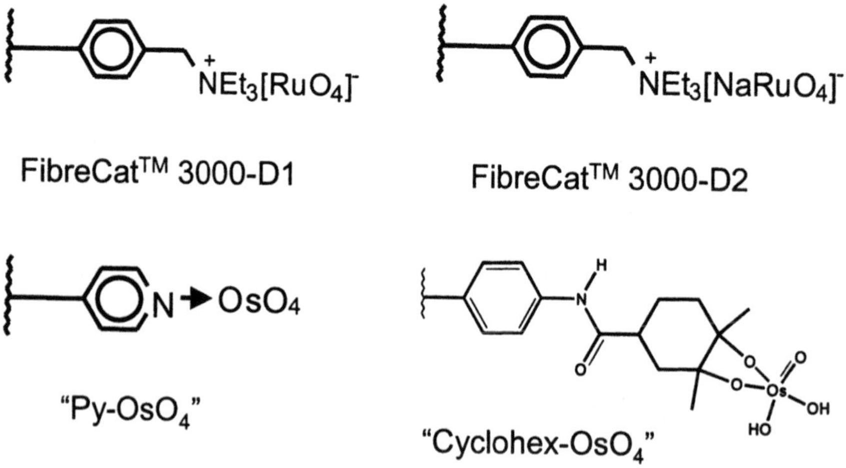

FibreCat™ 3000-D1

FibreCat™ 3000-D2

"Py-OsO₄"

"Cyclohex-OsO₄"

Figure 8 Ruthenium and osmium FibreCat™ for selective oxidation and dihydroxylation.

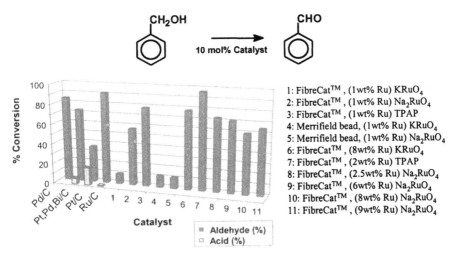

Figure 9 Oxidation of benzyl alcohol to benzaldehyde using ruthenium based Fibre-Cats[TM].

FibreCat[TM] samples showed good conversion to benzaldehyde, particularly the tetrapropylamine perruthenate (TPAP) loaded samples (for example, reaction 7 gave 95% conversion). The Merrifield bead catalysts gave only 10% conversion, while three of the heterogeneous catalysts gave over-oxidation to benzoic acid.

Figure 10 illustrates the use of osmium FibreCat[TM] catalysts for the cis-dihydroxylation of cyclooctene. Two FibreCat[TM] samples (pyridine fibre "Py-OsO₄" and cyclohexyl fibre "Cyclohex-OsO₄") and two co-oxidants (trimethyl-amine-N-oxide and 4-methylmorpholine-N-oxide) were tested for activity over 5 catalytic cycles. Between each cycle the fibre was filtered and washed twice in 4:1 t-butanol/water. The reason for the gradual loss in activity is not currently clear, except to say that it is not due to loss of osmium from the fibre. This is demonstrated in Figure 11, where we plot the amount of osmium found in solution after each of the reactions shown in Figure 10. Most of the samples showed around 5 ppm Os in solution with the highest being 10 ppm. This is no more than a few percent of the input Os. In general, the pyridyl FibreCat[TM] shows the lowest leaching and the best first cycle activity.

CONCLUSIONS

We have been able to demonstrate that polymer supported homogeneous catalysts are proving to be a credible alternative to homogeneous catalysts in a wide range of industrially important reactions. Similar reactivity and selectivity have been observed for the FibreCat[TM] catalysts in comparison with their homogeneous

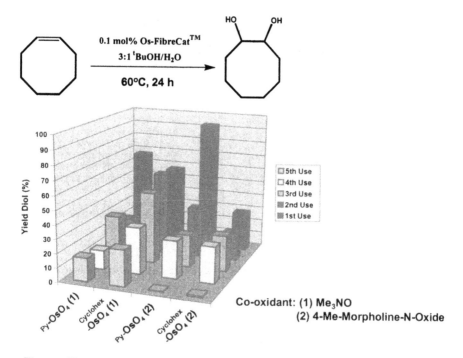

Figure 10 Cis-dihydroxylation of octene using osmium based FibreCats™.

counterparts. In particular they are easily separated from the reaction medium for metal recovery, and can frequently be recycled with little loss of metal into solution. In some circumstances, leaching of metal from the fibre has been observed. This can be due to species in the reaction medium such as amines, which compete with the fibre-bound ligands for coordination to the metal thus releasing it from the fibre. We are currently working on strategies to avoid this scenario. Furthermore, we are working on FibreCat™ systems with an increasing range of ligands, including chiral ligands, in order to enhance the reactivity and usefulness of the catalysts.

EXPERIMENTAL

Coupling Reactions

Suzuki coupling of 4-chloroacetophenone: 1 mol% of each catalyst (Pd) was mixed with 1.0 ml of 1M 4-chloroacetophenone in toluene, phenyl boronic acid (182 mg or 1.5 mmol) and 3mmol of base (174 mg potassium fluoride or 637 mg of potassium phosphate). The mixtures were stirred and heated to 100°C for 24 hours under argon. Samples were centrifuged and the supernatant analysed by GC.

Suzuki coupling of 4-bromoanisole: 1 mol% of catalyst (Pd) was mixed with 2.5 mg of 1M 4-bromoanisole in solvent, phenyl boronic acid (455 mg or 3.75 mmol) and 7.5 mmol of potassium carbonate base (1045 mg). The reaction mixtures were stirred and heated under argon to 60°C or to 100°C or the boiling point of the solvent if less than 100°C (methanol 65°C, ethanol 78°C, acetonitrile 82°C, THF 65°C). Samples were withdrawn and analysed as in the previous experiment.

Hydrogenation Reactions

Hydrogenation of carvone: Each reaction vessel was charged with a solution of carvone in THF (0.5M, 5ml) and the appropriate catalyst (containing 2.5×10^{-5} moles of Rh or Pd). The vessels were purged with nitrogen and then charged to 3 bar with hydrogen. The reaction was run at 70°C under 3 bar of hydrogen gas for 2 hours. At the end of the reaction, the system was depressurized and samples removed for centrifuging and the supernatant analysed by GC.

Hydrogenation of dimethylitaconate: Each reaction vessel was charged with a solution of dimethylitaconate in THF (0.5M, 5ml) and the appropriate catalyst (containing 1×10^{-5} moles of Rh). The vessels were purged with nitrogen and then

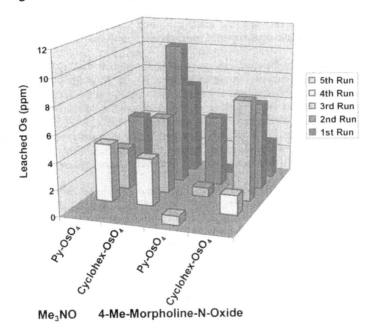

Figure 11 Cis-dyhydroxylation of octene using osmium based FibreCats™: leaching data.

charged to 3 bar with hydrogen. The reaction was run at 70°C under 3 bar of hydrogen gas for 1 hour. Samples were removed and analysed as in the previous example.

Selective Oxidations

10 mol% of each catalyst (Ru) was mixed with 1 mmol of benzyl alcohol in toluene (2 ml). The mixture was stirred and heated to 60°C for 6 hours under 3 bar of air. Samples from each reaction were withdrawn, centrifuged and the supernatant analysed by GC.

0.1 mol% of each catalyst (Os) was mixed with cyclooctene (5 mmol), and either trimethylamine-N-oxide or 4-methylmorpholine-N-oxide trimethylamine N-oxide co-oxidant (6.5mmol) in 17ml ᵗBuOH/Water (4:1). The mixture was heated to 60°C in air with stirring for 24 hours. Samples were then removed for centrifuging and the resultant clear supernatant analysed using GC.

REFERENCES

1. C.J.F. Barnard, K.G. Griffin, J. Froelich, K. Ekman, M. Sundell and R. Peltonen. Fibre-based catalysts: applications as heterogeneous and supported homogeneous catalysts for the fine chemicals industry. In Michael E. Ford, ed. Catalysis of Organic Reactions, Marcel Dekker, New York, 2001, pp 563–571.
2. F. R. Hartley. Gamma-radiation grafting: a novel approach for the preparation of robust and highly selective supported metal complex catalysts. British Polymer J., 16 (4), 199–204, 1984.
3. R. Stürmer. Take the right catalyst: palladium-catalyzed C-C, C-N, and C-O bond formation on Chloroarenes. Angew. Chem. Int. Ed., 38, 3307–3308, 1999.
4. a) D. W. Old, J. P. Wolfe and S. L. Buchwald. A highly active catalyst for palladium-catalyzed cross-coupling reactions: room-temperature Suzuki couplings and amination of unactivated aryl chlorides. J. Am. Chem. Soc., 1998, 120, 9722-9723. b) J. P. Wolfe and S. L. Buchwald. A highly active catalyst for the room-temperature amination and Suzuki coupling of aryl chlorides. Angew. Chem. Int. Ed., 1999, 38, 2413–2416. c) Ligands for metals and metal-catalyzed processes, patent application, WO 00/02887 (PCT/US99/15450).
5. a) R. H. Crabtree. The organometallic chemistry of the transition metals. John Wiley and Sons Inc., New York, 1988. ISBN 0-471-85306-2. b) S. Bhaduri and D. Mukesh. Homogeneous catalysis: mechanisms and industrial applications. John Wiley and Sons Inc., New York, 2000. ISBN 0-471-37221-8.
6. a) T. Mallat. J. Catal., 1995, 153, 131–143. b) B. Hinzen and S.V. Ley. Synthesis, 1995, 977.
7. a) J. Rocek and F. H. Westheimer. J. Am. Chem. Soc., 1962, 84, 2241. b) V. Van Rheenen, D. Y. Cha and W. M. Hartley. Org. Synth., Collective volume 1988, 6, 342.
8. P. A. Jacobs, A. Severeyns, D. E. De Vos, L. Fiermans, F. Verpoort, P. J. Grobet. A heterogeneous cis-dihydroxylation catalyst with stable, site-isolated osmium-diolate reaction centres. Angew. Chem. (Int. Ed.), 2001, 40 (3), 586–589.

6

Nickel Bromide as a Hydrogen Transfer Catalyst

Matthew D. Le Page, David Poon, and Brian R. James
Department of Chemistry, University of British Columbia, Vancouver, British Columbia, Canada

ABSTRACT

Catalyzed hydrogen transfer to a range of organics is achieved using $NiBr_2$ in alkaline, aqueous $NaOH/^iPrOH$ solutions under reflux, and the system seems of practical application for ketones and α,β-unsaturated ketones. The systems are homogeneous, and detailed kinetic studies on the catalyzed hydrogenation of cyclohexanone to cyclohexanol, particularly a 1st- to close to zero-order dependence on $[NiBr_2]$ with increasing nickel concentration and a 2nd-order dependence on $[NaOH]$, are interpreted in terms of an active monomeric Ni-isopropoxide species present in equilibrium with a high nuclearity Ni cluster. The catalytic cycle likely involves hydride transfer from coordinated isopropoxide to coordinated substrate, a conventional step in such catalyzed hydrogen transfer hydrogenations. At the optimum $[NaOH]$ of ~ 0.5 M, there is sufficient trace Ni salt impurity in the base to effect some catalytic transfer hydrogenation in the "base-only" blank systems.

INTRODUCTION

Catalyzed hydrogen transfer from a hydrogen donor (other than H_2) to an unsaturated organic substrate is attractive industrially because of safety, engineering, and economic concerns. The use of alcohols as both cheap and plentiful hydrogen sources under relatively mild conditions has been widely reported, although other donors, such as formic acid, amines, cyclic ethers, aldehydes, and hydrogenated aromatics, have also been used. The relevant literature can be traced through a review (1) and more recent references (2-7), when it becomes clear that most catalytic systems require the use of a base as cocatalyst to "activate" the system, and that the most widely used medium is ^{i}PrOH containing KOH.

During our studies on $NiX_2(PPh_{3-n}py_n)_2$ complexes (X =halogen, py = 2-pyridyl, n = 1-3), some of which are water-soluble (8), a report appeared on the use of a $NiCl_2(PPh_3)_2$ /NaOH/ ^{i}PrOH system for transfer hydrogenation of ketones and aldehydes (9), and we decided to test our Ni complexes under corresponding refluxing conditions. For hydrogenation of cyclohexanone to cyclohexanol (eq. 1), we showed that the Ni-pyridyl complexes were as active as the PPh$_3$ complex, but surprisingly found that NiCl$_2$ · 6H$_2$O was equally effective, while anhydrous NiBr$_2$ was about 6 times more effective in terms of conversion over 1 h (90 vs. 14% conversion). These findings were published in a communication (10), and here we present a more detailed account of the catalyzed hydrogen transfer, emphasizing kinetic and mechanistic aspects of the cyclohexanone reduction.

EXPERIMENTAL

The key components of the catalytic system were obtained as commercial, reagent certified grade compounds: anhydrous NiBr$_2$ (Alfa Chemicals), ^{i}PrOH and NaOH (Fisher Chemicals); the alcohol was distilled from CaO. The halide complexes $NiX_2(PPh_3)_2$ (X = Cl, Br) were made by a literature procedure (8). The commercially obtained organic substrates were distilled under reduced pressure to give materials of > 98% purity as determined by ^{1}H NMR (Varian XL300 or Bruker AC200 machines), and/or GC (Hewlett Packard 5890A instrument, FI detector, a 20 m HP-17 column, He carrier gas); hydrogenated organic products were similarly identified by NMR and GC. UV-Vis and IR spectra were recorded on a Hewlett Packard 8452A diode array and ATI Mattson Genesis FTIR spectrometers, respectively. Electrospray MS data were recorded on a Bruker Esquire-LC instrument in which the MS unit is coupled to a liquid chromatography set-up, so that components can be separated prior to MS analysis.

The hydrogen transfer reactions were performed in a screw-capped, thick-walled 35 mL test-tube, equipped with a Teflon-coated magnetic stir-bar. The tube was charged with NiBr$_2$, NaOH or other base, and ^{i}PrOH, and then heated to 95°C to give a clear pale-green solution; the tube was then cooled to ~

0°C, charged with the organic substrate, purged with N_2 (although this is not essential) and then capped. Up to 10 such vessels were then loaded simultaneously into a 95°C oil-bath to a depth just sufficient to immerse the reaction mixture (typically ~ 3 mL), thereby allowing the remainder of the tube to serve as an air-cooled condenser; ^{i}PrOH boils at 83°C, but the set-up ensured a safe internal pressure, and no significant loss of sample volume was observed over 24 h. Aliquots of solution samples (25 μL) were taken and vacuum distilled to extract the solvent, substrate, and hydrogenated product from the inorganic residue, prior to GC or NMR analysis. The data used to determine initial rates and % conversions were the average from 2-3 injections per sample. For determination of kinetic dependences on cyclohexanone ($C_6H_{10}O$) and ^{i}PrOH, sufficient ^{t}BuOH, a convenient inert diluent (11), was added to maintain a total volume of 3.0 mL, while KF was found to be an innocuous salt for maintaining a constant ionic strength (μ).

RESULTS AND DISCUSSION

Preliminary studies. The earlier work (10) showed that optimum conditions for reduction of cyclohexanone to cyclohexanol (eq. 1) were realized using 5 mM $NiBr_2$, 0.5 M NaOH, and 1.5 M ketone in 3.0 mL ^{i}PrOH, when complete conversion was achieved in 30 min. Under such conditions, other saturated ketones such as acetophenone, butan-2-one and pentan-2-one were correspondingly reduced but longer reaction times (24-48 h) were necessary (10). Cyclohex-2-en-1-one was converted initially to cyclohexanone but this was subsequently reduced at a rate comparable to the rate of appearance of the cyclohexanol; e.g. after 30 min there was 40% reduction of the enone to equal amounts of the saturated ketone and alcohol, while after 48 h there was 73% conversion of enone to essentially just cyclohexanol – no cyclohexen-1-ol was detected (10). Nitrobenzene was reduced selectively to aniline, but the conversion was only 20% after 48 h, while 4-nitrobenzaldehyde gave a mixture of mainly nitrobenzyl alcohol and smaller amounts of aminobenzyl alcohol and the aminobenzaldehyde. Heptan-1-al was also reduced to heptan-1-ol, while benzaldehyde under such basic conditions underwent the Cannizzaro reaction to give the alcohol and benzoate. The findings summarized in this paragraph are all reproducible, while a notable transfer hydrogenation of oct-1-ene to octane noted in the earlier communication has not been duplicated; this will be commented on later.

Reduction of Cyclohexanone (C6H10O)

We investigated mechanistically this "ligand-free" $NiBr_2$ system which, at least for ketone substrates, has activity comparable or greater than that reported for transfer hydrogenation using the more commonly used platinum metal complexes with phosphine and amine ligands (1). The readily reduced $C_6H_{10}O$ was selected as substrate. Of importance, the system is homogeneous. The initially light-colored solutions gradually became yellow and then orange with reaction time but, after filtration through a 0.22 μm pore, the filtrate showed no loss of activity; the addition of Hg(0), an inhibitor for colloidal activity (12), to the catalyst solution gave < 10% decrease in conversion in several ketone systems; and also the kinetic data for the cyclohexanone system were reproducible. Further, the catalyst system could be re-used several times (see below), while colloidal catalysts, unless "stabilized", tend to aggregate with re-use to inactive, heterogeneous metallic species (13).

The conversion vs. time profile depends on the $[C_6H_{10}O]$; from 0.1 to ~ 1.0 M, standard 1st-order behavior is seen (Figure 1), and such data yield a measured pseudo 1st-order rate constant (k_{obs}) of 8.5×10^{-4} s^{-1} for the conditions shown. Of note, corresponding data for 4-Me- and 2-Me-cyclohexanone yield k_{obs} values of 8.0 and 2.5×10^{-4} s^{-1}, the lower value being consistent with inhibited binding of the ketone at the Ni center because of the presence of the *o*-Me substituent. More complex kinetic behavior is seen at higher $[C_6H_{10}O]$, and the kinetic dependences generally are conveniently presented by the use of initial rates (r_o) given in M s^{-1}, determined by the concentration of cyclohexanol generated over an initial time period when the rate remains practically linear. In this way, the general dependence on $[C_6H_{10}O]$ was determined (Figure 2): the rate is 1st-order up to ~ 1 M, but then levels off, and then decreases to zero, e.g. no hydrogenation occurs when $[C_6H_{10}O]$ is > 3 M. A similar complex dependence

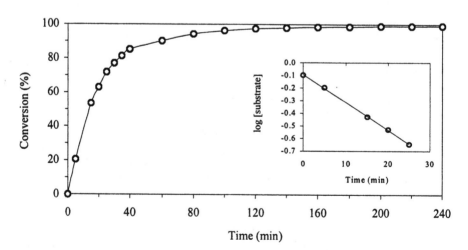

Figure 1 1st-order conversion plots for cyclohexanone; 0.4 mM $NiBr_2$, 0.26 M NaOH, 0.8 M $C_6H_{10}O$ in^{i}PrOH at 95°C.

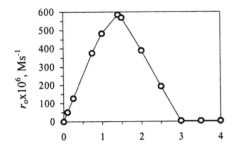

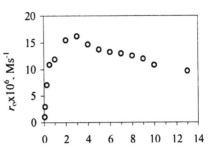

Figure 2 Dependence on [C₆H₁₀O]; 5.6 mM NiBr₂, 0.47 M NaOH, 9 M ʲPrOH at 95°C.

Figure 3 Dependence on [ʲPrOH]; 5.6 mM NiBr₂, 0.47 M NaOH, 0.1 M C₆H₁₀O at 95°C.

on [^{i}PrOH] (neat ^{i}PrOH is ~ 13 M) is shown in Figure 3. The existence of maxima in plots of r_o vs. [H-atom donor] or [H-atom acceptor] for transfer hydrogenations has precedence in both homogeneous and heterogeneous catalytic systems: e.g. Beaupère et al. (14) report an activity maximum at 1.0 M 1-phenylethanol for the reduction of 0.25 M cyclohexenone catalyzed by RhH(PPh$_3$)$_4$ in this alcohol, and Imai et al. (15) report a maximum activity at 0.4 M cycloheptene for its Pd/C catalyzed reduction using indoline as H-donor. The data were rationalized in terms of a competition between the donor and the substrate for the catalyst, and a similar explanation is offered here, with the coordination of ^{i}PrOH to Ni inhibited if [C$_6$H$_{10}$O] is > 1.4 M (Figure 2), and coordination of C$_6$H$_{10}$O inhibited if [^{i}PrOH] exceeds ~3 M (Figure 3). Attempts were made using IR spectroscopy to determine the relative binding affinities at Ni^{II} centers for the ketone substrate and the donor alcohol, although these experiments were limited for practical reasons to studies in C$_6$H$_6$ using NiBr$_2$(PPh$_3$)$_2$ (at 10 mM) as the Ni^{II} center. The addition of 0.2 M ^{i}PrOH, which itself gives bands at 3597 and 2893 cm^{-1}, to such a solution showed a new strong band at 3257 cm^{-1} attributed to a Ni^{II}(^{i}PrOH) species (a 3392 cm^{-1} band has been attributed to a Rh^I(^{i}PrOH) species (16)), while addition of 0.2 M C$_6$H$_{10}$O, which gives bands at 3690 and 1714 cm^{-1}, gives a new strong band at 2863 cm^{-1}, tentatively assigned to a Ni^{II}(C$_6$H$_{10}$O) species. Using an added mixture of 0.2 M ^{i}PrOH and 0.2 M C$_6$H$_{10}$O, the band at 2863 cm^{-1} is evident while that at 3257 cm^{-1} is not seen, implying that the ketone binds more strongly than ^{i}PrOH, albeit within a NiBr$_2$(PPh$_3$)$_2$ precursor. Coordination of the donor and acceptor are considered relevant in the reaction mechanism (see below).

That reaction (1) goes to completion, and that C$_6$H$_{10}$O conversion shows good 1st-order behavior, imply that, despite the production of cyclohexanol and acetone, there is no contribution from the back reaction of eq. 1, which is an example of the well known Meerwein-Ponndorf reversible equilibrium (11).

Consistent with this, addition of acetone or cyclohexanol to the reaction mixture prior to the refluxing procedure has little effect on the hydrogenation rates; in any case, much of the acetone will be in the gas headspace. A metal hydride will attack the ketone with the highest reduction potential (17), and calculations using literature $E°_{1/2}$ values for $C_6H_{10}O$ and acetone, 162 and 129 mV, respectively (18), show that thermodynamically (under standard conditions) the reduction of $C_6H_{10}O$ is favored by a factor of ~13.

Effect of base. Enhanced formation of $^iPrO^-$ is usually cited as the role played by added base in transfer hydrogenation using ^iPrOH (1, 5), but formation of active hydrides at high pH has been demonstrated for some Ru^{II}-PPh_3 systems (17). The kinetic dependence on NaOH was determined by two series of experiments, the first at constant μ of 0.05 M with [NaOH] = 0.00-0.04 M, and the second at μ = 0.60 M with [NaOH] = 0.00-0.60 M. The data reveal a 2nd-order dependence up to ~ 0.5 M (Figure 4); the sharp drop at [NaOH] > 0.5 M results from a solubility limitation of the base in the refluxing medium, as detected visibly. Before considering the kinetic dependence on [NiBr$_2$], the data can be discussed in terms of the mechanism outlined in Figure 5, one that is often postulated for H-transfer from ^iPrOH to a ketone (1, 6, 11, 19): the essential steps almost certainly involve: (a) coordination of $^iPrO^-$ to "Ni" to give **1**, (b) coordination of ketone to give **2**, (c)-(d) β-hydride migration via the metal (**3**) to the coordinated ketone to generate species **4** with coordinated acetone and the new alkoxide. The completion of the cycle, steps (e)-(f), formation and liberation of the product alcohol and loss of acetone, requires transfer of a proton for the overall stoichiometry. A direct 1st-order dependence on [NaOH] could result from a rapid pre-equilibrium involving formation of the $^iPrO^-$ needed for step (a) with K being small, with the $^iPrO^-$ then binding to the Ni, via again a small value equilibrium K_1 (eq. 2). If any of steps (b)-(f) were rate-determining, none would lead to a further 1st-order dependence on [NaOH]. The only

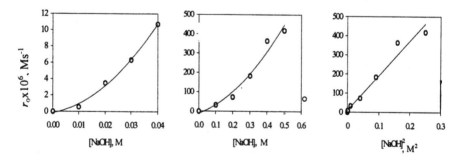

Figure 4 Plots showing dependence on[NaOH]; 5.6 mM NiBr$_2$, 1.41 M $C_6H_{10}O$, in ^iPrOH at 95°C.

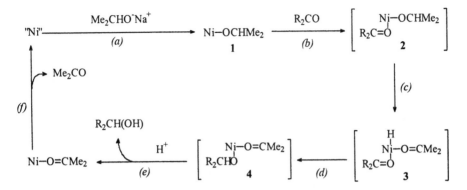

Figure 5 Proposed mechanism for transfer hydrogenation of a ketone, exemplified by the Ni-isopropxide system (auxiliary ligands and charges omitted).

possible source for the "second" [NaOH] dependence must come from a pre-equilibrium involving the Ni species, and one possibility is that **1** is a dibromo(hydroxo) species, eq. 3. If, for example, step (*b*) were rate-determining with rate constant k_b, the rate-law would be of the form shown in eq. 4. The numerator of the rate equation is at least qualitatively consistent with the

$$^iPrOH + NaOH \overset{K}{\rightleftharpoons} {}^iPrO^-Na^+ + H_2O \; ; \; \text{``Ni''} + {}^iPrO^- \overset{K_1}{\rightleftharpoons} Ni({}^iPrO) \quad (2)$$

$$NiBr_2 + OH^- \overset{K_2}{\rightleftharpoons} NiBr_2(OH) \quad (3)$$

$$Rate = k_b[Ni({}^iPrO)_2][R_2CO] = \frac{k_bKK_1K_2[\text{``Ni''}]_{Total}[R_2CO][{}^iPrOH][NaOH]^2}{H_2O} \quad (4)$$

observed 1st-order dependence on R_2CO (cyclohexanone) and on ^iPrOH (at least at lower [^iPrOH]), and 2nd-order in [NaOH]. Of note, step (*b*) was arbitrarily chosen as rate-determining; however, if step (*c*) were rate-determining, which is commonly postulated, the "same type" of rate-law would result as long as step (*b*) was now a rapid pre-equilibrium for binding the R_2CO substrate (again with a small binding constant, so that the kinetic dependence on [R_2CO] remains unity). This rationalization of the rate-law requires that the *dominant* Ni species in solution contains no liganded $^iPrO^-$ or $C_6H_{10}O$, with the *kinetically significant* species being **1**, **2**, or **3**, respectively, depending on whether step (*b*), (*c*), or (*d*) is rate-determining.

The above rate-law requires a direct inverse dependence on [H_2O]. Addition of 0.1% v/v H_2O to a system at optimum conditions decreased the conversion (after 10 min) from 28 to 20%; addition of 1.0% H_2O gave a 19% conversion, while 10.0% H_2O addition gave < 1% conversion; thus the effect is generally smaller than $[H_2O]^{-1}$, but variation of the $[H_2O]$ would almost certainly effect the nature, and distribution, of the Ni species present (see below), and the activity may well then result from contributions from more than one species. The formation of $Ni(OH)_2$ as a precursor catalyst, which could lead to 2nd-order in NaOH, is excluded as it is insoluble in refluxing ^iPrOH. Addition of NaBr has essentially no effect on reaction rates, this ruling out involvement of Br^- in any pre-equilibration of Ni species (e.g. Ni-Br + OH^- ⇌ Ni-OH + Br^-). A pre-equilibrium of the type "Ni" + 2 $^iPrO^-$ ⇌ "Ni($^iPrO)_2$" would also give rise to a 2nd-order dependence on $[OH^-]$, but would also require a 2nd-order dependence on $[^iPrOH]$, and a direct inverse 2nd-order on $[H_2O]$.

Addition of PPh_3 (up to 0.3 M) also has no effect on activity. Addition of PPh_3 to $NiBr_2$ in ^iPrOH readily yields the green $NiBr_2(PPh_3)_2$ species (20), but the species is unstable in alkaline medium (with dissociation of phosphine), implying that in the reported work on the $NiCl_2(PPh_3)_2$ system (9), the precursor catalyst may be simply $NiCl_2$. We found that the activity of $NiCl_2 \cdot 6 \ H_2O$ for cyclohexanone reduction was very similar to that of $NiCl_2(PPh_3)_2$, while $NiBr_2(PPh_3)_2$ was ~1/4 as active as $NiBr_2$ (conversions measured after 1 h) under corresponding conditions.

Nature of the Ni catalyst. The nature of the Ni species present in the strongly alkaline ^iPrOH solutions is unclear, but is of the utmost importance. Initial rate data for the $C_6H_{10}O$ system reveal a 1st- to fractional-order dependence with increasing $[NiBr_2]$ up to ~ 10^{-4} M, but then the rates become essentially independent of $[NiBr_2]$ at higher concentrations (Figure 6). A similar $NiBr_2$ dependence also holds for reduction of butan-2-one to butan-2-ol. An apparent zero-order dependence can result if the catalyst precursor reaches a solubility limit, but this was not evident. An alternative and favored explanation is that catalytically inactive multinuclear species are forming at the higher concentrations; this has been suggested for a Ru^{II}-PPh_3 hydrogen transfer system (17), although no direct experimental evidence was provided. If a monomeric catalyst is involved, and aggregation to inactive species occurs according to the process, n Ni ⇌ Ni_n, then the kinetic dependence on the total Ni concentration will become $(n)^{-1}$ at higher concentrations (e.g. 1/2 order for formation of a dimer, 1/3rd order for a trimer, etc.). For the determined 100-fold increase in $[NiBr_2]$ from ~7 x 10^{-5} to 7 x 10^{-3} M (cf. Figure 6), the essentially zero-order dependence would require n to be large (e.g. for $n = 50$, the rate should increase by about one tenth for the 100-fold increase).

The room temperature UV-Vis spectrum of $NiBr_2$ in ^iPrOH containing 0.47 M NaOH shows an absorption maximum at 432 nm, where Beer's Law is obeyed from (1.0 - 8.0) x 10^{-4} M with a molar extinction coefficient (ε) of ~ 1500; at higher concentrations (> 10^{-3} M), there was an "upward" deviation in the Beer's Law plot that could result from a nucleation process. At the lower [NiBr$_2$], 1.0 x 10^{-5} M, the 432 nm peak was still evident with ε still ~ 1500, but a new absorption maximum at 582 nm was seen (ε ~ 3100), and this could possibly be associated with a catalytically active, monomeric Ni species that aggregates to larger size, less active species at higher concentrations. These room temperature data are qualitatively consistent with the 95°C kinetic data on the Ni dependence. Preliminary ESI-MS data of a solution (preheated at 95°C for 1 h) comprising 5 mM NiBr$_2$, 1.5 M butan-2-one and 0.05 M NaOH (the practical maximum that could be used in the ES-instrument) revealed m/z peaks in the 600-900 range, at least consistent with formation of "clusters". Some of the peaks form an interesting pattern, where the mass units differ by 14: 736 - 750 - 764 - 778 - 792 - 806 - 820 - 834. As the ESI technique tends to give little or no fragmentation, one rationale is that the difference corresponds to the mass difference between Ni(^{i}PrO$^-$) and Ni(^{i}BuO); these species are readily envisaged in the reaction scheme of Figure 5 when using butan-2-one as ketone, i.e. species **1** and **4** (after loss of acetone). The only +1 ion peak that MS software could formulate a structure (with the ligands available) is one at 648, which corresponds to the tri-nickel cluster Br$_2$Ni(μ-OH)$_2$Ni(μ-Br)$_2$Ni(^{i}PrO$^-$)$_2$; if the ion is a +2 species, a related hexa-nickel cluster could be proposed. Thus, ESI data do reveal the presence of Ni clusters, but the nuclearity seems much less than that indicated by the kinetic data; however, the solution used for the MS measurements was much less strongly basic than the catalysis solutions. Use of NiI$_2$ · 6 H$_2$O was as effective as the bromide for transfer hydrogenation of C$_6$H$_{10}$O, but was hygroscopic and was more difficult to work with.

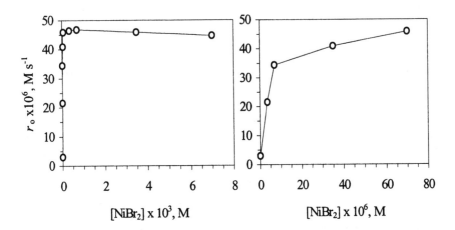

Figure 6 Dependence on [NiBr$_2$]; 0.47 M NaOH, 0.1 M C$_6$H$_{10}$O in ^{i}PrOH at 95°C.

We recently discovered that analytical grade NaOH when used at high concentrations will contribute trace amounts of transition metal salts, and notably Ni is the highest level impurity! Typical impurities are Na_2CO_3 (0.4%), K (0.012%), Cl (0.001%), Hg (< 0.01 ppm), Ag (< 0.002%), Fe (0.0002%) and Ni (0.007%). This implies that at ~ 0.5 M NaOH (optimal for the reduction of $C_6H_{10}O$), the [Ni] is ~ 3.5×10^{-5} M, in the region where the kinetic dependence on $NiBr_2$ is closer to zero- than 1st-order (see Figure 6). Thus, for the data of Figure 6, measured at 0.47 M NaOH, there perhaps should be an 'extra' [Ni] of 3.3×10^{-5} M added to the abscissa values, assuming that the impurity Ni contributes to the catalysis. This modification has little effect on the qualitative analysis of the data of Figure 6 that still points to a monomer/aggregate equilibrium. The kinetic analyses for the dependences on $C_6H_{10}O$, ^iPrOH, and NaOH are unaffected by the trace Ni, in that these experiments were carried out with $NiBr_2$ added in the 5 mM range. This recognition of trace Ni in analytical grade NaOH is important in that our earlier communication noted that transfer hydrogenation was evident also in the blank, base-only reactions, in the absence of added $NiBr_2$ (10); under optimum conditions for the Ni-catalyzed system for reduction of $C_6H_{10}O$ (see Figure 1), when cyclohexanone was fully reduced in 30 min, the base-only system gave ~ 30% reduction, and this almost certainly results from the trace Ni content of the NaOH. Acetophenone was just as effectively reduced in the absence of any added $NiBr_2$, although for most substrates, as with $C_6H_{10}O$, the Ni component significantly enhanced conversions. The quantitative effect of the trace Ni will depend on the kinetic dependence of Ni for the particular substrate (cf. Figure 6). Of note, there are early literature examples of base-only catalyzed hydrogen transfer from alcohols (21, 22); the suggested mechanisms were speculative, and the possibility of Ni impurities was not considered. Of interest, there is a report on a Grignard-type carbonyl addition to alkenyl halides mediated by $CrCl_2$-catalysis in systems that were subsequently found to be highly dependent on the presence of trace Ni salts, but here both metal components were necessary to effect the coupling reaction (23). It certainly seems that trace Ni in the NaOH does play a role in our "base-only" catalyzed systems, especially as preliminary experiments using different batches of NaOH give variable base-only conversions of $C_6H_{10}O$; more careful experiments using bases with well determined trace metal impurities are needed to clarify the situation. Use of high grade KOH (trace Ni concentration unknown) in place of NaOH under corresponding "base-only" conditions gave ~ 20% reduction of $C_6H_{10}O$ vs. 75% for the NaOH system; LiOH was ineffective but this has low solubility in ^iPrOH. Use of weaker bases such as [Me_4N]OH, NEt_3 and piperidine, that were soluble, gave minimal conversions. The use of $CoBr_2$ and CoI_2 under the optimum conditions for the $NiBr_2/C_6H_{10}O$ system resulted in conversions of ~ 60% after 30 min (vs. 100% for the $NiBr_2$ system).

As mentioned, use of base co-catalysts for metal complex catalyzed hydrogen transfer is common, and yet data on the base-only systems are rarely reported (10, 19, 24); our findings suggest that such blank reactions must be investigated within such systems.   The possibility that trace Ag or Fe salts contribute to the base-only catalysis cannot be fully excluded.

Despite the uncertain nature of the catalyst and mechanistic pathways, the simplicity of the system using commercially available $NiBr_2$ makes it attractive for some laboratory hydrogenations without the need for H_2. A further advantage is that the system operates under aerobic conditions, while the often used platinum metal, phosphine-containing complexes are usually air-sensitive in solution (1, 17). Further, the solid inorganic residue obtained after a reduction of $C_6H_{10}O$ could be used for repeat conversions, when only slow deactivation of the catalyst was noted: 1st run, 99.9% conversion after 30 min; 2nd, 95% (3 h); 3rd, 92% (7 h); 4th, 63% (3 h).

We were very disappointed to find that our earlier reported transfer hydrogenation of oct-1-ene, including its selective reduction in a mixture containing oct-1-ene and *trans*-oct-2-ene (10, 25), could not be duplicated. Transfer hydrogenation of olefins usually requires hydrogenated aromatics such as 1,2-dihydronaphthalene and indoline as donors, within homogeneous or heterogeneous systems (15, 26). The explanation eludes us. One possibility, considered remote, is that contaminated reaction tubes were used for this set of experiments; another project within our laboratories involves hydrogenation of aromatic compounds using stabilized Ru and Rh colloidal catalysts (13), and vessel contamination by such species might provide with ^iPrOH an environment for effective olefin reduction.

ACKNOWLEDGMENTS

We thank the Natural Sciences and Engineering Research Council of Canada for financial support.

REFERENCES

1.   G.Zassinovich, G.Mestroni and S.Gladiali, Chem. Rev. 97, 1051 (1992).
2.   K. Puntener, L. Schwink and P. Knochel, Tetrahedron Lett. 37, 8165 (1996).
3.   T. Sammakia and E.L. Strangeland, J. Org. Chem. 62, 6104 (1997).
4.   T.T. Upadhya, S.P. Katdare, D.P. Sabde, V. Ramaswamy and A. Sudalai, J. Chem. Soc., Chem. Commun. 1119 (1997).
5.   P. Barbaro, C. Bianchini, and A. Togni, Organometallics 16, 3004 (1997).
6.   C. de Bellefon and N. Tanchoux, Tetrahedron: Asymmetry 9, 3677 (1998).
7.   M. Yamakawa, I. Yamada and R. Noyori, Angew. Chem. Int. Ed, 40, 2818 (2001); R. Noyori and T. Ohkuma, Angew. Chem. Int. Ed, 40, 40 (2001); T. Ohkuma, H. Doucet, T. Pham, K. Mikame, T. Korenaga, M. Terada, R. Noyori, J. Am. Chem. Soc. 120, 1086 (1998).
8.   I.R. Baird, M.B. Smith and B.R. James, Inorg. Chim. Acta 235, 291 (1995).
9.   S. Iyer and J.P. Varghese, J. Chem. Soc., Chem. Commun. 465 (1995).

10. M.D. Le Page and B.R. James, J. Chem. Soc., Chem. Commun. 1647 (2000).
11. K. Haack, S. Hashiguchi, A. Fujii, T. Ikariya and R. Noyori, Angew. Chem. Int. Ed, 36, 285 (1997).
12. D.R. Anton and R.H. Crabtree, Organometallics 2, 855 (1983).
13. T.Y.H. Wong, R. Pratt, C.G. Leong, B.R.James and T.Q.Hu, Chem. Ind. (Marcel Dekker), 82, 255 (2001).
14. D. Beaupère, L. Nadjo, R. Uzan and P. Bauer, J. Mol. Catal. 20, 185 and 195 (1983).
15. H. Imai, T. Nishiguchi, Y. Hirose and K. Fukuzumi, J. Catal. 41, 249 (1976).
16. D. Beaupère, L. Nadjo, R. Uzan and P. Bauer, J. Mol. Catal. 18, 73 (1983).
17. Y. Sasson and J. Blum, J. Org. Chem. 40, 1887 (1975); A. Aranyos, G. Csjernyik, K.J. Szabo and J-E. Backväll, J. Chem. Soc., Chem. Commun. 351 (1999)
18. H. Adkins, R.M. Elofson, A.G. Rossow and C.C. Robinson, J. Am. Chem. Soc. 71, 3622 (1949).
19. R. Chowdhury and J-E. Backväll, J. Chem. Soc., Chem. Commun. 1063 (1991).
20. L. M. Venanzi, J. Chem. Soc., 719 (1958).
21. Y. Sprinzak, J. Org. Chem. 78, 466 (1956).
22. E.F. Pratt and A.P. Evans, J. Org. Chem. 78, 4950 (1956).
23. K. Takai, M. Tagashira, T. Kuroda, K. Oshima, K. Utimoto and H. Nozaki, J. Am. Chem. Soc. 108, 6048 (1986).
24. A. Mezzetti and G. Consiglio, J. Chem. Soc., Chem. Commun. 1675 (1991).
25. M.D. Le Page, Ph.D. Dissertation, University of British Columbia, Vancouver, 2000.
26. U. Gessner and A. Heesing, Chem. Ber. 118, 2593 (1985).

7

Small-Scale Reactors for Catalyst Evaluation and Process Optimization

Robert L. Augustine and Setrak K. Tanielyan
Center for Applied Catalysis, Seton Hall University, South Orange, New Jersey, U.S.A.

ABSTRACT

There are a number of different reactor systems which can be used to evaluate catalysts and optimize catalytic reactions, but most of the commercially available units are rather large. Reactions run in these systems, because of their size, usually need large amounts of catalyst and substrate so they are not very economical. Further, the time needed for reactor cooling, product removal and recharging for multiple use of the same catalyst can be rather long. It would seem that a more facile approach would be to use smaller reactors for these purposes.

We have developed several different types of small-scale reactors which are used routinely for studying catalytic processes both in the continuous and batch modes. Reaction volumes in these systems range from about 10mL to 200mL for the batch reactors with the continuous systems having a similar small scale. Some of these reactors are used at atmospheric pressure, some at low pressures (<60 psig), and others at high pressures (up to 1500 psig). All are interfaced with computers for continuous data acquisition. Some are computer controlled as well. The design and operation of these reactor systems will be discussed along with the presentation of some representative reaction data.

INTRODUCTION

A catalytic reaction can be run in two ways: using a batch process or in a continuous mode. A reactor used for batch processes is, essentially, a vessel containing the catalyst, the substrate, and the reacting gas in which these three components are placed in intimate contact by some form of agitation. Reactors are designed to be used at specific pressure ranges: atmospheric pressure, low pressure (15–60 psig) or high pressures (60–2000 psig). In a continuous process the substrate and hydrogen move continuously through the catalyst with the product removed at the same rate. There are two general types of continuous reactors; those in which the reactants pass through a slurry of the catalyst particles, and those in which the catalyst particles are packed into a fixed bed through which the reactant fluid (substrate and hydrogen) is passed. Here, too, the reactor design depends on the pressure used in the reaction.

A number of reviews have been published which describe the different types of reactors which are used in both batch and continuous catalytic hydrogenations (1–9). The reactor descriptions are generally generic in nature and are more suited for use in moderate to large-scale preparations than for use in catalyst screening or process optimizations. For these purposes small-scale equipment is usually more appropriate because such systems provide an economy of scale not possible with larger reactors. Not only do they use less catalyst, substrate and solvent, but they are also easier to use in repeated reactions on the same catalyst. Further, control of reaction variables is generally easier to accomplish in small reactors, so the data obtained can be more meaningful. While scaling up to larger reactors can introduce problems related to heat and mass transport, one can usually obtain sufficient data in the small

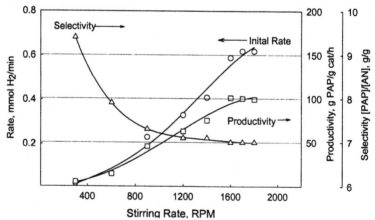

Figure 1 Effect of stirring on reaction rate, product selectivity and productivity in the partial hydrogenation of nitrobenzene to p-aminophenol (11).

reactors to determine how critical each of the reaction parameters are to obtaining the desired reaction product, and thus, which factors need more careful consideration in the scale-up. We have been using small-scale reactors for several years to develop catalytic processes for a number of industrial clients. While proprietary concerns prevent the presentation of any detailed data, there have not been any problems arising in using the data in large reactors.

In order to properly evaluate a catalytic procedure it is necessary that all pertinent reaction data be measured and recorded throughout the course of the reaction. This is usually accomplished by the use of computerized monitoring and data acquisition systems capable of measuring and recording at least the reaction temperature and pressure and the rate of gas uptake, where appropriate, over the entire reaction. With heterogeneously catalyzed reactions, the rate of agitation should also be recorded where possible. Many autoclaves have built-in tachometers which can be monitored. For other reactors, the very least that should be done is to use variable speed stirrers calibrated at several points with a tachometer. For reactions run using a magnetic stirrer a more appropriate solution would be to use a stirrer capable of maintaining a preset rpm regardless of the load in the reactor. Such a stirrer (10) was used to determine the effect of agitation on the rate and selectivity in the partial hydrogenation of nitrobenzene to p-aminophenol as depicted in Figure 1 (11).

Data obtained from a catalytic hydrogenation which is run at constant temperature, pressure and volume are most easily interpreted. With today's technology temperature control is relatively easily attained in most reactors, but maintaining a constant pressure within the reactor while still measuring hydrogen uptake requires special consideration. A common method used to

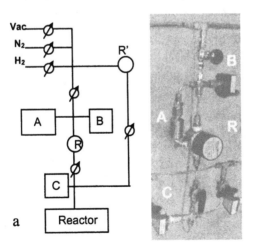

Figure 2 a) Diagram and b) picture of system for measuring pressure drip in a constant pressure reactor.

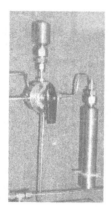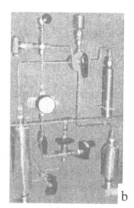

a   b

Figure 3 Multiple reservoir arrangements.

determine the rate of gas uptake while keeping the pressure in the reactor constant is shown in Figure 2a. The reaction gas is initially charged into the reservoir (A) at a pressure significantly higher than the pressure needed for the reaction. A gas regulator (R) is used to maintain the pressure in the reactor while a pressure transducer (B) measures the pressure drop in the reservoir and adjacent tubing and fittings. The volume of this high-pressure region is calibrated so the pressure drop can be used to calculate the moles of hydrogen consumed in the reaction. A second pressure transducer (C) is used to monitor the pressure inside the reactor. The regulator (R') controls the pressure in the reactor during the purging step before the actual reaction. Figure 2b shows a close-up of such an arrangement used in a low-pressure reactor system. Here the reservoir is a 10 cc pressure cylinder, but other sized reservoirs can also be used depending on the amount of gas needed for the particular reaction under study.

Figure 4 Low-pressure hydrogenation reactors.

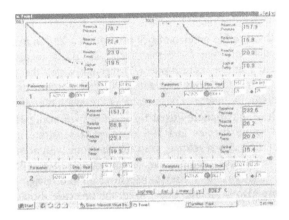

Figure 5 Computer screen showing progress of reactions run in reactors pictured in Figure 4.

Since the maximum pressure in the reservoir is fixed by the pressure rating of the cylinder and other safety considerations, the best way of using such a system for larger scale reactions is to change the size of the reservoir. A facile method for changing the reservoir size is shown in Figure 3 with different sized reservoirs connected to a selection valve. Figure 3a shows a unit with two reservoirs, and Figure 3b pictures one with four reservoirs.

LOW-PRESSURE REACTORS

For efficiency in data acquisition and recording it is possible to interface a single computer to several reactors. Figure 4 is a picture of a four-unit low-pressure reactor system which is somewhat similar to that described by Seif (12). In the present instance, however, the reactors are monitored individually by a single computer with reservoir and reactor pressure and reaction temperature recorded over the course of the reaction. In addition, the room temperature near the reservoirs is also monitored, and this value, along with the calibrated reservoir volume, is used to calculate the moles of gas consumed in the reaction. The computer screen (Figure 5) shows the progress of each reaction by depicting the pressure drop in each reservoir along with the reaction temperature, the moles of gas consumed and the elapsed time of the reaction.

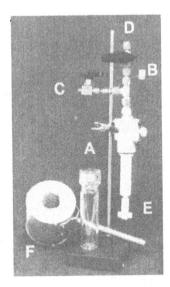

Figure 6 Glass reactor system.

The threaded pressure bottles available from Ace Glass (A in Figure 6) are readily adapted for use as reactors in low-pressure systems. They are connected to the manifold through the connector (C) present on a Teflon head. The fittings on these heads also provide for a thermocouple (B) to measure the internal reaction temperature and a septum port (D) through which the substrate can be introduced into the reactor after pretreatment of the catalyst or samples of the reaction mixture can be taken. These reactors are available in several different sizes. The smallest is commonly used for reactions run using 10–30 µmoles of the catalytically active species and 1–3 mmoles of substrate in 10–20 mL of solvent. The middle-sized reactor is used for reactions with about a 50 mL total liquid volume while the largest can accommodate 100–125 mL reaction volumes. These reactors can be used at pressures up to 60 psig. Larger threaded reactor vessels are also available, but if used, the pressure should be kept below 40 psig with them.

Normally in low-pressure reactors such as these, one would use a stir-bar at the bottom of the reactor along with a magnetic stirrer to provide agitation of the reaction mixture. However, such an arrangement can cause the catalyst particles to be crushed making reaction sampling and catalyst re-use difficult. One way of overcoming this problem is to use a magnetic stirrer with the stir-bar (E in Figure 6) suspended above the bottom of the reactor on a freely rotating shaft. The reactor can be heated using a heating mantle (F). Temperature control of ±1°C can routinely be attained using a control program in which there is a coarse adjustment based on the internal mantle temperature and fine adjustments based on the internal reactor temperature.

RAPID THROUGHPUT SCREENING

Figure 7 Apparatus for rapid screening.

There are a number of different types of commercial reactors available for the rapid screening of catalysts. Some are capable of evaluating several dozen catalysts at a time, but they are usually run at a preset temperature and pressure for a given time with the products analyzed to determine whether a specific catalyst was active or not. This type of reactor is usually not capable of measuring the gas uptake for the individual reactors.

Some others do have this capability but do not operate at constant pressure within the reactor. However, the biggest problem with most of these systems is the small size of the reactors. This may not be a problem with homogeneous catalysts where aliquots of a standard catalyst solution can be used to give identical amounts of the catalytically active species in each reactor. With heterogeneous catalysts, though, these small reactors can frequently necessitate the

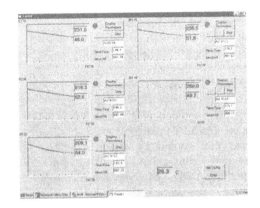

Figure 8 Computer screen showing reaction progress.

use of very small amounts of catalyst which can be difficult to weigh out equally. Another factor here is the possibility that very small portions taken from the same catalyst sample may have different activities.

One way of overcoming these problems is to use somewhat larger reactors set up using the reservoir system shown in Figure 2. The apparatus pictured in Figure 7 is such an arrangement. Each of the five reactors is connected, through a regulator, to a 10 cc reservoir with the pressure monitored by a computer. The pressure drop associated with each of the reactors is shown on the computer screen (Figure 8). The reaction temperature is controlled by the constant temperature bath in which the reactors are situated so all reactions are run at the same temperature. Each of the stirrers of the multi-stirrer is calibrated at several settings. In this way reaction parameters such as pressure, solvent, catalyst type and quantity and rate of agitation can be varied in each individual reactor. The computer monitors the temperature of the bath, the room temperature around the

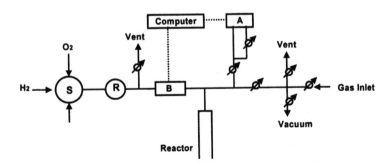

Figure 9 Diagram of an atmospheric pressure hydrogenator (11).

ballasts and the pressures of each reservoir and inside each reactor. The reactor vessels are the same as those pictured in Figure 6 but the heads do not have fittings for internal thermocouples. The small threaded reactors are used this system to screen reactions run using 10–20 umoles of the catalytically active species and 1–2 mmoles of substrate in about 10–15 mL of solvent.

Figure 10 Atmospheric pressure hydrogenator.

ATMOSPHERIC PRESSURE REACTORS

While these low-pressure reactor systems are useful for reactions run at pressures up to about 60 psig (depending on the size of the reactor) they are not adaptable for reactions run at atmospheric pressure. We have, however, designed an apparatus in which gas uptake can be automatically measured while keeping the pressure in the reactor constant at atmospheric (13). This is diagramed in Figure 9 with a picture shown in Figure 10.

The basic component in this system is a sensitive differential pressure transducer (A) with one side measuring atmospheric pressure and the other the internal reactor pressure. The computer monitors the transducer output, and when a slight pressure differential (about a half inch of water) is detected because of the gas uptake in the reactor, the pulse valve (B) is opened to introduce a known volume of gas into the reactor. The size of this pulse is dependent on the gas pressure at the pulse valve and the time it remains open. These factors are adjusted to provide calibrated pulses of near one half cc of reactor gas. The selection valve (S) provides a simple means of using different reaction gases. This apparatus can also be used to study reactions in which a gas is produced.

As discussed above, it can be more efficient to have a single computer interfaced with several of these systems. Figure 11 shows a picture of the screen of a computer interfaced with three of these automated atmospheric pressure reactor systems. Each data point represents the time of introduction of the pulse of reactant gas. For room temperature reactions the standard low-pressure reactors shown in Figure 6 can be used with these systems. When temperature control is needed, jacketed versions of these reactors are used with the temperature maintained by a constant temperature recirculating bath.

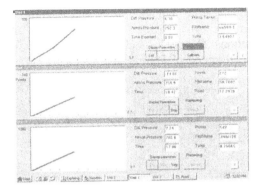

Figure 11 Computer screen showing reaction progress for three atmospheric pressure reactors.

Figure 12 High-pressure autoclaves.

HIGH-PRESSURE REACTORS

Reactions needing pressures between 100 and 1500 psig are run in small stainless steel autoclaves such as those shown in Figure 12. These reactors, which have reaction volumes of 15 mL to 150 mL, are run at constant pressure using the reservoir system described above. Reaction temperatures are maintained by external temperature controllers. A computer is interfaced with all three of the autoclaves to monitor reactor and reservoir pressures and the reaction temperature for each reactor. The stirring rates are also monitored throughout the reaction. The progress of the reaction is shown on the computer screen as the drop in reservoir pressure with time (Figure 13).

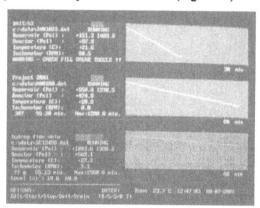

Figure 13 Computer screen showing progress of high-pressure reactions.

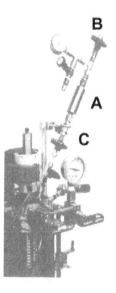

Figure 14 System for introducing substrate into reactor at high pressure.

It is frequently necessary in the use of high-pressure reactors to introduce the substrate after pretreatment of the catalyst without opening the reactor. This can be accomplished using the modification shown in Figure 14. Here the catalyst and some solvent are placed in the reactor and the substrate, either as a liquid or in solution, is placed in the pressure cylinder (A) through the septum port (B). After the pretreatment of the catalyst, the cylinder, A, is pressurized to a value somewhat higher than the pressure in the reactor. The valves (C) connecting the cylinder with the reactor are opened and the substrate is blown into the reactor. These valves are then closed and the reaction initiated. The pressure in the cylinder can be vented through the septum port.

Figure 15 Low-pressure reactor set up as a CSTR.

CONTINUOUS REACTORS

In a continuous process the substrate and hydrogen move continuously through the catalyst with the product removed at the same rate. There are two general types of continuous reactors: those in which the reactants pass through a slurry of the catalyst particles and those in which the catalyst particles are packed into a fixed bed through which the reactant fluid (substrate and hydrogen) is passed (14). Here, too, small-scale reactors can provide a considerable amount of information about a continuous catalytic reaction in a relatively short time. This is particularly true for catalyst activity and longevity studies as well as the effect which reaction parameters can have on product selectivity.

A continuously stirred tank reactor (CSTR) is an adaptation of a batch reactor in which the substrate is added continuously to the reactor while the reaction mixture is removed at the same rate. Since it is possible in such a system that some substrate molecules can move through the reactor unchanged, the product yield will usually be lower than in a batch process. These reactors are sometimes used in series with the reactant stream from one reactor passing into another so there is a stepwise increase in product formation on going from one reactor to the next. Figure 15 shows a low-pressure reactor set up as a CSTR. Substrate input and product removal are accomplished by the use of calibrated HPLC pumps.

Figure 16 pictures a high-pressure autoclave set up as a CSTR. In this arrangement the introduction of the substrate is done using an HPLC pump. The liquid level in the reactor is kept constant using a liquid level controller (A) which is monitored by the computer. When the liquid input reaches a set level an exit valve (B) is opened to drain some of the reaction liquid to maintain the liquid level in the reactor. With such a system there is no need for a second pump to remove the product stream.

One problem with this type of continuous reactor is the potential for catalyst loss during the removal of the reaction liquid. With a powdered catalyst this loss can be minimized by the use of an appropriate filter on the exit line inside the reactor. Catalysts made from granular or extruded supports can be placed in a basket

Figure 16 High-pressure autoclave in CSTR mode.

inside the reactor with the reaction liquid circulated through the catalyst bed by the proper mode of agitation. In this way the catalyst remains in place during the reaction. Figure 17 is a picture of a fixed basket modification of a standard low-pressure reactor head. The basket is made up of two concentric mesh walls a few millimeters apart with space in the center. The catalyst is placed between the mesh walls of the basket. With good agitation by an efficient stir-bar on the bottom of the reactor this arrangement allows the reaction liquid to circulate through the catalyst during the reaction. Catalyst baskets which are either fixed in the reactor or part of the stirrer are commercially available for most high-pressure autoclaves.

Figure 17 Fixed basket system for low-pressure CSTR.

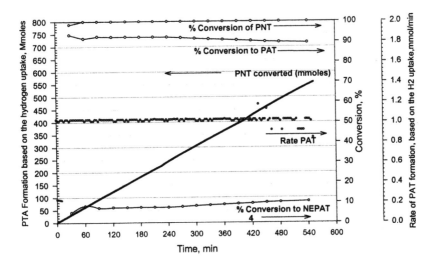

Figure 18 Reaction progress in the CSTR hydrogenation of p-nitrotoluene.

Figure 18 shows the reaction data collected over a six-hour period during the hydrogenation of p-nitrotoluene (PNT) to p-aminotoluene (PAT) using a proprietary catalyst. The reaction was run in EtOH at 400 psig and 140°C with a solution of 0.73 mmole PNT/mL EtOH pumped through the reactor at 1.38 mL/min. N-Ethyl aminotoluene (NEPAT) was formed as a by-product.

Figure 19 is a picture of a fixed bed reactor system. This is a small-scale unit capable of being used with different sized reactor tubes: The standard reactor is a twelve-inch length of heavy walled stainless steel tube, a half-inch in diameter. Other reactors ranging in diameter from one-quarter inch to one inch can also be used. With the smaller diameter tubes, reaction pressures of 1500 psig can be used. Larger reactors can only be used at lower pressures. This system is fully automated with all aspects controlled by a computer. Once the proper parameters are entered into the computer using the set-up screen, the computer takes over operation and cycles the system through purge, pre-hydrogenation, reaction and shut-down modes. A schematic of the system is displayed on the computer screen which indicates which mode is presently in progress and displays all pertinent reaction parameters.

This reactor has three-stage controlled heating with the computer monitoring these temperatures, the inlet and outlet pressures, the liquid flow rate and inlet and outlet reactor gas flows. All parameters have set limits which when exceeded automatically trigger complete system shut-down.

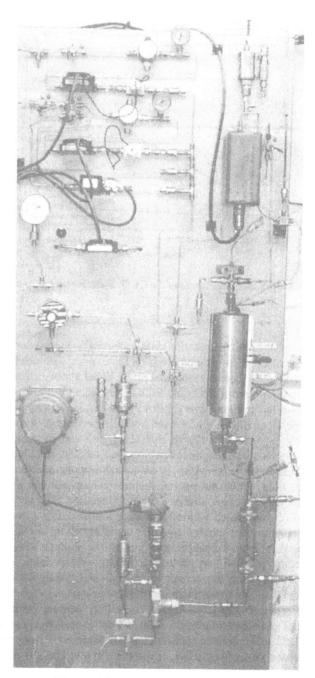

Figure 19 Fixed bed flow reactor.

After leaving the reactor, the reaction mixture goes through a gas-liquid separator with the gas going through the exit gas flow meter and the liquid through an automatic sampler and then into a product storage container.

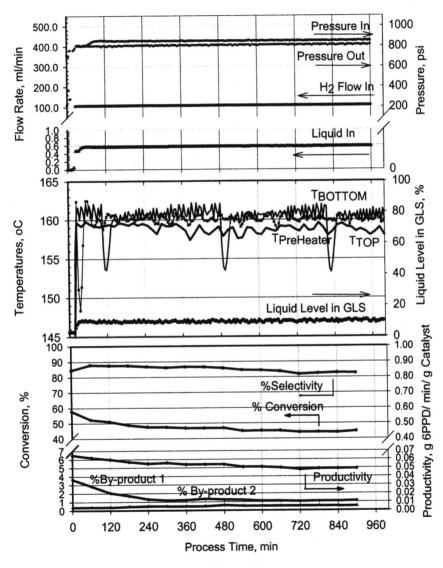

Figure 20 Fixed bed flow reactor data.

Figure 20 depicts the reaction data obtained from a proprietary reaction run in this fixed bed unit over a period of 7-8 hours. These graphs show the various gas and liquid flow rates, gas pressures at both the reactor inlet and the exit flow meter as well as the temperatures of the three reactor stages and the liquid level in the gas-liquid separator (GLS). The product composition is also measured by analyzing the samples taken by the automated sample collector. In this reaction, selectiviiy was a problem because of the formation of two by-products.

The use of continuous reactors introduces an additional reaction parameter which is of importance in optimizing reaction selectivity. In contrast to batch reactors in which the reactants are in contact with the catalyst throughout the entire reaction, in a continuous reactor the time of contact between the reactants and the catalyst can be regulated by changing the flow rate of the reactants through the reactor. The flow rate is usually described as the liquid hourly space velocity (LHSV) which is the number of times a volume of reactant equal to the volume of the catalyst passes through the reactor in an hour. The higher the number, the shorter the contact time. This dependency is shown graphically in Figure 21 which shows the relationship between the LHSV for the reaction with the data depicted in Figure 20 and the selectivities, conversions and formation of the two by-products. Also shown is the reaction productivity which is the amount of product formed per gram of catalyst per minute.

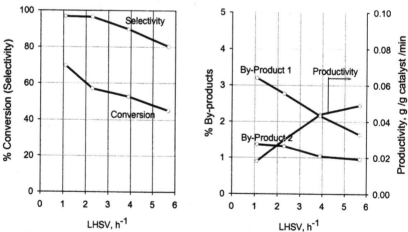

Figure 21 Changes in the reaction characteristics with changes in LHSV.

In a similar vein, Figure 22 shows the effect of reactant flow rate on the amount of PAT formed over time in the CSTR hydrogenation of PNT described above. At the fastest flow rate the substrate is going through the reactor too fast so less nitrobenzene is being hydrogenated and the amount of hydrogenated product accumulated decreases with time.

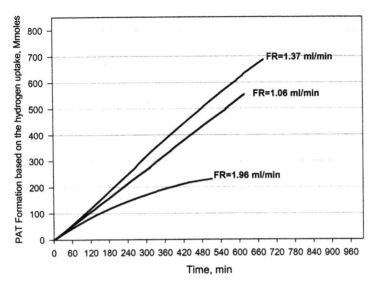

Figure 22 Effect of flow rate on the rate of PAT accumulation.

REFERENCES

1. K.G. Denbigh and J.C.R. Turner, *Chemical Reactor Theory. An Introduction,* Cambridge University Press, Cambridge, 1984.
2. J.F. Jenck, *Stud. Surf. Sci. Catal.,* **59** (Heterog. Catal. Fine Chem. II), 1 (1991).
3. P.L. Mills, P.A. Ramachandran and R.V. Chaudhari, *Rev. Chem. Eng.,* **8**, 1 (1992).
4. J. Shu, B.P.A. Grandjean, A. Van Nest and S. Kaliaguine, *Can. J. Chem. Eng.,* **69**, 1036 (1991).
5. T.T. Tsotsis and R.G. Minet, *Chem. Ind. (Dekker),* **51** (Computer Aided Design of Catalysts), 471 (1993).
6. G. Saracco and V. Specchia, *Catal. Revs.,* **36**, 305 (1994).
7. K.D. Cowan, D.L. Bymaster, H. Hall and E.J. Eisenbraun, *Chem. Ind. (London),* 105 (1986).
8. J.M. Berty, *Applied Industrial Catalysis, Vol 1,* Academic Press, New York, 1983, p. 41.
9. J.H. Rushton, *Chem. Ing. Prog.,* **46**, 395 (1950).
10. VWRbrand Stirrer, Model 400S.
11. J.J. Nair, S.K. Tanielyan, R.L. Augustine, R.J. McNair and D. Wang., *Chem. Ind. (Dekker),* This volume, chapter 52.
12. L.S. Seif, D.A. Dickman, D.B. Konopacki and B.S. Macri, *Chem. Ind. (Dekker),* **53** (Catal. Org. React.), 69 (1994).
13. R.L. Augustine and S.K. Tanielyan, *Chem. Ind. (Dekker),* **53** (Catal. Org. React.), 547 (1994).
14. T. Tacke, I. Beul, C. Rehren, P. Panster and H. Buchold, *Chem. Ind. (Dekker),* **82** (Catal. Org. React.), 91 (2001).

8

Catalytic Hydrogenation of Citral: The Effect of Acoustic Irradiation

J.-P. Mikkola, Tapio Salmi, Päivi Mäki-Arvela, J. Aumo, and J. Kuusisto
Process Chemistry Group, Åbo Akademi University, Åbo/Turku, Finland

INTRODUCTION

Citral (Figure 1) is a molecule with two double bonds and a carbonyl group, displaying a complex behavior upon hydrogenation: many intermediates and end-products - not to speak about the optical isomers - are formed during a typical hydrogenation experiment (1, 2). Citral can be used in the preparation of citronellol (refered as OL hereafter) or citronellal (refered as AL hereafter), to be utilized by the perfume industry etc. due to their pleasant aroma (Figure 2). Acoustic irradiation has been found to enhance heterogeneous catalysis in many cases (3, 4). In hydrogenation of various organic species, Ra-Ni is an often used catalyst. Raney nickel often performs quite well, in terms of both activity and selectivity. However, the deactivation is generally quite notable. Therefore, we tried to apply this technique for two sample systems in hope of retarded catalyst deactivation.

To study whether acoustic irradiation might have a beneficial effect on the hydrogenation velocity, or on the obtained selectivity to the desired products, a series of experiments with on-line unltrasonic treatment of the reaction mixture was performed.

Acoustic energy is mechanical by its nature. Cavitation bubbles are formed, provided that the intensity of ultrasonic field is sufficiently high. Cavitation close to the liquid-solid interface differs from that in pure liquid. Different mechanisms for the effects of cavitation in the vicinity of the surfaces have been proposed: microjet impact and shockwave damage (3, 4). Heterogeneous catalysis involving suspended solid particles in liquid is accelerated by the use of acoustic irradiation. The regenerating effect of acoustic irradiation is proposed to be primarily caused by the localized erosion and grinding generating newly exposed, highly active surface. In this work, we studied on-line acoustic irradiation in batchwise, three-phase hydrogenation of citral to citronellal and citronellol over a Raney nickel catalyst.

Figure 1 Citral molecule (Nemesis molecular modeling, Oxford Molecular Modeling Ltd.).

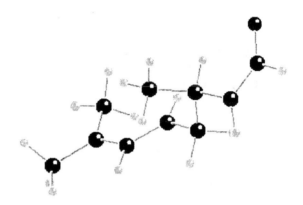

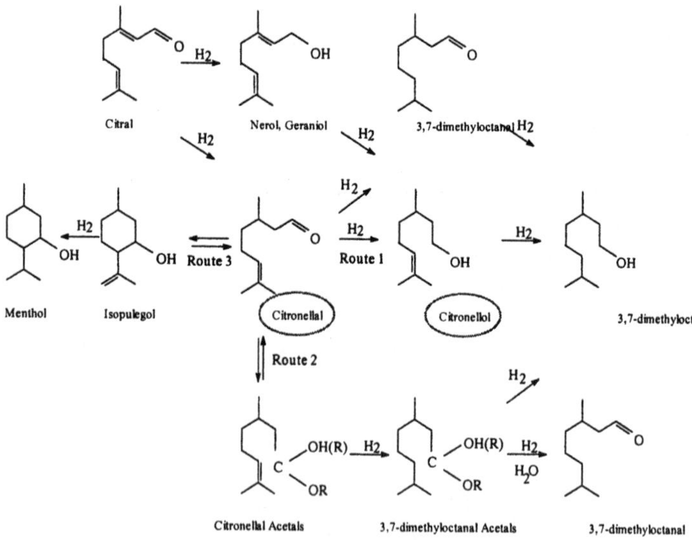

Figure 2 Hydrogenation of citral. The desired end-products, citronellal and citronellol are marked by the ellipses

EXPERIMENTAL

Hydrogenation experiments with simultaneous acoustic irradiation were carried out by using iso-propanol (2-propanol) as solvent in an automatic laboratory-scale autoclave (Parr 4560) with an effective liquid volume of 250 ml. The operating conditions were as follows: 50 bar hydrogen pressure and 70°C (343 K) as the reaction temperature. The catalyst-to-citral ratio was 1:25 (wt:wt) in the beginning of the reaction. A commercial molybdenum promoted Raney nickel catalyst with a mean particle size of approx. 22 μm and the specific surface area in the range of 80 m²/g was used in the experiments. The reactor contents were analyzed off-line with gas chromatography (GC).

A high-intensity ultrasonic liquid processing system was integrated to the reactor autoclave applied (5, 6). A tailor-made connection to the bottom of the pressurized autoclave was prepared. The apparatus had a generator with an adjustable power output (0-100 W, nominal operational frequency 20 kHz,) and a nominal effect of 50% of the maximum – which during the course of reaction gradually diminished to 20% – was used. The strength of the acoustic field was gradually diminished during the course of the experiments. The reason for the effect to diminish was that some of the fines (catalyst) gradually settled at the base of the ultrasonic horn, i.e. the connection between the horn and the reactor vessel and, therefore, the maximum amplitude of the vibrating horn was limited. Thus, the total input of acoustic energy supplied to the system was 0.2-0.08 W/ml, or related to the tip area (1.54 cm²) of the probe 32.5-13 W/cm². However, since the field inside the reactor is not evenly distributed but concentrated to a relatively small area on top of the horn, the molecules and catalyst particles were only periodically exposed to the strongest acoustic field. A schematic drawing of the equipment is displayed in Figure 3.

RESULTS AND DISCUSSION

The hydrogenation results indicated that the hydrogenation velocity, as well as the selectivity to the desired end-products, citronellol and citronellal, were enhanced by the use of on-line acoustic irradiation. Figure 4 illustrates the course of a hydrogenation batch in the presence and absence of acoustic irradiation. As revealed by the figure, in the case of acoustic irradiation the combined yield of OL and AL is very close to 100%, a value slightly higher than obtained by the conventional hydrogenation. Moreover, the optimum batch time is shorter (around 80 min) than in the absence of acoustic irradiation (close to 200 min) in the first batch with a fresh catalyst. The yields of AL / OL were in the presence of acoustic irradiation at various times as follows: 0.69/0.23 at 90 min.; 0.65/0.29 at 120 min. and 0.55/0.38 at 185 min. In the case of no acoustic irradiation the maximum yields were obtained at 205 min.: 0.72/0.16.

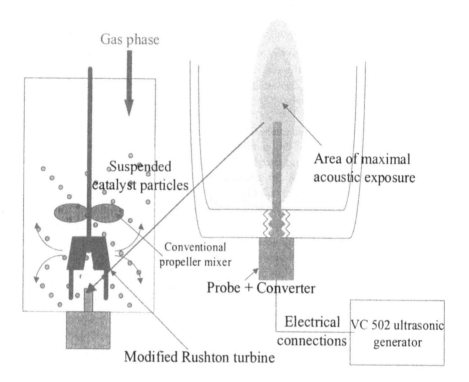

Gas phase

Suspended
catalyst particles

Area of maximal
acoustic exposure

Conventional
propeller mixer

Probe + Converter

Electrical
connections

VC 502 ultrasonic
generator

Modified Rushton turbine

Figure 3 A schematic illustration of the equipment for simultaneous hydrogenation and acoustic irradiation. The enlargement illustrates principally the nature of the acoustic field: it is at its strongest close to the tip of the vibrating horn.

When the catalyst was recycled for use in the second batch, the deactivation was found to be rather severe. This, however, was expected - mainly due to the very small amount of catalyst applied into the hydrogenation batch. Nevertheless, the results indicated that ultrasound enhanced the reaction rate. The maximal yields were 0.60/0.08 at 220 min, compared with those obtained by conventional hydrogenation technology, 0.42/0.06 at 345 min. It is probable that the yield of OL would eventually revert in the absence of acoustic irradiation, as well.

CONCLUSIONS

Catalytic three phase hydrogenation of citral to citronellol and citronellal under simultaneous acoustic irradiation was studied. The effect of acoustic irradiation was focused to the suspended solid particles, i.e. catalyst, residing in the liquid bulk. Thus, we aimed at retarding the catalyst deactivation, mainly. The results indicated that on-line acoustic irradiation during a hydrogenation batch has a significant effect on the hydrogenation velocity as well as the product composition obtained. The desired products were either citronellal or citronellol. The optimum,

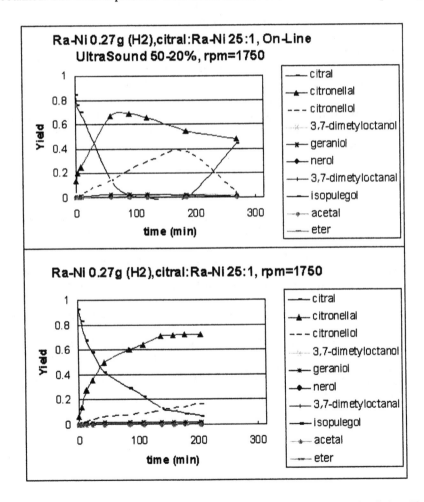

Figure 4 Hydrogenation of citral in the presence and absence of acoustic irradiation. The by-product emerging towards the end of the batch (batch with acoustic irradiation) is isopulegol.

looking for the maximum in combined yields of AL and OL was 0.88 (0.72 and 0.16), in the absence of acoustic irradiation. In the presence of acoustic irradiation the optimum yield was 0.94 (0.65 and 0.29, respectively). In the presence of the acoustic irradiation, the optimum reaction time was around 80-120 min, whereas when the system was not exposed to acoustic irradiation, the optimum was reached around 200 min.

The acoustic field is not homogeneous over the entire reactor volume. This fact, indeed, is one of the complications related to the acoustic irradiation technology: the results tend to be equipment dependent, since the field strength is never homogenous over the complete volume space. Moreover, the design and placement of the horn varies from equipment to equipment.

The experiments illustrated the beneficial effects of acoustic irradiation in the hydrogenation of citral to citronellal and citronellol in a polar solvent. This is in line with previous studies conducted with different chemical systems, such as hydrogenation of xylose to xylitol (5, 6). The future challenge lies in detailed understanding of the microscopic phenomena taking place in the system exposed to the acoustic irradiation: although numerous investigations have been conducted in the use of ultrasound in chemistry, the exact mechanisms still remain a secret to the scientific community.

REFERENCES

1. Mäki-Arvela, P., Tiainen, L.-P., Gil, R., Salmi, T., Kinetics of the hydrogenation of Citral over supported Ni Catalyst, Heterogeneous Catalysis and Fine Chemicals IV: 273-280, 1997
2. Salmi, T., Maki-Arvela, P., Toukoniitty, E., Kalantar Neyestanaki A., Tiainen, L.-P. Lindfors, L.-E., Sjoholm R., Laine, E., Immobile silica fibre catalyst in liquid-phase hydrogenation, Studies in surface science and catalysis, 130: 2033-2038, 2000
3. Suslick, K.S., Science, 247, March 23:1439-1445, 1990
4. Suslick, K.S., Casadonte, D.J., Green, M.L.H. and Thompson, M.E., Ultrasonics, 25, Jan: 1987
5. Mikkola, J.-P. and Salmi, T., Three-phase catalytic hydrogenation of xylose to xylitol - prolonging the catalyst activity by means of on-line ultrasonic treatment, the 16th Canadian Symposium on Catalysis, Banff 2000, Catalysis Today, 64: 271-277, 2001
6. Mikkola, J.-P. and Salmi, T., Proc. of the 7th Meeting of the European Society of Sonochemistry (PROGEP), Biarritz-Guéthary, France, May 2000

9

Catalytic Applications of Aluminum Isopropoxide in Organic Synthesis

James E. Jerome and Rodney H. Sergent
Chattem Chemicals, Inc., Chattanooga, Tennessee, U.S.A.

ABSTRACT

The well-established utility of aluminum isopropoxide (AIP) as a catalytic agent in numerous areas of organic synthesis is discussed. Classic Meerwein-Ponndorf-Verley (MPV) reactions for the reduction of aldehydes and ketones to alcohols are summarized. This review emphasizes improved catalytic activity through AIP derivatives (chloro-, TFA) and co-catalysts (AlCl₃, aluminum *sec*-butoxide, ZnCl₂). The use of aluminum isopropoxide for the preparation of other catalysts is also discussed where AIP is the aluminum alkoxide of choice for preparing higher alkoxide and acylate Al-catalysts. The use of AIP and its derivatives has unfortunately been supplanted in many cases by the use of hydrides such as LiAlH₄ and NaBH₄, which may cause it to be overlooked by modern practitioners of the synthetic organic art. This review provides a look at the catalytic versatility of AIP and an extensive bibliography of the aluminum isopropoxide literature as an aid to those current researchers who wish to explore its utility in their own work. Additionally, we present some results from current work on alternative approaches for catalysis with AIP as a comparison with the existing literature work on some well-known systems.

INTRODUCTION TO ALUMINUM ISOPROPOXIDE AND OTHER ALKOXIDES

Aluminum alkoxides, particularly those formed from secondary alcohols, have been of interest to synthetic chemists since the mid-1920s due to their catalytic activity. Examples of these trialkoxides include aluminum isopropoxide (AIP) and aluminum sec-butoxide (ASB). They are easily prepared at lab or plant scale and provide highly selective reductions and oxidations under mild conditions. These reductions are termed Meerwein-Ponndorf-Verley (MPV) reactions after the chemists (1-3) who first investigated their utility. Because a MPV reaction are accuratelybe described as an equilibrium process, the reverse reaction (oxidation) can also be exploited. These associated reactions are termed Oppenauer oxidations (4). Meerwein-Ponndorf-Verley reductions and Oppenauer oxidations as well as other reaction types and applications will be discussed, but first some background is provided concerning structure, preparation, and characterization of aluminum isopropoxide and related compounds.

Aluminum alkoxides have been historically prepared (5-7) from aluminum metal and the corresponding alcohol usually catalyzed by a Hg-compound with or without the presence of iodine. One of the earliest preparative methods (5) in the literature from 1922 utilized aluminum turnings from commercial pipe as Al- feedstock, and the reaction was purported to have considerable violence (8). This reaction violence was shown later to be an artifact of the alloying metals incorporated in commercial Al piping stock available at that time (6). Higher $Al(OR)_3$ compounds (where R= n-butyl, sec-butyl, tert-butyl, longer chain alkyls, aryl, and aryls substituted with alkyl, alkoxy, or halogens) are routinely prepared by transalcoholysis reactions with simpler Al-alkoxides such as $Al(O^iPr)_3$, $Al(OEt)_3$, or $Al(OMe)_3$. Current manufacturing processes have been developed and are being utilized that avoid mercury catalysts completely to eliminate the health risks and regulatory demand associated with mercury compounds. Researchers also have available the route where alkyl-Al compounds are reacted with the desired alcohol to form the Al-alkoxide directly or *in situ*. Some more exotic preparative methods for aluminum alkoxides are also known. One such method involves heating the selected alcohol to convert it to vapor and passing the organic reactant gas flow through an aluminum metal packing (9,10) then flushing out the liquid AIP product from the metal packing with the recirculated flow of unreacted alcohol vapor. A method(11) has also been patented which employs aluminum dissolved in gallium or gallium alloys to make the desired trialkoxide. Of course, this method will never achieve widespread use as the cost of gallium is prohibitive.

Commercially, aluminum isopropoxide is available as a solid, or as a solution in an inert solvent like toluene or hexane, and, rarely, as a liquid. The world production of aluminum isopropoxide is approximately 5000 metric tonnes. Approximately 40% of the world market is in the United States. The largest consumers of aluminum isopropoxide are manufacturers of rheology modifiers for printing inks and industrial greases, and producers of high purity alumina and aluminum salts. The largest usage for catalyst applications is in the synthesis of agricultural, flavor and fragrance, and pharmaceutical compounds. Similarly, the liquid aluminum sec-butoxide is used in large-scale applications and is available in several grades with the double-distilled form being the highest. Aluminum isopropoxide is less expensive than complex metal hydrides like LiAlH$_4$, NaBH$_4$, and derivative compounds like sodium bis(2-methoxyethoxy) Al hydride (Vitride®) (12) and does not require the use of peroxide-forming solvents. Aluminum isopropoxide is readily hydrolyzed by the presence of moisture, but does not have the additional sensitivity to oxygen (air), or carbon dioxide that is seen in complex metal hydrides.

Compounds of the entire series of aluminum trialkoxides are often represented as monomers displaying three alkoxide ligands around the trigonal planar aluminum center. This is done more for convenience in viewing and performing stoichiometric calculations than anything else. However, in truth the simple tris-alkoxides are not monomeric, but dimeric, trimeric and tetrameric in nature. Aluminum isopropoxide has never been isolated as a monomer, even in the vapor phase (13,14) An extensive kinetics study of the interconversion of the various forms of aluminum isopropoxide has shown that the compound exists in a dimeric form in the vapor phase and in liquid phase at elevated temperatures (14). The dimeric form of aluminum isopropoxide shown in Figure 1 has

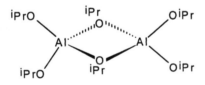

dimer

Figure 1 Aluminum isopropoxide dimer.

the same bridging structure seen for Al(O^{t}Bu)$_3$ and is correctly described as [μ_2-(O^{i}Pr)Al(O^{i}Pr)$_2$]$_2$. Aluminum isopropoxide, when freshly made, is a liquid containing primarily a trimeric species in which Al centers exist in tetrahedral environments. This liquid has been termed (15) a super-cooled melt due to the temperature dependent interconversion of oligomeric units. Pertinent data from thermogravimmetric methods (TGA, DTG, DSC, etc.), which would lend further insight and confirmation of these thermal conversions, is not available. These studies are inexplicably missing from the body of research concerning aluminum isopropoxide. Slowly over several days to several weeks (16-18), this same material undergoes a structural unit conversion to a solid, tetrameric species with one 6-coordinate (octahedral environment) Al center and three 4-coordinate (tetrahedral environment) Al centers per repeat unit. The tetramer as a crystalline solid has a reported melting point of 127-128°C (15), which the authors have confirmed (127.7°C) through DSC measurements. Interestingly, the solid AIP remains in the tetrameric form when dissolved until heated, whereafter a temperature dependent conversion back towards the trimeric form is initiated. To counter this known conversion to the solid tetrameric form, methods of stabilizing small alkyl, liquid aluminum alkoxides such as the tri-*n*-butoxide, triisopropoxide, tri-*n*-propoxide, triethoxide, and trimethoxide have been developed. One stabilization approach (19) involves mixing the Al alkoxide with aluminum tri-*sec*-butoxide to prevent solidification. Alternately, the inclusion of small molar percents of a longer chain alcohols like decyl, nonyl, iso-octyl, or 2-ethylhexanol during the synthesis of aluminum isopropoxide gave clear, stable liquids even after standing for five months (20). Another approach avoids the lower alkoxides entirely by reacting mixtures of n-butyl and isobutyl alcohol to produce stable liquid solutions in concentrations of up to 90% aluminum alkoxide (21). Greco and Triplett (22) patented still another stabilization method where aluminum trialkoxides are mixed in an organic cosolvent with phosphate compounds where the phosphorus center has alkyl, alkoxy, or aryloxy ligands.

The structure of solid-state aluminum isopropoxide was determined at 25°C in 1979 by Turova et al. (23) confirming Bradley's assertion of a possible non-cyclic tetrameric form (24). Folting and coworkers studied crystalline aluminum isopropoxide by X-ray diffraction as well as proton and ^{27}Al NMR. They determined it to be a tetrameric material having the following formula, $Al[(\mu_2\text{-}O^iPr)_2Al(O^iPr)_2]_3$, which supports earlier structural proposals (24,25) and the spectroscopic data of other researchers (15,18). The tetrameric form of aluminum isopropoxide is depicted in Figure 2 where the octahedral, central Al

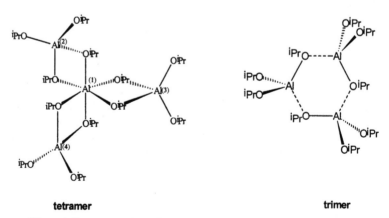

Figure 2 Tetrameric and trimeric forms of aluminum isopropoxide

atom is μ_2-bridged to three tetrahedral Al centers through the isopropoxy oxygen atoms. The terminal isopropoxy groups show the shortest Al-O (1.70Å) distances in the molecule. The six-coordinate Al surrounded by four tetrahedral Al atoms seen in Figure 2 is the structural unit confirmed for tetrameric solidified 'Al(O^{i}Pr)$_3$', while the "melt" form of aluminum isopropoxide is composed of primarily 4- and 5-coordinate Al atoms (25-27).

Cryoscopic and NMR investigation (14) indicates that starting from the pure dimer, aluminum isopropoxide undergoes a very fast equilibrium shift to a mixture containing mainly the trimeric form followed by a quite slow conversion to the tetrameric form. Conversely, starting from the purely tetrameric form, an equilibrium mixture is slowly obtained consisting of all three forms.

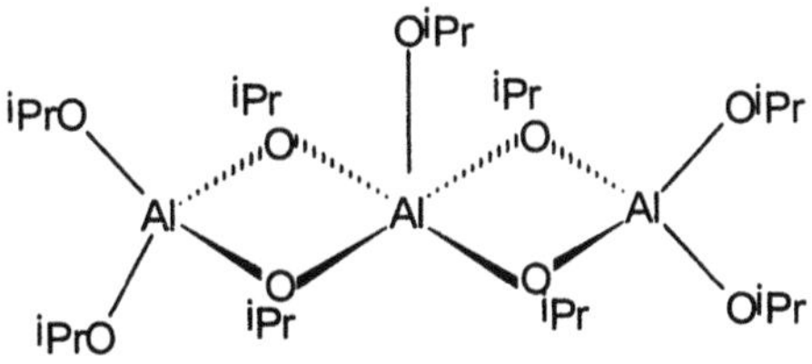

Figure 3 Proposed alternate trimeric form of AIP (From Ref. 14).

Kleinschmidt (14) postulated an alternative trimeric form containing two 4-coordinate Al centers and one penta-coordinate aluminum as shown in Figure 3. He proposed a tetramer to dimer conversion mechanism (Scheme I) and a trimer to hexamer conversion mechanism (Scheme II). Both of these proposed mechanisms involve species containing 5-coordinate Al centers.

CLASSIC MEERWEIN-PONNDORF-VERLEY-OPPENAUER (MPVO) REACTIONS

In 1925, Meerwein and Schmidt (1) published their work on reducing an aldehyde with ethanol in the presence of aluminum tris-ethoxide. A few months later, Verley reported (3) the reduction of butryaldehyde by geraniol using the same aluminum alkoxide. In 1926, the versatility and selectivity of the reduction was illustrated by Ponndorf's investigation (2). He extended the use of aluminum alkoxide reductions of aldehydes to ketones by employing easily oxidized secondary alcohols and their corresponding Al-alkoxides, particularly isopropanol and Al(O^{i}Pr)$_3$. A salient point he brought to the attention of chemists was that the reversibility of this reaction could be exploited via simple Le Chatelier manipulation whereby evaporative removal of the resulting aldehyde product caused a corresponding increase in the reductive conversion percent. Ten years later, Lund built on the previous MPV work by demonstrating the reaction successfully on a large number of carbonyl-containing substrates (8,28).

By the time comprehensive reviews by Wilds (29) in 1944 and Djerassi (30) in 1953 became available, MPV reactions were a standard reductive technique in the organic chemistry community. For example, a 1945 patent (20) describes the utility of using aluminum alkoxides in the presence of an organic nitrogen as a weak base for the reduction of carbonyl groups on oxo compounds such as 7-hydroxy-cholesterol acetate, benzaldehyde, cinnamic aldehyde, and citronellal.

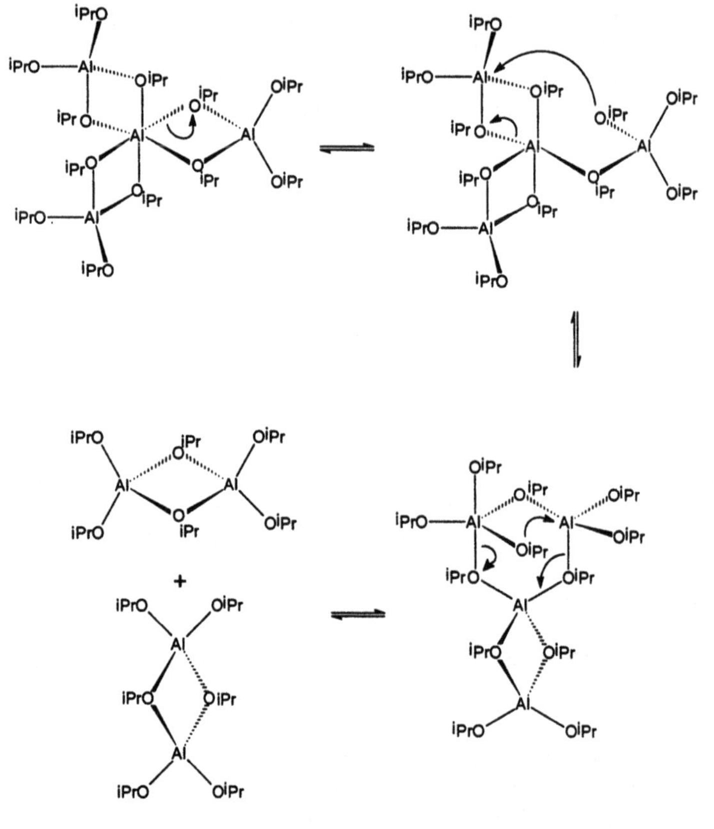

Scheme I Proposed tetrameric to dimeric AIP conversion mechanism (From Ref. 14).

Successful MPV reduction of aliphatic β-aminoaldehydes to their corresponding β-aminoalcohols was reported by Hayes and Drake (31). However, since the introduction of LiAlH$_4$ in 1947 and the growth in usage of complex metal hydride reducing agents, aluminum alkoxides have fallen out of favor in organic synthesis.

In the same year that Lund's work was published, Oppenauer reported (4) the oxidation of steroids containing secondary alcohol moieties in high yield and without harsh conditions using acetone in benzene as the reductant, with aluminum *tert*-butoxide as the catalyst. Oppenauer also investigated using aluminum isopropoxide and aluminum phenoxide as catalysts for the same purposes. The application of oxidants like benzophenone, cyclohexanone, and quinones that are readily reduced expanded the use of Oppenauer oxidations to the conversion of primary alcohols into their corresponding aldehydes. A great number of oxidative methods for alcohols are available, but a distinct advantage of Oppenauer reactions over these alternative methods is that carboxylic acid products from overoxidation are avoided. More recently, Lehman (32) reviewed the literature on Oppenauer oxidations up to 1974 with a particular emphasis on reactions involving steroids and alcohols containing nitrogen, which are of great utility in natural products synthesis.

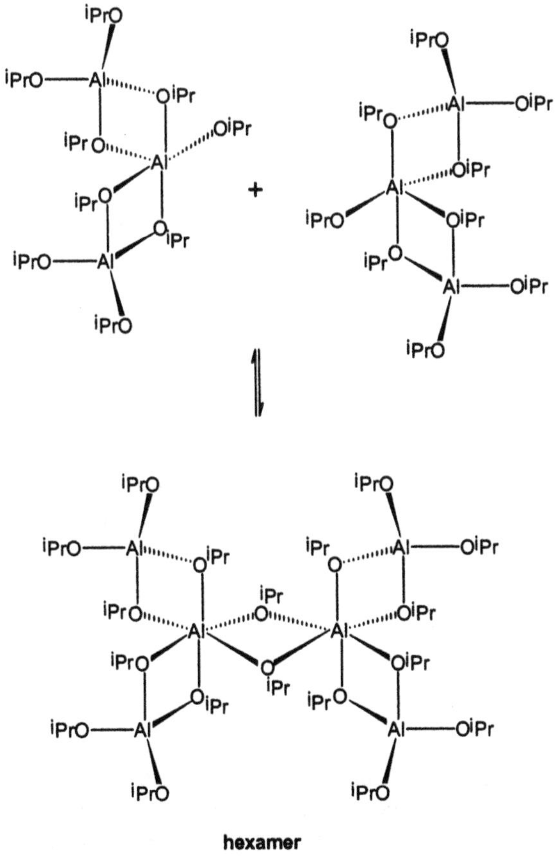

hexamer

Scheme II Proposed trimeric to hexameric AIP conversion mechanism (From Ref. 14).

MECHANISTIC DISCUSSION

In support of the 6-membered transition state mechanism (28) for MPV reductions depicted in Scheme III, deuterium tracer studies by von Doering and Ashner (33,34), and by Williams and coworkers (35) show that the source of hydride transferred to the carbonyl substrate is indeed the carbinol carbon. Later data from Moulton, et al. (34), in 1961 did not preclude the existence of a non-

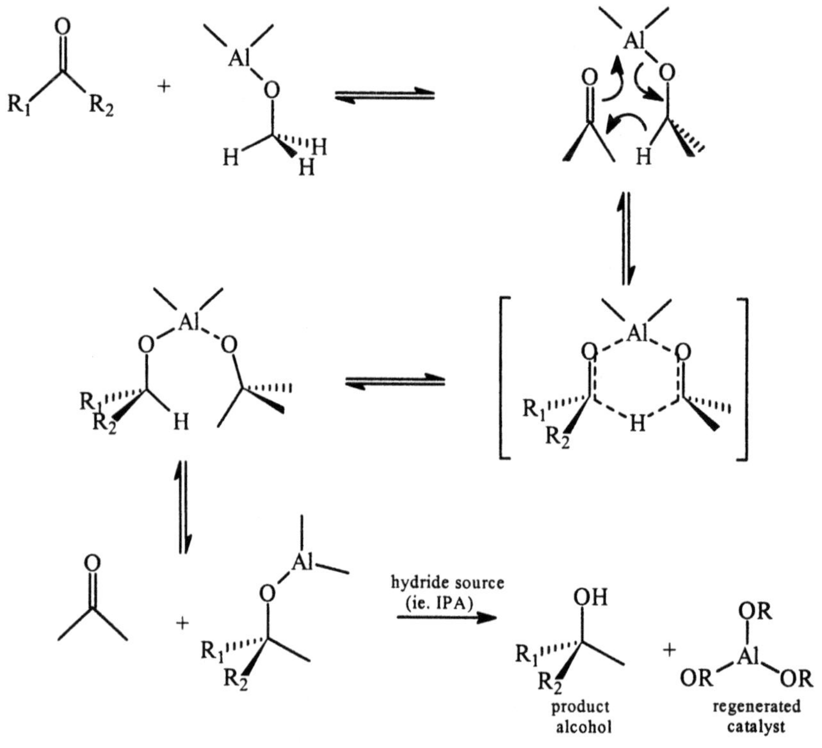

Scheme III Accepted Cyclic transition State Mechanism for MPV Reduction.

cyclic transition state shown in Figure 4 and its alternate mechanistic pathway. Their work does point to the conclusion that both mechanisms may well occur simultaneously. The drawback in organic catalysis reactions employing AIP or similar aluminum alkoxides is that to achieve excellent conversion of the substrate the AIP or similar compounds must be present in stoichiometric or higher ratios to the substrate. Thus aluminum alkoxides generally give less than optimum results in MPV reactions at catalytic levels. This has been attributed to the slow ligand exchange rate of traditional Al-alkoxide catalysts. Shiner and Whittaker (15) have shown that the exchange rate of free isopropyl alcohol with the tetrameric form of aluminum isopropoxide to be quite slow. This is in keeping with the general observation that commercially available granular aluminum isopropoxide (tetrameric) is not as catalytically active as the freshly prepared or

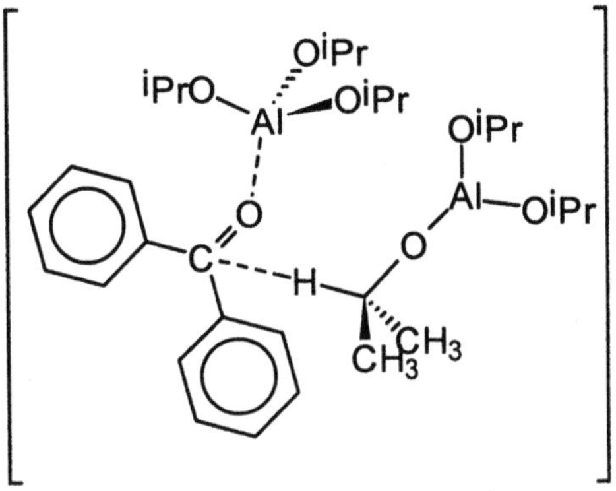

Figure 4 Proposed Non-cyclic Transition State for MPV Reduction of Benzophenone with Aluminum Isopropoxide (From Ref. 34).

freshly distilled AIP. This is understandable considering that the tetramer undergoes this slow conversion to the trimer before the catalytic transition state can be formed with the secondary alcohol and substrate. When the liquid form is used, the rate-limiting, tetramer to trimer step is excluded or significantly diminished, allowing for better catalytic efficiency. By adding amounts of AIP well above those percentages considered normal for a catalytic system, however, the alternate proposed mechanism described in Scheme II, if present, may well have an additive effect with that of the six-membered transition state mechanism because of a higher localized concentration of AIP molecules per substrate carbonyl moiety.

Unfortunately, MPV reactions are not applicable to substrates that are easily enolizable (28) such as β-diketones and β-diketoesters. With metal alkoxides, such substrates readily form stable β-enolate complexes that do not undergo reduction by hydride attack. A detailed kinetic study of the Meerwein-Ponndorf-Verley reduction of benzophenone was presented in 1961 by Moulton, van Atta, and Ruch (34) and provided some mechanistic insight into these types of reactions. They reacted a dilute solution of benzophenone in isopropanol with aluminum isopropoxide and monitored the reaction by sampling at intervals during a constant temperature regime. They quenched the reaction samples by hydrolysis of the AIP and then measured the concentrations of benzophenone and acetone through polarographic means giving results indicating pseudo-first order reduction rate of benzophenone and having an approximate ~1.5 rate order in AIP. For the reduction to proceed entirely by the accepted, cyclic transition state mechanism, the rate would have to be first order in benzophenone and first order in aluminum isopropoxide. They suggest from analysis of their data that the reaction may follow two concurrent mechanisms. The predominant process is the accepted mechanism involving a six-membered, cyclic transition state that incorporates one AIP molecule. They propose that the second mechanism proceeds through a non-cyclic transition state that requires two separate aluminum isopropoxide molecules. This alternate mechanism would be expected to be first order in benzophenone and second order in AIP. They were able to determine through Arrhenius equation plots approximate energy of activation and activation entropy values for data points of 25°C and 45°C (ΔH_{act} of 14 kcal, ΔS of - 20 entropy units); 25°C and 35°C (ΔH_{act} of 11 kcal, ΔS of – 31 entropy units); and 35°C and 45°C (ΔH_{act} of 17 kcal, ΔS of – 42 entropy units). A qualitative summary of their Arrhenius plot data is that this MPV reduction proceeds with relatively small energies of activation and corresponding large, negative entropic values.

A recent exploration of MPV reductions by Campbell, Zhou, and Nguyen (36) provides comparative data for simple aluminum alkyl and chloro-alkyl pre-catalysts (with ^{i}PrOH), and aluminum isopropoxide catalysts in the same reaction systems (Table 1). These reactions were performed at room temperature in toluene with four equivalents of isopropanol. They confirmed through ^{1}H NMR that the alkyl aluminum pre-catalyst species they investigated were entirely converted in situ to AIP in 2 hours at room temperature. The AIP used for comparison was obtained from commercial sources as a solid. Their results indicate a substantial advantage in reduction efficiency when the aluminum isopropoxide catalyst is formed *in situ*. They postulate that this higher activity seen in Al-alkoxide species derived from simple alkylaluminum precursors in solution is due to low aggregation state of these species. This conclusion supports the previously discussed correlation between catalytic activity and the oligomeric state (tetrameric versus trimeric/dimeric) of aluminum isopropoxide. The solid, commercial AIP they evaluated can be assumed to be comprised substantially of tetrameric species and thus have a higher aggregate character. Their route to AIP from labile methylaluminum compounds presumably avoids the rate limiting interconversion process from tetramer to trimer, and thus explains the very rapid reduction reactions.

Mills and coworkers (37) performed a series of MPV reductions of cyclic ketones to the corresponding epimeric mixtures of alcohols using aluminum isopropoxide as the catalyst in isopropyl alcohol. A summary of their results is given in Table 2. They conclude that the predominant influence in these reductions pertinent to the cis/trans ratio of the resulting epimeric alcohol mixtures is the degree of steric hindrance of the carbonyl moiety. They also separately investigated (38) a series of MPV reductions of aldehydes and unstable ketones where slow, drop-wise addition of the carbonyl substrate to the reactions improved the yields.

They also provide examples of this technique whereby convenient solvent-less MPV reductions are afforded.

Table 1 MPV reductions with alkylaluminum and AIP catalysts (From Ref. 36).

Substrate	Pre-catalyst /catalyst (10 mol%)	Time (hrs)	% Reduction
2-hexanone	$Al(CH_3)_3$	12	11
2-hexanone	$Al(CH_3)_2Cl$	12	50
2-hexanone	$Al(O^iPr)_3$	12	0
cyclohexanone	$Al(CH_3)_3$	3	82
cyclohexanone	$Al(CH_3)_2Cl$	2	96
cyclohexanone	$AlCH_3Cl_2$	12	5
cyclohexanone	$Al(O^iPr)_3$	12	7
benzaldehyde	$Al(CH_3)_3$	2	91
benzaldehyde	$Al(CH_3)_2Cl$	1	60
benzaldehyde	$Al(CH_3)_2Cl$ (in neat ^iPrOH)	4	85
benzaldehyde	$AlCH_3Cl_2$	12	6
benzaldehyde	$Al(O^iPr)_3$	12	3
2-chloroacetophenone	$Al(CH_3)_3$	12	99
2-chloroacetophenone	$Al(CH_3)_2Cl$	12	65
acetophenone	$Al(CH_3)_3$	12	51
acetophenone	$Al(CH_3)_3$ at 65°C	12	80
acetophenone	$AlCH_3Cl_2$	12	55
acetophenone	$Al(O^iPr)_3$	12	0

Table 2 MPV reductions of cyclic ketones with AIP catalysts (From. Ref. 37).

ketone substrate	% reduction	% cis-alcohol in product mixture
(±)-2-methylcyclohexanone	95	50
(±)-3-methylcyclohexanone	93-95	55
4-methylcyclohexanone	93-95	33
4-isopropylcyclohexanone	93-95	40
4-isopropylcyclohex-2-en-1-one	95	39
(-)-menthone	92-96	70
(+)-camphor	>99	70

Aluminum isopropoxide has been described in a 1958 US patent (39) as an MPV reductant to selectively convert the 7-ethylenedioxy derivative of a substituted dodecahydrophenanthrene to its 4b-methyl-7-ethylenedioxy-1,2,3,4,4a,4b,5,6,7,8,10,10a-dodecahydrophenanthrene-4-ol-1-one. The MPV reduction of pyrimidin-2(1H)-ones to the corresponding 3,4-dihydro- pyrimidin-2(1H)-ones has been accomplished with zirconium tetraalkoxide and with AIP, although the Zr-alkoxide was more effective (40). Aluminum isopropoxide has proven useful in the reduction of 5β-hydroxy-8-oxo-1,4,4aα,5,8,8aα-hexahydronapthalene-1β-carboxylic acid to a precursor used in the synthesis of reserpine and related comnpounds (41,42).

Asymmetric, borane-modified MPV reduction of a variety of aromatic ketones to their corresponding alcohols has recently (43) been reported using a chiral aluminum alkoxide catalyst shown in Figure 5. This compound was formed *in situ* from aluminum isopropoxide and (R)-1,1'-binapthyl-diol in

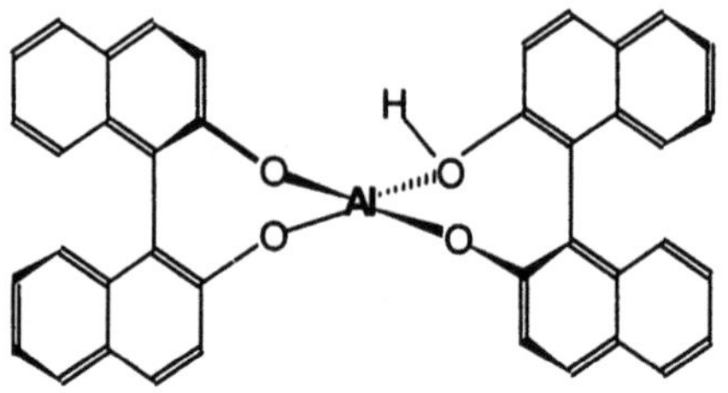

Figure 5   Chiral aluminum alkoxide catalyst (From Ref. 43).

methylene chloride at room temperature. The ketone substrates subjected to re-duction included 2-bromoacetophenone, 1-indanone, 2'-acetonapthone, propio-phenone, acetophenone, 4'-chloroacetophenone, α-tetralone, isobutyrophenone, and 2,2-dimethylpropiophenone. The alcohol products were obtained in high yields with ee values up to 83%. They probed the system by preparing a similar catalyst complex where the hydroxy moiety seen in Fig. 5 was replaced with a methoxy group and obtained similar reduction yields but greatly decreased enan-tiomeric excess. This indicates that the catalyst system can be tuned for asym-metric selectivity. The use of aluminum catalysts with chiral alkoxide ligands that allow enantioselective synthesis represents an entirely new area of MPVO reaction investigation.

NEW COMPARATIVE WORK

In an effort to provide some comparative data to previously published work, a series of experiments were conducted using molten AIP in lieu of the granular form commonly available commercially. This 'molten' material was not freshly prepared and neither was it granular material re-melted. It was from AIP that had been prepared at industrial scale more than six months previously and kept in the liquid state by heated (~115-120°C) storage in a sealed metal can. The reason for this was to achieve a relatively stable and consistent equilibrium of the complex oligomeric interconversions of aluminum isopropoxide discussed earlier. Solid (granular) AIP is primarily tetrameric and less active than the liquid trimeric form obtained from a fresh preparation. With few exceptions, the literature has focused on work involving the tetrameric form. Well-documented MPV reductions on common substrates were chosen to employ molten AIP (80-85°C, free-flowing liquid) as the catalyst (44). All reactions performed under dry conditions in toluene at room temperature using 10 mol% molten AIP (80-85°C at time of addition) with aliquots removed at 24 hours and quenched with dilute HCl prior to analysis. The acetone produced by these reactions was not removed through distillation, but allowed to remain in the mixture over the course of the reaction. This decision was supported by the reduction results reported (45) by Truett and Moulton with AIP in refluxing isopropanol for a variety of substrates including benzophenone (96%), cyclopentanone (60%), benzaldehyde (85%), benzoin (94%), dl-camphor (92%), benzosuberone (94-99%), and benzalacetophenone (76%).

Table 3 Reduction comparison using molten AIP, ASB, and AIP/ASB.

Reaction	Substrate	% Reduction of Substrate[a] quenched at 24 hrs[b]		
		Molten AIP	ASB	AIP/ASB
A	2-hexanone	0.6	9.85	58.2
B	cyclohexanone	87.5	33.4	96.2
C	benzaldehyde	78.4	36.0	93.5
D	acetophenone	12.2	14.0	38.6
E	benzophenone	30.8	47.9	69.2

[a]Conversion percents determined by gas chromatography[46]; [b]No attempt was made to recover or regenerate the AIP after the reaction.

The five test reductions A through E described in Figure 6 were also performed using 10 mole % aluminum sec-butoxide (ASB) at room temperature in lieu of molten AIP, and alternatively, a mixed aluminum alkoxide catalyst system (AIP/ASB, total 10 mole %) at room temperature was explored in an effort to provide comparative data to reactions employing AIP or ASB alone. It

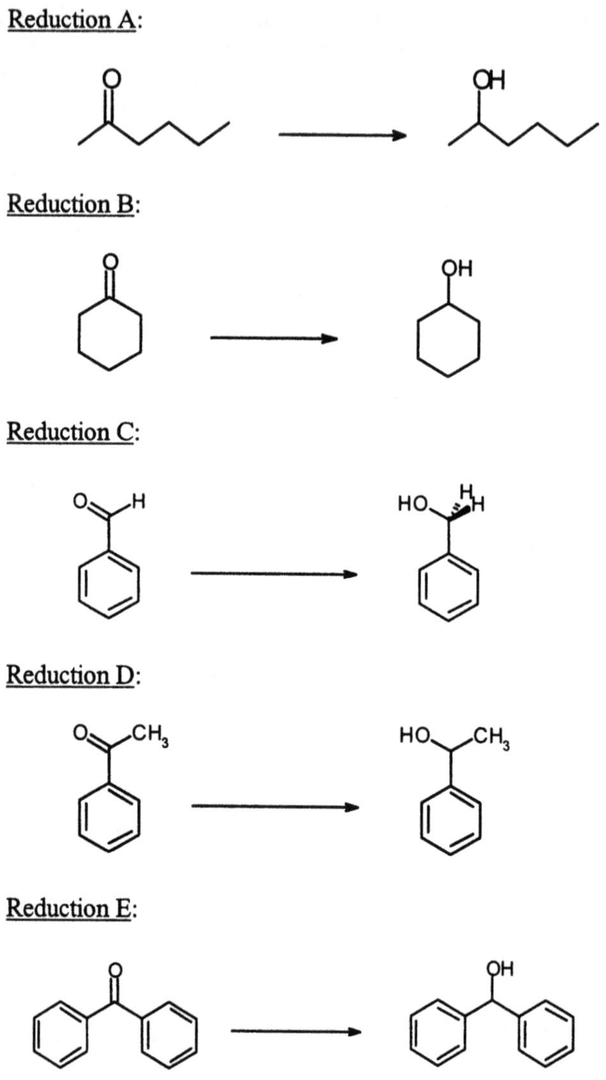

Reduction A:

Reduction B:

Reduction C:

Reduction D:

Reduction E:

Figure 6 MPV reductions selected as test reactions.

is established that freshly prepared, liquid AIP is more desirable for catalysis due to its more labile mixture of dimeric and trimeric forms existing in the liquid state. Aluminum isopropoxide can be stabilized in its more active liquid state when combined with aluminum sec-butoxide (ASB). Using a weight ratio of 70:30 AIP to ASB provided a satisfactory co-solution of the two pure aluminum alkoxides that was liquid at room temperature with long term stability. This approach is based on Mesirow's 1954 patent (19) for the stabilization of lower Al alkoxides in liquid phase by the addition of aluminum sec-butoxide. In theory, this stabilization effect is attributed to alkoxide exchange upon mixing to produce mono-sec-butoxyaluminum diisopropoxide and/or di-sec-butoxyaluminum isopropoxide *in situ* as equilibrium species. Mesirow indicated that an AIP mixture containing 40% ASB is permanently stable as a liquid, but that lesser percentages of ASB would give long-term stability particularly if the mixture was subjected initially to several hours of a heat treatment. The 70:30 AIP/ASB mixture employed by the authors was liquid and stable for up to four months at room temperature.

The results of these comparative reactions are given in Table 3 and indicate a trend of improved reduction from molten for the 2-hexanone, acetophenone, and benzophenone systems. This trend increases from AIP to ASB to AIP/ASB mixture. In the benzaldehyde and cyclohexanone system, the use of aluminum sec-butoxide alone gave much poorer reduction than that seen with molten AIP or with 70:30 AIP/ASB. This behavior is as yet unexplained. The most significant observation from the results of these comparative reactions is that the mixed alkoxide catalyst provides a greatly improved reduction yield over the same reactions performed with either AIP or ASB. The implication is that the mixed AIP/ASB catalyst should be employed with previously reported MPV reactions that proceed slowly or poorly using either of these aluminum trialkoxides separately.

The preliminary experimental work described herein has led us to multiple avenues of research into MPVO reactions using simple and modified aluminum alkoxide catalyst systems. The authors are continuing these various efforts and will publish the detailed results of these investigations elsewhere in the literature.

CONCLUSIONS

The authors have endeavored herein to refresh the collective memory of the organic catalysis community about the usefulness of aluminum isopropoxide and related or derivative compounds, and to present limited comparative results from preliminary efficiency studies. The diversity of applications of aluminum isopropoxide and other alkoxides described herein demonstrates the continued value of this class of compounds to organic catalysis. We desired to stimulate new consideration of this catalytic tool among current investigators as it awaits the full exploration of its potential. Our object was to consolidate the extensive AIP and aluminum alkoxide literature for easy review by researchers to facilitate their usage of this body of knowledge. Hopefully, our efforts have served that purpose. The authors are indebted to Kerry Rickerd (Chattem Chemicals) for GC method development and analysis of the reduction experiment samples.

REFERENCES AND NOTES

1. Meerwein, H. and Schmidt, R., Ein neues Verfahren zur Reduktion von Aldehyden und Ketonen. *Justus Liebigs Ann. Chem.*, **39**, 221-238, 1925.
2. von Ponndorf, W., Der reversible Austausch der Oxydations-stufen zwischen Alde den oder Ketonen einerseits und primären oder sekundären Alkoholen anderseits. *Angew. Chem.*, **39**, 138, 1926.
3. Verley, A., *Bull. soc. chim. France*, Sur l' change de groupements fonctionnels entre deux mol cules. Passage des c tones aux alcohols et inversement. 37, 537, 1925.
4. Oppenauer, R. V., Dehydration of secondary alcohols to ketones. I. Preparation of sterol ketones and sex hormones. *Recl. Trav. Chim. Pays-Bas*, **56**, 137, 1937.
5. Adkins, J., *J. Am. Chem., Soc.*, **44**, 2178, 1922.
6. Doumani, T. F., Process for the production of aliphatic alcohols. US Patent 2,394,848, February 12, 1946.
7. Young, W. G.; Hartring, W. H.; and Crossley, F. S., *J. Am. Chem. Soc.*, **58**, 100, 1936.
8. Lund, H., Aluminum isopropylate as a reducing agent. A general method for carbonyl reduction. *Ber. Dtsch. Chem. Ges.*, **70**, 1520, 1937.
9. Towers, R. S., Process for preparing organometallic compounds. US Patent 3,094,546., June 1963.
10. Anderson, A. R. and Smith, W., Process for preparing metal alkyls and alkoxides. US Patent 2,965,663, December 20, 1960.
11. Cuomo, J. J.; Leary, P. A.; and Woodall, M., Process for producing aluminum alkoxide or aluminum aryloxide. US Patent 4,745,204 May 17, 1988.
12. Jaroslov Vit, VITRIDE®: hydride reductions come of age. Technical Information Bulletin, HEXCEL Chemical Products, 1991.
13. Mehrota, R. C., *J. Indian Chem. Soc.*, **31**, 85, 1954.
14. Kleinschmidt, D. C.; Shiner, V. J.; and Whittaker, D., Interconversion reactions of aluminum isopropoxide polymers. *J. Org Chem.*, **38:19**, 3334-3337, 1973.
15. Folting, K.; Streih, W. E.; Caulton, K. G.; Poncelet, O.; and Huber-Pfalzgraf, L. G., Characterization of aluminum isopropoxide and aluminosiloxanes. *Polyhedron*, **10:14**, 1639-1646, 1991.
16. Mehrota, R. C., *J. Indian Chem. Soc.*, **30**, 585, 1953.
17. Whitaker, G. C., Aluminum alcoholates and the commercial preparation and uses of aluminum isopropylate, in Metal Organic Compounds, Advances in Chemistry Series, No. 23, pp. 184-189, Sept. 1959.
18. Wengrovius, J. H.; Garbauskas, M. F.; Williams, E. A.; Going, R. C.; Donahue, P. E.; and Smith, J. F., Aluminum alkoxide chemistry revisited: synthesis, structures, and characterization of several aluminum alkoxide and siloxide complexes. *J. Am. Chem. Soc.*, **108**, 982-989, 1986.
19. Mesirow, R., Preparation of liquid aluminum alkoxides. US Patent 2,687,423, August 24, 1954.
20. Pratt, C., Modified aluminum tri-alkoxide compounds- replacing part of isopropyl alcohol of aluminum triisopropoxide with a higher alcohol, storage stability. US Patent 4,525,307, June 25, 1985.
21. Lerner, R. W.; Towers, R. S.; and Flasch, R., Stable aluminum alkoxide solutions. US Patent 4,052,428, October 4, 1977.
22. Greco, C. C. and Triplett, K. B., Stabilized liquid aluminum alkoxides and process. US Patent 4,596, 881, June 24, 1986.

23. Turova, N. Y.; Fozunov, V. A.; Yanovskii, A. I.; Bokii, N. G.; Struchkov, Y. T.; and Tarnoplskii, B. L., Physico-chemical and structural investigation aluminum isopropoxide. *J. Inorg. Nucl. Chem.*, **41**, 5, 1979.
24. Bradley, D. C., Metal alkoxides. *Adv. Chem. Ser.*, **23**, 10, 1959.
25. Shiner, V. J.; Whittaker, D.; and Vernandez, V. P., The structures of some aluminum alkoxides. *J. Am. Chem. Soc.*, **85**(15), 2318-2322, 1963.
26. Anwander, R.; Palm, C; Gerstberger, G.; Groeger, O.; and Engelhardt, G, Enhanced catalytic activity of MCM-41-grafted aluminum isopropoxide in MPV reductions. *Chem. Commun.*, **17**, 1811-1812, 1998.
27. Shiner, V. J. and Whittaker, D., *J. Am. Chem. Soc.*, **91**, 394, 1969.
28. de Graauw, C. F.; Peters, J. A.; van Bekkum, H.; and Huskens, J., Meerwein-Ponndorf-Verley reductions and Oppenauer oxidations: an integrated approach. *Synthesis*, **10**, 1007-1017, 1994.
29. Wilds, A. L., Reduction with aluminum alkoxides. *Org. React.*, **2**, 178, 1944.
30. Djerassi, C., *Org. React.*, **6**, 207, 1953.
31. Hayes, K. and Drake, G., The reduction of some aliphatic beta-amino aldehydes. *J. Org. Chem.*, **15**, 873-876, 1950.
32. Lehman, H., In Houben-Weyl, 4th ed., Vol. 4/1b; Müller, E., Ed.; Thieme: Stuttgart, 1975, p. 901.
33. von Doering, W. and Ashner, T. C., *J. Am. Chem. Soc.*, **75**, 393, 1953.
34. Moulton, W. N.; van Atta, R. E.; and Ruch, R. R., Mechanism of the Meerwein-Ponndorf-Verley reduction. *J. Org. Chem.*, **26**, 290-292, 1961.
35. Williams, E. D.; Krieger, K. A.; and Day, A. R., The mechanism of the Meerwein-Ponndorf-Verley reaction. A deuterium tracer study. *J. Am. Chem. Soc.*, **75**, 2404-2407, 1953.
36. Campbell, E. J.; Zhou, H.; and Nguyen, S. B. T., Catalytic Meerwein-Pondorf-Verley reduction by simple aluminum complexes. *Org. Lett.*, **3:15**, 2391-2393, 2001.
37. Jackman, L. M.; Macbeth, A. K.; and Mills, J. A., Reductions with aluminum alkoxides. Part I. The relative proportion of epimers in alcohols derived from cyclic ketones. *J. Chem. Soc.*, 2641-2646, 1949.
38. Macbeth, A. K. and Mills, J. A., Reductions with aluminum alkoxides. Part II. Modified procedure. *J. Chem. Soc.*, 2646-2649, 1949.
39. Arth, G. E.; Poos, G. I.; and Sarelt, L. H., Dodecahydrophenanthrene compounds and methods of preparing same. US Patent 2,842,557, July 8, 1958.
40. HØseggen, T.; Rise, F.; and Undheim, K., Tetraisopropoxyzirconium and tri-isopropoxyaluminum in regioselective reduction of pyrimidinones. *J. Chem. Soc.*, Perkin Trans. I, **5**, 849-850, 1986.
41. Montmorency, R. J. and Warnant, J., Process of producing a brominated lactone useful in the preparation of reserpine and related compounds. US Patent 2,951,852, September 6, 1960.
42. Muller, G.; Nominé, G.; and Warnant, J., Polycyclic compounds and process of preparing same. US Patent 2,952,682, September 13, 1960.
43. Fu, I.-P. and Uang, B.-J., Enantioselective borane reduction of aromatic ketones catalyzed by chiral aluminum alkoxides. *Tetrahedron Asymmetry*, **12:1**, 45-48, 2001.
44. Experimental- Al[OCH(CH$_3$)$_2$]$_3$ was obtained from commercial manufacture (Chattem Chemicals) and maintained in liquid form by storage at 85°C in a sealed can for >6 months. Double distilled Al[O(CH$_3$)CH$_2$CH$_2$CH$_3$]$_3$ was used as supplied from commercial production (Chattem Chemicals). 70:30 wt% AIP/ASB was prepared from freshly made commercial aluminum isopropoxide and aluminum sec-butoxide. Tolu-

ene (≥99.9%), 2-hexanone (99.9% GC assay%), cyclohexanone (99.9% GC assay), benzaldehyde (99.7% GC assay), and benzophenone (99.9 % GC assay) were used as supplied from Aldrich. All glassware used was oven-dried overnight then flushed with dry nitrogen immediately before usage. All reactions were performed in sealed 250 mL flasks, with aliquots quenched with 0.05M HCl upon collection.

45. Truett, W. L. and Moulton, W. N., A modified method for the Meerwein-Ponndorf Verley reduction. *J. Am. Chem. Soc.*, **73**, 5913-5914, 1951.
46. Substrates and reduction products were identified on a Hewlett Packard HP6890 gas chromatograph equipped with an FID detector and employing a 30m Restek RTX-200 capillary column (trifluoropropylmethyl polysiloxane packing). Yields of the resulting alcohols from these reductions were based on area percents correlated to known quantities of high purity standards (Aldrich).

10

Liquid-Phase Methylation of m-Cresol Catalyzed by Mixed Oxides Derived from Hydrotalcite Precursors: Effect of Acid-Base Features on Catalytic Performance

M. Bolognini, F. Cavani[*]**, C. Felloni, and D. Scagliarini**
Dipartimento di Chimica Industriale e dei Materiali, Università di Bologna, Bologna, Italy

C. Flego and C. Perego
EniTecnologie SpA, S. Donato Milanese (MI), Italy

The reactivity of Mg-Fe mixed oxides obtained from hydrotalcite-like precursors in m-cresol alkylation with methanol was studied, and compared with that of Mg-Al mixed oxides. Samples were characterized using CO_2 adsorption and Thermal-Programmed-Desorption (TPD) technique to obtain information on the relationship between surface properties and catalytic performance. The Mg-Fe mixed oxides were found to be less basic than the reference Mg-Al mixed oxide. Particular features of Fe-containing catalysts, such as (i) high selectivity towards C-alkylated products rather than O-alkylated products, (ii) formation of side-chain alkylated products (mainly thymol) and (iii) the high amount of heavy compounds obtained, were explained by hypothesizing a synergic contribution of acid and basic centers, which affected the reaction pathway. Tuning the composition of MgO-based mixed oxides makes it possible to control the catalytic performance in alkylation reactions of phenol derivatives with methanol.

[*]*E-mail*: cavani@ms.fci.unibo.it

INTRODUCTION

Considerable interest has been growing in recent years concerning the possibility of developing solid basic catalysts to replace currently employed homogeneous systems [1,2]. The benefits of heterogeneous systems are well-known and concern the lowering of the environmental impact by reduction of the amount of inorganic salts in the effluents and corrosion, and by easier catalyst separation. Amongst the base-catalyzed reactions which have been studied, the alkylation of phenol derivatives represents a reaction of industrial interest for the synthesis of intermediates for the fine chemicals and polymers industry [3]. For example, 2-methylphenol is a monomer for the synthesis of epoxy cresol novolack resin; 2,5-dimethylphenol is an intermediate for antiseptics, dyes and antioxidants; 2,6-dimethylphenol is used for the manufacture of polyphenylenoxide resins, and 2,3,6-trimethylphenol is a starting material for the synthesis of vitamin E. Starting from m-cresol, and using 2-propanol as the alkylating agent, a commercially interesting product can be synthesized: thymol (2-isopropyl-5-methylphenol) which is an important intermediate for the industrial production of menthol.

Methylation can be carried out with either acid or basic catalysts, and catalyst properties affect the distribution of the products, especially when different positions for methylation are present. Acid zeolites are very active catalysts in alkylation of phenol derivatives; however, the considerable formation of heavy compounds leads to a fast deactivation of the catalyst. Basic catalysts, such as single oxides (MgO) and Mg-Al mixed oxides, have been found to be less active than the acid ones, but did not form heavy products [4]. Mg-Al mixed oxides, prepared starting from hydrotalcite precursors, have shown the best basic features, and indeed in recent years these materials have been reported as catalysts for different basic reactions, such as the Claisen-Schmidt condensation, the Knoevenagel condensation, and many others [5-9].

With the aim of studying the acid-base properties of Mg-Fe mixed oxides we chose the alkylation of m-cresol with methanol as a model reaction. The distribution of products obtained may take account of the different surface properties of the catalysts [3,4]. The catalytic performances were compared with the acid and basic sites distributions, as determined by pyridine and CO_2 adsorption and desorption, respectively.

EXPERIMENTAL

Mg-Fe hydrotalcite-precursors having different Mg/Fe ratios were prepared following the conventional procedure as described elsewhere [10]. The precursors obtained were then calcined at 450°C, for 8 hours. Activation of the catalysts before reaction was carried out at 450°C for 3 hours in a gas-chromatographic N_2 flow. A Mg-Al mixed oxide (Mg/Al = 2.0) was synthesized and used as the reference compound. Single oxides MgO and Fe_2O_3 were prepared following the same procedure as for the preparation of the hydrotalcite precursors, in the absence of the second element. The catalysts were characterized by (i) X-ray diffraction (Philips PW 1050/81), (ii) surface area measurements (single point BET, Sorpty 1700 Carlo Erba), (iii) adsorption at 21°C and thermal-programmed-desorption of CO_2 from 21 to 450°C (PulseChemisorb 2705, Micromeritics; 300 mg of samples, which were preliminarily treated at 450°C to desorb adsorbed species), and (iv) adsorption at 21°C and stepwise desorption (1h, dynamic vacuum) of pyridine, followed by FT-IR spectroscopy (Perkin-Elmer, mod. 2000).

The catalytic tests were carried out in a 300 mL Parr 4560 autoclave reactor loading 32 g of m-cresol, 19 g of methanol (m-cresol/methanol molar ratio=0.5) and 5.7 g of catalyst. The initial molar ratio between m-cresol and methanol in the liquid phase at reaction conditions was equal to 2.3 (in the gas phase 0.31). The reaction temperature was 300°C; the initial autogeneous pressure at 300°C was 38 atm and the final pressure was around 35 atm. The reaction was carried out for six hours. The stirring rate was equal to 700 min^{-1}. The testing procedure can be summarized as follows: The catalyst was loaded first, the reactor was sealed and evacuated, and the m-cresol and methanol mixture was then loaded. Heating up was started, and the beginning of the reaction time was taken as being when the temperature of 300°C was reached.

RESULTS AND DISCUSSION

Structural Characterization of the Calcined Samples

The surface areas of the samples after calcination at 450°C are reported in Table 1. All the Mg-Fe mixed oxides show a surface area around 150 m^2/g. The MgO and Mg/Al=2.0 samples have a slightly higher surface area (around 200 m^2/g). Fe_2O_3 has a surface area much lower than that of the other samples.

Table 1 Surface area and Mg/Fe(Al) ratios of the calcined samples

Sample	Surface area, m^2/g
Fe_2O_3	32
Mg/Fe = 1.5	150
Mg/Fe = 2.0	149
Mg/Fe = 2.5	144
Mg/Fe = 3.0	142
Mg/Fe = 4.0	147
MgO	206
Mg/Al = 2.0	185

The X-ray diffraction patterns show the presence of the hydrotalcite-like structure in all the dried Mg-Fe and Mg-Al samples. Figure 1 reports the X-ray diffraction patterns of Mg/Fe=2.0, Mg/Al=2.0 and MgO samples after calcination. The patterns of the calcined Mg-Fe and Mg-Al samples correspond to that of a poorly crystallized MgO. The shift of the main reflections in the Mg/Al sample towards higher 2θ values with respect to MgO is due to the decrease in the volume of the crystallographic cell of MgO (periclase), in agreement with an isomorphic replacement of Mg^{2+} cations with the smaller Al^{3+} cations (Mg^{2+} = 0.65 Å; Al^{3+} = 0.50 Å) [7]. The shift is not observed for the Mg-Fe sample, because of the similar size of the Mg^{2+} and Fe^{3+} cations (Fe^{3+} = 0.69 Å). The sample prepared in the absence of magnesium corresponds to Fe_2O_3 hematite.

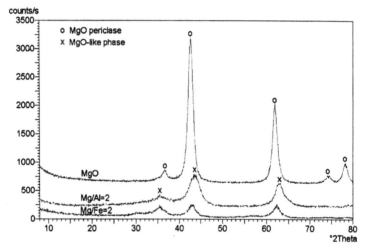

Figure 1 XRD patterns of Mg/Fe=2.0, Mg/Al=2.0 and MgO calcined samples.

Acid-Base Properties of the Calcined Samples

Information about the basicity distribution was obtained by CO_2 adsorption/TPD (Table 2). The overall density (μmol/g) of the basic sites was evaluated from the adsorption of CO_2 at 21°C. The basic strength distribution was evaluated from the capacity of the material to retain the probe molecule during desorption at increasing temperature (TPD). The TPD profiles of Mg/Fe=2.0 and Mg/Al=2.0 mixed oxides, and of MgO and Fe_2O_3 single oxides, are shown in Figure 2. The complex CO_2 desorption profiles are clearly related to the presence of high heterogeneity of the basic strength. The profiles for all samples consist of i) one low temperature peak with a maximum of desorption at temperatures lower than 100°C, attributed to the interaction with sites having weak basic strength, ii) one desorption peak with a maximum in the range 140-170°C, related to desorption of CO_2 from sites with medium basic strength, and iii) one broad desorption area, which covers the temperature range from 200 to 450°C, attributed to CO_2 desorption from sites with strong basic strength.

In Mg-Al calcined hydrotalcites the weak basic sites are proposed to be the surface OH⁻ groups, while the medium strength sites are related to Mg^{2+}-O^{2-}, but also to Al^{3+}-O^{2-} pairs [11-13]. Isolated O^{2-} anions, common in pure oxides, are also responsible for the strong basic sites in calcined hydrotalcites [14,15].

Table 2 Basicity distribution from CO_2 adsorption/TPD experiments

Sample	Overall amount adsorbed, μmol / g	Overall amount desorbed, μmol / g	Weak sites, μmol/g	Medium sites, μmol/g	Strong sites, μmol/g
Fe_2O_3	151	132	7	21	104
Mg/Fe = 1.5	124	120	28	41	51
Mg/Fe = 2.0	159	143	33	64	46
Mg/Fe = 3.0	146	142	33	55	54
Mg/Fe = 4.0	161	158	26	77	55
MgO	356	346	52	50	244
Mg/Al = 2.0	263	274	57	115	102

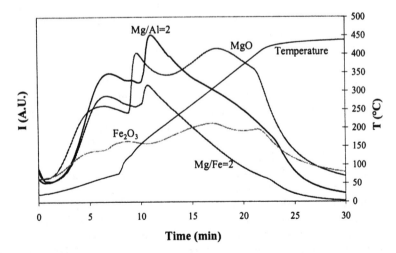

Figure 2 TPD profiles of Mg/Fe=2.0 and Mg/Al=2.0 mixed oxides, and of MgO and Fe_2O_3 single oxides.

In order to distinguish the different components, a deconvolution procedure was applied, the results of which are the numbers reported in Table 2. Curve fit quality was measured by the correlation coefficient (R^2), standard deviation (σ) and the Levenberg-Marquardt algorithm (χ^2). The quality of the deconvolution was high in all experiments: $R^2>0.996$, $\delta<0.004$ and $\chi^2<1.133$. The pure oxides (Fe_2O_3 and MgO) have a large number of strong basic sites (respectively 79 and 71%) and a small number of basic sites with weak (5 and 15%) or with medium (16 and 14%) strength. The overall amount of basic sites is much greater for MgO: 356 μmol/g are compared with 151 μmol/g for Fe_2O_3. However, in the case of Fe_2O_3, the difference observed between the amount of CO_2 adsorbed and desorbed, higher than the experimental error, shows a further quantity of strong basic sites, that retain CO_2 up to 450°C.

Within the series relative to Mg-Fe mixed oxides, the distribution of the basic strength is quite similar for all components, with an amount of the weak-strength sites between 11 and 23%, and a comparable amount of the medium and strong basic sites: 34-49% for the former and 32-40% for the latter. With respect to the single oxides, the number of strong basic sites is much lower, while that of medium strength sites is higher.

Comparing the Mg/Fe=2.0 sample with the reference Mg/Al=2.0 one, the replacement of Fe^{3+} for Al^{3+} does not produce relevant effects in the distribution of the basic strength, whereas the total density of the basic sites changes considerably. The overall amount of CO_2 adsorbed is equal to 159 μmol/g for the Mg-Fe sample and 263 μmol/g for the Mg-Al sample, despite the comparable surface area.

To investigate the acid properties of these materials, adsorption/desorption experiments with pyridine as the probe molecule were carried out. FT-IR spectra of samples were recorded after adsorption of pyridine and desorption at increasing temperatures. The acid strength is subdivided into: i) very weak, i.e., acid sites able to retain pyridine only after evacuation at 21°C, ii) weak, i.e., acid sites able to retain pyridine after evacuation at 100°C, and iii) medium, i.e., acid sites able to retain pyridine after evacuation up to 200°C. The results are reported in Table 3.

Within the series of Mg-Fe mixed oxides, no Brønsted acid sites are detected. Pyridine interacts essentially with Lewis acid sites and disappears almost completely after evacuation at temperatures higher than 21°C, indicating the weak nature of this bonding. The acid sites are coordinatively unsaturated Fe^{3+} (octahedral) cations; pairs of coordinatively unsaturated Fe^{3+} in octahedral and tetrahedral coordination are also present in the samples at lower Mg/Fe atomic ratio. With increasing Mg/Fe ratio, a decrease in the total Lewis acid sites density is observed. The MgO single oxide, like the Mg/Al=2.0 mixed oxide, does not show the presence of any acid sites able to chemisorb pyridine. The acidity in the Fe_2O_3 samples was not detected because of the poor transparency of this material in the IR region.

Table 3 Acidity of the samples from pyridine adsorption/FT-IR measurements.

Sample	Lewis acid sites			
	Total sites, μmol/g	Very weak sites, μmol/g	Weak sites, μmol/g	Medium sites, μmol/g
Fe$_2$O$_3$	Not detectable			
Mg/Fe = 1.5	210	101	65	44
Mg/Fe = 2.0	204	142	62	0
Mg/Fe = 3.0	113	63	50	0
Mg/Fe = 4.0	46	44	2	0
MgO	0	0	---	---
Mg/Al = 2.0	0	0	---	---

Reactivity in M-Cresol Methylation

In order to verify the reaction scheme, tests were made using different reaction times, with the catalyst having a Mg/Fe ratio equal to 1.5. The results obtained are reported in Figure 3.

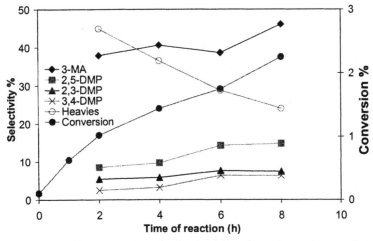

Figure 3 m-Cresol conversion and selectivities to the main products as functions of the reaction time. Catalyst: Mg/Fe=1.5.

The main reaction products are 3-methylanisole (3-MA, the product of O-alkylation on m-cresol), 2,3-dimethylphenol and 2,5-dimethylphenol (2,3-DMP and 2,5-DMP, products of ortho C-alkylation on m-cresol), 3,4-dimethylphenol (3,4-DMP, the product of para C-alkylation on m-cresol), and "heavies". Other

products formed in minor amounts (not reported in Figure 3) are side-chain alkylation products (mainly thymol: 2-isopropyl-5-methylphenol) and 2,3,6-
trimethylphenol (2,3,6-TMP, the product of ortho C-dialkylation on m-cresol). In
our reaction conditions one of the prevalent products is 3-MA, with a selectivity
between 38 and 46%. The amount of "heavies" is high, with selectivity between
24 and 45%. These are compounds obtained by dehydrogenation of methanol to
formaldehyde, followed by reaction of formaldehyde with methanol to yield hemiformal and formal which undergo nucleophilic attack by m-cresol or its polyalkyl-
ated products. Another possible mechanism of "heavies" formation consists in a
direct attack of m-cresol or methylated m-cresol on the formaldehyde and successive alkylation/hydroxyalkylation on the aromatic ring or on the O atom. Figure 3
shows an increase in 3-MA and in DMPs with increasing m-cresol conversion,
whereas the selectivity to "heavies" decreases. This may be attributed either to
consecutive transformations occurring on "heavies", or to the deactivation of the
strongest basic sites (responsible for the dehydrogenation of methanol and the
formation of condensation products [16]) in the first stages of reaction. Heavies
might thus form during the first stages of reaction and then their amount remain
constant; therefore the corresponding selectivity decreases with increasing relative
concentration of the other products.

As shown in Figure 3, the O- and C-alkylation products form through parallel reactions, and none of these products apparently undergo consecutive transformation. This may be due to the relatively low m-cresol conversion achieved under
these conditions, and it can not be excluded that a consecutive contribution of 3-
MA transformation to C-alkylated products might occur under conditions of
higher m-cresol conversion.

The following reaction parameters can be used to compare the catalyst properties: i) conversion of m-cresol, ii) ratio between the selectivity to products of
ortho C-alkylation and that of para C-alkylation (ortho/para ratio), and iii) ratio
between the selectivity to 3-MA and the selectivity of the products of C-alkylation
(O/C ratio). The importance of each of the above mentioned parameters for the
reaction of phenol alkylation has been discussed in a previous paper [4].

The values of m-cresol conversion and of normalized m-cresol conversion
(i.e., referred to the specific surface area of the catalyst) as a function of the Mg/Fe
atomic ratio are reported in Figure 4, while Figure 5 shows the corresponding values of selectivity to the main products. The conversion is maximum for the single
oxides Fe_2O_3 and MgO (both 3.1 %), while, within the series of Mg/Fe mixed oxides, the conversion slightly decreases as the amount of Mg increases. However,
due to the low surface area of Fe_2O_3, the trend of the normalized conversion shows
a maximum for iron oxide. With Fe_2O_3 the selectivity to the C-alkylated products
(2,5-DMP, 2,3-DMP and 3,4-DMP) is the highest, while the selectivity to the O-
alkylated product (3-MA) is the lowest. Therefore, the O/C ratio for Fe_2O_3 is very
low (Figure 6). On the contrary, on MgO the selectivity to 3-MA is the highest and
that to C-alkylated products is the lowest; this means that MgO has the highest
O/C ratio. With Mg-Fe mixed oxides, increasing the amount of Mg leads to an

increase in the selectivity to 3-MA and a decrease in that to C-alkylated products. The O/C ratio becomes higher as the Mg/Fe ratio increases. The amount of side-chain (S-C) alkylation products is maximum for Fe_2O_3 (6.7 %) and minimum for MgO (0.5 %); within the Mg-Fe series the selectivity is between 1.5 and 4.5%. The selectivity to "heavies" is the highest for Fe_2O_3, while no "heavies" form with MgO. For the Mg-Fe mixed oxides the amount of "heavies" is almost constant and does not depend on the Mg/Fe ratio.

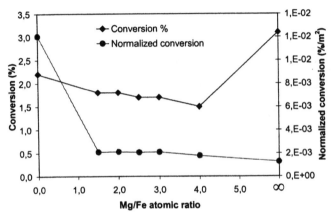

Figure 4 m-cresol conversion and normalized conversion as a function of the Mg/Fe atomic ratio.

Also reported in Figure 6 are the ortho/para ratio and the 2,5-DMP/2,3-DMP ratio as a function of the Mg/Fe atomic ratio. These ratios may give useful indications about the absorbed state of the m-cresol on the surface of the catalysts. In all cases the adsorption of the m-cresol is carried out via the interaction between bulk oxygen ions and the hydrogen atom of the hydroxy group, with abstraction of either a proton or a hydride depending on the properties of the bulk oxygen anion. The position of the aromatic ring with respect to the reactive surface depends on the acid-base features of the catalyst [17]. On acid catalysts the interaction between the electron-rich aromatic ring of phenol and the Lewis acid sites makes the ring plane parallel to the catalytic surface (in this case the ortho/para ratio will be close to 2). On basic catalysts the repulsion between the aromatic ring and the Lewis basic sites makes the molecule plane perpendicular to the catalytic surface (ortho/para ratio much higher than 2). Figure 6 shows that the ortho/para ratio is maximum for Fe_2O_3 and MgO single oxides. This confirms the stronger basic strength (Table 2) of these oxides with respect to the Mg-Fe mixed oxides.

The 2,5-DMP/2,3-DMP ratio takes account of possible interaction or repulsion between the methyl group of m-cresol and the catalyst surface. With Fe_2O_3 this ratio is equal to 2.7 but, when magnesium is present (Mg-Fe series), the ratio decreases as the Mg/Fe ratio increases.

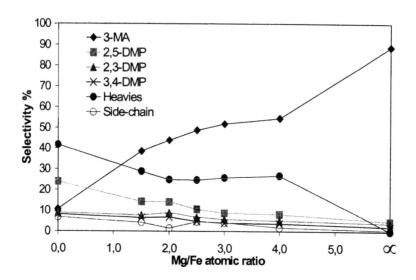

Figure 5 Selectivity to the main products as a function of Mg/Fe atomic ratio.

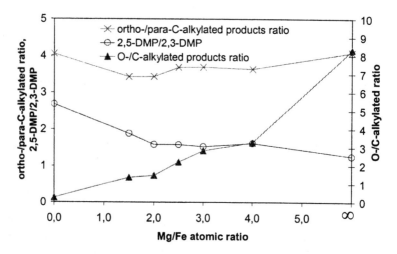

Figure 6 Ortho/para ratio, O/C ratio and 2,5-DMP/2,3-DMP ratio as a function of the Mg/Fe atomic ratio.

These data can be explained by considering the possible contribution of Lewis acid sites, the relative amount of which increases with increasing Fe content in the samples (Table 3). The interaction between methanol, more basic than m-cresol, and the Lewis acid sites makes the alcohol more electrophilic than on fully basic catalysts, and thus able to give not only attack on the O atom of m-cresol but also on the aromatic ring. As a consequence, the O/C ratio decreases with increasing Fe content. Moreover, the interaction between the aromatic ring and the surface Lewis acid site makes the para position (which otherwise would not be accessible, being too far from the surface) also available for C-alkylation, with a decrease in the ortho/para ratio in correspondence with an increase in the Fe content. Concerning the increase in the 2,5-DMP/2,3-DMP ratio, a planar position of the aromatic ring makes the ortho position adjacent to the methyl group less accessible for methanol attack due to steric hindrance, differently from a vertical adsorption of the ring, which would make the two ortho positions virtually with similar accessibility. For the same reason, the main side-chain alkylated product is thymol (2-isopropyl-5-methylphenol): the methyl group in position 2 is close to the methanol adsorbed and susceptible to further attack until the formation of the isopropyl substituent.

Another possible explanation for the observed phenomena is based on the hypothesis formulated by Grabowska et al. [18], who suggested that on basic metal oxides, methoxy intermediates decompose via β-hydride elimination leaving positively charged carbon atoms [19-21], with generation of an electrophilic metoxide species. The coordination properties of Fe make it possible to adsorb both methoxide and methylphenoxide at the same site, and consequently C-alkylation becomes the easiest reaction to occur. Moreover, in accordance with the proposed reaction pathway, position 5 of the aromatic ring is favored because, to reduce the steric hindrance, the m-cresol methyl group will settle in the opposite direction with respect to the methanol adsorbed. As a consequence, the selectivity to 2,5-DMP increases with increasing Fe content and becomes maximum for Fe_2O_3.

Figure 7 shows a comparison between Mg/Fe=2.0 and Mg/Al=2.0 samples. The Mg/Al=2.0 catalyst reaches a greater conversion (12.9 %), against 2.2 % for Mg/Fe=2.0. Moreover, one relevant difference is represented by the ortho/para ratio which is equal to 3.4 for Mg/Fe=2.0 and 26.1 for Mg/Al=2.0. These samples are quite similar in the basic strength distribution, but the Mg-Al sample shows a greater density of basic sites, and does not possess acid sites. The fact that the overall amount of basic sites for MgO is higher than that of Mg-Al, but the ortho/para ratio for the former sample is equal to 4.1, indicates that the distribution of the basicity is also important. These data can be explained with the hypothesis that sites having intermediate basic strength are the active ones for the alkylation of m-cresol under our reaction conditions [4]. Therefore, the higher value of the ortho/para ratio for the Mg-Al sample is due to the larger number of sites having medium basic strength with respect to the Mg-Fe and MgO samples. On the other hand, the presence of Lewis acid sites is likely responsible for the lower O/C ratio in the Mg-Fe catalyst as compared to the Mg-Al catalyst.

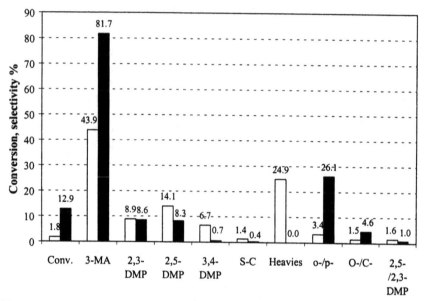

Figure 7 Conversion of m-cresol, selectivity to the main products, ortho-/para-C-alkylated products (o-/p-), O-/C-alkylated products (O-/C-) and 2,5-DMP/2,3-DMP ratio for Mg/Fe=2.0 (white bars) and Mg/Al=2.0 (black bars) catalysts.

CONCLUSIONS

Mg-Fe mixed oxides obtained starting from hydrotalcite-like precursors were characterized by means of CO_2 adsorption/TPD and IR experiments with pyridine as the probe molecule, in order to check the acid-base properties. The basicity of Mg-Fe mixed oxides is lower than that of Fe_2O_3 (in terms of the strength of the basic sites), MgO (number and strength of basic sites) and Mg-Al mixed oxides (density of basic sites). Within the Mg-Fe series the presence of weak Lewis acid sites was detected, the relative amount and strength of which increased with increasing Fe content. Particular features in the m-cresol methylation were observed for the Fe-containing catalysts as compared to Mg-Al mixed oxides: i) higher selectivity towards C-alkylated products, ii) formation of side-chain alkylated products (mainly thymol), and iii) formation of heavy products. The data can be explained by hypothesizing a contribution of surface acidity, which activates methanol (enhancing its electrophilicity) and favors the planar adsorption of the aromatic ring. The results obtained demonstrate that it is possible to control the distribution of the products in the methylation of phenol derivatives by changing the catalyst composition, and introducing metal cations having specific acid-base properties.

REFERENCES

1. H. Hattori, Chem. Rev., 95 (1995) 527
2. Y. Ono, T. Baba, Catal. Today, 38 (1997) 321
3. S. Velu, S. Sivasanker, Res. Chem. Intermed., 24(6) (1998) 657
4. M. Bolognini, F. Cavani, D. Scagliarini, C. Flego, C. Perego, M. Saba, Catal. Today, in press;
5. A. Guida, M.-H. Lhouty, D. Tichit, F. Figueras, P. Geneste, Appl. Catal., A: General, 164 (1997) 251
6. A. Corma, V. Fornés, R.M. Martin-Aranda, F. Rey, J. Catal., 134 (1992) 58
7. A. Corma, V. Fornés, F. Rey, J. Catal., 148 (1994) 205
8. R.J. Davis, E.G. Derouane, J. Catal., 132 (1991) 269
9. S. Velu, C.S. Swamy, Appl. Catal., A: General, 162 (1997) 81
10. F. Cavani, F. Trifirò, A. Vaccari, Catal. Today, 11(2) (1991) 173
11. D. Tichit, F. Fajula, Stud. Surf. Sci. Catal. 125 (1999) 329
12. D. Tichit, M. Hassane Llhouty, A. Guida, B. Huong Chiche, F. Figueras, A. Auroux, D. Bartalini, E. Garrone, J. Catal. 151 (1995) 50
13. J. Shen, M. Tu, C. Hu, J. Solid State Chem. 137 (1998) 295
14. J. Di Cosimo, V.K. Diez, M. Xu, E. Iglesia, C.R. Apesteguia, J. Catal. 178 (1998) 499
15. T.M. Jyothi, T. Raja, M.B. Talawar, B.S. Rao, Appl. Catal. 211 (2001) 41
16. E Palomares, G.Eder-Mirth, M. Rep, J. A. Lercher, J. Catal., 180 (1998) 56
17. K. Tanabe, Stud. Surf. Sci. Catal. 20 (1985) 1
18. H. Grabowska, W. Mista, L. Syper, J. Wrzyszcz, M. Zawadzki, J. Catal, 160 (1996) 134
19. N. Kizhakevariam, E. M. Stuve, Surf. Sci. 286 (1993) 246
20. B.C. Wiegand, P. Uvdal, J. G. Serafin, C. M. Friend, J. Phys. Chem. 96 (1992) 5063
21. B.-R. Shen, S. Chaturvedi, D. R. Strongin, J. Phys. Chem. 98 (1994) 10258

11

2002 Paul Rylander Plenary Lecture

Selective Heterogeneous Catalysis: Opportunities and Challenges

Jerry R. Ebner
Retired from Monsanto Company, St. Louis, Missouri, U.S.A.

ABSTRACT

In chemical synthesis today, high value can be achieved through the design and application of highly selective catalytic chemistry. Very high selectivity provides a means to eliminate undesirable pollutants and byproducts and achieve the required economics for commercial production. Both homogeneous and enzyme catalytic reaction chemistries are considered primary sources of very high selectivity catalysis. Because of the robustness of reaction conditions often required for heterogeneous catalytic systems, they are not usually the systems of first choice for high selectivity, especially for oxidation reactions. The development of heterogeneous surfaces of defined structure is hindered by the inherent challenges of characterizing the structure of the needed active sites, and also is constrained by the limited range of economically viable synthesis methods for the scaled synthesis of surfaces with the desired structural components. Added to this is the inevitability of surface reconstruction of the catalyst upon use in the chemical reaction! Not surprisingly, the empirical methods of catalyst development are still heavily relied upon. There is plenty of challenge for the heterogeneous catalyst designer seeking high selectivity chemistry. Thus, a great opportunity and need continues to exist for discovery of better surface structural characterization methods, especially *in situ* methods, and a much expanded synthetic toolbox for surface structure design and control.

129

Over the past decade, our heterogeneous catalysis research in oxidation chemistry has been aimed at achieving high selectivity to reduce waste and improve profitability. One new technology developed in this effort is a copper on carbon catalyst system useful in the dehydrogenation of alcohols in the presence of strong base to form carboxylic acid salts. Catalysts with supported copper have a rich history for hydrogenation and dehydrogenation catalysis. However, the preparation of copper metal on carbon supports is not well studied. Overcoming the inherent challenges of preparing and characterizing *stable, high selectivity* copper on carbon catalysts will be the scientific focus of this presentation. It will also serve to illustrate the opportunities for discovery of new synthesis and structural methods, as discussed above.

Editor's Note: Due to proprietary information, Dr. Jerry Ebner was unable to publish the contents of his presentation. The abstract of his talk is included in these proceedings.

12

Diisopropyl Ether One-Step Generation from Acetone-Rich Feedstocks

John F. Knifton
Shell Global Solutions, Houston, Texas, U.S.A.

Robert J. Taylor and Pei-Shing E. Dai
Texaco Global Products, The Woodlands, Texas, U.S.A.

SUMMARY

A one-step integrated process for the generation of the high-octane fuel ether, diisopropyl ether (DIPE), from acetone-rich feedstocks has been demonstrated. Three continuous, downflow, reactor configurations have been considered, including: a) A two-bed reactor design containing one multi-functional catalyst, but with each bed separated by inerts, b) a gradient, multi-catalyst, two-reactor design, containing three distinct transition-metal/solid acid catalyst combinations, and c) an integrated, two-zone layout, incorporating two, multi-component, transition-metal catalyst combinations on zeolite, or oxide supports. Typically, the bifunctional catalysts have both hydrogenation and etherification/dehydration capabilities and may comprise Group IB, VIB, and VIII metals incorporated into acidic, large and medium-pore zeolites, Group III or IV metal oxides, as well as heteropoly acid structures. DIPE syntheses are typically conducted at 100 to 165°C, under hydrogen pressure. The gradient reactor design, with prudent choice of hydrogenation and etherification catalyst compositions, allows DIPE to be generated in exit concentrations approaching 40%, with quantitative acetone conversions and little or no $C_3/C_6/C_9$ formation.

INTRODUCTION

We have previously described the synthesis of diisopropyl ether (DIPE) (1,2) (as well as isopropyl *tert*-butyl ether (3) from acetone-rich feed streams involving initial hydrogenation of the acetone fraction to isopropanol (IPA), followed by dehydration of the intermediate IPA to DIPE, as depicted in equation (1). Such a two-stage DIPE process normally requires the interstage separation of hydrogen in order for the etherification catalyst to remain effective, yet removal of said hydrogen may cause any co-produced propylene to oligomerize in the etherification unit. In a commercial process this could significantly increase the costs associated with DIPE purification, as well as necessitate frequent catalyst regeneration.

Here we disclose procedures for the one-step conversion of acetone-rich feed streams to DIPE using novel, dual-functional, catalyst systems that incorporate both ketone hydrogenation capabilities in addition to dehydration activity. DIPE one-step synthesis has been demonstrated in continuous unit equipment using three types of catalyst bed/reactor configurations.

The excellent octane blending properties of DIPE for reformulated gasoline are now well established (4,5). Alternative routes to diisopropyl ether synthesis include propene hydration followed by IPA dehydration – pioneered by Mobil Corporation (6). With low-cost, crude acetone feedstocks as the basic building block, we believe that this new, one-step, route to DIPE (equation 2) enjoys very attractive, long-term, economics (2).

EXPERIMENTAL

Catalyst Screening: All catalyst screening studies were performed in continuous microreactor units operated in the downflow configuration. For experiments conducted using a two-bed catalyst design, each bed comprised 4 cc of catalyst having the same composition, separated by 4 cc of inert material. Internal thermocouples were positioned at the bottom of each catalyst bed. The acetone-rich liquid feed was charged to the unit using a high pressure pump and hydrogen was metered through a mass flow controller. Liquid products were collected in a chilled receiver at -15°C and 20 bar, and analyzed by gas-liquid chromatography (gc).

The catalyst beds were activated by heating slowly to 315°C over a period of 6 hours, under flowing nitrogen, at 5 bar. The unit pressure was then raised to 35 bar with hydrogen and the catalyst bed held at 315°C for 10 hours under flowing hydrogen. The bed was then allowed to cool to < 90°C, and crude (97%) acetone charged to the unit at 1.5 liquid hourly space velocity (LHSV), based upon the total catalyst volume. The hydrogen flow rate was adjusted to give a hydrogen to acetone molar feed ratio of ca. 5:1, at a total unit pressure of 35 bar. The typical operating temperature range was 100 to 165°C.

For the gradient reactor studies the microreactor unit consisted of two reactors in series separated by a quench zone. The top reactor was loaded with 4 cc of catalyst with the highest hydrogenation activity and the lowest dehydration/etherification activity. This first reactor was operated adiabatically. The second reactor had two catalyst beds of 4 cc each separated by a 4 cc bed of inert material. The top bed in the second reactor contained catalyst having an intermediate hydrogenation and etherification activity. The bottom bed in the second reactor contained the catalyst with the highest dehydration/etherification activity. The total catalyst charge was 12 cc. A typical gradient reactor configuration is illustrated in Figure 1.

For the integrated, two-zone, reactor design with differing catalyst compositions, Figure 2 illustrates the unit set-up. Here the first and second catalyst beds (A and B) are of different composition.

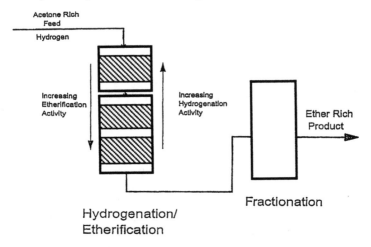

Figure 1 Gradient reactor configuration, three catalysts.

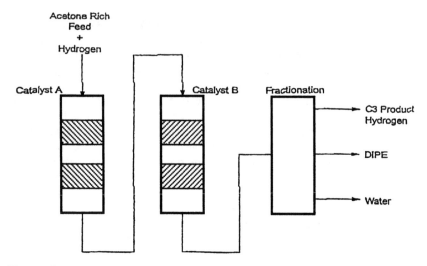

Figure 2 Two-zone reactor configuration, two catalysts.

RESULTS AND DISCUSSION

The one-step conversion of acetone-rich feed streams to diisopropyl ether (equation 2) has been demonstrated in this work using dual-functional catalyst systems that incorporate both ketone hydrogenation and dehydration/ etherification activities. These catalyst combinations have been tested in three types of reactor configurations, namely:

- A two-bed reactor design containing one multi-functional catalyst, each bed separated by inerts, and the same catalyst composition in both beds (7,8).
- A gradient, multi-catalyst, two-reactor design, containing three distinctly different transition-metal/solid acid catalyst compositions (9).
- An integrated, two-zone layout, incorporating two, multi-component, transition-metal catalyst compositions on zeolite, or oxide supports (10).

All catalyst screening for DIPE synthesis was conducted in continuous microreactor units, operated in the downflow mode as described in section 3. The bifunctional catalysts comprised Group 1B, VIB, and VIII metals incorporated into acidic, large and medium-pore zeolites, Group III and VI oxides, as well as supported heteropoly acid structures. Typical catalyst preparations have been described previously (11). After loading into the microreactors, each catalyst bed was pretreated in a stream of hydrogen, at elevated temperatures, then fed a mix of acetone (97%, also containing methanol plus water) and hydrogen (the H_2-to-$(CH_3)_2CO$ molar feed ratio being ca. 5:1) over a range of operating temperatures. The typical operating temperature range was 100 to 160°C. Experimental details may be found in section 3.

Product composition data for DIPE production using the first reactor configuration – the two catalyst beds with the same composition – are illustrated in Table 1. Five catalyst compositions were initially tested, three 32% nickel/copper on 10, 50, and 60% Beta samples, a nickel/copper on 80% ZSM-5, and a 12-tungstophosphoric acid on silica-alumina (3). It is intended that for each catalyst formulation, the metals component provides the ketone hydrogenation activity, and that the acidic zeolite ensures subsequent IPA etherification. The first four examples each gave quantitative acetone conversions over a range of operating temperatures. In the case of example 3, at 124°C and an acetone feed rate of 1.5 LHSV, DIPE selectivities of 25 wt% were realized, where the combined DIPE plus IPA yields were ca. 90 wt%. Liquid recovery in this case was quantitative. All other product sample recoveries in experiments 1 to 4 were >99 wt%. Comparing the data of examples 1-3, it appears that for this particular reactor design, DIPE productivity increases with increase in catalyst β–zeolite content.

DIPE production as a function of operating temperature is illustrated in Figure 3 for the nickel-copper/50% Beta catalyst of example 2.

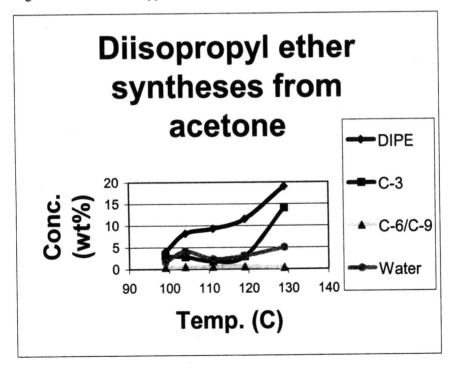

Figure 3 DIPE generation as a function of catalyst bed temperature, single catalyst.

TABLE 1 Two-Bed Reactor Configuration, Single Catalyst

Ex.	Catalyst	Temp (C)	Product Composition (wt%)						Liquid Recovery (wt %)
			C-3	Acetone	IPA	DIPE	C-6/C-9	Water	
1	NiCu/10% Beta	101	0.9	0	97.4	0.5	0.5	0.9	99
		113	1.8	0	94.6	1.9	0.5	1.1	98
		130	0.5	0.1	89.6	6.6	0.9	2.4	100
		143	6.8	0.1	71.4	14.6	1.8	5.1	93
		163	22.1	0.2	49.2	17.1	2.6	8.6	80
2	NiCu/50% Beta	99	2.9	0	91	4	0.4	1.6	97
		104	2.8	0	84.3	8.2	0.5	4.1	99
		111	1.8	0	86	9.3	0.5	2.4	99
		119	2.9	0.1	81.7	11.5	0.6	3.1	97
		129	14.1	0.1	61.1	19	0.4	5	86
3	NiCu/60% Beta	124	0.3	0	65	24.8	1.8	7.9	100
4	NiCuCr/ZSM-5	104	1.8	0	82.7	9.8	1.3	4.5	100
5	PW12/SiO2-Al2O3	141	1.6	4.6	82.5	10.9	0.5	3.8	100

A significant improvement in DIPE productivity has been realized by running the acetone hydrogenation/dehydration sequence using a gradient reactor configuration. Such a unit has a series of fixed-bed catalyst zones loaded along the reactor axis so that there is a hydrogenation activity gradient along the reactor in one direction and an etherification activity gradient in the opposite direction. A typical layout is illustrated in Figure 1. In the DIPE synthesis examples illustrated here, there are three distinct catalyst beds of differing compositions, and the metal loadings in each of the beds range from 40 – 25%, to 20 – 10%, to 10 – 0% metals, while the zeolite content increases from 0 – 5%, to 40 – 60%, to 70 – 90%. Table 2 includes gc product analyses data for two such experiments. In both runs 6 and 7, the catalyst charge has been loaded so that the initial zone (first seen by the fresh acetone feed) contains a high hydrogenation function and a low etherification function, while successive zones display decreasing hydrogenation catalyst activity and more etherification activity. In the first (top) catalyst bed, the catalyst composition is selected to provide high conversion of the acetone to IPA, without significant condensation to undesirable C-6 and C-9 products, or the dehydration of IPA intermediate to C-3 gas plus water. As the acetone/IPA stream passes further down the unit, the remaining acetone is reduced and the alcohol fraction is etherified in those subsequent catalyst zones that display increasing quantities of acid functionality. In the case of example 6, the precise catalyst compositions are:

- Top bed: 32% Ni plus Cu on alumina
- Middle bed: 16% Ni plus Cu on 50% Beta/50% alumina
- Bottom bed: 80% Beta/20% alumina

Again the acetone conversions are essentially quantitative over the full temperature range considered (see column three, Table 2, for the typical temperatures in each of the three catalyst zones). However in this case, it is noteworthy that the IPA + DIPE molar selectivities exceed 98% and we are able to minimize the coproduction of C-3 gas, as well as C-6/C-9 condensation products (DIPE to C-3/-6/-9 wt ratio >30). Consequently, DIPE effluent concentrations reach ca. 38% when the catalyst temperature sequence is 104-159-146°C. Liquid recoveries in this experimental series were again essentially quantitative.

Somewhat similar data are realized in experiment 7 where the catalyst configuration is:

- Top bed: 32% Ni plus Cu on 10% Beta/90% alumina
- Middle bed: 16% Ni plus Cu on 50% Beta/50% alumina
- Bottom bed: 8% Ni plus Cu on 80% Beta/20% alumina

TABLE 2 Gradient Reactor Configuration, Three Catalysts

Ex.	Catalyst	Temp (C)	Product Composition (wt%)						Liquid Recovery (wt %)
			C-3	Acetone	IPA	DIPE	C-6/C-9	Water	
6	NiCu/Al2O3 +NiCu/50% Beta + 80% Beta	101 - 131 - 133	0.2	0	81.6	13.6	0	3.5	99
		99 - 134 - 128	0.3	0	74.9	20	0	4.9	100
		102 - 138 - 142	0.5	0.5	64.5	26.3	0	6.6	98
		100 - 143 - 139	0.7	0	58.4	32.7	0	8.1	100
		102 - 150 - 143	0.7	0	51.1	37.4	0	10.3	100
		104 - 159 - 146	1.1	1	42.5	37.5	0	11.9	94
7	NiCu/10% Beta + NiCu/50% Beta + NiCu/80% Beta	106 - 129 - 129	0	0	86.5	9.9	0.3	3.3	100
		105 - 142 - 136	0.7	0	64.9	26	0.3	7.7	99
		105 - 141 - 132	0.5	0	71.6	21.9	0	6	100
		107 - 146 - 139	0.6	0	65.3	24.4	0	6	96

Table 3 data illustrate the production of DIPE using the third reactor configuration – a two-zone design, outlined in Figure 2. Here the acetone feed is hydrogenated to an isopropanol-rich effluent by multi-metal catalyst A, and then said IPA is passed directly to the second reactor, without separation of liquid and gas, where, in the presence of a strongly acidic zeolite-based catalyst, B, it is etherified to DIPE. Example 8 uses a 32% nickel-copper on alumina catalyst in combination with a 32% nickel-copper on 80% Beta/20% alumina support (11). Again, near complete acetone conversions were achieved under these test conditions, with IPA and DIPE co-generated as desired products. Small quantities of propene (from IPA dehydration) provided the major co-products. DIPE yields increased substantially with increasing temperature of the second reactor. Raising the acetone feed space velocity, and then adjusting the catalyst bed temperatures, is demonstrated in example 9.

DIPE yields of up to ca. 36% were achieved in example 10 using a 32% nickel-copper on 60% β-zeolite/40% alumina as the second catalyst bed, at 146°C. Unfortunately, in this case, there is also a 9.2% coproduction of light (C-1 plus C-2) gases (see column six), and if the etherification temperatures are raised still further (> 150°C), there is a deleterious effect on the combined IPA + DIPE yields (to < 80%) through the formation of additional undesired lights coproduct.

Further extensions of this chemistry (equation 2) include the addition of alkanols – particularly *tert*-butanol or methanol – to the acetone feed streams, thereby generating, in one step, a mix of DIPE with IPTBE, MTBE, etc (9). Again, these one-step, multi-ether syntheses can be conducted in the gradient reactor system without the co-production of large quantities of undesirable C-3 and C-4 gaseous co-products. Typical alkanol feed concentrations may be in the 10 to 70% range (9).

CONCLUSIONS

We conclude that of the three unit configurations considered in this research program, the gradient reactor design has the triple advantages of:

- The highest exit concentrations of desired DIPE, approaching 40% (see Table 2, ex. 6).
- Very little co-product C-3/C-6/C-9 formation.
- Quantitative acetone conversion over a broad range of operating temperatures.

Additional improvements in DIPE productivity may be anticipated through further optimization of the hydrogenation/dehydration activities of the three catalyst compositions, as well as the temperature profiles selected for DIPE generation.

Table 3 Two-Zone Reactor Configuration, Two Catalysts

Ex.	Catalyst	Temp (C)	LHSV	Time-on-stream (hr)	C-1/C-2	C-3	Acetone	IPA	DIPE	Water	Liquid Recovery (wt %)
8	NiCu/Al2O3 +	115 - 126	1	9	1.4	0.2	0.6	83.9	8.7	4.8	99
	NiCu/80% Beta	118 - 137	1	20	4.6	0.5	0.5	54.3	30	10	95
9	NiCu/Al2O3 +	115 - 127	1	9	0	0	0	82.9	12.2	4.9	100
	NiCu/30% Beta	115 - 135	1	14	5.3	0.3	0.5	64.1	25.2	4.6	95
		151 - 126	2	22	0	0	0	90.4	4.7	4.9	100
10		152 - 134	2	30	4	0.3	0.5	74.3	16.2	4.7	96
	NiCu/Al2O3 +	116 - 117	1	5	0	0	0	90.5	4.6	4.9	100
	NiCu/60% Beta	117 - 135	1	9	4.2	0.3	0	66.1	24.7	4.7	96
		118 - 146	1	17	9.2	0.3	0	50.3	35.8	4.5	91
		119 - 147	1	19	18.9	0.8	0	43.1	33.2	4	81

Product Composition (wt%)

REFERENCES

1. JF Knifton and PE Dai. Diisopropyl ether syntheses from crude acetone. *Catal. Lett.* **57**:193-197, 1999.
2. JF Knifton and PE Dai. High octane fuel ethers from alkanol/crude acetone steams via inorganic solid acid catalysis. In: M. E. Ford, ed. Catalysis of Organic Reactions. New York: Marcel Dekker, 2001, pp 145-151.
3. JF Knifton, PE Dai, and JM Walsh. Isopropyl *tert*-butyl ether from crude acetone streams. *Chem. Commun.* 1521-1522, 1999.
4. WJ Piel. Expanding refinery technology leads to new ether potential. *Fuel Reform.* 34-40, November/December 1992.
5. WJ Piel. Diversify future fuel needs with ethers. *Fuel Reform.* 28-33, March/April 1994.
6. A Wood. Mobil cuts the alcohol out of oxygenate production. *Chem. Week* **7**, April 15 1992.
7. PE Dai, RJ Taylor, JF Knifton, and BR Martin. US Patent 5 476 972 (1995) to Texaco Chemical Inc.
8. RJ Taylor, PE Dai, JF Knifton, and BR Martin. US Patent 5 637 778 (1997) to Texaco Chemical Inc.
9. RJ Taylor, PE Dai, and JF Knifton. US Patent 5 550 300 (1996) to Texaco Chemical Inc.
10. RJ Taylor, PE Dai, and JF Knifton. US Patent 5 583 266 (1996) to Texaco Inc.
11. RJ Taylor, PE Dai, and JF Knifton. Diisopropyl ether one-step generation from acetone-rich feedstocks. *Catal. Lett.* **68**:1-5, 2000.

13

Highly Selective Catalytic Oxidation of Alkylbenzenes to the Corresponding Hydroperoxides

Isabel W. C. E. Arends, Manickam Sasidharan, Sandrine Chatel, and Roger A. Sheldon*
Delft University of Technology, Delft, The Netherlands

Carsten Jost, Mark Duda, and Adolf Kühnle
CREAVIS Gesellschaft für Technologie und Innovation mbH, Marl, Germany

INTRODUCTION

The autoxidation of alkylbenzenes constitutes the industrial route for the production of the corresponding hydroperoxides. The two well-known examples are the production of 1-phenylethyl hydroperoxide in the Shell and ARCO processes for the co-production of styrene and propene oxide, and the production of cumene hydroperoxide for the production of phenol via the Hock process (1,2) A disadvantage of the Hock-process is the co-production of acetone. One possible alternative involves the use of cyclohexylbenzene (CHB) in Scheme 1.

---

* *Corresponding author,* r.a.sheldon@tnw.tudelft.nl

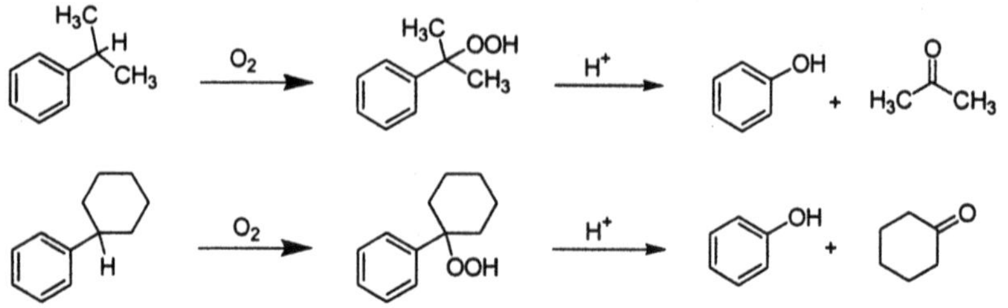

Scheme 1 Production of phenol via decomposition of cumene hydroperoxide and cyclohexylbenzene-1-hydroperoxide, respectively.

Analogous to cumene, the corresponding tertiary hydroperoxide can be easily converted into phenol and cyclohexanone. However, in this case the cyclohexanone coproduct can be dehydrogenated to give a second molecule of phenol (3). CHB is readily obtained from benzene via (a) selective hydrogenation to cyclohexene, using a ruthenium catalyst (4), followed by Friedel-Crafts alkylation or (b) oxidation to biphenyl (5) followed by selective hydrogenation (6). This provides for the overall conversion of two molecules of benzene to two molecules of phenol, i.e. a coproduct-free route to phenol (Scheme 2).

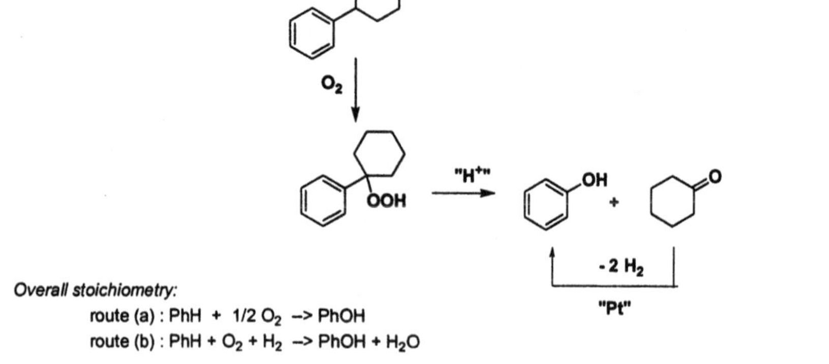

Overall stoichiometry:
    route (a) : PhH + 1/2 O_2 –> PhOH
    route (b) : PhH + O_2 + H_2 –> PhOH + H_2O

Scheme 2 Process scheme for the production of phenol.

However, we anticipated that there could be a selectivity problem. Cumene has one reactive tertiary bond and six relatively unreactive primary C-H bonds. Cyclohexylbenzene in contrast has one tertiary C-H bond and 10 secondary C-H bonds. Earlier reports from patents indicated, however, that in the presence of azo-initiators air-oxidation of cyclohexylbenzene resulted in a selectivity for cyclohexylbenzene-1-hydroperoxide (CHBHP) as high as 90%, albeit at low conversion levels of 3-5% (7). This encouraged us to embark on the study of the selective autoxidation of cyclohexylbenzene. For practical utility the autoxidation of CHB should afford CHBHP in >90% selectivity at conversions of ca. 30%.

The free radical chain mechanism of hydrocarbon autoxidation is well documented (2,8). The susceptibility of any substrate to autoxidation is determined by the ratio $k_p/(2k_t)^{1/2}$ – referred to as the oxidizability (9) – which determines the length of the propagating chain and, thus, the rate of the reaction (see Scheme 3).

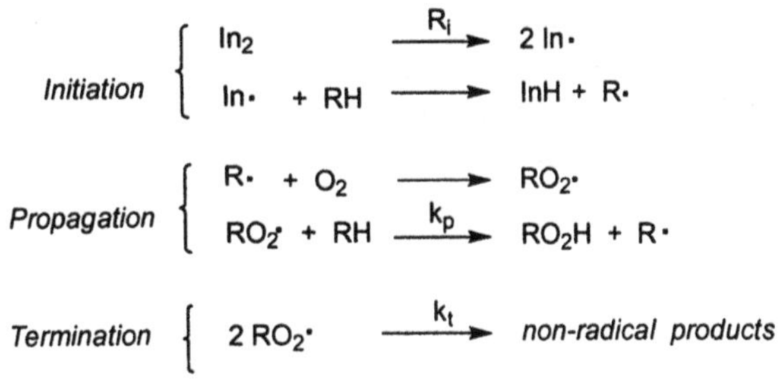

$$
\begin{array}{ll}
\text{Initiation} & \left\{
\begin{array}{l}
In_2 \xrightarrow{\ R_i\ } 2\ In\cdot \\[4pt]
In\cdot + RH \longrightarrow InH + R\cdot
\end{array}
\right. \\[18pt]
\text{Propagation} & \left\{
\begin{array}{l}
R\cdot + O_2 \longrightarrow RO_2\cdot \\[4pt]
RO_2\cdot + RH \xrightarrow{\ k_p\ } RO_2H + R\cdot
\end{array}
\right. \\[18pt]
\text{Termination} & \left\{
\begin{array}{l}
2\ RO_2\cdot \xrightarrow{\ k_t\ } \text{non-radical products}
\end{array}
\right.
\end{array}
$$

Scheme 3 Standard mechanism autoxidation.

A vital role is played by alkylperoxy radicals, which are the active propagating species. In order to improve the performance in autoxidation processes, a large variety of additives has been demonstrated to exhibit (minor) advantages, among which HBr/Br⁻ is probably most well understood (10). This effect is ascribed to the replacement of the chain propagating alkylperoxy radicals by bromine atoms. The latter react more selectively than alkyperoxy radicals with hydrocarbon substrates (11). The resulting HBr efficiently scavenges alkyperoxy radicals, thereby increasing the ratio of propagation to termination rates and, hence, the rate and/or selectivity of the autoxidation (see Scheme 4).

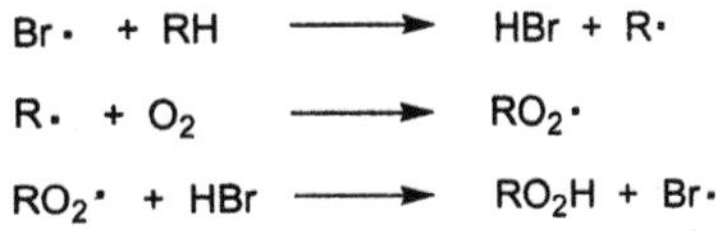

$$
\begin{array}{l}
Br\cdot + RH \longrightarrow HBr + R\cdot \\[4pt]
R\cdot + O_2 \longrightarrow RO_2\cdot \\[4pt]
RO_2\cdot + HBr \longrightarrow RO_2H + Br\cdot
\end{array}
$$

Scheme 4 Effect of HBr on hydrocarbon autoxidations.

However, HBr is not a suitable alternative for cyclohexylbenzene autoxidation as it would catalyze the facile rearrangement of the hydroperoxide to cyclohexanone and phenol. Although this is the final goal, the in-situ formation of phenol would inhibit the autoxidation reaction. Hence, we needed a neutral additive which would not promote the (acid-catalyzed) decomposition of hydroperoxide.

In this context our attention was drawn to reports by Ishii et al. on the selective oxidation of a large variety of hydrocarbon substrates using the combination of N-hydroxyphthalimide (NHPI) and metal salts (12). Under these conditions, alcohols, ketones and carboxylic acids are the main products, via alkyl hydroperoxides as putative intermediates. The use of NHPI/metal resulted in high yields and selectivities under mild conditions. For example, using NHPI (10 mol%) and $Co(OAc)_2$ (0.5 mol%) the oxidation of toluene in acetic acid for 20 h would afford benzoic acid and benzaldehyde in 81% and 3% yield, respectively (13). The metal salt functions as an initiator for the autoxidation reaction, and also catalyzes the decomposition of the intermediate hydroperoxides. Since the metal salt acts primarily as an initiator, we envisaged that it could be replaced by organic initiators leading to an NHPI catalyzed oxidation which would afford the hydroperoxide in high selectivity. The validity of this idea was given support by a recent publication of Ishii's group which described the preparation of hydroperoxides in 75% yield, using 10 mol% NHPI as the catalyst (14). This publication prompted us to report our own results on the selective oxidations of cyclohexylbenzene, cumene and ethylbenzene, using NHPI as a catalyst (15). We will show that when using NHPI concentrations as low as 0.1 mol%, alkylbenzene hydroperoxides can be prepared with excellent selectivities at relatively high conversions.

The role of NHPI is analogous to that of HBr in metal-catalyzed autoxidation processes. The proposed reaction sequence for cyclohexylbenzene is shown in Scheme 5. NHPI efficiently traps the intermediate alkylperoxy radicals (reaction 4) thereby suppressing competing termination. The thus formed phthalimide N-oxy radical (PINO) abstracts a hydrogen from the substrate to regenerate NHPI (reaction 2). Obviously, the ratio of propagation to termination and, hence, the hydroperoxide selectivity will be directly dependent on the NHPI concentration. A decrease in NHPI concentration should lead to a decrease in both rate and selectivity of RO_2H formation.

RESULTS AND DISCUSSION

Cyclohexylbenzene

In initial experiments we investigated the oxidation of HB in the presence of added hydroperoxide and azo compounds as initiators at different temperatures. Oxidation reactions were performed using neat CHB with bar of pure oxygen. The results in this case were not satisfactory: selectivity decreased rapidly with

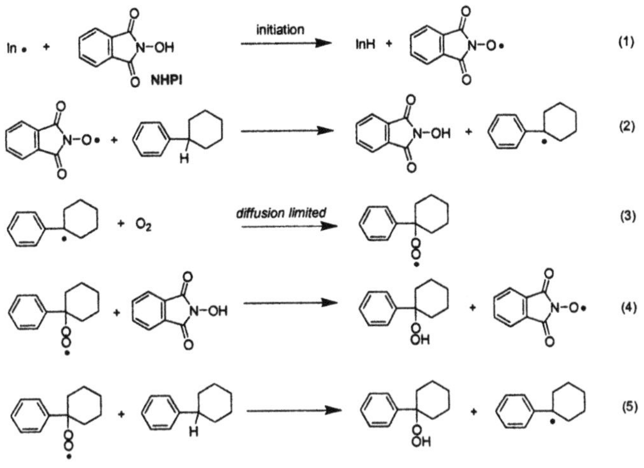

Scheme 5 Reaction scheme for the NHPI catalyzed oxidation of cyclohexylbenzene

increasing conversion. At 5% CHB conversion (95°C), a selectivity to CHBHP of 87% was obtained while at 9% conversion the selectivity decreased to 65%. However, upon adding NHPI to the system with CHBHP as the initiator, high conversions and selectivities could be obtained. Table 1 shows the results. At 100°C after 8 h a conversion of 29% was reached with a selectivity to CHBHP of 96%. We studied the influence of the NHPI concentration on conversion and selectivity. Even at NHPI concentrations of 0.05%, the selectivity for CHBHP formation remained high although the conversion decreased. The optimum NHPI concentration proved to be 0.5 mol%. Higher amounts of NHPI had no further beneficial effect owing to the limited solubility of NHPI in cyclohexylbenzene. Noteworthy is that poor results were obtained with NHPI but without added CHBHP in the beginning of the reaction (see Table 1).

The effect of NHPI is threefold. (1) It increases the reaction rate; (2) It increases the selectivity of initial attack at the 1 versus the 4 position and (3) It suppresses the formation of the byproducts A and B. These main byproducts, derived from intramolecular (transannular) abstraction of hydrogen from the 3-position in the intermediate peroxy radical 1, are illustrated in Scheme 6. Transannular H-abstraction results in the formation of a di-hydroperoxide (A) which is relatively unstable and decomposes to product B. In the presence of NHPI but without extra CHBHP (second entry Table 1) this transannular hydrogen abstraction was still significant. However, in the presence of both NHPI and CHBHP this latter rearrangement was largely suppressed. The preference for oxidation at the 1-position -

Table 1 Oxidation of cyclohexylbenzene catalyzed by NHPI [a]

NHPI	Temp. (°C)	Initiator	Conv. (%) CHB	Selectivity Products (%)[b]				
				1-ROOH	2-ROOH	4-ROOH	A	B
No NHPI	100	CHBHP	3.2	86	0.9	6.0	2.9	-
1%	95	ACCN[c]	17	58	1.0	0.8	9.8	12
1%	95	CHBHP	8.4	98	0.2	1.8	-	-
1%	100	CHBHP	29	96	-	2.1	1.6	0.1
1%	110	CHBHP	37	94	0.2	2.5	2.7	0.2
0.5%	100	CHBHP	32	96	-	2.1	1.6	-
0.1%	100	CHBHP	19	97	0.1	3.2	-	-
0.05%	100	CHBHP	14	94	0.3	4.0	1.7	-

[a]Conditions: no solvent, 60 mmol CHB, 0.6 mmol NHPI (1 mol%), 0.03 mmol azo-initiator (0.05%) or 1.2 mmol CHBHP (2 mol%), 8 h reaction time, 1 atm O_2, internal standard used is naphthalene; Hydroperoxide determined by GC as well as iodometric titration. [b] Selectivity of products (see Scheme 6) formed, missing balance is accounted for by several unidentified products in low quantities. [c] 1,1'-Azobis(cyclohexane carbonitrile) (ACCN) was used as initiator, $t_{1/2}$ = 150 min at 100°C.

relative to the other positions in the cyclohexane ring – in the presence of NHPI is apparent when comparing the last entries in Table 1. The ratio for initial H-attack at the 1-position versus attack at other positions in the ring increases from 11 without NHPI, to 47 with 0.1 mol% NHPI. This effect results from the fact that at sufficiently high concentrations of NHPI, PINO radicals are responsible for propagation (reaction 2; Scheme 5). Apparently, PINO radicals are more selec-

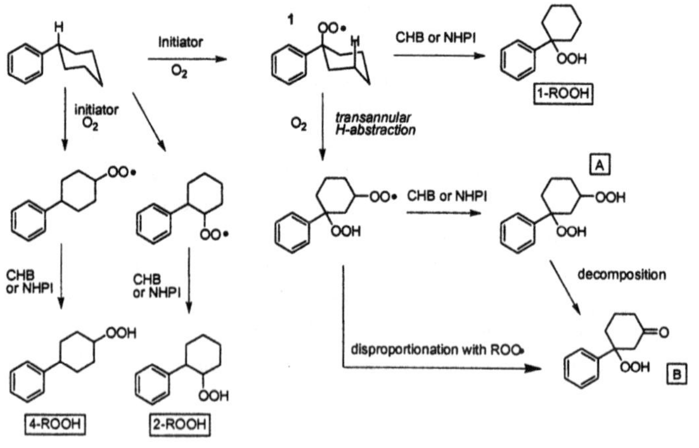

Scheme 6 Formation of products in the radical oxidation of cyclohexylbenzene.

tive than the phenylcyclohexyl peroxy radicals, resulting in H-abstraction at the thermodynamically more favourable 1-position.

The effect of rate acceleration by NHPI (32% conversion with 0.5% NHPI, vs. 3.2% conversion without NHPI after 8 h at 100°C) can be ascribed to the lower rate of termination as discussed above.

Cumene

The excellent results obtained with cyclohexylbenzene prompted us to apply the combination of NHPI and alkyl hydroperoxide to cumene and ethylbenzene autoxidation. The results for cumene autoxidation using the NHPI / alkyl hydroperoxide method are shown in Table 2 and Figure 1. In this case the autoxidation was carried out in an autoclave (at 10 bar using a flow of 8% oxygen in nitrogen), as well as in glassware at atmospheric pressure. Also, there is a clear influence of NHPI on both the selectivity and rate of the reaction. The formation of cumyl alcohol as well as acetophenone decreases with increasing NHPI concentration. In this way a selectivity towards cumene hydroperoxide of 94% was obtained, at a conversion as high as 37%.

Table 2 NHPI catalyzed autoxidation of cumene at 100°C[a]

NHPI	Reaction time	Conv. (%) cumene	Selectivity products (%)[b]		
			ROOH	Cumyl alcohol	acetophenone
No NHPI	6 h	27	84	12	2.2
0.05%	5 h	21	92	6.7	1.1
0.1%	5.5 h	30	92	6.4	1.1
0.5%	5.5 h	37	94	4.5	0.7

[a] Conditions: no solvent, flow of 8% O_2 and 92% N_2 (100 ml/min) is led through the reactor. Autoclave: Hastelloy C; 148 mmol cumene, 3 mmol cumene-hydroperoxide, internal standard used is naphthalene; Hydroperoxide determined by GC as well as iodometric titration and corrected for hydroperoxide added. [b] Selectivity of products formed, missing balance is accounted for by several unidentified products in low quantities.

In order to confirm that NHPI influences, the relative rates of propagation and termination, rather than the rate of initiation, we examined the addition of varying amounts of NHPI to cumene autoxidations that were already in progress. As can be seen in Figure 2, the addition of varying amounts of NHPI (0.05 to 1 mol%), after 3 hours, resulted in increases in the rate of oxygen uptake.

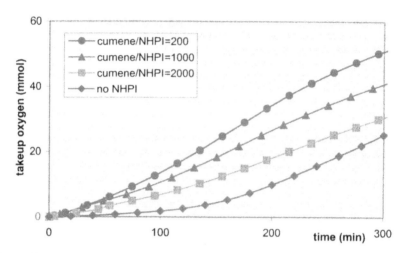

Figure 1 Cumene autoxidation followed in time, as a function of NHPI concentration (conditions as in Table 2).

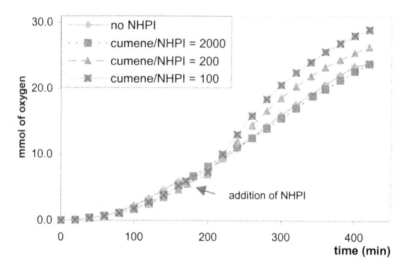

Figure 2 NHPI catalyzed autoxidation of cumene, in a glass reactor: NHPI was added after 3 h, to evaluate its effect on the rate of propagation. Conditions: 115°C, 1 bar O_2, 2 mol% cumene hydroperoxide, 60 mmol cumene.[†]

---

† Data obtained after 7 h. No NHPI: 40% conv., 89% CHP, 9.1% cumyl alcohol, 2.0% acetophenone. 0.05% NHPI: 42% conv. 90% sel. CHP, 8.2% cumyl alcohol, 1.6% acetophenone; 0.5% NHPI 46% conv., 92% CHP, 6.6 % cumyl alcohol, 1.1% acetophenone.

Ethylbenzene

We turned our attention next to the autoxidation of ethylbenzene (EB) to the corresponding hydroperoxide (EBHP) which constitutes the first step in the SMPO (styrene monomer propene oxide) process for the co-production of styrene and propene oxide from ethylbenzene and propene (Scheme 7). The overall selectivity to propene oxide obviously depends on the selectivity to EBHP in the first step, which is believed to be 80-85% in the commercial process. This is lower than for cumene as a result of secondary (in the case of EB) versus tertiary (in the case of cumene) C-H bond oxidation. The main byproduct in the autoxidation of ethylbenzene is acetophenone (16). From an economic viewpoint the production of acetophenone should be kept as low as possible.

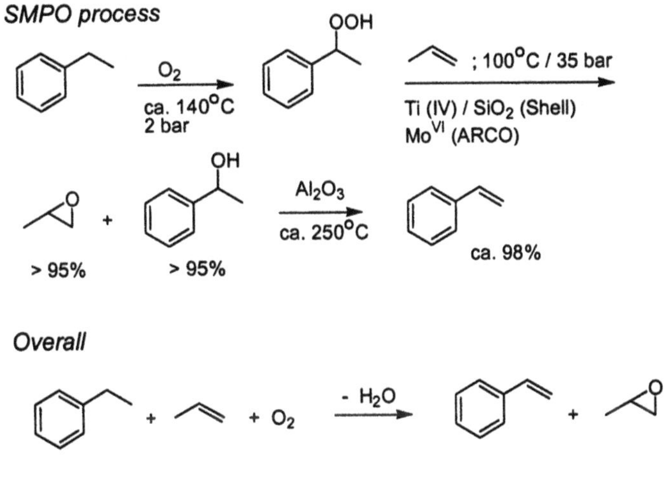

Scheme 7 Process scheme for the coproduction of styrene and propene oxide from ethylbenzene and propene.

When we added NHPI in combination with EBHP reasonable selectivities were obtained up to 12 % conversion (Table 3). Although yields and selectivities obtained in this case were not so high as in the case of cumene and cyclohexylbenzene, the same trend can be discerned. At low conversions selectivities are always much higher then at higher conversions where – due to the number of terminations which have taken place – the amount of byproducts due to disproportionation of alkylperoxy radicals is substantial. NHPI is able to substantially change the ratio of selectivity versus conversion. In Figure 3, the selectivity is plotted as a function of conversion.

Table 3 NHPI catalyzed autoxidation of ethylbenzene at 115°C. [a]

Entry	Ratio EB/NHPI	Ratio EB/EBHP	Conv. EB (%)	Selectivity Products (%)[b]	
				EBHP	acetophenone
1	No NHPI	No EBHP	5	88	12
2	No NHPI	50	14	27[c]	48
3	1000	1000	10	91	8
4	400	1000	10	92	7
5	200	1000	12	89	9
6	100	1000	12	88	9
7	100	No EBHP	12	86	11

[a]Conditions: no solvent, 40 mmol EB, 0.4 mmol NHPI (1 mol%), 0.04 mmol EBHP (0.1 mol%), 8 h reaction time,1 atm O_2, internal standard used is 1,2-dichlorobenzene; Hydroperoxide determined by GC as well as iodometric titration. [b] Selectivity of products formed, missing balance is accounted for by several unidentified products in low quantities. [c] 1-phenylethanol was found to be present with 25% selectivity.

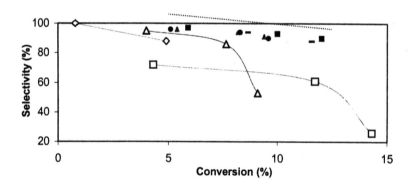

Figure 3 Autoxidation of ethylbenzene catalyzed by NHPI.
Conditions: 115°C, 1 bar O_2, 40 mmol EB. Open symbols: no NHPI and   2 mol% EBHP; Δ 0.1 mol% EBHP;   no EBHP. Closed symbols with NHPI and 0.1 mol% EBHP   0.1 mol% NHPI   0.25 mol% NHPI   0.5 mol% NHPI   1.0 mol% NHPI.

The selectivity to EBHP remained high at conversions above 10%, while in the absence of NHPI the selectivity for EBHP decreased rapidly with increasing conversion.

EXPERIMENTAL

Safety caution: the production of hydroperoxides should always be performed with caution. In our case with 100% molecular oxygen, we worked above the explosion limit, and monitored the oxygen uptake by a burette, making sure that the conversion stayed below 40%. Laboratory glassware was used behind safety screens, and the scale was limited to 10 ml solutions. Hydroperoxides were stored at 4°C as dilute solutions, with the corresponding alkylbenzenes as solvent.

Materials:

Cyclohexylbenzene was distilled before use. Cyclohexylbenzene-1-hydroperoxide was prepared from the corresponding alcohol using H_2O_2 and H_2SO_4, adopting a procedure similar to a reported one (17).

Typical Experimental Procedure:

Procedure in glassware. Reactions were performed in a stirred three-necked flask equipped with a condenser. The outlet of the condenser was connected to a gasburette filled with 1 bar oxygen. In this way oxygen take-up could be followed with time.

Procedure in autoclave. Experiments were performed at 10 bar using a 180 ml Hastelloy C Parr autoclave. A continuous stream of 8% O_2 in N_2 was passed through the solution. The amount of oxygen consumed during the reaction was monitored by an oxygen meter and plotted against time.

Analysis:

Products were analyzed by GC using an internal standard, molar responses were based on pure samples. Either a CP-WAX 52 B (50 m x 0.53 mm) or a CP-Sil-5 CB (50m x 0.53 mm) column was used. HPLC analysis was performed with a reverse phase column using 50/50 methanol/water as an eluens.

CONCLUSIONS

We have developed an effective method for the selective autoxidation of alkylaromatic hydrocarbons to the corresponding benzylic hydroperoxides using 0.5 mol% NHPI as a catalyst and the hydroperoxide product as an initiator. Using this method we obtained high selectivities to the corresponding hydroperoxides, at commercially viable conversions, in the autoxidation of cyclohexylbenzene, cumene and ethylbenzene. The highly selective autoxidation of cyclohexylbenzene to the 1-hydroperoxide product provides the basis for a coproduct-free route to phenol and the observed improvements in ethylbenzene hydroperoxide production provide a basis for improving the selectivity of the SMPO process for styrene and propene oxide manufacture.

REFERENCES

1.   H Hock, B Lang. Autoxydation von kohlen-wasserstoffen, IX Mitteil: über peroxide von benzol-derivaten. Chem. Ber. 77: 257-264, 1944.
2.   RA Sheldon, JK Kochi. Metal-catalyzed oxidations of organic compounds, New York: Acad. Press, 1981.
3.   W Jordan, H van Barneveld, O Gerlich, M Kleine-Boymann, J Ullrich in Ullmann's encyclopedia of industrial chemistry, Weinheim: VCH, 1991, A19, p. 307.
4.   J. Struijk, JJF Scholten, Appl. Catal. A Gen. 82: 277-287, 1992; RL Augustine, Heterogeneous catalysis for the synthetic chemist, New-York: Marcel Dekker, 1996, p. 405.
5.   S Mukhopadhyay, G Rothenberg, G Lando, K Agbaria, M Kazanci, Y Sasson. Air oxidation of benzene to biphenyl – a dual catalytic approach. Adv. Synth. Catal. 343: 455-459, 2001.
6.   PN Rylander, Catalytic hydrogenation in organic syntheses, New York: Acad. Press, 1979, p. 179; PN Rylander, DR Steele US 3387048, 1968.
7.   SHA Dai, CY Lin, FA Stuber. US Pat. 4,282,383, 1981, to Upjohn Company.
8.   L Reich, SS Stivala. Autoxidation of hydrocarbons and polyolefins, New York: Dekker, 1969; NM Emmanuel, ET Denisov, ZK Maizus, Liquid phase oxidation of hydrocarbons (BJ Hazzard, transl.) New York: Plenum, 1967.
9.   JA Howard. Absolute rate constants for reactions of oxyl radicals. Adv. Free-Radical Chem. 4: 55-173, 1972.
10.   C Walling. In: Active oxygen in chemistry. CS Foote, JS Valentine, A Greenberg, JL Liebman, Eds. , London: Blackie Acad. & professional, 1995, pp. 24-65.
11.   The relative reactivities for pimary vs. secondary vs. tertiary C-H bonds in case of RO_2 are 1:9.3:15.9 (from JA Howard and KU Ingold, Can. J. Chem. 46: 1017, 1968) while for Br differences are larger resp. 1 :17:37 (from GA Russell, CJ de Boer, J. Am. Chem. Soc. 85:3136, 1963).
12.   Y Ishii, S Sakaguchi, T Iwahama. Innovation of hydrocarbon oxidation with molecular oxygen and related reactions. Adv. Synth. Catal. 343: 393-427, 2001 and refs. cited herein.
13.   Y Yoshino, T Hayashi, T Iwahama, S Sakaguchi, Y Ishii. Catalytic oxidation of alkylbenzenes with molecular oxygen under normal pressures and temperatures by N-hydroxyphthalimide combined with Co(OAc)$_2$. J. Org. Chem. 62: 6810-6813, 1997.
14.   O Fukuda, S Sakaguchi, Y Ishii. Preparation of hydroperoxides by N-hydroxyphthalimide-catalyzed aerobic oxidation of alkylbenzenes and hydroaromatic compounds and its application. Adv. Synth. Catal. 343: 809-813, 2001.
15.   A Kühnle, M Duda, U Tanger, RA Sheldon, S Manickham, T Nadu, IWCE Arends, DE 10015874, 2001 patent to Degussa; A Kühnle, M Duda, U Tan-

ger, RA Sheldon, IWCE Arends, S Manickam, WO 01/74767 A, 2001, patent to Degussa.

16. The mechanism of the autoxidation of ethylbenzene has been thoroughly studied, see NM Emanuel, D Gál, Modelling of Oxidation Processes, prototype: the oxidation of ethylbenzene, Budapest: Akademiai Kiado, 1986.

17. AG Davies, RV Foster, AM White, J. Chem. Soc. 1541-1547, 1953.

14

Vapour Phase Oxidation of Methyl Aromatics to the Corresponding Aldehydes

Andreas Martin, Ursula Bentrup, Bernhard Lücke, and Gert-Ulrich Wolf
Institut für Angewandte Chemie Berlin-Adlershof e.V., Berlin, Germany

ABSTRACT

Alkali cation-containing vanadia solids (M = Li, Na, K, Rb, Cs) were synthesised and used as catalysts in the vapour phase oxidation of toluene, *p*-methoxytoluene and *p*-chlorotoluene to the corresponding aldehydes. Moreover, the effect of the addition of a non-oxidisable base (e.g. pyridine) to the reaction mixture to increase basicity of the reaction mixture was studied. Parent catalyst samples and used specimens were characterised by X-ray diffractometry and infrared spectroscopy. Vanadates and bronze phases of the more bulky alkali metal cations (K, Rb, Cs) are generated during catalyst synthesis and catalytic reaction that may lead to new surface arrangements with vanadia. Aldehyde selectivity up to 60-70 % was reached at high conversions. It could be shown that the formation of the aldehydes strongly depends on nature, strength and concentration of acidic and basic surface sites of the catalysts, the acid-base properties of the reaction mixture and, additionally, on the electronic properties, i.e. nucleophilicity of the reactants.

INTRODUCTION

Heterogeneously catalysed partial oxidation of hydrocarbons to the corresponding aldehydes, acids and anhydrides are reactions of great industrial importance in the production of intermediates for the manufacture of diverse

157

bulk chemicals like polymers, resins, fibres etc. and fine chemicals, respectively. More than 20 % of all organic chemicals and intermediates produced world wide are synthesised by such heterogeneous catalysed reactions. The production of aromatic aldehydes belongs to this sector and an increasing number of reports in the literature deals with their heterogeneously catalysed synthesis proving this development.

For example, benzaldehyde is produced in the vapour phase at rather low toluene conversion rates (10-20 % per pass) at short residence time (< 1 s); even then, it is only 40-60 % of the theoretical yield [1]. The reaction processing in the liquid phase shows no significantly improved results [1]. Recent developments using ultrasound, supercritical media as solvents, photochemical activation and enzyme-catalysed processing seem to set new standard in liquid-phase oxidations [2]. Best benzaldehyde selectivity of ca. 40-80 % is reported at about 25 % conversion using Co-containing catalysts in liquid phase so far [1]. Due to that rather poor performance of direct oxidation, various substituted aromatic aldehydes are still manufactured in traditional way of the organic chemistry, for example by acid hydrolysis of benzal chlorides [1]. Besides, other synthesis routes became known; e.g. the indirect oxidation by electrochemical assistance using Co, Mn or Ce redox couples in the presence of a suitable solvent [e.g. 3,4]. However, these processes produce many waste causing environmental problems and, additionally, the efficiencies of these routes seem to be not so high. Nevertheless, the vapour phase oxidation route is of current interest due to the advantageous handling of reactants and products, the use of solid catalysts and the environmentally benign reaction processing.

Some work has been done in the last two decades in the field of the oxidation of methyl aromatics to their corresponding aldehydes in the vapour phase to get a deeper knowledge and insight into reactant-catalyst interaction by application of various *in situ*-methods [e.g. 5-7] and to find promising new catalyst compositions. Activities were also focused to increase the catalytic performance of known systems, respectively; for example by using of *(i)* K_2SO_4 impregnated V_2O_5 solids or V_2O_5-TiO_2 catalysts in the synthesis of benzaldehyde from toluene, [e.g. 8,9], *(ii)* V-P-Cu-O catalysts doped with potassium sulphate [10,11] or V_2O_5-CaO-MgO compositions [12] developed for the manufacture of *p*-methoxybenzaldehyde from *p*-methoxytoluene and *(iii)* Cs-Fe-O compounds doped with various transition metals for the production of *p*-chlorobenzaldehyde from *p*-chlorotoluene [13]. Obviously, the applied catalyst compositions combine acidic, basic and redox functions. Mainly Japanese groups reported on the role of acid and base properties of vanadium oxide catalysts, containing basic metal oxides (M = K, Rb, Cs, Tl, Ag) for partial oxidation of methyl aromatics to aldehydes [e.g. 14]. In general, they stated that the selectivity to substituted benzaldehydes is closely related to the basic properties of the catalyst whereas the activity strongly depends on the amount and strength of acid sites.

This finding matches the results of our recent studies on the oxidation of toluene to benzaldehyde on the rather acidic $(VO)_2P_2O_7$ catalyst [15]. Mainly Brønsted acid sites (OH-groups formed during reaction by interaction of V-O-P bonds with water) are responsible for a restricted aldehyde desorption that leads to poor aldehyde selectivity by consecutive overoxidation. Interestingly, the addition of the non-reactive base pyridine to the feed significantly improves the aldehyde selectivity by a blockade of these acidic sites [16]. This principle was also proven for the synthesis of pyridin-4-carbaldehyde using the basic 4-picoline as feedstock and $(VO)_2P_2O_7$ as catalyst. An aldehyde selectivity of ca. 55 % was reached at a 4-picoline conversion of about 80 mol-% [17]. 4-Picoline acts in a twofold way: *(i)* it is converted to the desired aldehyde and *(ii)* it can block acidic Brønsted sites that significantly influence the catalyst properties.

The aim of the present paper is concerned with the oxidation of toluene, *p*-methoxytoluene and *p*-chlorotoluene using vanadia catalysts doped with alkali metal cations. Furthermore, the influence of a pyridine admixture to the feed was studied. Changes in catalyst composition were followed by XRD and FTIR-spectroscopic investigations.

EXPERIMENTAL

Catalysts

Alkali metal-containing V_2O_5 catalysts (MVO) were prepared [e.g. 18-21] by incipient wetness method with an aqueous M_2SO_4 solution (M = Li, Na, K, Rb, Cs). 50 ml of such a solution (0.01 mol M_2SO_4) were added to V_2O_5 (0.1 mol) and evaporated using a rotary evaporator at 343 K for 1 h. A further evaporation to dryness was carried out under vacuum at this temperature. The obtained product was dried overnight at 403 K. Table 1 gives a survey of the catalysts denotation and the BET surface areas.

Additionally, a potassium vanadate (KV_3O_8) was synthesised according to a method described by Kelmers [22]. 45.49 g (0.25 mol) V_2O_5 were introduced stepwise into a solution of 28.14 g (0.5 mol) KOH in 500 ml distilled water during 30 min. 3 ml H_2O_2 (30%) were added to get complete vanadium(V). After filtration the solution was heated up to 353 K under vigorous stirring and 20.5 g (ca. 0.2 mol) H_2SO_4 were slowly added dropwise. The orange-brown solid was further stirred for 72 h at 353 K, then filtered, washed and dried.

Sample Characterisation

The XRD-patterns were recorded using the transmission diffractometer STADI P (Stoe) and the infrared spectra of the samples were recorded with the FTIR-spectrometer Galaxy 5020 (Mattson) using KBr technique.

The surface areas were determined by N_2-physisorption using the BET method (Gemini III 2375, Micromeritics). The samples (ca. 0.5-1 g) were pretreated at 423-473 K for 1 h under vacuum.

Table 1 Survey on used catalysts, denotation and BET surface area

Catalyst	Denotation	S_{BET} $(m^2\,g^{-1})$
Li_2SO_4/V_2O_5	LiVO	7.22
Na_2SO_4/V_2O_5	NaVO	6.52
K_2SO_4/V_2O_5	KVO	5.18
Rb_2SO_4/V_2O_5	RbVO	6.07
Cs_2SO_4/V_2O_5	CsVO	4.69

Catalytic Measurements

The catalytic properties of the synthesised solids were determined during the partial oxidation of toluene (TO) to benzaldehyde (BA), p-methoxytoluene (MTO) to p-methoxybenzaldehyde (MBA, anisaldehyde) and p-chlorotoluene (CTO) to p-chlorobenzaldehyde (CBA). A microreactor set-up that contains a metering system for liquids and gases and a fixed bed quartz-glass reactor was used. The catalysts were introduced into the reactor as sieve fractions (1-1.25 mm) and mixed prior to oxidation runs with the equal portion of quartz glass (1 mm) to avoid local overheating. The product stream was analysed by on line-GC or it was trapped in aqueous ethanol and determined by off line-capillary GC. The formation of carbon oxides was continuously followed by non-dispersive IR photometry.

RESULTS AND DISCUSSION

First partial oxidation tests of various toluenes as feed using KVO catalyst show drastically increased aldehyde selectivity compared to the runs on VPO catalysts reported recently [15,23]. These tests were also repeated in the presence of pyridine to increase the basicity of the reaction mixture. Table 2 shows some selected results.

Surprisingly, in the case of CTO the admixture of a small amount of pyridine to the reaction mixture had no significant influence. Moreover, the increase in the pyridine amount leads to a decrease of the CTO conversion, maybe by blocking of sorption sites. A similar effect can also be seen when pyridine is added to a TO feed. The selectivity is only slightly increased. A significant improvement of the catalytic performance is obtained when pyridine is added during the conversion of MTO. This could be due to differences in electronic properties, i.e. basicity or nucleophilic properties of the reactant and the competition of pyridine and the reactant molecules on the surface for acidic Lewis sorption sites. The basicity of the reactant molecules increases in the following order: CTO < TO < MTO (see Scheme 1); the basicity of the reaction products shows the same ranking. CBA desorbs easily because electron density is additionally withdrawn by the formed carbonyl group, whereas for a fast removal of MBA additional surface basicity or a competitor for sorption sites (e.g. pyridine) is needed.

Table 2 Comparison of partial oxidation of p-chlorotoluene (CTO), toluene (TO) and p-methoxytoluene (MTO) to the corresponding aldehydes on KVO catalyst (molar ratio CTO : O_2 : N_2 = 1 : 6 : 220, TO : O_2 : N_2 = 1 : 83 : 295, MTO : O_2 : N_2 = 1 : 72 : 256) $vs.$ similar runs carried out in the presence of aqueous pyridine (*; molar ratio CTO : O_2 : N_2 : py : H_2O = 1 : 6 : 220 : 0.4 : 40, TO : O_2 : N_2 : py : H_2O = 1 : 84 : 298 : 5 : 24, MTO : O_2 : N_2 : py : H_2O = 1 : 72 : 256 : 5 : 21).

Feed	Temperature K	$X_{toluene}$ mol%	$Y_{aldehyde}$ mol%	$S_{aldehyde}$ mol%
CTO	678	24.4	17.9	73.4
TO	673	70.2	14.4	20.5
MTO	633	51.8	12.5	24.1
CTO*	679	23.9	15.1	63.2
TO*	673	43.9	10.4	23.7
MTO*	635	61.2	35.5	58.0

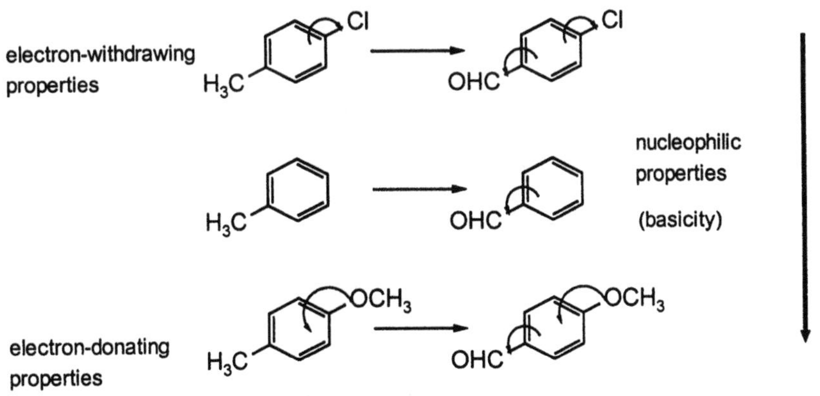

Scheme 1

162   **Martin et al.**

The chemisorption of pyridine should not be too strong, otherwise, the reaction comes to a standstill. Such observation was described by Nag et al. [24] using Mo oxide catalysts for toluene partial oxidation. The authors found out that catalyst acidic surface sites being responsible for non-selective reaction path can be poisoned by pyridine pulses that lead to strong increase in the benzaldehyde selectivity but conversion drops to zero. Otherwise, our own studies have shown that conversion could not be further increased by temperature increase because the effect of the pyridine addition vanished. The acidic sites seem to be too weak to adsorb pyridine to a sufficient extent and for a defined period at higher temperatures (T > 653 K). This behaviour is also known from the usage of pyridine or other amines as a probe molecule in determination of nature, strength and amount of acidic sites of solids by spectroscopic as well as temperature-programmed methods [e.g. 25,26]. Thus, the selectivity-enhancing effect of such a base admixture depends on various factors like nature, strength and amount of surface acidic sites, reaction temperature, nature of the reactant molecule as well as its concentration.

Figure 1 depicts MBA selectivity at similar MTO conversion on LiVO, KVO and CsVO with and without pyridine under comparable reaction conditions. In general, selectivity increases with increasing size of the alkali cation. Furthermore, significant differences in catalytic performance were obtained using pyridine as admixture (*) compared to runs without pyridine; for example, nearly fourfold MBA selectivity was observed.

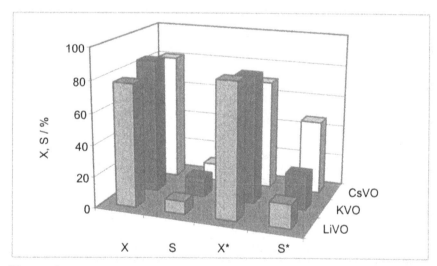

Figure 1 Selectivity of *p*-methoxybenzaldehyde formation on various MVO catalysts at similar *p*-methoxytoluene conversion (* - in the presence of pyridine).

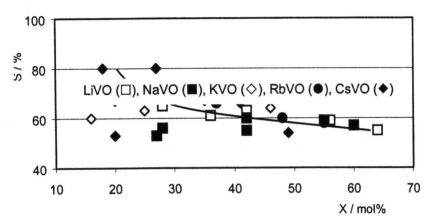

Figure 2 Selectivity of *p*-methoxybenzaldehyde formation *vs.* *p*-methoxytoluene conversion on several MVO catalysts (T = 603 - 648 K, *p*-methoxytoluene : O_2 : N_2 : py : H_2O = 1 : 85 : 300 : 145 : 32, catalyst weight = 0.5 g, W/F = ca. 0.4 ghmol^{-1})

All alkali metal-containing catalysts were checked in a second test running at higher water vapour and pyridine content in the reaction mixture (Fig. 2). A similar selectivity *vs.* conversion dependence was found. LiVO and NaVO catalysts show best MBA selectivity of ca. 60 % at a conversion of ca. 25-30 mol%. However, KVO as well as CsVO catalysts reveal significant higher MBA selectivities of ca. 80 % at the same conversion. *p*-Methoxybenzoic acid (anisic acid) and mainly carbon oxides were the by-products. The selectivity increasing effect could be explained if we assume that the increasing size of the alkali metal cation would enlarge the distances between the V_2O_5 layers leading to a deterioration in its electronic interaction and oxygen transport capacity in the bulk. Consequently, this would lead to a decrease in the catalyst activity and an increase in selectivity. Moreover, with the increasing size of the cation the surface basicity is enhanced and an easier desorption of the formed aldehydes might be realised. Otherwise, new crystalline phases could be formed that may significantly influence catalytic performance.

As recently reported [27], the parent KVO catalyst contains proportions of a crystalline KV_3O_8 phase beside crystalline V_2O_5. The formation of the KV_3O_8 phase and its crystallinity strongly depend on the synthesis and the drying conditions. Figure 3 depicts a comparison of the XRD patterns of the fresh MVO catalysts. The patterns always show the V_2O_5 reflections but also those of MV_3O_8 phases. Such vanadates could be observed in parent KVO, RbVO and CsVO. Interestingly, the as-synthesised LiVO and NaVO samples do not show such phase. The existence of the vanadate species in KVO, RbVO and CsVO parent samples could also observed by FTIR spectroscopy showing a characteristic band at ca. 960 cm^{-1} beside the typical V=O band of V_2O_5 at ca. 1020 cm^{-1} (Fig. 4).

The XRD patterns of the used samples indicate no changes for LiVO and NaVO, only reflections of V_2O_5 can be detected. In contrast, the used KVO, RbVO and CsVO samples show significant changes. The reflections of the vanadate phases are vanished but new phases are formed beside the remaining V_2O_5 as depicted in Figure 5.

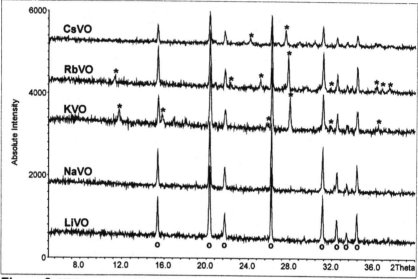

Figure 3 XRD patterns of fresh MVO catalysts (o V_2O_5, * MV_3O_8).

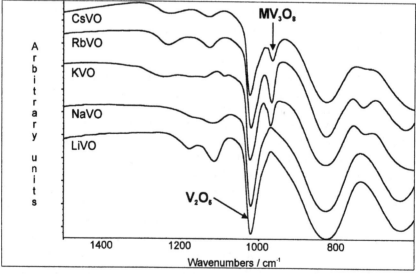

Figure 4 FTIR spectra of fresh MVO catalysts.

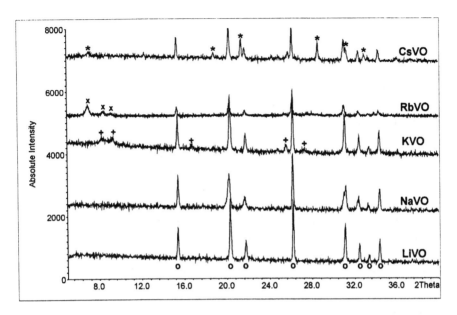

Figure 5 XRD patterns of MVO catalysts used during partial oxidation of *p*-methoxytoluene (o V_2O_5, * $Cs_{0.3}V_2O_5$, + $K_xV_2O_{5-x}$, x $Rb_xV_2O_{5-x}$).

The KVO derived solid demonstrates some additional weak reflections, especially at $2\Theta = 8.0 - 9.8$. These reflections were also observed during earlier investigations [27] and it was supposed that they could be assigned to a vanadium(IV)/vanadium(V) mixed-valent $K_{0.5}V_2O_5$ ($K_{0.5}V_{1.5}^{5+}V_{0.5}^{4+}O_5$) phase that would be involved in catalytic reaction as active or selective surface species, probably. In dependence on the reaction conditions (mainly oxygen content) these crystalline phases show varying potassium contents (x = 0.24 - 0.5, for KVO samples as identified by XRD) and, therefore, different vanadium reduction degrees. A similar transformation process could be seen for RbVO and CsVO; the parent RbVO solid is transformed during the oxidation reaction into a material that contains crystalline V_2O_5 and proportions of a phase that is similar to $Rb_{0.4}V_2O_5$ ($Rb_{0.4}V_{1.6}^{5+}V_{0.4}^{4+}O_5$). The same process was found for the parent CsVO material that is transformed into a catalyst that contains V_2O_5 and a phase that is known as an hexagonal bronze ($Cs_{0.3}V_{1.7}^{5+}V_{0.3}^{4+}O_5$). The partial reduction of V(V) in catalytically used MVO samples (K, Rb, Cs) was also validated by IR-spectroscopic investigations.

It is supposed that such mixed-valent phases may be formed during catalytic runs from MV_3O_8 phases and that they can influence the catalytic performance. For validation, a pure KV_3O_8 sample was separately synthesised as mentioned and used as catalyst during MTO oxidation to MBA. Interestingly, the catalytic performance of this solid was rather poor, MTO conversion up to 10 mol% and

MBA selectivity below 20 % were obtained. These results lead to the conclusion that the V_2O_5 phase still acts as an active phase whereas the formation of alkali metal-containing bronze-like species effects an increase in basicity of the catalyst surface and a separation of the active vanadyl sites and, hence, leads to the increase in aldehyde selectivity.

CONCLUSION

Alkali metal-containing vanadia catalysts show a graduated catalytic activity in dependence on the nature of the cation. The formation of crystalline K, Rb and Cs-containing mixed valent vanadium oxides beside V_2O_5 lead to a separation of vanadyl active centres that also increases aldehyde selectivity. Li and Na-containing solids do not show the formation of such phases. An additional increase of the basicity of the reaction mixture by admixture of pyridine to the feed results in significantly increased aldehyde selectivity. Furthermore, the catalytic performance of the catalysts is strongly influenced by substituent effects of the reactants.

ACKNOWLEDGEMENT

The authors thank Ms. H. French and Ms. E. Oliev for experimental assistance and the Federal Ministry of Education and Research, Germany for financial support (project No. 03 C 0279 0).

REFERENCES

1. F. Brühne and E. Wright in: *Ullmann's Encyclopedia of Industrial Chemistry*, Vol A3, VCH Weinheim, 1985, p. 463.
2. A.K. Suresh, M.M. Sharma and T. Sridhar, Ind. Eng. Chem. Res., 39 (2000) 3958.
3. T. Tzedakis and A.J. Savall, Ind. Eng. Chem. Res. 31 (1992) 2475.
4. D. Bejan, J. Lozar, G. Falgayrac and A. Savall, Catal. Today 48 (1999) 363.
5. D.A. Bulushev, L. Kiwi-Minsker and A. Renken, Catal. Today, 57 (2000) 231.
6. G. Centi, S. Perathoner and S. Tonini, Catal. Today, 61 (2000) 211.
7. U. Bentrup, A. Brückner, A. Martin and B. Lücke, J. Mol. Catal. A: Chemical, 162 (2000) 383.
8. M. Pontzi, C. Duschatzky and A. Carrascull, E. Ponzi, Appl. Catal. A 169 (1998) 373.
9. A.O. Rocha, Jr., A.L. Chagas, L.S.V.S. Suné, M.F.S. Lopes and J.A.F.R. Pereira, Stud. Surf. Sci. Catal., (1997) 1193.
10. US Patent 4,054,607 (1977), Tanabe Seiyaku Co.
11. H. Seko, Y. Tokuda and M. Matsuoka, Nippon Kagaku Kaishi 1979, 558.
12. B.M. Reddy, M.V. Kumar and K.J. Ratnam, Appl. Catal. A: General, 181 (1999) 77.

13. EP 0 723 949 (1996), Hoechst AG.
14. M. Ueshima, N. Saito and N. Shimizu, Stud. Surf. Sci. Catal. 90 (1994) 59.
15. A. Martin, U. Bentrup, A. Brückner and B. Lücke, Catal. Lett. 59 (1999) 61.
16. A. Martin, U. Bentrup, B. Lücke and A. Brückner, Chem. Commun. 1999, 1169.
17. A. Martin, B. Lücke, H.-J. Niclas and A. Förster, React. Kinet. Catal. Lett. 43 (1991) 583.
18. M.S. Wainwright and N.R. Foster, Catal. Rev.-Sci. Eng. 19 (1979) 211.
19. D.W.B. Westerman, N.R. Foster and M.S. Wainwright, Appl. Catal. 3 (1982) 151.
20. M. Baerns, H. Borchert, R. Kalthoff, P. Käßner, F. Majunke, S. Trautmann and A. Zein, Stud. Surf. Sci. Catal., 72 (1992) 57.
21. P. Concepción, S. Kuba, H. Knözinger, B. Solsona and J.M. López Nieto, Stud. Surf. Sci. Catal., 130 (2000) 767.
22. A.D. Kelmers, J. Inorg. Nucl. Chem., 21 (1961) 45.
23. A. Martin, U. Bentrup, G.-U. Wolf, Appl. Catal. A: General, 227 (2002) 131.
24. N.K. Nag, T. Fransen and P. Mars, J. Catal., 68 (1981) 77.
25. G. Busca, Phys. Chem. Chem. Phys., 1 (1999) 723.
26. A. Martin, U. Wolf, H. Berndt and B. Lücke, Zeolites, 13 (1993) 309.
27. A. Brückner, U. Bentrup, A. Martin, J. Radnik, L. Wilde and G.-U. Wolf, Stud. Surf. Sci. Catal., 130 (2000) 359.

15

Use of Platinum Group Metal Catalysts for the Selective Oxidation of Alcohols to Carbonyls in the Liquid Phase

K. G. Griffin and P. Johnston
Johnson Matthey plc, Royston, Hertfordshire, United Kingdom

S. Bennett and S. Kaliq
Johnson Matthey, Reading, Berkshire, United Kingdom

ABSTRACT

This work illustrates that supported heterogeneous Platinum Group Metal (PGM) catalysts are effective for the selective oxidation of alcohols to aldehydes and ketones using air as the oxidant. These reactions can be performed under mild reaction conditions (<70EC and <5 bar), giving high selectivities to the carbonyl when using organic solvents such as toluene and xylene. In general Pt/C catalysts have the highest activity for aliphatic alcohols, but Ru/C catalysts offer higher selectivities if the alcohol is prone to over oxidation to the carboxylic acid. However, for aromatic alcohols, although Pt/C and Ru/C catalysts are still effective, the best activities and selectivities are obtained with Pd,Pt,Bi/C mixed metal catalysts. In all cases the presence of base in the system is beneficial for both activity and selectivity, which is preferably a property of the catalyst rather than base added to the reaction mixture.

INTRODUCTION

The selective oxidation of hydrocarbons by heterogeneously catalysed routes still presents a major challenge for current technologies within the Fine Chemical Industry (1, 2). The use of stoichiometric inorganic reagents (e.g. potassium dichromate, potassium permanganate etc) although decreasing is still widespread (3, 4, 5), such that catalytic oxidation methods are highly desired by the Fine Chemical Industry (6).

The aim of this work is to demonstrate the effectiveness of heterogeneous PGM catalysts in the selective oxidation of alcohols to carbonyls without over oxidation to carboxylic acid or gaseous products in non-aqueous solvent systems, whereas prior work (7, 8) has concentrated on the use of aqueous solvent with the addition of base to control the selectivity. A range of monometallic and bimetallic PGM catalysts were studied under varying conditions of temperature, pressure and solvent using air as the oxidant.

EXPERIMENTAL

Catalysts

A range of carbon and oxide supported PGM catalysts as shown in Table 1 were used in the initial High Throughput Screening studies.

These catalysts were screened with a range of aliphatic and aromatic alcohols, namely 1, 2 & 3-pentanol, geraniol, benzyl alcohol, 2 & 4-hydroxybenyl alcohol, 2-thiophene methanol and sec-phenethyl alcohol. Selected catalysts from the range were also used in the larger scale autoclave test programme.

Reactors and Analytical Procedures

Catalyst testing was carried out in two different reactor systems. Initial tests were done in High Throughput Screening systems and scale up testing was done in a traditional stirred autoclave.

The High Throughput Screening experiments were performed in a stainless steel multi-well reactor capable of testing up to 40 catalysts at a time, with a maximum working temperature of 100EC and pressure of 6 Bar.

In all cases the experiments were run for 6 hours at 60EC and 3 bar with 0.1g of reactant dissolved in 2 ml of either toluene or acetonitrile (unless otherwise stated) at a reactant:metal molar ratio of 80:1 using air as the oxidant. The reactor was agitated by means of an orbital shaker running at 200rpm.

Scale up testing of selected catalysts was performed in a 250ml stirred autoclave using 5g of reactant dissolved in 120ml of the appropriate solvent at the same reactant:metal molar ratio of 80:1. Again the reactions were performed at 60EC and 3 bar air, employing a stirrer speed of 800 rpm. In this case the oxygen partial pressure in the reactor was maintained at a constant value by topping up the reactor with oxygen rather than air.

Analysis was performed by GC in all cases using mesitylene as an internal standard.

Table 1 Range of catalysts used in the High Throughput Screening studies

No.	Catalyst	pH	No.	Catalyst	pH
1	20% Ag/C	Basic	13	5% Pt/C(128)	Basic
2	5% Pd/C(487)	Basic	14	5% Pt/C(18)	Neutral
3	5% Pd/C(87L)	Neutral	15	5% Pt/C(117)	Acidic
4	10% Pd/C(87L)	Neutral	16	5%Pt/Graphite(289)	Basic
5	5% Pd/C(38H)	Acidic	17	5%Pt/Al$_2$O$_3$(94)	Basic
6	5% Pd/C(331)	Acidic	18	5% Rh/C(592)	Acidic
7	5%Pd/Graphite(450)	Basic	19	5% Rh/C(20A)	Acidic
8	2% Pd/Al$_2$O$_3$	Basic	20	5% Rh/Al$_2$O$_3$(524)	Acidic
9	0.2%Pd,0.1%Au/Al$_2$O$_3$	Basic	21	5% Ru/C(619)	Basic
10	8% Pd,2% Pt/C(464)	Neutral	22	5% Ru/C(97)	Basic
11	3% Pd,2% Bi/SiO$_2$	Neutral	23	5% Ru/Al$_2$O$_3$(698)	Basic
12	4%Pd,1%Pt,3%Bi/C(430)	Neutral			

pH = pH measurement of the aqueous extract of the catalyst suspension.

RESULTS AND DISCUSSION

Initial screening of the full range of PGM catalysts in the selective oxidation of 1, 2 & 3-pentanol to the corresponding carbonyl (Figures 1 and 2) showed that the most active catalysts were based on Platinum, particularly when supported on activated carbon. Ruthenium was a close second in activity, but offers greater selectivity to the aldehyde with 1-pentanol than is the case with Platinum, which tends to give measurable amounts of pentanoic acid. In all three cases good mass balances were achieved (only alcohol substrate, carbonyl and acid products observed) indicating that over oxidation to gaseous products did not occur.

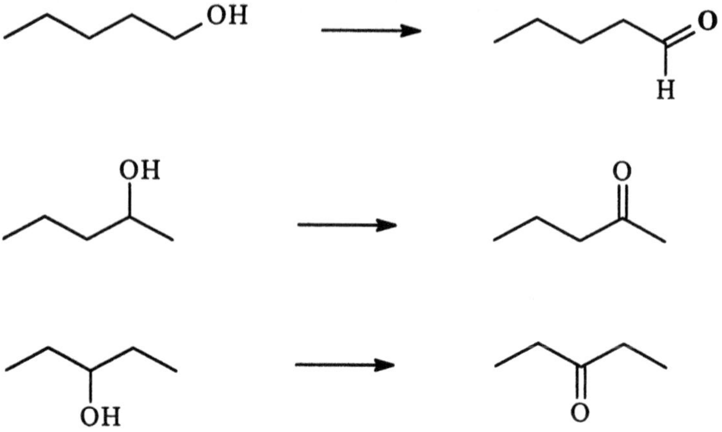

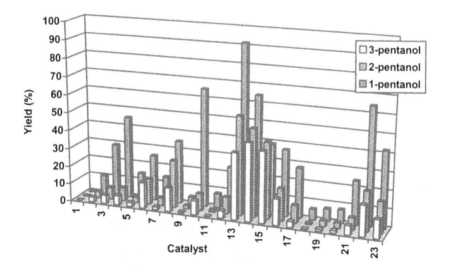

Figure 1 Selective Oxidation of 1, 2 & 3-pentanol to the carbonyl.

Figure 2 Results of catalyst screening in selective oxidation of 1, 2 & 3-pentanol after 6 hours reaction time at 60EC and 3 bar air in toluene solvent with reactant:metal molar ratio of 80:1.

Table 2 Yields obtained in the oxidation of 1, 2, & 3-pentanol after 6 hours reaction time at 60°C and 3 bar air in toluene solvent with reactant:metal molar ratio of 80:1.

No.	Catalyst	pH	Pentanal (acid)	2-pentanone	3-pentanone
13	5%Pt/C	Basic	92 (6)	53	37
14	5%Pt/C	Neutral	65 (6)	49	43
15	5%Pt/C	Acidic	39 (5)	42	40
8	2%Pd/Al$_2$O$_3$	Basic	35 (4)	26	13
10	8%Pd,2%Pt/C	Neutral	65 (9)	10	8
22	5%Ru/C	Neutral	64 (1)	22	17

No other side products i.e. only alcohol substrate, carbonyl and acid detected.

The most active Platinum catalysts are the basic ones (e.g. catalyst 13) supported on activated carbon and the trends on activity are the same for all three of the pentanols as shown in Figure 2 and Table 2. As expected the ease of oxidation for the three alcohols follows the trend 1-pentanol > 2-pentanol > 3-pentanol as shown by the conversion levels in Figure 2 and Table 2.

Results obtained for the other alcohols used in the catalyst screening studies are shown in Table 3, again with yields obtained after a 6 hour reaction time at 60°C and 3 bar air in toluene solvent, except for 2 & 4-hydroxybenzyl alcohol where acetonitrile was employed due to their poor solubility in toluene.

Pt/C catalysts tend to be too active with benzyl alcohol, such that a high level of benzoic acid was observed. This can be overcome by using Ru/C catalysts, which give no over oxidation to benzoic acid. In the case of the 2 & 4-hydroxybenzyl alcohols, again the Pt/C catalysts tend to be too active and give over oxidation to the acid. The most selective catalyst in this instance was 4% Pd, 1% Pt, 3% Bi/C (catalyst 12).

Both Pt/C and Ru/C are effective for the oxidation of sec-phenethyl alcohol to acetophenone with yields > 85% in the 6 hour reaction time.

It is interesting to note that sulfur containing systems, such as 2-thiophene methanol are successfully oxidised to the aldehyde with little evidence of catalyst poisoning by the sulfur (activities similar to non sulfur containing systems). Both Pt/C and Ru/C are active, but Ru/C retains a higher activity on recycle than the Pt/C catalyst (yields of 85/70/45 compared to 65/45/15 for 3 cycles), thus showing better poison resistance than the Pt/C catalyst.

Table 3 Results illustrating catalyst effectiveness with a range of alcohols after 6 hours reaction time at 60°C and 3 bar air in toluene solvent with reactant:metal molar ratio of 80:1.

Alcohol	Product	Catalyst	% Yield Carbonyl (acid)
Geraniol	Citral	13 Pt/C	100 (trace)
Benzyl Alcohol	Benzaldehyde	13 Pt/C	67 (33)
		22 Ru/C	100 (0)
2-hydroxybenzyl alcohol	2-hydroxybenzaldehyde	13 Pt/C	42 (0)
		12 Pd, Pt, Bi/C	98 (1)
4-hydroxybenzyl alcohol	4-hydroxybenzaldehyde	13 Pt/C	30 (0)
		12 Pd, Pt, Bi/C	74 (0.1)
2-thiophene methanol	2-thiophene carboxalde-hyde	13 Pt/C	65 (0)
		22 Ru/C	82 (0)
Sec-phenethyl alcohol	Acetophenone	14 Pt/C	85 (0)
		22 Ru/C	90 (0)

Solvent was acetonitrile for 2 & 4-hydroxybenzyl alcohol due to poor solubility in toluene.

The effect of solvents was studied using geraniol and 2-thiophene methanol as substrate with the 5% Pt/C (catalyst 13) and 5% Ru/C (catalyst 22) catalysts at 60°C and 3 bar air with the standard reaction time of 6 hours and reactant:metal molar ratio of 80:1.

The results in Table 4 show that toluene and xylene gave the highest conversions and selectivities to the aldehyde for both substrates. This appears to be a general trend and solvents, which are immiscible with water tend to give the highest selectivities to the aldehyde with little over oxidation to the carboxylic acid. However, water appears to play a role in the reaction as the best results were obtained when using paste rather than dry powder catalysts, which were also basic in nature.

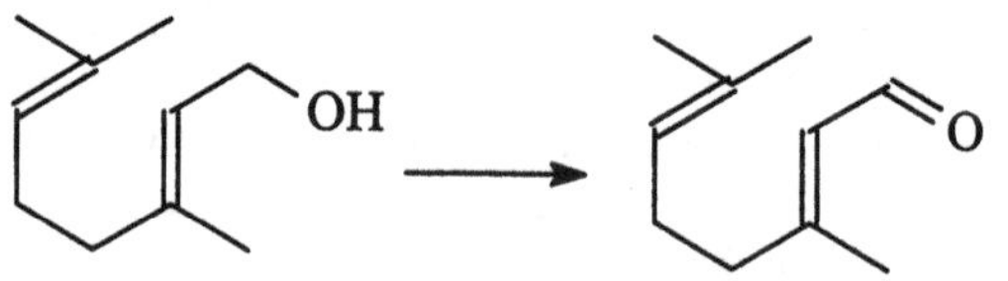

Figure 3 Selective oxidation of geraniol to citral

Table 4 The effect of solvent on aldehyde yield with Pt/C (catalyst 13) after 6 hours reaction time at 60°C and 3 bar air with reactant:metal molar ratio of 80:1.

Solvent	Geraniol	2-thiophene methanol
Toluene	80	55
Xylene	90	45
n-hexane	40	-
Cyclohexane	42	-
DMF	10	5
Acetonitrile	20	10

These studies were performed on the oxidation of geraniol to citral (Figure 3) in toluene solvent using dry powder 5% Pt/C catalysts 12 and 15 (in all previous screening experiments the carbon supported catalysts were used as water wet pastes) with the addition of either water or water containing sodium hydroxide as the base. As usual the reactions were run for 6 hours at 60°C, 3 bar air with a reactant:metal molar ratio of 80:1.

The results in Table 5 show that the addition of water and base together have a marked effect on the yield of citral. For both the neutral and acidic catalysts the yield of citral is increased when base is present in the solvent system.

This indicates that both the water and/or the base play a significant role in the mechanism of the reaction to give either the aldehyde or the acid as the product in the oxidation of primary alcohols.

Table 5 Effect of water and base addition on Citral yield after 6 hours reaction time at 60°C and 3 bar air in toluene solvent with reactnat:metal molar ratio of 80:1.

Reaction mix	5% Pt/C (cat 12)	5% Pt/C (cat 15)
Dry	27	71
Addition of 1% water	28	-
Addition of 5% water	24	73
Addition of 1% water and base	45	-
Addition of 5% water and base	71	83

Table 6 Results of scale up studies on oxidation of geraniol to citral in toluene solvent at 60°C and 3 bar air with reactant:metal molar ratio of 80:1.

Scale (g)	Time (hours)	Citral Yield (%)	Geranic Acid Yield (%)
0.1	6	100	Trace
5	6	41	0.2
5	16	85	1.9
Neat	6	29	0.1

Table 6 shows results obtained when using the 5% Pt/C (catalyst 13) catalyst in the oxidation of geraniol to citral at various scales from the initial HTS studies employing 0.1g of geraniol to autoclave tests employing 5g of geraniol (using toluene or no solvent).

At the larger scale, particularly without solvent, reaction rates are lower, but even when the reaction is run for an extended period to give higher conversions of the geraniol, selectivities remain very good with little over oxidation to geranic acid.

EFFECT OF WATER AND BASE ON REACTION MECHANISM

It is proposed that the alcohol adsorbs onto the metal surface and undergoes dehydrogenation leading to the formation of a metal hydride and liberating the carbonyl. The metal hydride then reacts with oxygen to generate water and regenerate the metal surface to react with further alcohol to continue the catalytic cycle. This is consistent with published mechanistic studies of alcohol oxidation reactions (7, 9, 10). Base promotes this reaction by assisting the abstraction of the hydrogen from the alcohol and it is proposed that the promoting effect of the water is due to it acting as a weak base.

If the product is a ketone there is no further reaction and the ketone is the only product. However, if the product is an aldehyde (i.e. when R_1 is H), this can react further with water in the system to generate an aldehyde hydrate, which can react with oxygen to produce the carboxylic acid.

In aqueous systems the reaction between the aldehyde hydrate and oxygen is fast, so high yields of the carboxylic acid are formed as the equilibrium between the aldehyde and the hydrate is pulled to the hydrate as it is removed from the system by forming the acid with oxygen.

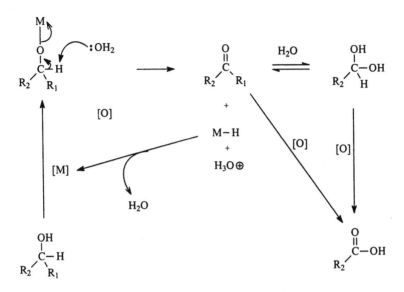

Figure 4 Proposed mechanism for oxidation of alcohols to carbonyls and acids.

In non aqueous systems low yields of acid are observed as the direct oxidation of the aldehyde to the acid is a slow step and only occurs under forcing conditions.

Addition of a strong base such as sodium hydroxide enhances the reaction rate as it abstracts the hydrogen from the alcohol at a greater rate than the weak base water.

CONCLUSIONS

This work demonstrates that supported heterogeneous Platinum Group Metal catalysts are very effective for the selective oxidation of primary alcohols to aldehydes and secondary alcohols to ketones using air as the oxidant.

These reactions can be performed under mild reaction conditions (< 70EC and < 5 bar) using non aqueous solvents such as toluene and xylene.

In general the most active catalysts for aliphatic alcohols are Pt/C, but Ru/C catalysts are also effective and are preferred if the alcohol is prone to over oxidation to the carboxylic acid as Ru generally has a lower activity than Pt.

For aromatic alcohols Pt/C and Ru/C catalysts are also effective, but the highest activities and selectivities are obtained with the mixed metal catalysts containing Bismuth e.g. 4% Pd, 1% Pt, 3% Bi/C.

In all cases studied the presence of base in the system (particularly if incorporated in the catalyst) has a beneficial effect on activity.

These results compare well with those reported by Mallat (7) in his studies on the oxidation of cinnamyl alcohol to cinnamaldehyde using a 5% Pt, 3% Bi/C catalyst in an aqueous solvent. Careful control of the system pH was critical to achieve high selectivities to the aldehyde. In our system using organic solvents rather than the water/base solvent no such control of the pH is necessary and a simpler monometallic catalyst can be used rather than the Bi promoted bimetallic catalyst.

For the oxidation of geraniol to citral the basic 5% Pt/C (catalyst 13) catalyst is very effective giving good conversion rates and very high selectivities to citral. Even if water and base are added to the system the high selectivity is maintained.

REFERENCES

1. R. A. Sheldon, J. Dakka. Catal Today 19: 215, 1994.
2. R. A. Sheldon, I. C. E. Arends, A. Dijksman. Catal Today 57:157, 2000.
3. J. H. Clark. Green Chemistry 1:1, 1999.
4. B. M. Choudry, M. L. Kantam, P. L. Santhi. Catal Today 57:17, 2000.
5. I. L. Finar, Organic Chemistry Vol 1, 6th ed, 1973, pp183-186
6. R. A. Sheldon. Chem Ind 12, 1997.
7. T. Mallat, Z. Bodnar, P. Hug, A. Baiker. J Catal 153:131-145, 1995.
8. K. Deller, Catalysis of Organic Reactions, Marcel Dekker, 1992, pp 261-265.
9. R. DiCosimo, G. M. Whitesides. J Phys Chem 93:768-775, 1989.
10. Y. Schuurman, B. F. M. Kuster, K. van der Wiele, G. B. Martin. Stud Surf Sci Catal 72:43-55, 1992.

16

Destruction of PTA Offgas by Catalytic Oxidation

Baoshu Chen, Jason Carson, Jaime Gibson, Roman Renneke, and Thomas Tacke
Degussa Corporation, Calvert City, Kentucky, U.S.A.

Martin Reisinger, Ralf Hausmann, Andreas Geisselmann, Guido Stochniol, and Peter Panster
Degussa AG, Hanau, Germany

ABSTRACT

The emission streams from purified terephthalic acid (PTA) plants commonly contain carbon monoxide, methyl bromide, and various volatile organic compounds (VOC's). Before the vent gas (offgas) is exhausted to the atmosphere, these contaminants (often regulated) must be destroyed, normally by the catalytic oxidation process. Currently, most commercially available catalysts are used at an inlet temperature higher than 350°C. The improvement of the catalyst activity is desired to increase the catalyst life-time and to reduce the operational cost. Additionally, the catalyst selectivity needs to be improved to minimize or eliminate the formation of polybromobenzenes (PBB's) which can cause plugging or blockage in process lines.

Having inherited the original Halohydrocarbon Destruction Catalyst (HDC) technology from AlliedSignal, Degussa Catalysts has been continuously working on developing a superior catalyst formulation for PTA offgas abatement. Starting with powder catalysts we prepared and tested a variety of supported Pt catalysts. We studied the effect of the following factors on the catalyst performance: different supports (such as ZrO_2 and TiO_2), mixtures of various oxides, and Pt loading.

After we identified a new superior washcoat, a monolithic catalyst (HDC-25) was prepared with this washcoat and was tested for activity and selectivity. For comparison, our existing commercial PTA offgas catalyst (HDC-7) was also tested under the identical condition.

Degussa's new generation catalyst for PTA offgas abatement (HDC-25) has activity higher than HDC-7. The light-off temperature (T50) on HDC-25 for benzene, methyl bromide and methyl acetate is reduced by more than 30°C compared to HDC-7. The complete conversion temperature (T99) for benzene and methyl acetate improved by 70 and 90°C, respectively. HDC-25 also produces less by-products compared to HDC-7.

INTRODUCTION

Catalytic incineration (complete air oxidation) for the purification of gas streams is now quite commonly used in many applications (1-7), being preferred in these over thermal (non-catalytic) incineration and adsorption methods. It can offer advantages over thermal incineration in terms of costs, size, efficiency of destruction, and minimization of thermal NO_x by-product formation. The catalytic incineration systems are now commonly employed in such applications as exhaust emission purification from a variety of industrial processes (including manufacture of organic chemicals and polymers) and air-stripping catalytic processes used to clean contaminated water or soil.

The purification of streams containing halogenated hydrocarbons originally presented a formidable challenge for catalytic incineration as the halogen typically poisoned, deactivated, or otherwise damaged the catalyst (8, 9). However, a catalyst capable of effectively purifying such streams was introduced to the market in the early 1990's by Allied-Signal (10-14); it was termed the halohydrocarbon destruction catalyst or HDC. This was a coated honeycomb monolith (similar to those used in automotive emission control catalysts) offering low pressure drop (and thus low energy loss). Other commercial catalyst suppliers soon followed with their catalysts for halohydrocarbon-containing gas streams. Various open literature studies have appeared examining catalysts for the destruction of halohydrocarbons (1, 3, 15, 16).

The purification of the exhaust gas of PTA plants is one application where halohydrocarbon destruction catalysts have found use at a global scale (12, 13, 17, 18). Typically the untreated exhaust contains a mixture of volatile organic components including methyl bromide, carbon monoxide, hydrocarbons, methyl acetate, and organic acids. The presence of the methyl bromide sets forth the requirement that a catalyst such as the HDC be used. Additionally, the catalyst must be able to effectively destroy all the other organic components (with their widely different intrinsic reactivities toward air oxidation) of the mixture at reasonably low temperatures. Currently most PTA offgas remediation catalysts are used at an inlet temperature higher than 350°C. An improvement of catalyst activity is desired to increase the catalyst life-time and to reduce the operational cost. Additionally, the catalyst selectivity needs to be improved to minimize or eliminate the formation of polybromobenzenes (PBB's), which can cause plugging or blockage in process lines.

Degussa Catalysts has worked on obtaining a more active and durable PTA offgas treatment catalyst since the acquisition from Allied-Signal of the HDC catalyst-related business unit. One earlier improved version of the original HDC catalyst was HDC-7. Here is presented the Degussa HDC-25 catalyst, showing a significant improvement in activity and selectivity characteristics over HDC-7 and other state-of-the-art catalysts.

EXPERIMENTAL

Both powder and monolithic catalysts were evaluated for their activity and selectivity for PTA offgas destruction. For powder catalyst screening, a total of 1 ml of catalyst powder was placed on a quartz frit in a U-shaped, downflow reactor. A reactant mixture, which included carbon monoxide, methyl bromide, benzene, and methyl acetate, was used to represent an industrial PTA offgas. The concentrations of these compounds were 3500 ppm CO, 35 ppm CH_3Br, 9 ppm C_6H_6, and 410 ppm $CH_3CO_2CH_3$, respectively. The gas hourly space velocity (GHSV) was 60,000 h^{-1}, and ambient pressure was used for the test. The gases exiting the catalyst test cell were introduced into a gas chromatograph for analysis. The conversion of CO was measured by an NDIR analyzer (Figure 1).

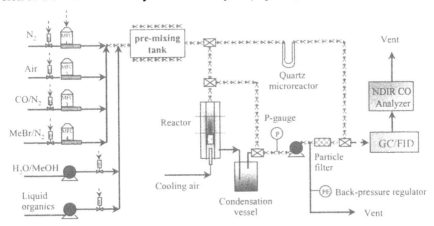

Figure 1 Schematic of catalyst testing apparatus.

For the monolithic catalyst testing the monolith was first cut to a core with a dimension of 1" diameter and 2" in length. The core was placed in a stainless steel reactor.   The specific components were added through mass flow controllers (gases) and micro-HPLC pumps (liquids). The composition, GHSV, and pressure are similar to the typical PTA offgas. The concentrations of CO, CH_3Br, C_6H_6, and $CH_3CO_2CH_3$ were the same as that used in the experiments for powder catalyst testing. Other compounds such as methanol and p-xylene were also present in the inlet gas stream to simulate the typical PTA offgas. The gases exiting the catalyst test cell were introduced into a gas chromatograph for analysis (Figure 1).

RESULTS AND DISCUSSIONS

Effect of Varying Washcoats

Both TiO_2 and ZrO_2 powder (washcoats) were used to support Pt by impregnating the powder with $Pt(NH_3)_4(OH)_2$ solution. The standard incipient wetness method was used, and the Pt loading was 1 wt%. The powder Pt/TiO_2 and Pt/ZrO_2 catalysts were then tested for the activity and selectivity. The conversion of various compounds (CO, benzene, methyl bromide, and methyl acetate) were then plotted versus the catalyst bed temperature, as shown in Figures 2 and 3.

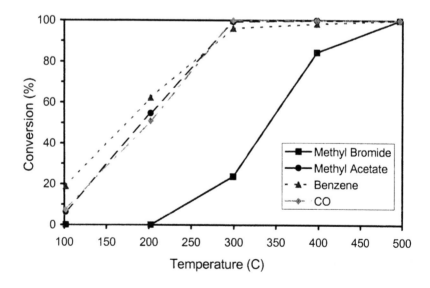

Figure 2 Activity test over 1 wt% Pt/TiO_2.

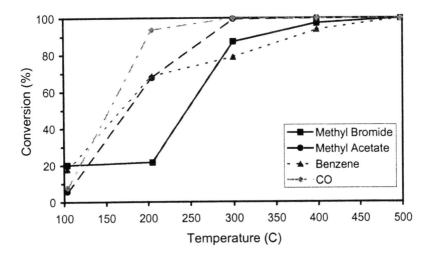

Figure 3 Activity test over 1 wt% Pt/ZrO$_2$.

The TiO$_2$-supported Pt catalyst is more active for the oxidation of CO and non-halogenated VOC's than for methyl bromide. As shown in Figure 2, at 300°C almost complete conversion was observed for CO, methyl acetate, and benzene. However, only about 20% methyl bromide was converted at 300°C.

On the other hand, the ZrO$_2$-supported Pt catalyst has a significantly higher activity for methyl bromide oxidation. Figure 3 shows that almost 90% of methyl bromide was converted at 300°C. Higher conversion was also observed for CO and methyl acetate. However, benzene conversion was lower on Pt/ZrO$_2$ compared to Pt/TiO$_2$. Some by-products such as monobromobenzene were also observed on 1% Pt/ZrO$_2$ at 300°C.

Effect of Precious Metal Loading

Since 1 wt% Pt/ZrO$_2$ had superior performance for PTA offgas oxidation (Figure 3), the Pt loading was increased from 1% to 1.5% using the same preparation method to see if this would further increase its activity. As shown in Figure 4, methyl bromide conversion at low temperature was significantly increased. More than 70% of methyl bromide was converted at 200°C. Conversions for other compounds also increased. At 200°C, 100% of CO, 80% of methyl acetate, and about 60% of benzene were destroyed.

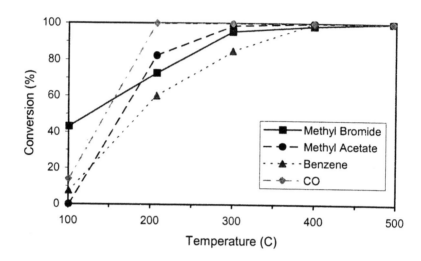

Figure 4 Activity test over 1.5 wt% Pt/ZrO$_2$.

When the Pt loading was decreased from 1% to 0.5%, 0.3%, and 0.1%, activity test results showed that reducing the Pt metal loading on the support decreased the catalytic activity, as expected. However, 0.5% Pt/ZrO$_2$ had an activity only slightly less than that of 1% Pt/ZrO$_2$. As shown in Figure 5, about 75% CO was oxidized at 200°C. More than 60% of methyl bromide and benzene were converted at 300°C.

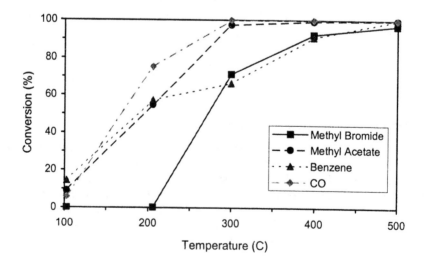

Figure 5 Activity test over 0.5 wt% Pt/ZrO$_2$.

Effect of Physically Mixed Washcoats

When TiO_2 was physically mixed with Al_2O_3 (with a ratio of 4:1) as a support, the resulting catalyst had activities for CO and non-halogenated VOCs much higher than pure TiO_2-supported Pt catalysts. At 300°C both CO and non-halogenated VOC's were almost completely oxidized (Figure 6). However, less than 40% of methyl bromide was converted on this catalyst at 300°C.

By Mixing ZrO_2 and TiO_2 together (1:1) as a support, the catalyst showed intermediate performance between Pt/TiO_2 and Pt/ZrO_2, as expected. At 300°C, more than 90% of benzene was converted, and higher than 60% conversion was observed for methyl bromide (Figure 7).

When Pt was deposited on Degussa's proprietary washcoat, overall superior results were obtained for PTA offgas abatement. More than 80% of all measured compounds were converted at 300°C (Figure 8).

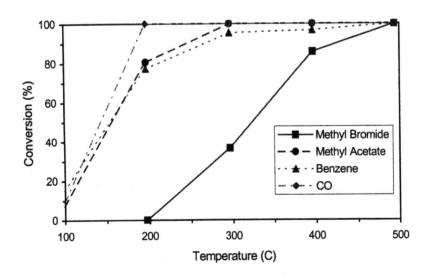

Figure 6 Activity test over 1 wt% $Pt/(80\% \ TiO_2 + 20\% \ Al_2O_3)$.

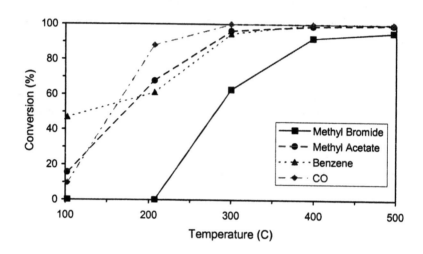

Figure 7 Activity test over 1 wt% Pt/(50% TiO$_2$ + 50% ZrO$_2$).

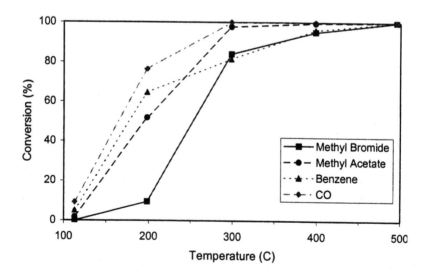

Figure 8 Activity test on 1%Pt / (Degussa's proprietary washcoat).

Monolithic Catalysts

After we identified a new superior washcoat, a monolithic catalyst (HDC-25) was prepared with this washcoat and was tested. The honeycomb monolith with 400 cpsi was used as a support, and powder catalysts were coated on the monolith by Degussa's proprietary washcoating technology. For comparison, our existing commercial PTA offgas catalyst (HDC-7) was also tested under identical conditions. The results are shown in Figures 9 and 10, and Table 1.

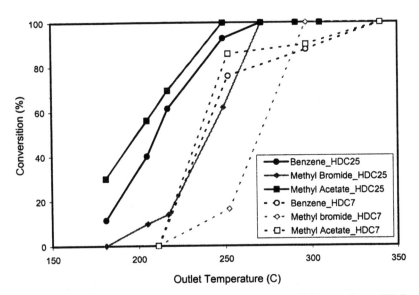

Figure 9 Activity tests over Degussa commercial monolithic catalysts: HDC-7 and HDC-25.

Table 1 Comparison between HDC-7 and HDC-25 monolithic catalysts.

Catalysts	HDC-7		DHC-25	
	T50[a] (°C)	T99[b] (°C)	T50 (°C)	T99 (°C)
Benzene	240	340	210	270
Methyl Bromide	270	300	240	270
Methyl Acetate	235	340	200	250

[a] T50: the light-off temperature or the temperature with 50% conversion.
[b] T99: the temperature with 99% conversion.

Figure 9 compared conversion of benzene, methyl bromide, and methyl acetate for both HDC-25 and HDC-7. Degussa's new generation catalyst for PTA offgas abatement (HDC-25) had activity higher than HDC-7. The light-off temperature (T50) on HDC-25 for benzene, methyl bromide and methyl acetate was reduced more than 30°C compared to HDC-7. The complete conversion temperature (T99) for benzene and methyl acetate was improved by 70 and 90°C, respectively (Table 1).

HDC-25 also produced less by-products compared to HDC-7. As shown in Figure 10, about 0.5 ppm monobromobenzene was detected between 250 and 300°C when HDC-7 was tested. However, no by-products (polybromobenzenes or PBBs) were detected with HDC-25.

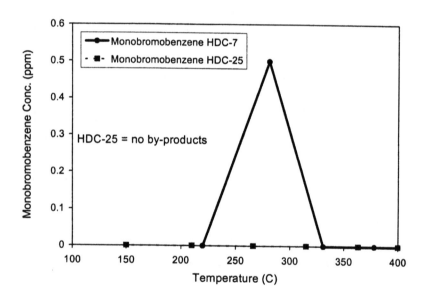

Figure 10 By-product formation on catalysts: HDC-7 and HDC-25.

SUMMARY

The monolithic precious metal catalyst is widely used in purified terephthalic acid plants for their offgas treatment. Currently most commercially available catalysts are used at high inlet temperature, thus causing high operational costs. With continuous R&D at Degussa Catalysts, a new generation of PTA offgas catalyst (HDC-25) has been developed with the significantly improved performance, including lower inlet temperature and less by-product formation.

REFERENCES

1. G. C. Bond & N. Sadeghi. Catalyzed destruction of chlorinated hydrocarbons. J. Appl. Chem. Biotechnol., 25: 241-248, 1975.
2. D. Pope, D. S. Walker & R. L. Moss. Evaluation of platinum-honeycomb catalysts for the destructive oxidation of low concentrations of odorous compounds in air. Atmosph. Env., 12: 1921, 1978.
3. J. J. Spivey. Complete catalytic oxidation of volatile organics. Ind. Eng. Chem. Res., 26: 2165-2180, 1987.
4. B. A. Tichenor & M. A. Palazolo. Destruction of volatile organic compounds via catalytic incinerator. Environmental Progress, 6 (3): 172-176, 1987.
5. W. Chu & H. Windawi. Control VOCs via catalytic oxidation. Chemical Engineering Progress, March: 37-42, 1996.
6. P. Papaefthimiou, T. Ioannides, & X. E. Verykios. Catalytic incineration of volatile organic compounds present in industrial waste streams. Applied Thermal Engineering, 18: 1005-1012, 1998.
7. K. J. Herbert. Catalysts for volatile organic compound control in the 1990's. Presented at the 1990 Incineration Conference, San Diego, California, May 14-18, 1990.
8. M. M. Farris, A. A. Klinghoffer, J. A. Rossin, & D. E. Tevault. Deactivation of a Pt-alumina catalyst during the oxidation of hexafluoropropylene. Catalysis Today, 11: 501-516, 1992.
9. B. Mendyka, A. Musialik-Piotrowska, & K. Syczewska. Effects of chlorine compounds on the deactivation of platinum catalysts. Catalysis Today, 11: 597-6810, 1992.
10. G. R. Lester. Catalytic destruction of hazardous halogenated organic chemicals. Paper No. 89-96A.3. AWMA 82nd Annual Meeting and Exhibition, Anaheim, California, June 29, 1989.
11. G. R. Lester. Catalytic destruction of hazardous halogenated organic chemicals. Catalysis Today, 53(3): 407-418, 1999.
12. G. R. Lester, R. F. Renneke, & K. J. Herbert. Activity, stability, and commercial applications of the Allied-Signal halohydrocarbon destruction catalyst. Paper No. 93/109.05. AWMA 85th Annual Meeting & Exhibition, Kansas City, Missouri, June 21-27, 1992.
13. M. Freidel, A. C. Frost, K. J. Herbert, F. J. Meyer, & J. C. Summers. New catalyst technologies for the destruction of halogenated hydrocarbons and volatile organics. Catalysis Today, 17(1-2): 367-382, 1993.
14. G. R. Lester. Mechanisms of catalytic destruction of halogenated organic compounds. Book of Abstracts, 214th ACS National Meeting, Las Vegas, Nevada, September 7-11, 1997.
15. B. Chen, C. Bai, R. Cook, J. Wright, & C. Wang. Gold/cobalt oxide catalysts for oxidative destruction of dichloromethane. Catalysis Today, 30: 15-20, 1996.
16. S. Chatterjee & H. L. Greene. Oxidative catalysis of chlorinated hydrocarbons by metal-loaded acid catalysts. Journal of Catalysis, 130: 76-85, 1991.
17. T. G. Otchy & K. J. Herbert. First large scale catalytic oxidation system for PTA plant CO and VOC abatement. Paper No. 92/109.02. AWMA 85th Annual Meeting & Exhibition, Kansas City, Missouri, June 21-27, 1992.
18. M. Moreton. Catalytic oxidation for PTA plant emissions control. Int. J. Hydrocarbon Eng., 3(7): 57-59, 1998.

17

2002 Murray Raney Plenary Lecture

Asymmetrically Modified Nickel Catalyst (MNi): A Heterogeneous Catalyst for the Enantio-Differentiating Hydrogenations

Akira Tai
Professor Emeritus, Himeji Institute of Technology, Hyoga, Japan

INTRODUCTION

The asymmetrically modified nickel catalyst (MNi) is the simplest heterogeneous catalyst for the enantio-differentiating hydrogenation of prochiral ketones into the corresponding chiral alcohols. It can easily be prepared by soaking activated nickel black in an aqueous solution of an optically active modifying reagent. The unique functions of MNi are those associated with the cooperation between an active nickel surface for the hydrogenation and the chiral modifying reagent implanted on the surface for the enantio-face-differentiation of the substrate. In this respect, studies aiming to improve the enantio-differentiating ability (e. d. a.) of MNi contribute not only to the development of a significant catalyst for the synthesis of optically active compounds, but also to an understanding of the characteristic phenomena taking place on the surface of the nickel catalysts. In the main part of this lecture, I would like to show the evolution of the MNi catalyst systems from a relatively low e.d.a. in the early days to the excellent e.d.a.'s that have been achieved in recent years. The lecture concludes with an outline of optical enrichment of the hydrogenation products and some applications of the available optically enriched compounds for organic chemistries.

DEVELOPMENT OF A HIGHLY EFFICIENT MRNi CATALYST SYSTEM BASED ON HYPOTHETICAL MODELS

Prolog

Izumi and Akabori first invented MNi in 1963.[1] They modified freshly prepared Raney nickel (W–1 type) with a mono-sodium salt of glutamic acid (Gul-MRNi). This catalyst gave a 15% e.d.a. for the hydrogenation of methyl acetoacetate (MAA) to methyl 3–hydroxybutanoate (MHB). It was sheer luck that MAA was employed as the substrate. Even today MAA is one of the best substrates, and it has long been employed as a standard. Immediately after the discovery by Izumi et al, Klabunovskii and coworkers started studies of this type of catalyst[2], and the groups led by Izumi and Klabunovskii carried out almost all the early research.[3]

Important results of these early studies are summarized as follows: 1) Among almost 100 water soluble optically active compounds examined as a modifying reagent, tartaric acid (TA) was clearly superior as a modifying reagent. Figure 1 shows the stereochemistry and e.d.a. for some representative modifying reagents. 2) As a precursor RNi catalyst for TA-modification, RniH obtained at a high temperature digestion of Raney alloys was better than RNiL obtained at a low digestion temperature. 3) Modification of RNiH with a slightly acidic aqueous 1% solution of TA (pH adjusted to 3.5–5.5 with NaOH) at 100°C

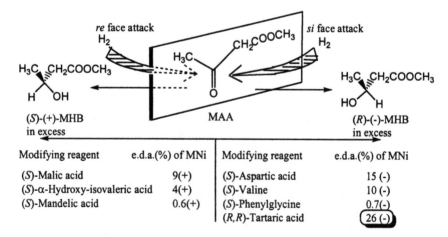

Modifying reagent	e.d.a.(%) of MNi	Modifying reagent	e.d.a.(%) of MNi
(S)-Malic acid	9(+)	(S)-Aspartic acid	15 (-)
(S)-α-Hydroxy-isovaleric acid	4(+)	(S)-Valine	10 (-)
(S)-Mandelic acid	0.6(+)	(S)-Phenylglycine	0.7(-)
		(R,R)-Tartaric acid	26 (-)

Figure 1 Stereochemistry of the enantio-differentiating hydrogenation of MAA over various MRNi catalysts. (From Ref. 4a and selected data from Refs.2 and 4a).

(TA–MRNi) gave the optimum e.d.a. (44-50%). 4) Although the hydrogenation of MAA proceeds well without solvent (neat), the use of methyl propionate (MP) or THF as a solvent is favorable. During the hydrogenation the infiltration of water significantly decreases the e.d.a., while small amounts of acid such as acetic acid raise the e.d.a. Both H_2 pressures (10~100 Kg/cm^2) and reaction temperatures (50–100°C) have little effect on the e.d.a. 5) The enantio–differentiating hydrogenation was possible with a TA-modified metal black catalyst such as Ru, Co, Fe, and Cu. The e.d.a.'s of these modified catalysts were far less than that of TA–MRNi.

MNi catalysts can be prepared from RNi as well as from various types of Ni catalysts such as supported Ni on SiO_2, AlO_3, Kieselguhr, and metal black obtained by the reduction or thermal decomposition of the Ni derivative. In the 70s' and 80s', TA–MNi catalysts prepared from various types of Ni catalysts have been studied in detail by numerous catalyst scientists from Europe, the USA, and Japan, mostly based on inorganic and physicochemical approaches. Their results disclosed many features of TA-MNi and provided important information for our further studies. However, advances in the e.d.a.'s of these catalysts were not significant.

The total process of the enantio–differentiating hydrogenation with MNi is not as sophisticated as shown in Figure 2. However, many variables mutually interact and affect the performance of the system. Thus, an approach to optimize

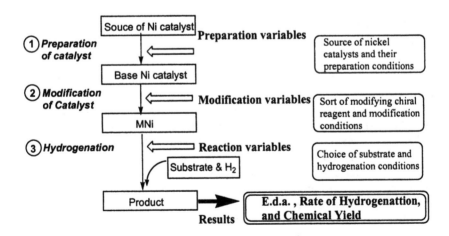

Figure 2 The total procedure for the enantio-differentiating hydrogenation with MNi and the functions of the active site based on various types of indirect information.

each variable one by one is practically impossible. The other approach, a logical design of the catalyst system based on the true reaction mechanism is much more difficult. To gain insight into the mechanism, the states of all the molecules on the Ni surface are simultaneously determined under the reaction conditions. Even today, this is too ambitious a goal. As the third approach to overcome such difficulties, we took an approach similar to the one that had been employed for the studies of enzymes by organic chemists in the early 1950's. Organic chemists made hypothetical models for each enzyme in order to simulate and predict the structure and the functions of the active site based on various types of indirect information such as the enzyme's response to the action of various chemicals. Later, most enzyme models were found to be compatible with the real situation revealed by spectroscopic and X–ray crystallographic studies. Realizing the limitation of conventional approaches, we started to build our own hypothetical models for TA-MNi based on the accumulation of a wide variety of fragmentary information obtained from various research groups and by ourselves. Three models, the "Reaction Process Model", the "Catalyst Region Model", and the "Stereochemical Model" were proposed in order to answer three simple questions: 1) Which elementary reaction step is decisive for the enantio-differentiation? 2) Which region on the TA-MNi catalyst contributes to the enantio-differentiation?, and 3) How does the adsorbed modifying reagent differentiate between the enantio-faces of the substrate? These models not only well simulate the functions of TA-MNi, but also have afforded important clues to improve the e.d.a. of MNi catalyst.

Reaction Process Model (Enantio-differentiation and hydrogenation steps in the reaction path)

The MNi possesses two functions, hydrogenation and enantio-differentiation. In order to simulate the relation of these two functions, we initially studied the rates and e.d.a.'s for the liquid phase hydrogenation of MAA with various MRNi's at atmospheric H_2 pressure.[4] Examples of the Arrhenius plots are shown in Figure 3. For all the MRNi's, parallel lines are obtained with an apparent activation energy (E_A) of 44.2 KJ mol^{-1}, regardless of the e.d.a. value of each MRNi. There is no evidence for a systematic relationship between the e.d.a. and the reaction temperature, e.g., ln([S-product]/[R-product]) is porportional to (1/T). Secondly, the relation between the rate and the MAA concentration was investigated for TA-MRNi and malic acid MRNi (MA-MRNi).[5] From the initial rate (v), the order with respect to the substrate was found to be 0.2-0.3 in both cases. Plots of $1/v$ *versus* 1/[MAA] were linear and overlapped each other. These results indicate that 1) the reaction has a Langmuir-Hinshelwood mechanism, which is generally proposed for the liquid phase hydrogenation of ketones, and 2) the nature of the modifying reagent does not affect the rate determining step. Hence, the e.d.a. must be independently determined at a non-rate determining step such as an adsorption step. Various groups have also carried out kinetic studies of the hydrogenation of MAA with TA-MNi in the liquid phase[6,7] and gas phase.[8,9] All results were essentially compatible with our results.

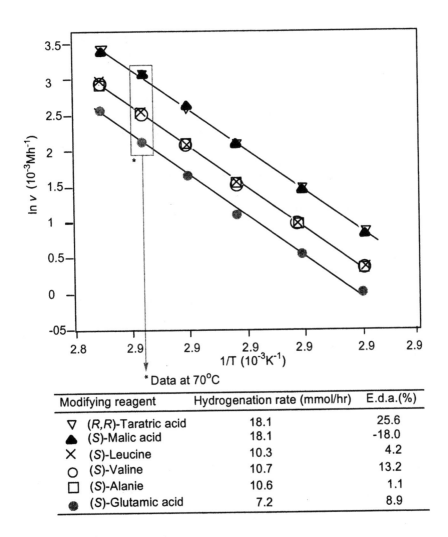

* Data at 70°C

Modifying reagent	Hydrogenation rate (mmol/hr)	E.d.a.(%)
▽ (R,R)-Taratric acid	18.1	25.6
▲ (S)-Malic acid	18.1	-18.0
✕ (S)-Leucine	10.3	4.2
○ (S)-Valine	10.7	13.2
☐ (S)-Alanie	10.6	1.1
⬤ (S)-Glutamic acid	7.2	8.9

Figure 3 Arrhenius plots of the hydrogenation rate of MAA with various MRNi. (From Chem. Soc. Jpn., (ed.) "Kagaku Sousetsu, No 34", Japan Scientific Societies Press, Tokyo, 1981, p.153).

To obtain further information about the e.d.a. determining step, the hydrogenation of methyl 2-methyl-3-oxobutanoate (1) was attempted.[10] The configuration of the C-2 position of 1 is highly susceptible to racemization. However, as soon as 1 is hydrogenated to methyl 2-methyl-3-hydroxybutanoate (2), no further epimerization occurs at the C-2 position. Hence, the stereochemical changes at the C-2 position provide an internal probe for the elucidation of the enantio-differentiation step. The results of the hydrogenation of 1 with (R,R)-TA-MNi are shown in Figure 4. Although racemic 1 is used as a substrate, the (2S)-isomers [(2S,3R)-2 +(2S,3S)-2] are formed in excess compared to the (2R)-isomers [(2R,3R)-2+ (2R,3S)-2]. This indicates that the interaction between 1 and (R,R)-TA leads to an excess adsorbed (2S)-1 on the catalyst surface. Since more (3R)-isomers are found than (3S)-isomers, the carbonyl group of 1 must be hydrogenated by the predominant si face attack of hydrogen in the same manner as in the MAA hydrogenation on (R,R)-TA-MNi. These results indicate that the configuration of the C-2 and C-3 positions of 2 are simultaneously determined, in an adsorption step that precedes the rate-determining step. From the similarity between the structure of MAA and that of 1, it is obvious that 1 as well as MAA and its analogs are hydrogenated through the same enantio-differentiating process.

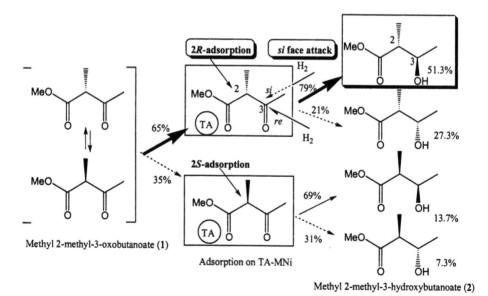

Figure 4 Stereo-differentiating hydrogenation of methyl 2-methyl-3-oxobutanoate. (Selected data from Ref. 10).

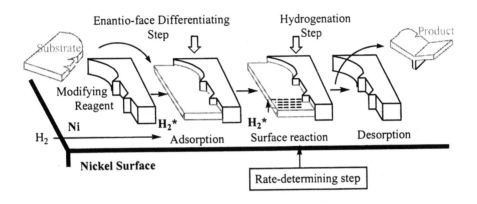

Figure 5 Schematic illustration of "Reaction Process Model" (From Chem. Soc. Jpn., (ed.) "Kagaku Sousetsu, No. 34", Japan Scientific Societies Press, Tokyo, 1981, p.154.

Thus, the enantio-face differentiation of the substrate occurs in an adsorption step, which is followed by a rate determining hydrogenation step. This hypothesis, which is based on stereochemical and kinetic information, is the focus of the "Reaction Process Model" as illustrated in Figure 5. An important implication is that the activated complex of the rate-determining step need not be considered in order to explain or improve the stereo-control of the reaction.

Catalyst Region Model (Enantio-differentiating and non-enantio-differentiating regions on MNi)

State of the Adsorbed TA: In a modifying solution prepared from TA and NaOH, TA can exist as a free acid (TAH_2), monosodium salt (TAHNa), or disodium salt ($TANa_2$) depending on the pH. Since contaminants on the Ni surface might influence the adsorption species of TA, pure Ni black prepared from pure NiO by hydrogenolysis (HNi) was employed to study the effect of the pH of the modifying solution.[11] Figure 6 plots the amount of adsorbed TA *versus* the pH and the e.d.a. *versus* the pH. The plot of the adsorption amount *versus* the pH comprises two levels, a high value for a pH between 2 and 4.5, and a low value for a pH over 6 with a transition at pH 4.5-5.5. The surface concentration is low when both carboxyl groups of TA are neutralized with NaOH ($TANa_2$). At low pH the adsorption species must be TAHNa and TAH_2. If Ni corrosion takes place, (TAH_2) Ni and (TANa)(TAH)Ni may also be present on the surface. While the e.d.a. is extremely low at pH<2.5, it sharply increases to 50% at pH 3.5, and gradually rises to a plateau at pH 5.5. These results indicate that the amount of adsorbed TA does not directly determine the e.d.a. of the catalyst. The results also show that the TA must be present as TANa$_2$, TAHNa when a high e.d.a. is obtained. In the early work, adjustment of the pH with an alkaline hydroxide other than NaOH, such as LiOH, KOH, RbOH, and NH$_4$OH, was found to result in a poor e.d.a.[12] When this fact is taken into account, the presence of Na$^+$ on the catalyst is indispensable. It seems that the cation radius may affect the mode of TA on the Ni.

XPS studies on TA-MNi (modifying pH 5.1) show that the C(1s) binding energy resembles that of TANa$_2$ or TANi rather than TAH$_2$,[13] in agreement with the prevalence of dianionic TA^{-2}. The Ni (2p$_{3/2}$) spectrum shows the characteristic peak of pure Ni metal, with some peaks corresponding to NiO. The former peak is preserved after exposure of the TA-MNi catalyst to air, while the unmodified Ni catalyst surface is completely oxidized upon air contact.[14] This result proves that despite the corrosion of the Ni surface, the adsorption of TA^{-2}/Na$^+$ protects the zero valent Ni metal surface.

The degree of coverage of the Ni surface with TA is an important parameter in order to imagine the TA-MNi surface. From electrochemical measurements, Fish and Ollis determined the amount of adsorbed TA as 3.0 x 10^{-10} mol/cm^2 at pH 5 and 5.3 x 10^{-10} mol/cm^2 at pH 4.[15] Based on a molecular model, the adsorption of a complete mono-layer was estimated to 8.8 x 10^{-10} mol/cm^2. Hence, fractional coverages are 0.34 at pH 5 and 0.6 at pH 4. Based on adsorption isotherms as in Figure 6, and based on BET surface areas, we have estimated the optimum surface coverage of TA-MHNi at 2.7 x 10^{-10}mol/cm^2, or a fractional coverage of 0.3. For TA-MNi/SiO$_2$ (silica supported Ni), Webb and

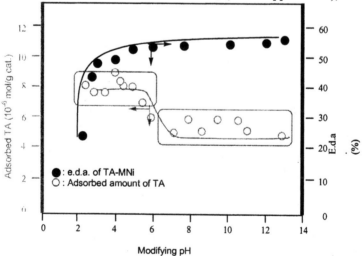

Figure 6 Effect of the modifying pH on the e.d.a. of TA-MHNi and the adsorbed amount of TA on TA-MHNi. (From Ref. 11).

Keane estimated that for various modifying conditions, a fractional surface coverage of 0.20 corresponds to the optimum e.d.a.[16] A sub-mono-layer coverage of metallic Ni surface with TA is expected to be required in order to allow the adsorption of MAA.

Enantio-differentiating and non-enantio-differentiating region on MNi: During our kinetic studies with TA-MRNi and MA-MRNi,[5] it was found that the rate of hydrogenation (v) decreased with an increasing conversion and reached a plateau, while the e.d.a. increased with increasing conversion and reached a plateau. (Figure 7). Since the relationship v is proportional to $[MAA]^{0.2-0.3}$ applies under the reaction conditions, the rate decrease is mainly caused by deactivation of part of the catalyst rather than by the change in the MAA concentration. The coinciding rate plots for both the TA- and MA-MRNi catalysts indicate that this deactivation process does not depend on the type of modifying reagent, even though the e.d.a. values are different for the two catalysts.

These results suggest the presence of an enantio-differentiating region (E-region) and non-enantio-differentiating region (N-region) on the active surface of Ni during hydrogenation, and led to the following hypotheses: 1) The increase in

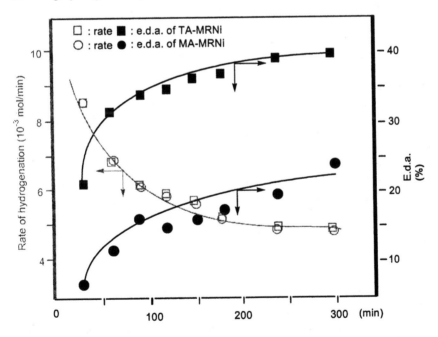

Figure 7 Time dependence of the rate and e.d.a. during the hydrogenation of MAA over TA-MRNi and MA-MRNi. (From Y. Iwasawa (ed.), Tailored Metal Catalysts, D. Reidel, Dordrect, 1986, p.287).

the e.d.a. is related to the disappearance of the N-region which is initially present on the catalyst, 2) The e.d.a. of the E-region is the intrinsic e.d.a. and does not depend on the ratio of the N- and E-regions.   This intrinsic e.d.a. (i) is determined by the nature of the modifying reagent and substrate.   Bringing together these hypotheses, we formulated the "Catalyst Region Model" as follows:

$$\text{e.d.a. (\% e.e.)} = i \times [E] / ([E] + [N]) \qquad (1)$$

where [E] and [N] are the contributions of the E-region and N-region to the hydrogenation, respectively, and i is the intrinsic e.d.a. (% e.e.) for the modifier and substrate.

That the N-region of a freshly prepared MRNi is liable to gradually loose its hydrogenation activity suggests that the N-regions consist of dispersed fine metal particles, with a disordered Ni surface, and with substantial amounts of contaminants such as Al. In contrast, the E -region is probably stable crystalline metal particles with an ordered surface. These ideas on the N- and E-regions are well supported by the studies of Nitta and coworker.[17]   With XPS and XRD techniques, these authors carefully investigated the relationships between the e.d.a., surface contaminants, and the mean crystalline size (Dc) of TA-MNi's prepared from various Ni blacks. The results are that the contaminants on the Ni surface directly lower the e.d.a. to some extent and they also decrease the Dc of the catalysts during the preparation process, which is even more detrimental to the e.d.a. For a Dc > 20 nm, the e.d.a. is always excellent. However, for a Dc smaller than 10 nm, the e.d.a. steeply decreases.

From the characteristics of the E-region, HNi prepared by the reduction of pure NiO with H_2 is expected to be a promising precursor. Indeed, TA-MHNi prepared under suitable preparation and modification conditions gives a good e.d.a. (more than 80% e.e.).[18] However, its hydrogenation activity is inferior to that of TA-MRNi.   As a result, long reaction times are required with low chemical yields (around 50 %) and the formation of oligomeric by-products. Although TA-HNi showed an attainable e.d.a. of MNi, its hydrogenation activity was not practical.   Practical enantio-differentiating hydrogenation catalysts should possess both a high hydrogenation activity and a high e.d.a. Along this line, we decided to develop an improved catalyst using RNi as the precursor, relying on its high activity and allowable size of Dc to achieve a good e.d.a.

Enhancement of the e.d.a. of TA-MRNi based on the "Catalyst Region Model":
The "Catalyst Region Model" states that the removal of the N-region from the RNi catalyst is essential to obtain a high e.d.a. Fortunately, the N-region is expected to be sensitive to chemical or mechanical treatment.

1) Treatment with acids: It is well known that certain components in
Raney alloys, e.g., NiAl or NiAl$_3$, are difficult to dissolve with NaOH. However,
they can be dissolved by aqueous solutions of α-hydroxy acids. Selective
elimination of Al, together with the removal of disordered Ni results in an
improved precursor Ni catalyst, denoted as the acid-treated RNi (RNiA). Thus,
treatment of a conventional RNiH (6% Al) with a solution of glycolic acid (GA)
or TA at pH 3.5 and 100°C yields RNiA(GA) or RNiA(TA) with an Al content
less than 3%. "Soft" TA modification (pH 5; 0°C) does not eliminate Al, but
nevertheless gives MNi with a better e.d.a. than RNiH (Table 1).[19] The optimum
modifying conditions stated in the early studies (pH 3.5-5; 100°C) thus not only
provide the favorable adsorption species (i.e., TAHNa) but also partly remove
the N-regions by corrosion with acid (a one pot preparation of TA-MRNiA).

2) Modification with Na:.   Although the acid treatment was effective, the
maximum e.d.a. of 72% was still insufficient from a practical viewpoint. As an
alternative approach, we tried to eliminate the activity of the N-region by partial
poisoning. Using sulfur and nitrogen compounds resulted in failure. Although
Keane has recently reported that thiophene is effective,[20] this compound is too
strong a catalyst poison to be utilized as a selective poisoning reagent. While
studying partial poisoning, we notice an appreciable increment of e.d.a. when
water contaminated with NaCl and Na$_2$SO$_4$ was accidentally used in preparing
the modifying solution. Subsequent screening of various sodium salts revealed
that NaBr most enhanced the e.d.a. of TA-MRNi (Table 2).[11, 21] This increase
was observed only when the TA and NaBr modification were carried out
simultaneously with one solution (TA-NaBr-MRNi).

Table 1 TA-MNi prepared from various Ni black[a)]. *Source*: Selected data from
Refs. 11 and 19.

No	Base Catalyst	Al Content (%)	Modifying Condition		E.d.a. of TA-MNi
			pH	Temp(°C)	
1	RNiL	6	5.0	0	35
2	RNiH	5	5.0	0	40
3	RNiH	5	4.1	100	44
4	RNiA(GA)	3	5.0	0	72
5	RNiA(TA)	2	5.0	0	62
6	HNi	-	7.3	0	71
7	HNi	-	3.2	100	75
8	HNiA(TA)	-	7.3	0	82

a)Reaction conditions: MAA, 11.5 ml; MNi, 0.9 g; Methyl propionate, 23 ml; AcOH, 0.2 ml;
initial H$_2$ pressure: 90 Kg/cm^2, 100 °C.

Table 2 Enantio-differentiating hydrogenation of MAA over TA-inorganic
salt-MRNi[a]. *Source*: Selected data from Refs. 11 and 19.

Inorganic salt	Amount of salt in modifying solution (100 ml containing 1 g TA)	E.d.a. (%)
None	-	39
NaBr	10	83
NaF	3	61
NaCl	10	72
NaI	5×10^{-4}	51
Na_2SO_4	10	56
LiBr	3	62

a) Reaction conditions: MAA, 11.5 ml; MNi, 0.9 g; Methyl propionate, 23 ml;
AcOH, 0.2 ml; initial H_2 pressure: 90 Kg/cm^2, 100°C

On the TA-NaBr-MNi, the amount of adsorbed NaBr is found to directly affect the e.d.a. as shown in Figure 8. The amount of adsorbed NaBr increases with the increasing concentration of NaBr in the modifying solution up to a saturation value $[NaBr]_{sat}$. Similarly, the e.d.a. increases until it reaches maximum at $[NaBr]_{sat}$. A comparison of the e.d.a. of MRNiH and MRNiA indicated that NaBr modification enhances the e.d.a. to the same extent, but it does not change the difference in e.d.a. between these two catalysts. This result suggests that the functions of the acid treatment and the NaBr modification may be different.

Although NaBr modification has been discovered during the study of partial poisoning, the effect of NaBr is now explained either in terms of enhancement of the *i* factor in the E-region by interaction with adsorbed TA or in terms of partial poisoning of the N-region as originally expected. When the fact that TA-MRNiH and TA-NaBr-MRNiH give almost the same hydrogenation activities, the former idea is more plausible. However, the real mode of action is still open to discussion.

A typical one-pot preparation method of TA-NaBr-MRNi follows: RNi prepared from 2 g Ni-Al alloys (Ni/Al = 42/58) is modified with a 100 ml solution of TA (1 g) and NaBr (10 g), adjusted to pH 3.2 with 1 mol NaOH solution. By the 90's, TA-NaBr-MRNi was the best MNi catalyst. The TA-NaBr-modification is also effective for HNi[11] and various supported Ni catalysts.[22, 23, 24]

3) Mechanical treatments: As an additional approach, mechanical elimination of the N-region from RNi is promising, since the N-region is

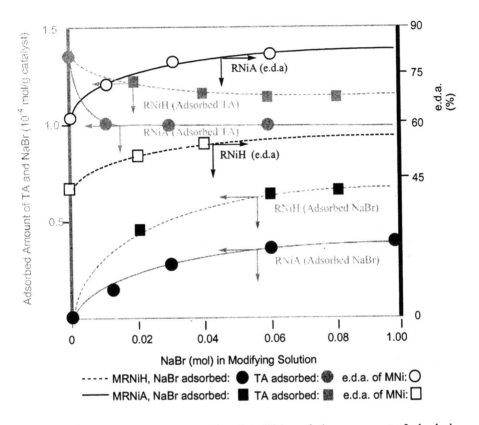

Figure 8 Effect of NaBr concentration of modifying solution on amount of adsorbed NaBr, adsorbed TA, and e.d.a. of resulting MRNi. (From DE De Vos, IFJ Vankelecom, PA Jacobs (ed.), Chiral Catalyst Immobilization and Recycling, Wiley-VCH, Weinheim, 2000, p.187.

expected to contain fragile, disordered Ni domains with substantial amounts of residual Al.   Ultrasonic irradiation was successfully applied to remove unfavorable parts from the RNi.[25]   A freshly prepared RNiH was suspended in water, and was subjected to ultrasonic irradiation (48 KHz, 600W, 5 min) after which the turbid supernatant was removed by decantation from paramagnetic nickel powders fixed with a magnet. The same treatment was repeated until the supernatant becomes transparent. EPMA (SEM-EDX) observations also proved the effectiveness of ultrasonic irradiation for removing disordered nickel domain from RNi.[26]   In the original RNiH, small particles (less than 1 μm) and large particles (10-20 μm) form an aggregate-like cluster.   Ultrasonic treatment crushes the cluster and reduces the amount of small particles.   Thus, the resulting ultrasound irradiated RNi (RNiU) mostly consists of medium size particles (5-10 μm) with a smooth, pure Ni surface. The solid in the supernatant is a mixture of small particles (less than 1 μm) and rather large particles (5-10 μm) of flat plates with rough surfaces. The latter is expected to fall in flake from the cluster in RNi. Both particles contain a large amount of Al. The similar X-ray diffraction patterns of RNiH and RNiU indicate that the ultrasonic irradiation does not change the Dc but merely homogenizes the particle size distribution, with removal of the Al-enriched particles.

Under optimum modifying conditions, TA-NaBr-MRNiU gives 86% e.d.a. in the hydrogenation of MAA, with a hydrogenation activity 8 times higher than that of conventional TA-NaBr-MRNiH. The progress of the e.d.a., as guided by the "Catalyst Region Model" is summarized in Table 3.

Table 3 Improvement of MNi monitored by the enetio-differentiating hydrogenation of MAA. *Source*: Selected data from Refs. 11, 19, 26.

Catalyst	E.d.a. (%)	Hydrogenation activity	Age of improvement
TA-NRNi	52	high	
TA-MRNiA(GA)	72	high	1977
TA-MHNi	80	fair	1978
TA-NaBr-MRNi	83	high	1978
TA-NaBr-MRNiU	86	very high	1991

Stereochemical Model (Interaction Between Substrate and TA on Mni)

As mentioned previously in Catalyst Region Model, the adsorbed TA is present on the Ni surface as TA^{-2} with less than a monolayer coverage. Early studies on the hydrogenation of MAA with TA-MNi indicated that the addition of extra TA to the reaction decreases the e.d.a. and that there is a linear relationship between the optical purity of the TA employed for modification and the e.d.a. of the resulting MNi.[27] Yasumori studied the thermal desorption of MAA from TA-MNi and unmodified Ni, and proposed the presence of an association between MAA and TA through hydrogen bonds.[14] Sachtler and coworker,[8] and Keaned[20] found in the hydrogenation of MAA in gas and liquid phase, respectively, that the E_A values for TA-MNi is less than those for unmodified Ni. The association between MAA and TA on the catalyst surface also explained the decrease in E_A after TA-modification. Summarizing these facts, an enantio-face-differentiation of MAA takes place with a one-to-one interaction between the adsorbed MAA and TA^{-2} on the Ni surface through hydrogen bonds.

"Stereochemical Model" based on interaction through two hydrogen bonds (2P Model): This model was first proposed to explain the distribution of diastereomers in the hydrogenation of methyl 2-methyl-3-oxobutanoate over TA-MNi (Reaction Process Model, Figure 4).[10] Later, it was generalized to understand the stereochemistry of the hydrogenation of MAA and its analogs 'based on the information listed in Table 4 and Table 5.[28] In Table 4, the e.d.a.'s are listed for the hydrogenation of MAA with the MRNi catalyst prepared from

Table 4 The e.d.a.'s of TA (and analogs)-modified RNi's determined by the hydrogenation of MAA[a)]. *Source*: Selected data from Ref. 28.

Modifying reagent	$\begin{array}{c} H \quad X_2 \\ \mid \quad \mid \\ Y-C-C-COOH \\ \mid \quad \mid \\ X_1 \quad H \end{array}$			E.d.a. of MNi (%)
No.	X_1	X_2	Y	
1	-OH	-OH	-COOH	83
2	-OH	-H	-COOH	61
3	-OH	-OCOPh	-COOH	65
4	-OH	-OMe	-COOH-	63
5	-OH	-OH	-CH₃	1.2
6	-OH	-OH	-H	0
7	-OCOPh	-OCOPh	-COOH	8
8	-OCH₃	-OCH₃	-COOH	0.2

a) Reaction conditions; 11.5 ml MAA, 0.9 g MNi, 23ml MP, 0.2 ml AcOH, initial H_2 pressure; 90 Kg/cm^2, 100°C.

Table 5 Enantio-differentiating hydrogenation of prochiarl ketones with TA-NaBr-MNi[a)]. *Source*: Selected data from Ref. 28.

Substrate	E.d.a. (%) of hydrerogenation and configuration of product
1 $H_3C-C-C-OCH_3$ (O O)	2 (R)
2 $H_3C-C-CH_2-C-OCH_3$ (O O)	83 (R)
3 $H_3C-C-CH_2CH_2-C-OCH_3$ (O O)	38 (R)
4 $H_3C-C-CH_3CH_2CH_2-C-OCH_3$ (O O)	0 -
5 $H_3C-C-CH_2CH_2CH_2-CH_3$ (O)	8 (S)

a) Reaction conditions; 0.09-0.10 mol substrate, 0.9 g MNi, 23ml MP, 0.2 ml AcOH,

initial H_2 pressure; 90 Kg/cm^2, 100°C.

TA and its analogs and NaBr. Of the examined modifiers, only TA had a higher than 80% e.d.a. When one of the hydroxyl groups of TA is replaced by hydrogen, a benzoyloxy or methoxy group, an appreciable decrease in the e.d.a. is observed (Table 4, entries 2,3,4). When one of the carboxyl groups of TA is replaced by H or by a methyl group, the e.d.a. is completely lost (Table 4, entries 5, 6). Apparently, the interactions of two carboxyl groups with the Ni surface are essential to immobilize the modifying reagent on the surface and to create a uniform chiral environment on the catalyst. A drastic decrease in e.d.a. is also observed when both hydroxyl groups of TA were acylated or alkylated (Table 4, entries 7, 8). The results clearly show that the two hydroxyl groups and two carboxyl groups in TA are vital for the effective enantio-face differentiation of MAA.

In the hydrogenation of the methyl esters of the α-, β-, γ- and δ-keto acids and 2-hexanone with (R,R)-TA-MNi, an excellent e.d.a. is obtained only for the β-ketoester (MAA) (Table 5). Thus, the distance between the two carbonyl groups in the substrate is critical. When the distance of two hydroxyl groups in TA is considered, it is clear that only MAA can be bound with its two carbonyl groups *via* two hydrogen bonds to the hydroxyl groups of the adsorbed TA^{-2}. Such a precise fit is not possible with the other substrates. A pictorial model of the interaction between MAA and (R,R)-TA *via* two pairs of hydrogen bonds is shown in Figure 9 (2P-model). Construction of a space-filling CKP model reveals that one of the hydroxyl groups of TA comes close to the catalyst surface (site 1), whereas the second hydroxyl group is somewhat remote from the surface (site 2). In Figure 9a, the carbonyl group of MAA to be hydrogenated is fixed on site 1 and comes within 0.1 nm from the surface with its *si*-face toward the catalyst. When MAA is adsorbed in this fashion, it is ready to be hydrogenated by the supply of an activated hydrogen from the Ni surface, leading to (R)-MHB (hydrogen attack from *si* face; Figure 9b).

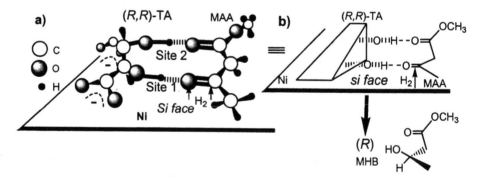

Figure 9 Schematic illustration of the interaction between MAA and TA adsorbed on the Ni surface through two hydrogen bonds, "2P model". (From Ref. 28).

The 2P stereochemical model rationally explains the stereochemistry of the hydrogenation of MAA over TA-MNi. It also predicts that the TA-MNi system should be effective for the enantio-differentiating hydrogenation of the β-functionalized prochiral ketones as listed in Tables 6 and 7. As shown in Table6, the hydrogenation of a series of β-ketoesters and ketones with an electronic structure similar to that of MAA gives an excellent e.d.a. value.[29,30,31]

Table 6 Enantio-differentiating hydrogenation of prochiral ketons suitable for 2P model[a]. *Source*: Selected data from Refs. 25, 28–30.

No.	Substrate R=	Catalyst	Solvent /Additve	Product E.e (%)	Configuration
1	CH₃-	(R,R)-TA-MRNi	neat	33	R
2		(R,R)-TA-NaBr-MRNi	MP/AcOH	83	R
3		(R,R)-TA-NaBr-MRNiU	MP/AcOH	86	R
4	CH₃CH₂-	(R,R)-TA-MRNi	neat	66	R
5		(R,R)-TA-NaBr-MRNi	MP/AcOH	87	R
6		(R,R)-TA-NaBr-MRNiU	MP/AcOH	92	R
7	CH₃(CH₂)₆-	(R,R)-TA-NaBr-MRNi	MP/AcOH	83	R
8		(R,R)-TA-NaBr-MRNiU	MP/AcOH	89	R
9	CH₃(CH₂)₇-	(R,R)-TA-NaBr-MRNi	MP/AcOH	86	R
10		(R,R)-TA-NaBr-MRNiU	MP/AcOH	94	R
11	AcO-(CH₂)₇-	(R,R)-TA-NaBr-MRNiU	MP/AcOH	85	R

12	H-	(R,R)-TA-MRNi	THF	27	R
13		(R,R)-TA-NaBr-MRNi	THF	70	R
14	CH₃-	(R,R)-TA-NaBr-MRNi	THF	68	-

15	CH₃CH₂-	(R,R)-TA-NaBr-MRNi	THF	71	-
16	CH₃(CH₂)₄-	(R,R)-TA-NaBr-MRNi	THF	68	-
17	CH₃(CH₂)₇-	(R,R)-TA-NaBr-MRNi	THF	67	R

a) Reaction conditions; 0.09-0.10 mol substrate, 0.9 g MNi, 23ml solvent, 0.2 ml AcOH, initial H₂ pressure; 90 Kg/cm², 100°C

Acetylacetone (AA) and its analogs are also suitable substrates. The double hydrogenation of AA to 2,4-pentanediol (PD) proceeds stepwise. In a first and fast step, the enantio-face-differentiating reaction over (R,R)-TA-MNi gives (R)-4-hydroxyl-2-pentanone (HP) in large excess. In the second step, the enantio-face differentiation controlled by (R,R)-TA-MNi and the diastereo-face differentiation induced by the chiral center of HP simultaneously participate. Fortunately, both mechanisms work in the same direction and contribute to produce (R,R)-PD in large excess. In this step, the third stereo-differentiating mechanism, the enantiomer-differentiation (kinetic resolution) of the two HP enantiomers is also important, particularly if the reaction is stopped rather early. Indeed, the hydrogenation of racemic HP indicated that the consumption of (R)-HP was much faster than (S)-HP. As a result, the formation of (R,R)-PD is even more favored. In the overall reaction, the productivity is mostly governed by the e.d.a. of the first step, while the e.e. of (R,R)-PD can be further enriched by kinetic resolution if the reaction discontinues an appropriate conversion . The major product of the reaction, (R,R)-PD, is readily separated from the meso-isomer, $(R*, S*)$-PD by recrystallization from ether at -20°C. As the isolated crystals contain (R,R)-PD of 100% e.e., the isolated (R,R)-PD yield can be used as a measure for the stereo-differentiating ability of the catalyst. Results of the

Table 7 Stereo-differentiating hydrogenation of 1,3-diketones (1) over MNi catalyst[a]. *Source*: Selected data from Refs. 26, 34.

No.	Substrate(1) R=	Catalyst	Products composition[b]				E.e. of 4 (%)	Yield (%) of optically pure 4 based on 1[c]
			1	2	3	4		
1	-CH₃	(R,R)-TA-NaBr-MRNi	0	0	13	87	86	41(R,R)
2		(R,R)-TA-NaBr-MRNi	0	20	10	70	90	21(R,R)
3		(R,R)-TA-NaBr-MRNiU	0	7	7	86	91	60(R,R)
4	-CH₂CH₃	(R,R)-TA-NaBr-MRNi	0	0	20	80	-	25(R,R)
5	–CH⟨CH₃ CH₃	(R,R)-TA-NaBr-MRNi	1	17	16	66	85	32(S,S)
6		(R,R)-TA-NaBr-MRNiU	0	6	22	72	90	59(S,S)
7	-(CH₂)₂CH₃	(R,R)-TA-NaBr-MRNi	0	0	15	85	-	13(R,R)
8	-(CH₂)₅CH₃	(R,R)-TA-NaBr-MRNi	0	0	20	80	-	20$R,R)$
9	-Ph	(R,R)-TA-NaBr-MRNi	0	0	23	77	-	20(S,S)

a) Reaction conditions; 100 g substrate, 19 g MNi, 100 ml THF, initial H₂ pressure; 90 Kg/cm², 100°C. b) Determined by gravimetry after separation of each compound by MPLC. c) Two successive recrystallization of the hydrogenation products from diethyl ether gave optically pure 4.

hydrogenation of AA and its homologues are summarized in Table 7.[32,33,34] Particularly TA-NaBr-MRNiU is an excellent catalyst to obtain optically pure 1,3-diols with high chemical yield.[25,26]

"Stereochemical Model" based on interaction through one hydrogen bond (1P Model): Hydrogenation of a series of prochiral ketones reveals that (*R,R*)-TA-MNi can differentiate to some extent between the enantio-faces of simple ketones (Table 5, entry 5). In the case of 2-hexanone, the enantio-face differentiation can only be governed by the relative size of the methyl and butyl groups connected to the carbonyl group. The carbonyl group may adsorb on the catalyst surface and interact with one of the hydroxyl groups of TA (site 1) in the same manner as MAA. Meanwhile, the other hydroxyl group (site 2) will repel the large hydrophobic butyl group to the other side. Consequently, the small methyl group is accommodated in the inside of the cavity. This interaction model comprising a pair of hydrogen bonds and a steric repulsion (1P model) is depicted in Figure 10. The model is consistent with the stereochemistry of the reaction, since (*R,R*)-TA-MNi gives an excess of the (*S*) isomer *via re*-face hydrogen attack.[28]

Initially, 2-alkanones were hydrogenated under the optimum conditions for MAA hydrogenation. The reaction mixture also contains a small amount of acetic acid as an additive, which later turned out to be essential for the enantio-differentiating hydrogenation of 2-alkanones. A survey of the carboxylic acid additives indicates that carboxylic acid with branching at the α-position were very effective in increasing the e.d.a. of the system. The best results are obtained with pivalic acid (PA) in a concentration more than twice that of the substrate.[35,36] The effect for PA can also well explained by the 1 P model, taking into account an association between PA and TA (Fig 10b). PA associates with the hydroxyl group of TA at site 2, forming a bulky, polar fence towering over the

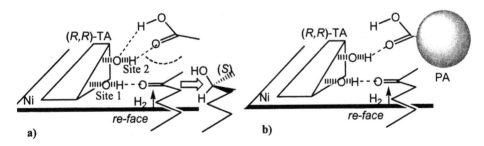

Figure 10 Schematic illustration of the interaction between 2-alkanone and TA adsorbed on the Ni surface through a hydrogen bond and steric repulsion, "1P model". (From Ref. (28) and A. Tai, Shokubai (Catalyst & Catalysis), 37:527, 1995).

long-chained alkyl group of the adsorbed substrate. As a result, the discrimination between methyl and alkyl group becomes far more pronounced than the presence of acetic acid.[37]

Results from a series of 2-alkanone hydrogenations are listed in Table 8.[36] In all cases, the e.d.a.'s significantly increase by decreasing the reaction temperature from 100° to 60°C. Almost the same e.d.a.'s are obtained for straight chain 2-alkanones as for 2-hexanone. As expected from the 1P model, branching in the α-position of the carbonyl group increased the e.d.a. up to 85 % (Table 8, entry 5). For 2-butanone, the e.d.a. is 60 % under these conditions (Table 8, entry 1), and it can be raised to 72% by further optimizing the reaction conditions.[38] This is the first example of an effective differentiation between methyl and ethyl groups, with the exception of enzymatic reactions.

The e.d.a's for the 2-octanone hydrogenation with various MNi's are listed in Table 8. The best catalyst for the MAA hydrogenation always brings about the best e.d.a. in the alkanone reaction. This proved that the MAA and 2-alkanone hydrogenation occur in the same E-region of the catalyst.

Table 8 Enantio-differentiating hydrogenation of 2-alkanone over (R,R)-TA-MNi[a]. *Source*: Selected data from Ref. 36.

	Substrate RCOCH₃	E.d.a. (%) Hydrogenation Temperature	
	R=	100°C	60°C
1	CH_3CH_2-	49	63
2	$CH_3(CH_2)_3-$	66	80
3	$CH_3(CH_2)_5-$	66	80
4	$CH_3(CH_2)_{10}-$	65	75
5	CH_3CH- CH_3	63	85
6	CH_3CHCH_2- CH_3	-	75
7	$CH_3CHCH_2CH_2-$ CH_3	-	77
8	$CH_3CHCH_2CH_2CH_2-$ CH_3	-	66

a) Reaction conditions: substrate, 10g; MNi, 1.6 g; PA, 20 ml; THF, 20 ml; initial H_2 pressure: 90 Kg/cm²

"Extended Stereochemical Model" (Merging the 2P and 1P models): Study of
the 2P and 1P models teaches that both enantio-differentiating reactions take
place in the same region on the catalyst. Their relative contribution depends on
the type of substrate. In order to prove this idea, hydrogenation of a series
ketoesters and 2-octanone was reexamined with TA-NaBr-MRNiU in the
absence and presence of PA. These results are shown in Figure 11.[39] In the
absence of PA, the reaction of the β-ketoester leads to the highest (R)-excess.
The e.d.a. gradually decreases in going from the γ-ketoester to the δ-ketoester.
Eventually, with the ε-ketoester, a slight excess of (S)-configuration is obtained.
The addition of a large amount of PA brings a characteristic change in the e.d.a.
for all substrates. Lower (R)-excesses are obtained for the β- and γ-ketoesters,
while appreciably high (S)-excesses are observed for the δ-ketoester and ε-
ketoester as well as for 2-octanone. The e.d.a. for the δ-ketoester and ε-ketoester
has the same temperature dependence as that of 2-octanone. From these
observations, it can be deduced that the relative contributions of 2P and 1P in the
absence of PA are 2P>>1P for the β-ketoester, 2P>1P for the γ-ketoester, 2P=1P
for the δ-ketoester, 2P<1P for the ε-ketoester. The addition of PA evidently
decreases the contribution of the 2P mode in favor of the 1P mode resulting in
the relative contributions; 2P>1P for β-, 2P=1P for ?γ-, 2P<<1P for δ- and ε-
ketoester.

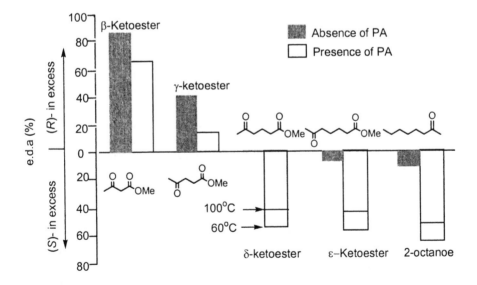

Figure 11 Enantio-face-differentiating hydrogenation of various prochiarl ketones over
TA-NaBr-MRNiU in the absence and presence of PA in the system. (From A. Tai,
Yuukigousei Kyoukaisi (J. Synthetic Organic Chemistry Japan), 58: 573, 2000).

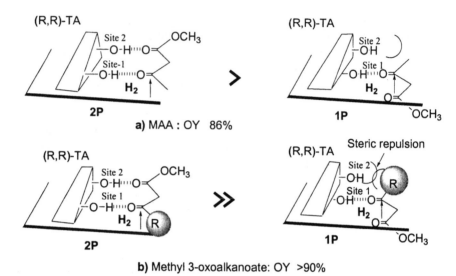

Figure 12 Schematic illustration of the relative contribution of 2P of 1P in the hydrogenation of MAA and its homologous. (From A. Tai, Syokubai (Catalyst & Catalysis), 41:270, 1999).

The results mentioned above brought about a modified hypothetical model called the "Extended Stereochemical Model". In view of this model, the intrinsic e.d.a. (i), as proposed in equation (1), may be replaced by the intrinsic e.d.a. resulting from the 2P and 1P contributions. The intrinsic e.d.a. is eventually determined by the interaction between the modifying reagent and substrate.

In early studies, it was observed that the hydrogenation of the higher homologues of MAA such as methyl 3-oxodecanoate afforded better e.e. than of MAA (Table 6). The increase in the e.d.a. is not easily explained based on the 2P model alone, but became clear with the "Extended Stereochemical Model". Figure 12 shows a sketch of the 2P and 1P models for MAA (a) and its long chain homologue (b). For MAA, a good 2P fit is expected, but the 1P mode of adsorption is also possible because of the small size of the methyl group on the acyl side. For long chain homologues, the contact between site 2 and the long alkyl groups on the acyl side impedes the 1P adsorption mode. This increases the contribution of the 2P mode and eventually gives a higher (R)-excess than for MAA.

The "Extended Stereochemical Model" also suggests that MAA, which has long been considered to be the standard β-ketoester, may not be the best substrate for TA-MNi. For the MAA/TA-MNi systems, the *i* factor of 100% is hard to reach because of a small 1P contribution. Based on the extended stereochemical model, the β-ketoester carrying a relatively large alkyl group on the acyl side should be a more favored substrate due to the steric repulsion. If steric repulsion plays some role, its efficiency is expected to be temperature-dependent. Accordingly, a series of β-ketoesters was hydrogenated over TA-NaBr-MRNiU at 60° and 100°C (Table 9).[39] TA-NaBr-MRNiU is an excellent catalyst for low temperature hydrogenation because its high activity.

Table 9   Enantio-differentiating hydrogenation of various -ketoesters with (R,R)-TA-NaBr-MRNiU[a]. *Source*: Selected data from Refs. 39, 40.

No.	Substrate	Reaction Temp. 100°C E.d.a.	60°C	No.	Substrate	Reaction Temp. 100°C E.d.a.	60°C
1		86	86	9		88	96
2		85	87	10		88	96
3		84	88	11		90	95
4		91	94	12		84	96
5		90	93				
6		87	90	13		80	Slow rate
7		80	88				
8		95	**98**	14		No reaction	

a) Reaction conditions: substrate, 4-10g; MNi, 0.9g; THF, 20 ml; AcOH, 0.2 ml; initial H2 pressure: 100 Kg/cm2

Except for acetoacetates, all substrates gave a systematically higher e.d.a. at 60° than at 100°C (Table 9, entries 4-9, 12). For acetoacetates, neither the type of alkoxy group nor temperature has much effect on the e.d.a. In going from a methyl group on the acyl side (MAA) to an ethyl group (methyl 3-oxopentanoate), the e.d.a. increases (Table 9, entries 1 *vs.* 4). Further increase in the alkyl chain length on the acyl side had little effect (Table 9, entries 5, 6). Branching of the alkyl group (isopropyl, cyclopentyl, or neopentyl) further increases the e.d.a. (Table 9, entries 7-9), but a too bulky alkyl groups slowed down the hydrogenation itself (Table 9, entries 10, 11). Thus, an alkyl group of medium size gives the best e.d.a.'s. An excellent result is the 96% e.e. for the hydrogenation of methyl 4-methyl-3-oxopentanoate (Table 9, entry 7). The e.d.a. of 96% had been the highest record achieved by a heterogeneous catalyst until we recently broke the record ourselves.

This breakthrough comes from a vague notion that the isopropyl group of methyl 4-methyl-3-oxopentanoate may even be a little too large and the ethyl group of methyl 3-oxopentanoate may be a little too small. Hence we decided to test a cyclopropyl group, which has a size between that of the isopropyl and ethyl groups (eq. 2).

$$(2)$$

The hydrogenation of methyl 3-cycloporopyl-3-oxopropionate with TA-NaBr-MRNiU proceeds smoothly giving more than a 98% e.d.a. and an almost quantitative chemical yield (Table 9, entry 12).[40] Fortunately, the cyclopropyl group remains intact under the hydrogenation conditions.

Epilog

The e.d.a. of more than 98% strikingly illustrates that even a simple compound such as TA adsorbed on Ni can differentiate between the enantio-faces of a substrate with more than 99% accuracy. Therefore, the present study counters the *a priori* idea that a heterogeneous asymmetric catalyst cannot yield an excellent e.d.a. because the surface is not uniform. or because steric constriction of the stereo-controlling active site is hard to achieve on a solid. In terms of the "Catalyst Region Model", the contribution of the N-regions in TA-NaBr-MRNiU is less than 2%. While we had this catalyst in hand for sometime, it was the prediction based on the "Extended Stereochemical Model" that led us to using the very best substrate. This again, underlines the importance of sufficient tuning of the reaction system. The "Reaction Process Model" which is well compatible with the above-mentioned two models, also clearly simulates the dual functions of MNi ; i.e., the hydrogenation and enantio-face-differentiation of the substrate. The reaction and stereochemical models of TA-MNi have been proposed by several research groups.[41, 42, 43] However, in comparison with our work, these models are less instructive, because they insufficiently incorporate information from organic stereochemistry.

In conclusion, our models allow simulating the function of MNi and predicting ways to improve the e.d.a. of the catalyst system without claiming to imply a real mechanism. Full understanding of MNi will require further significant advances in the physicochemical approach.

OPTICAL ENRICHMENT OF HYDROGENATION PRODUCTS AND APPLICATIONS OF OPTICALLY PURIFIED COMPOUNDS FOR SYNTHETIC ORGANIC CHEMISTRY

Optical Enrichment

In the enantio-differentiating hydrogenation of substrates that meet with the "Stereochemical Model", certain types of substrates give almost prefect e.d.a.'s. but all others give around a 75 to 90% e.d.a. This section briefly deals with a supplementary procedure, which allows for the imperfect nature of MNi and provides optically pure compounds from the reaction products.

The so-called preferential crystallization is the simplest and the most practical method for optical enrichment of chiral compounds. Applicability of this method for the hydrogenation products with MNi has been systematically investigated. As summarized in Table 10, all the hydro-genation products are able to be enriched to an optically pure state, without using special and expensive reagents or devices.

A series of methyl 3-hydroxyalkanoic acids; the hydrogenation products of β-ketoesters, was saponified and the resulting acids were converted to crystallized salts. Among the various salts examined, a dibenzyl ammonium salt was found to be favorable for short chain acids (Table 10, entries1,2,3),[44] while a dicyclohexyl ammonium salt was better for long chain acids (Table 10, entries 4-7).[31] When the salt derived from the hydrogenation products is crystallized, a part of the racemic modification crystallizes as a racemate but not conglomerate, while a part of the excess enantiomer gives crystals of a single enantiomer. In all cases, the former crystals are much soluble than the latter. Hence crystals of a single enantiomer are effectively isolated from racemic modification. This optical enrichment procedure is effectively applied for 3-hydroxyalkanoic acids of more than 80% e.e. on a practical scale.

The hydrogenation products of the 2-akanones, the 2-alkanols are converted to crystalline esters with pthalic or 3,5-dinitrobenzoic acid. Optically pure esters are obtained by recrystallization of the resulting crystals. The following saponification of the purified esters gives optically pure 2-alkanols (Table 10, entries 8-12).[36] 3-Hydroxy sulfones, hydrogenation products of β-ketosulfones, are cystallizable compounds and can be isolated as crystals of a single enantiomer (Table 10, entries 14-16).[34] As mentioned in Stereochemical 2P Model, optically pure pentanediol, one of the hydrogenation products of actylactone, is crystallized from an ether solution of the reaction products. A series of symmetric β-diols obtained by the hydrogenation of homologues of acetylacetone can be purified in the similar way (Stereochemical 2P Model, Table 7).

Table 10 Optical enrichment of hydrogenation products with MNi.
Source: Selected data from Refs. 31, 34, 36, 44.

Optical enrichment of 3-hydroxyalkanoic acids *via* crystalline ammonium salts

$$R\text{-}CH\text{-}CH_2\text{-}COOCH_3 \longrightarrow R\text{-}CH\text{-}CH_2\text{-}\overset{-}{C}OO\overset{+}{N}HR'_2 \longrightarrow R\text{-}CH\text{-}CH_2\text{-}COOR''$$

	OH	1)aq. NaOH	OH	1)aq. NaOH	OH	
		2) H$^+$	Recrystallization	2) H$^+$ or R"Br		
		3) R'$_2$NH				

No	Hydrogenation Products	Crystalline derivatives	Recrystallization		Final products'	
	R (e.e)	R'	Solvent, (Number)		R"	Yield (%)
1	CH₃- (83)	-CH2Ph	CH₃CN/H2O=2/1	(2)	n-Bu-	77
2	CH₃CH₂- (84)	-CH2Ph	CH₃CN	(2)	Me-	65
3	CH₃(CH₂)₄- (89)	-CH2Ph	CH₃CN	(2)	H-	65
4	CH₃(CH₂)₆- (87)	-cyclohexyl	CH₃CN	(3)	H-	50
5	CH₃(CH₂)₈- (88)	-cyclohexyl	CH₃CN	(3)	H-	73
6	CH₃(CH₂)₁₀- (85)	-cyclohexyl	CH₃CN	(3)	H-	73
7	CH₃(CH₂)₁₂- (87)	-cyclohexyl	CH₃CN	(3)	H-	75

Optical enrichment of 2-alkanols *via* crystalline esters

$$R\text{-}CH\text{·}CH_3 \xrightarrow{\text{Esterification}} R\text{-}CH\text{-}CH_3 \xrightarrow[\text{Recrystallization}]{\text{saponification}} R\text{-}CH\text{·}CH_3$$

with OH → O-R' → OH

R'= -C(=O)-C₆H₄-COOH   Acid phtalate (APH)

-C(=O)-C₆H₃(NO₂)₂   2,4-dinitrobenzoate (DNB)

No	R (e.e)	Crystalline derivatives	Recrystallization Solvent (Number)		Yield (%)
8	CH₃(CH₂)₃- (80)	DNB	MeOH	(4)	23
9	CH₃(CH₂)₄- (71)	APH	Hexane	(4)	21
10	CH₃(CH₂)₅-(77)	APH	CH₃CN	(3)	37
11	CH₃(CH₂)₇-76)	DNB	EtOH	(3)	28
12	CH₃(CH₂)₁₀- (75)	DNB	EtOH	(3)	56

Optical enrichment of 2-hydroxysulfones by direct crystallization
R-CH(OH)-SO₂-CH₃

No	R (e.e)	Solvent	(Number)	Yield (%)
15	CH₃CH₂- (71)	EtOAc	(3)	30
16	CH₃(CH₂)₄-(68)	Et₂O	(3)	20
17	CH₃(CH₂)₇- (67)	EtOAc	(3)	40

Applications of Optically Purified Hydrogenation Products

Optically purified hydrogenation products listed in Table 7 and 10 are promising compounds as a chiral synthetic block for the synthesis of various natural products as well as a chiral auxiliary for diastereo-differentiating reactions. Using available optically purified reaction products, we have carried out several application studies. Among them, two major works, in the field of bioactive natural products and stereo controlled synthetic organic chemistry, will briefly be mentioned.

The sex pheromone of pine sawflies from optically pure 3-hydroxybutyric acid: Pine feeding sawflies are widely distributed in the coniferous forests of the Eurasia and North American continents, and are known to be severe pests of conifers. It is said that there more than 100 species, including old world genera (*Genus Diprion*) and new word genera (*Genus Neodiprion*). In 1976, Jewett and co-worker at the University of Wisconsin isolated a major pheromone component from the *Neodiprion* species and determined its chemical structure as acetate or propionate of 3,7-dimethly-2-pentadecanol (**3**) (a compound having three chiral centers).[45] Since non stereo-controlled synthetic compounds they prepared did not show sufficient pheromone activity, it was anticipated that the stereochemistry of this compound could be an important factor to describe the pheromone system of various sawflies. At the request of the Wisconsin group, we have synthesized a series of stereochemically specified compounds employing a optically purified hydrogenation products with MNi as the starting materials. Results of cooperative studies with Wisconsin group are summarized in Table 11.[46] The pheromone component of the *Neodiprione* species is described by two stereoisomers, (2S,3S,7S)-**3** and (2S,3R,7R)-**3**. The former isomer is the main pheromone and the latter is an auxiliary pheromone, which functions as a synergist or inhibitors depending on the blend ratio with the main pheromone. These results lead to an important ecological finding that the difference in the acyl groups as well as the difference in the blend ratio of stereoisomers play an essential role in the pheromone activity to avoid interbreeding and to maintain reproductive isolation of each species, especially for sympatric *Neodiprion* species. In these studies, the pheromone of *Diprion simils,* an old world species, was also found to be the propionate of (2S, 3R, 7R)-**3**, which corresponds to the auxiliary pheromone for *Neodiprion* species. Later, the propionate of (2S,3R,7R)-dimethy-2-tridecanol (**4**) and propionate of (2S,3R,9S)-dimethyl-2-undcanol (**5**) are determined to be the pheromone of *Diprion pini* and *Diprion nipponica*, respectively. Although pheromones of old world species are more diverse than those of new world species, structure and stereochemistry of 2 and 3 positions are conserved (Table 11).[47,48]

The syntheses of two important pheromone components; (2S,3S,7S)-**3** and (2S,3R,7R)- **3**, are carried out by the coupling of the C12 block and C5 block as

Table 11 Pheromone system of pine sawflies. *Source*: Selected data from A. Tai, Yukigosei Kyoukaichi (J. Synthetic Organic Chemistry Japan), 58:576, 2000, and Refs. 47, 48.

Species	Flying season	Acyl group (X)	Optimum blend ratio Main pheromone / Auxiliary pheromone	
N.sertifer	Autumn	Acetate	1	0.001-0.0001
N.p.banksianae	Autumn	Acetate	1	0.1-0.01
N.nanulus	Autumn	Acetate	1	-
N.lecontei	Summer	Acetate	1	-
N.pinetum	Late spring	Acetate	1	2
N.swainei	Summer	Propionate	1	2

Diprion Species

D.similis — Propionate of (2S,3R,7R)-3,7-Dimethyl-2-pentadecanol (3)

D.pini — Propionate of (2S,3R,7R)-3,7-Dimethyl-2-tridecanol (4)

D. nipponica — Propionate of (2S,3R,9S)-3,7-Dimethyl-2-undecanol (5)

is shown in Figure 13. Optically pure (S)- and (R)-C12 blocks are derived from a natural product, (R)-pulegone *via* (R)-citlloneric acid.   (S)-3-hydroxybutyric acid obtained by the optical enrichment of hydrogenation product of MAA over (S,S)-TA-MNi was employed for the preparation of stereochemically pure key intermediates of the C5 blocks; methyl (2R,3S)- and (2S,3R)- 2-methyl-3-hydroxybutanoate (**2**).   The stereochemistry of these compounds have coincidentally been established in the study of "Reaction Process Model..[49] These stereochemically well specified and pure intermediates are converted to C5 blocks and subjected to reaction with a Grignard reagent derived from C12 blocks. The pheromones of *D. pini* and *D. nipponic* have also been synthesized from the same C5 blocks.

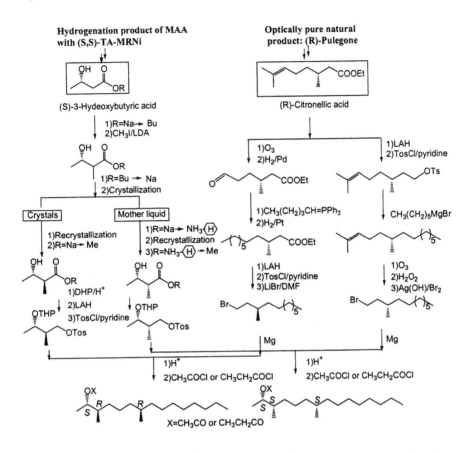

Figure 13 Preparation of the pine sawflies sex pheromone components from 3-hydroxybutyric acid obtained by the hydrogenation of MAA with MRNi. (From A. Tai, Yuukigousei Kyoukaishi (J. Synthetic Organic Chemistry Japan), 58: 575, 2000).

Application of 2,4-pentanediol (PD) and its homologues for new diastereo-differentiating reactions as chiral auxiliaries: In the early 80's, optically pure PD prepared by our simple procedures became available in a practical scale and various types of use were proposed shortly thereafter. One of the most extensively studied uses is as a chiral auxiliary built in the prochiral ketone or aldehyde as an acetal. The diastereo-differentiating attacks of various nucleophiles on the PD acetals activated by the Lewis acid, proceeds effectively giving the PD mono-ether of chiral alcohols in 80-95 % d.e. After the reaction, removing the PD unit from the mono-ether can easily be carried out by oxidation of the hydroxy group and mild base treatment. Because of good d.e.'s. of the reaction system and easier removablity of the PD auxiliary from the products, PD acetals have been widely applied to the preparation of optically active sec and tert alcohols (Eq. 3).[50]

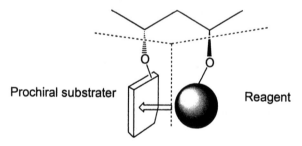

As the other important application of PD, we have developed a series of new diastereo-differentiating reactions basing on the structural nature of PD, a chiral bidentate compound with C_2 symmetry. The idea is that a prochiral substrate and a reagent are linked by PD as a chiral tether, and that various kinds of reactions are allowed to proceed *via* an intermolecular pathway as illustrated in Figure 14. This reaction design has been successfully applied to a variety of substrate-reagent combinations as summarized in Table 12.[51] All the reactions listed have resulted in high diastereo-differentiation to give products with over a 99% diastereomer excess (d.e.) without special attempts for optimization of the reaction conditions.

2,4-Pentanediol (PD) chiral tether

Prochiral substrater   Reagent

Figure 14 Intra molecular diastero- and enantio-differentiating reaction by PD tether as a chiral auxiliary. From T. Sugimura, Yuukigousei Kyoukaisi (J. Synthetic Organic Chemistry Japan), 55: 270, 1997).

The excellent ability and the wide applicability of the PD-tethered reaction have thus established a handy diastereo-differentiating reaction for the preparation of optically active compounds. However, the outstandingly high stereo-control in this system is not explained by a conventional mechanism. The chiral PD-tether part has only two methyl groups and leads to a flexible medium ring in the transition states, where the methyl groups are not expected to introduce enough steric-repulsion nor structural strain to perform strict stereo-

Table 12 The PD-tethered stereo-differentiating reactions. *Source*: A. Tai, Yukigosei Kyoukaishi (J. Synthetic Organic Chemistry Japan), 58:576, 2000.

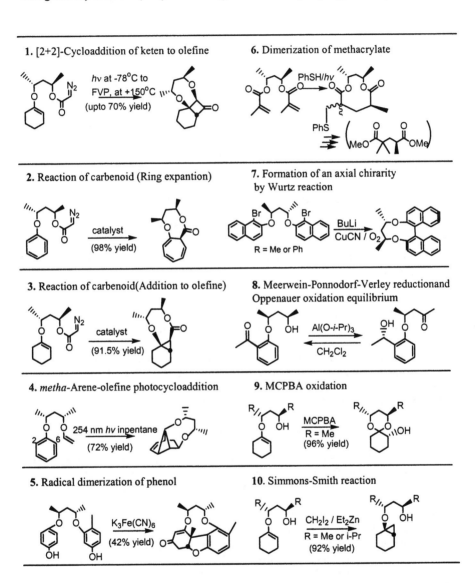

1. [2+2]-Cycloaddition of keten to olefine	6. Dimerization of methacrylate
2. Reaction of carbenoid (Ring expantion)	7. Formation of an axial chirarity by Wurtz reaction
3. Reaction of carbenoid(Addition to olefine)	8. Meerwein-Ponnodorf-Verley reductionand Oppenauer oxidation equilibrium
4. *metha*-Arene-olefine photocycloaddition	9. MCPBA oxidation
5. Radical dimerization of phenol	10. Simmons-Smith reaction

control. Our recent study on the [2+2] cycloaddition (Table 12, entry 1) shows that the e.d. of this reaction always maintained 100% over a wide range of reaction temperatures (185-423°K).[51] These results strongly suggest that the diastereo-differentiating ability of the PD tether is responsible for the difference in entropy. A series of PD tethered reactions reveals a new aspect for designing stereo-differentiating reactions, i.e., the Entropy Controlled Asymmetric Reaction.

REFERENCES

1)   Izumi, Y.; Imaida, M.; Fukawa, H.; Akabori, S. *Bull. Chem. Soc. Jpn.* **1963**, 36, 21.
2)   Izumi, Y. *Agew. Chem. Int. Ed. Engl.*, **1971**, 10, 871. m
3)   Klabunovskii, E. I. ; Vedenyapin, A. A. Asymmetricheski Kataliz Hidrogeizatsya no Metalakh, Izdatel'stov Nauka, 1980.
4)   a) Ozaki, H;Tai, A; Izuimi, Y; Chem. Lett. **1975**, 935
     b)Harada, T.; Hiraki, Y.; Izumi,Y.; Muraoka, J.; Ozaki, H.; Tai, A. Proc. 6th Int. Congr. Catal., p. 1204 (London, 1976).
5)   Ozaki, H.; . Tai, A.; Kobatke, S.; Watanabe, S.; Izumi, Y. *Bull. Chem. Soc. Jpn.* **1978**, 51, 3559.
6)   Nitta, Y.; Imanaka, T.; Teranishi, S. *J. Catl.* **1983**, 81 31.
7)   Keane, M.A. *J. Chem. Soc., Farady trrans.*, **1977**, 93, 2001.
8)   Sachtler, W.H.M. in R. L. Augastin (ed.) Catalysis for Organic Reaction, p. 189 (Mercel Dekker, 1985),
9)   Yasumori, I.; Inoue, Y.; Okabe, K. in B. Delmon (ed.),Catalysis, Heterogeneous, Homogeneous, p. 41 (Elsevier, 1975).
10) Tai, A.; Watanabe, H.; Harada, T. *Bull. Chem. Soc. Jpn.* **1979**, 52, 1468.
11) Harada, T.; Yamamoto, M.; Onaka, S.; Imaida, M.; Tai, A.; Izumi, Y. *Bull. Chem. Soc. Jpn.* **1981**, 54, 2323.
12) Tanabe, T.; Okuda, K.; Izumi, Y. *Bull. Chem. Soc. Jpn.***1973**, 46, 514.
13) Inoue, Y.; Okabe, K.; Yasumori, I. *Bull. Chem. Soc. Jpn.* **1981**, 54, 613.
14) Yasumori, I. *Pure Appl . Chem.* **1978**, 50, 971.
15) Fish, M. J.; Ollis, D. F. *J. Catal.* **1977**, 50, 353.
16) Keane, M.; Webb, G. *J. Catal.* **1992**, 136, 1.
17) Nitta, Y.; Swekine, F.; Imanaka, T.; Teranishi, S. *Bull. Chem. Soc. Jpn.* **1981**, 54, 980.
18) Harada, T.; Imachi,Y.; Tai, A.; Izumi, Y. in B. Imelik (ed) Metal-Support and Metal-Additive Effects in Catalysis, p. 377 (Elsevier, Amsterdam 1982).
19) Harada,T.; Tai, A.; Yamamoto, M. H. Ozaki, and Y. Izumi, Proc. 7th Int. Congr. Catal. p. 364 (Tokyo,1980).
20) Keane, M.A. *Langmuir* **1997**, 13, 41.
21) Harada, T.; Izumi, Y. *Chem. Lett.*, **1978**,1195.
22) Botelaar, L. J.; Sachtler, W. H. M. *J. Mol. Catal.* **1984**, 27, 387.
23) Brunner, H.; Muschiol, M.; Wischert, T.; Weil, J.; Heeeraeus, W. C. *Tetrahedron Asymmetry* **1990**, 1, 159.
24) Keane, M.A.; Webb, G. *J.Catal.* **1992**, 80, 136.
25) Tai, A.; Kikukawa, T.; Sugiura, T.; Inoue, Y.; Osawa, T.; Fujii, S. *Chem. Commun.* **1991**,795.
26) Tai, A.; Kikukawa, T.; Sugiura, T.; Inoue, Y.; Abe, S.; Osawa,T.; Harada, T. *Bull. Chem. Soc. Jpn.* **1994**, 67, 2473.

27) Tatsumi, S. *Bull. Chem. Soc. Jpn.* **1968**, 1, 402.

28) Tai, A.; Harada, T.; Hiraki, Y.; Murakami, S. *Bull. Chem. Soc. Jpn.* **1983**, 56, 1414.

29) Murakami, S.; Harada,T.; Tai, A. *Bull. Chem. Soc. Jpn.* **1980**, 53, 1356.

30) Hiraki,Y.; Ito, K.; Harada,T.; Tai, A. *Chem. Lett.* **1981**, 131.

31) Nakahata, M.; Imaida, M.; Ozaki, H.; Harada, T.; Tai, A. *Bull. Chem. Soc. Jpn.* **1982**, 55, 2186.

32) Ito, K.; Harada, T.; Tai, A. *Chem. Lett.* **1980**, 1049.

33) Ito, K.; Harada, T.; Tai, A. *Bull. Chem. Soc. Jpn.*, **1980**, 53, 3367.

34) Tai, A.; Ito, K.; Harada, T. *Bull. Chem. Soc. Jpn.* **1981**, 54, 223.

35) Osawa, T.; Harada, T. *Bull. Chem. Soc. Jpn.* **1984**, 57,1618.

36) Osawa, T.; Harada, T.; Tai, A. *J. Catal.* **1990**, 21, 7.

37) Osawa, T. *Chem. Lett.*, **1985**, 1069.

38) Harada, T.; Osawa, T. in J. Gannes and V. Duois (eds) Chiral reaction in Heterogeneous catalysis, p. 75 (Plenum Press, 1995).

39) Sugimura, T.; Osawa, T.; Nakagawa, S.; Harada, T.; Tai, A. Proc. 11th Int. Congr. Catal. p. 231. (Baltimore 1996),

40) Nakagawa, S.; Sugimura,T.; Tai, A. *Chem. Lett.* **1997**, 859.

41) Hoek, A.; Woerde, M. H.; Sachtler, W. H. M. Proc. 7th Int. Congr. Catal., p. 376 (Tokyo 1980).

42) Klabunovskii, E. I.; Vedenyapin, A. A.; Karpeiskaya, E. I.; Pavlov,A. V. Proc. 7th Int. Congr. Catal. p. 390 (Tokyo 1980).

43) Yasumori, I.; Yokozeki, M.; Inoue, M. *Faraday Discuss. Chem. Soc.* **1981**,72, 385.

44) Kikuawa, T.; IIzuka,Y.; Sugimura, T.; Harada, T.; . Tai, A. *Chem. Lett.* **1987**, 267.

45) Jewett, D. M.; Matsumura, F.; Coppel, H. C. *Science*, **1976**, 192, 51.

46) Tai, A.; Morimoto, N.; Yoshikawa, M.; Uehara, K.; Sugimura, T. *Agric. Biol. Chem.*, **1990**, 54, 1753, and references therein

47) Bergström, G.; Wassgern, A. –B.; Anderbrant, O.; Fägerhag, J.; Edlund, H.; Hendenström, E.; Högberg, H. –E.; Geri, C.; Auger, M. A.; Varsma, M.; Hansson, S. B.; Löfqvist, J. *Experientia*, **1995**, 51, 370.

48) Tai, A.; Syouno, E.; Tanaka, K.; Fujitra, M.; Sugimura, T.; Higasiura, Y.; Kakizaki, M.; Hara, H.; Naito, T. *Bull. Chem. Soc. Jpn.*, 75:111-121, 2002.

49) Tai, A.; Imaida, M.. *Bull. Chem. Soc. Jpn.* **1978**, 51, 1114.

50) a) Bartlett, P. A.; Johonson, W. J. *J. Am. Chem. Soc.*, **1983**, 105, 2088, b) Alexakis,A.; Magneney, P. *Tetrahedron Asymmetry*, **1990**, 1, 477.

51) Sugimura, T.; Tei, T.; Mori, A.;Okuyama, T.; Tai, A. *J. Am. Chem. Soc.*, **2000**, 122, 2129, and references therein.

18

Improved Stabilities of Skeletal Cu Catalysts for Dehydrogenation and Hydrogenolysis Reactions

Liyan Ma and Mark S. Wainwright
School of Chemical Engineering and Industrial Chemistry, The University of New South Wales, Sydney, Australia

ABSTRACT

This study reports improved stabilities of skeletal Cu catalysts for use in organic synthesis reactions. The promoted skeletal Cu catalysts have been characterised by measuring their resistance to structural rearrangement in caustic solutions, thermal stabilities and activities for the reactions of methanol dehydrogenation and methyl formate hydrogenolysis. Comparisons have been made with an unpromoted skeletal Cu catalyst and a commercial coprecipitated copper chromite catalyst.

Preliminary results showed that skeletal Cu, after promotion with chromia, has a very high surface area and a stable structure. The Cu crystallite sizes and surface areas remained almost the same as the initial values after being exposed to 6.1M NaOH at 323K for 170 hours, at 373 for 120 hours and at 423K for 100 hours respectively. However, unpromoted skeletal Cu, under the same conditions, showed a significant loss of surface area. When the temperature was increased the BET surface area decreased even further. This significant decrease of surface area was attributed to dissolution and reprecipitation of copper in this highly corrosive environment.

Activities and stabilities of the promoted and unpromoted skeletal Cu catalysts were studied using methanol dehydrogenation and methyl formate hydrogenolysis as test reactions. The results showed that there was almost no deactivation observed for the promoted skeletal Cu catalyst during 7 hours on stream at 453K. However, significant decreases in activity were observed for the unpromoted skeletal Cu catalyst. Similar results were also obtained when using both catalysts for the hydrogenolysis of methyl formate. These results indicate that promotion of skeletal Cu by chromia suppresses fouling by polymerized formaldehyde, an intermediate during these reactions, and therefore mitigates the problem of deactivation. The promoted skeletal Cu catalyst was shown to have higher activity at low temperatures than that of the commercial copper chromite catalyst and similar stability.

These novel skeletal copper catalysts have significant potential for use in a wide range of organic synthesis reactions including those in which high caustic concentrations are involved such as in the dehydrogenation of primary alcohols to form carboxylic acid salts.

INTRODUCTION

Dehydrogenation of alcohols and hydrogenolysis of esters are very important reactions in industrial organic synthesis processes. The hydrogenolysis of natural fatty acid esters for production of fatty alcohols [1-3] has been operated in large-scale industrial processes for more than five decades [1]. The hydrogenolysis of formates to methanol is another typical application in a methanol synthesis route in which an alcohol reacts with carbon monoxide to generate a formate which is then hydrogenolysed into methanol and the parent alcohol [1]. There has been particular interest in the hydrogenolysis of diesters to diols [4,5], which are mainly used in the production of polyesters. The liquid phase dehydrogenation of amino alcohols to carboxylic acid salts [6-11] is widely used to produce glycine, iminodiacetic acid and nitrilotriacetic acid etc. which are the raw materials for preparation of pharmaceuticals, agricultural chemicals, pesticides and detergents etc.. The gas phase dehydrogenation of methanol to methyl formate [12-15] is a first step in the synthesis of acetic acid, N,N-dimethylformamide, and hydrogen cyanide [16].

The use of copper-based catalysts for the reactions of ester hydrogenolysis and alcohol dehydrogenation is well established [2-26, 30,31]. Raney®-type skeletal copper catalysts are widely used in these reactions [5-11,17-26,30,31]. However, significant catalyst deactivation is experienced [5-6,25-29]. Evans et al studied the hydrogenolysis of formates in the gas phase over a variety of copper-based catalysts, including skeletal copper and copper chromite [17,18,25] and reported that skeletal Cu is the most active and selective catalyst for the reaction [17] but subject to severe catalyst deactivation [25]. Similar deactivation behavior has been observed for the reverse reaction, dehydrogenation of methanol to methyl formate [26]. Deactivation has been explained to be due to formaldehyde polymer fouling the copper surface [26]. Catalyst deactivation due to polymer fouling the surface has also been observed when skeletal Cu catalysts were used for the liquid-phase hydrolysis of acrylonitrile to acrylamide [27-29].

Studies of the hydrogenolysis of dimethyl succinate [23,24,30,31] have shown that, at moderate pressures and hydrogen/diester ratios, γ-butyrolactone is the major product that is further hydrogenolysed to tetrahydrofuran [31]. Very high hydrogen pressure is necessary to form 1,4-butanediol [31]. Skeletal Cu and skeletal Cu-Cr catalysts, prepared by leaching a Cu-Cr-Al alloy in a NaOH solution, provide the best selectivity and high activity leading to improved 1,4-butanediol yields [23]. Copper-chromite is the most active catalyst for the hydrogenolysis of γ-butyrolactone to tetrahydrofuran [24]. However, the presence of water formed by this reaction leads to a very large reduction in the reaction rate [30]. Franczyk has reported that copper catalysts, including Raney® copper, used in the process for the dehydrogenation of amino alcohols to amino carboxylic acid salts in the presence of highly concentrated caustic solutions, agglomerate and therefore lose activity [6]. The presence of small amounts (50-1000 ppm) of transition metals, such as Cr, Ta, Nb, Zr,V, Mo, Mn, W, Co, Ni, in Raney® Cu can extend the activity to a significant degree [6].

Köhler et al have investigated surface species formed by adsorption of ethyl acetate on copper using infrared spectroscopy and found that ethyl acetate initially dissociates into adsorbed ethoxy and acetyl fragments [32]. The amounts of the fragments produced correspond approximately to one, two carbon fragment for every twelve surface copper atoms [32], indicating that catalyst deactivation through fouling is related to the surface copper state.

Recent attention has been paid to developing new preparation processes to improve the performance of skeletal Cu catalysts [33-35]. In this paper we show that Cr_2O_3 promoted skeletal Cu catalysts have very high surface areas and are very stable in highly concentrated caustic NaOH solutions at moderate temperatures (up to 400K), which shows that they have great potential in use for the process of dehydrogenation of amino alcohols to amino carboxylic acid salts. The improved Cu surface areas and pore structures in these catalysts provide high stability for methanol dehydrogenation to methyl formate and the reverse reaction.

EXPERIMENTAL

Six Cu catalysts, namely two skeletal Cu catalysts, three promoted skeletal Cu catalysts and a commercial copper chromite catalyst, were used in this study. The skeletal copper catalysts (Cu1 and Cu2) were prepared by leaching $CuAl_2$ alloy particles (210-350 μm) in a large excess of 6.1M NaOH for 24 hours at 273K and 323K respectively. The promoted skeletal CuCr1 and CrCr2 catalysts were prepared under the same procedures as the skeletal Cu1 catalyst (at 273K) except that the 6.1M NaOH solutions contained 0.01M and 0.1M sodium chromate respectively. The promoted skeletal CuCr3 catalyst was prepared by leaching the same alloy particles in a 6.1M NaOH solution containing sodium chromate (0.01M) at 323K for 24 hours. Details of the precursor alloy and of the preparation procedures have been described elsewhere [33]. The commercial copper chromite catalyst was Harshaw Cu-0203T catalyst consisting of 80 wt% CuO and 20 wt% Cr_2O_3.

The chemical compositions of the catalysts were determined using atomic absorption spectroscopy (Varian/Spectr AA-20 plus) on acid-digested samples. The chemical compositions and surface areas of the catalysts are presented in Table 1.

Catalysts were characterized by measuring their stabilities in highly concentrated caustic solution, their thermal stabilities and catalytic properties for the reactions of methanol dehydrogenation and methyl formate hydrogenolysis.

Table 1 Catalyst characterization

Catalyst	Composition, wt%				Surface area*, m^2g^{-1}		Cu
	Cu	CuO	Cr_2O_3	Al	BET area	Cu area	crystallite size*, nm
Cu1	98.17	-	-	1.83	26.9	16.7	13.7
Cu2	99.31	-	-	0.69	12.4	5.1	14.7
CuCr1	94.54	-	1.59	3.87	51.6	20.0	9.9
CuCr2	94.81	-	2.38	2.81	52.2	23.5	10.5
CuCr3	94.27	-	3.12	2.61	41.8	11.7	9.6
Harshaw Cu-0203T	-	80	20	-	11.8	6.2	-

* In reduced form.

The stabilities in caustic solutions were determined by measuring the BET and Cu surface areas as well as the Cu crystallite sizes of the catalyst after being aged in 6.1M NaOH solutions at temperatures ranging from 323K to 423K for different periods. The aging experiments were conducted mostly in thick wall polyethylene containers (made by Nalgene) that were initially loaded with a 6.1M caustic solution (100ml) and preheated in an air oven at the aging temperature (for temperatures below 373K) for 48 hours before addition of the catalyst. For the aging temperatures above 373K, weldless cylindrical stainless steel containers (150mm length × 9.5mm i.d. and 12.7mm o.d.), sealed by stainless steel caps at the ends, were employed. The catalysts were loaded into these containers that were filled up with a 6.1M NaOH solution. Samples were taken at different periods of aging and washed in distilled water to a pH of 7.

The measurements of BET and Cu surface areas were carried out in a flow system (Flowsorb [36]). Samples of around 0.15 g were wet loaded into a sample tube and dried at 423K in a flow of H_2 initially for 30 minutes and then for another 30 minutes at 513K, before being flushed and cooled to 363K in flowing high purity helium. The Cu surface area was determined at 363K using a N_2O decomposition technique [24], followed by total surface area measurements using a single point BET method (nitrogen adsorption-desorption at 77K). The Cu crystallite sizes were determined by X-ray diffraction (Philips Expert System) using Ni-filtered CuKα (wavelength=1.543Å) radiation. The sample (covered by distilled water to eliminate oxidation) was uniformly spread on the surface of a silica sample holder prior to XRD analysis and scanned from 40° to 47° (2θ). The Cu(111) peak was employed to calculate Cu crystallite size using the Scherrer equation.

The thermal stabilities were determined by thermally treating the catalysts at temperatures ranging from 513 to 633K in flows of H_2 in the Flowsorb system for one hour, followed by measurements of the Cu and BET surface areas after thoroughly purging with pure helium. After pretreatment the samples were covered by distilled water to avoid oxidation and then characterized by X-ray diffraction by scanning from 10° to 75° (2θ).

The catalytic properties were determined by the reactions of methanol dehydrogenation and methyl formate hydrogenolysis, conducted in a conventional stainless steel microreactor system. About 0.15 grams of wet catalyst were loaded into the reactor and dried at 423K in a flow of H_2 for 30 minutes, followed by another 4 hours at 513K. The catalyst was then flushed with high purity helium and cooled to 453K for the reactions. The feedstock (methanol for dehydrogenation and methyl formate for hydrogenolysis) was introduced into the reactor by passing carrier gas (He for methanol and H_2 for methyl formate) through two saturators connected in series (at 288K for methanol and at 289K for methyl formate). Compositions of the reactants and products were analyzed by an on-line gas chromatograph (Shimadzu 9A) fitted with a thermal conductivity detector after separation using a Porapak Q column (3.2m) at 383K. The dehydrogenation of methanol was conducted at 101.3 kPa, 453K and a space velocity (GHSV) of c.a. $3.87 \times 10^4 h^{-1}$ using 9.21mol% methanol in helium as the feedstock. The hydrogenolysis of methyl formate was carried out at 101.3 kPa, 453K and a space velocity (GHSV) of ca. $1.17 \times 10^5 h^{-1}$. The mole ratio of H_2/methyl formate in the feedstock was 6.3:1.

The activity of methanol dehydrogenation was expressed as moles of methanol converted to methyl formate per gram of catalyst per hour. The activity of methyl formate hydrogenolysis was defined as moles of methyl formate converted to methanol per gram catalyst per hour.

RESULTS AND DISCUSSION

Catalyst Stability in Caustic Solutions

Some industrial organic synthesis reactions take place in the presence of aqueous caustic soda. A typical example is the dehydrogenation of amino alcohols to amino carboxylic acid salts, which is typically conducted at 1.0 MPa and 393K-483K in a concentration of caustic up to 50wt%. Under such harsh conditions, most supported copper catalysts cannot be used due to dissolution of the support material in caustic solution. Raney®- type skeletal copper catalysts are the most preferable catalysts for this application. However, agglomeration of copper in the concentrated caustic solution, after several reaction cycles, is also observed [6].

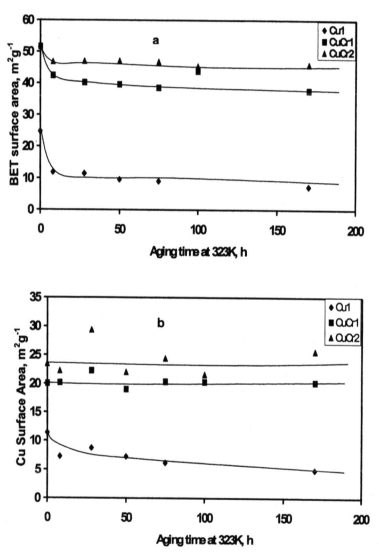

Figure 1 BET (a) and Cu (b) surface areas of the promoted (CuCr1&2) and unpromoted (Cu1) skeletal copper catalysts aged in 6.1M NaOH at 323K for the period up to 170 hours.

Figure 1 shows the surface areas of the promoted skeletal Cu catalysts (CuCr1&2) and the unpromoted skeletal Cu (Cu1) catalyst after different times of exposure to caustic solution at 323K. It is obvious that CuCr1 and CuCr2 catalysts were very stable in a 6.1M NaOH solution at 323K. After being aged for 170 hours, both the BET and Cu surface areas of these catalysts remained at very high levels (BET:40-46 m^2g^{-1}; Cu:20-23 m^2g^{-1}) and the maximum loss of the BET surface area was around 20%. There was no detectable loss of Cu surface area for these catalysts (Fig1b). However, the Cu1 catalyst lost about two thirds of its original BET surface area and around half of the original Cu surface area after 80 hours. The significant decrease of surface area was due to dissolution and reprecipitation of copper in this highly corrosive environment. It is clear that Cr_2O_3 in the skeletal Cu catalysts hindered the dissolution/redeposition of copper during aging and hence stabilized the surface area.

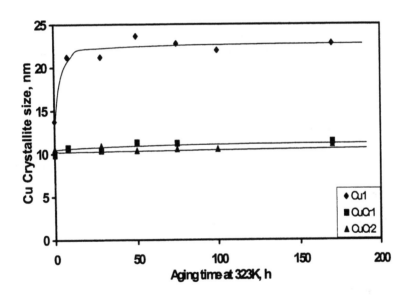

Figure 2 Copper crystallite sizes of the promoted (CuCr1&2) and unpromoted (Cu1) skeletal Cu catalysts corresponding to Figure 1.

It is also seen from Figure 1 that significant decreases in the surface areas appeared during the first 8-hour period, especially for the unpromoted skeletal Cu1 catalyst. XRD analysis (Figure 2) shows that the Cu crystallite size of the Cu1 catalyst grew from 14 nm to 23 nm, leading to the decreased surface areas. On the other hand, for the CuCr1 catalyst containing 1.59 wt% Cr_2O_3, the Cu crystallite size only increased from 9.9 nm to 10.7 nm. There was no crystallite size increase for the CuCr2 catalyst containing 2.94 wt% Cr_2O_3 after being aged for 170 hours. This indicates that the presence of Cr_2O_3 in the skeletal Cu catalyst suppresses Cu crystallite growth and stabilizes the Cu structure, thereby increasing the stability in highly corrosive caustic environments.

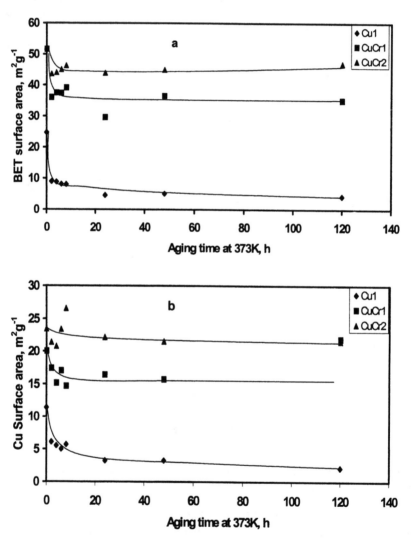

Figure 3 BET (a) and Cu (b) surface areas of the promoted (CuCr1&2) and unpromoted (Cu1) skeletal copper catalysts aged in 6.1M NaOH at 373K for a period up to 120 hours.

The effect of chromate additive on formation of skeletal Cu catalysts has recently been comprehensively studied [37, 38]. Chromate was found to deposit on the copper as chromium (III) oxide, hindering the leaching reaction as well as the dissolution/redeposition of copper [37]. The deposition of Cr_2O_3 resulted in a finer structure, with a corresponding larger surface area.

Figures 3 and 4 show the results of the same catalysts (in Figure 1) after aging at 373K. Although increasing the aging temperature resulted in further decreases to the surface areas, the two promoted skeletal Cu catalysts (CuCr1&2) still had very high BET and Cu surface areas (BET: 35-45 m^2g^{-1}; Cu: 16-21 m^2g^{-1}), compared to the unpromoted skeletal Cu catalyst (Cu1) which had a BET area of 4 m^2g^{-1} and a Cu area of 2 m^2g^{-1} after 120 hours aging at 373K. The CuCr1 and CuCr2 catalysts lost around 30% and 10% of their original BET surface areas as well as 20% and 8% of their original Cu surface areas respectively. The Cu1 catalyst lost more than 80% of both BET and Cu surface areas.

Corresponding to the decrease in the Cu surface area with the aging time (Fig3b) was an increase in the Cu crystallite size. Higher aging temperatures resulted in larger Cu crystallite sizes. It can be seen from Figure 4 that at 373K the crystallite size of the Cu1 catalyst increased from 14 nm to 31 nm, which was 8 nm larger than the catalyst aged at 323K for 170 hours. It is interesting

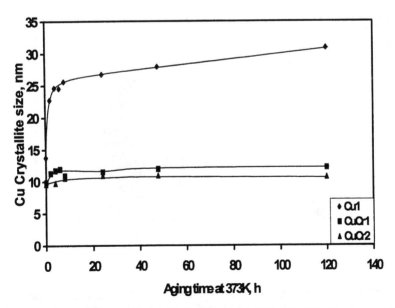

Figure 4 Copper crystallite sizes of the promoted (CuCr1&2) and unpromoted (Cu1) skeletal Cu catalysts corresponding to Figure 3.

to find that the effect of aging temperature on the CuCr1 and CuCr2 catalysts
was insignificant. The Cu crystallite sizes of these catalysts aged at 373K for
120 hours maintained at almost the same levels as those aged at 323K for 170
hours.

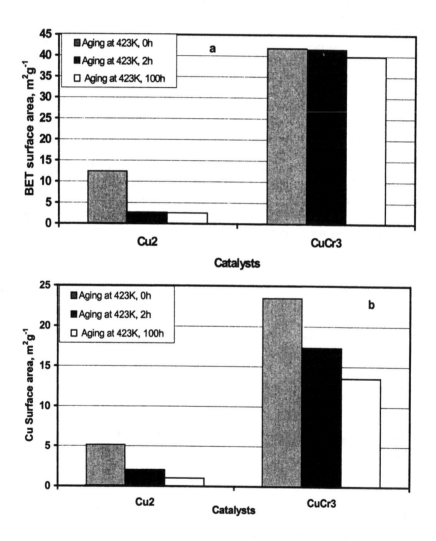

Figure 5 BET (a) and Cu (b) surface areas of the promoted (CuCr3) and unpromoted
(Cu2) skeletal copper catalysts aged in 6.1M NaOH at 423K for the period up to 100.

Further examination of the level of stability to caustic solution of both promoted and unpromoted skeletal Cu catalysts was carried out at 423K. The Cu2 and CuCr3 catalysts were used for these tests. As a result of the higher leaching temperature (at 323K) the catalysts had lower surface areas (Table 1 and Fig.5) than the corresponding catalysts prepared at 273K as expected [38].hours.

Figure 5 shows the BET and Cu surface areas of CuCr3 and Cu2 catalysts before and after being aged at 423K for the period up to 100 hours. The BET surface area of the CuCr3 catalyst only decreased from $42m^2g^{-1}$ to 40 m^2g^{-1}. However, the BET surface area of the Cu2 catalyst decreased from 12 m^2g^{-1} to 2 m^2g^{-1}. The Cu surface area of the promoted skeletal Cu catalysts after aging seemed to decrease significantly ($\sim10m^2g^{-1}$). However, the Cu crystallite size only increased from 9.6 nm to 11.9 nm which is a small change, compared to that observed for Cu2 catalyst for which the Cu crystallite size increased from 13.8 nm to 39 nm under the same aging conditions (Fig. 6). Figures 5 and 6 indicate that significant rearrangement of Cu occurred during the first two hours of aging and that Cr_2O_3 in the skeletal Cu catalyst retarded the Cu crystallite growth.

The stability of the chromia promoted skeletal Cu in highly concentrated caustic solution at temperature around 400K indicates significant potential for use in organic synthesis reactions such as the dehydrogenation of amino alcohols to carboxylic acid salts.

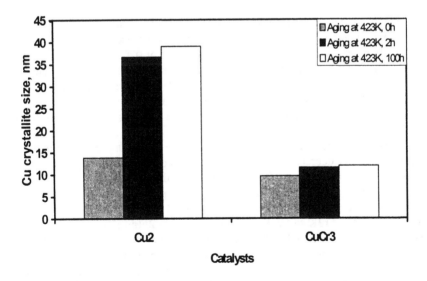

Figure 6 Copper crystallite sizes of the promoted (CuCr3) and unprompted (Cu2) skeletal Cu catalysts corresponding to Figure 5.

Thermal stability of the Catalysts

It is well known that most supported Cu catalysts cannot be used at temperatures above 573K because of significant Cu surface area losses caused by sintering of the finely dispersed copper particles on the surface of the support. Improvement to thermal stability is very important for the application of Cu catalysts in gas-phase reactions including hydrogenation and dehydrogenation. In this study, the thermal stability of the CuCr1 catalyst was examined. Comparison was made with the unpromoted Cu1 catalyst and a commercial copper chromite catalyst (Harshaw Cu-0203T). Details of the results are listed in Table 2 and Figure 7.

Table 2 BET and Cu surface areas of different catalysts at various thermal pretreatment temperatures.

Catalyst	Cu1		CuCr1		Harshaw Cu-0203T	
Surface area, $m^2 g^{-1}$	BET	Cu	BET	Cu	BET	Cu
Pretr. T, K						
513	18.7	12.7	51.2	19.5	10.6	8.2
563	16.4	11.7	48.1	22.6	11.8	6.2
593	18.1	7.6	49.8	28.3	17.5	5.5
623	17.2	9.7	49.3	27.9	23.0	4.0

Under the temperatures used for the thermal pretreatment (513-623K), the total surface areas (BET) of the tested catalysts were maintained at the same level as their original values, except the Harshaw Cu-0203T catalyst for which the BET area increased with increasing temperature. This seems to indicate that the structure of the Cu-0203T catalyst changed with increasing temperature. However, XRD analysis (Figure 7) shows that the XRD spectrum of the Cu-0203T catalyst pretreated at 513K was almost the same as that of pretreated at 623K.

It is interesting to note from Table 2 that the increasing temperature of the thermal treatment decreased the Cu surface areas of the Cu1 catalyst and of the Harshaw Cu-0203 catalyst (Table 1) but that the promoted CuCr1 catalyst showed no decrease in its Cu surface area at temperatures up to 623K. On the contrary, the Cu surface area appeared to gradually increase with increasing the temperature from 513K to 593K as seen from Table 2. Initially, it was suspected that some Cr_2O_3 was reduced to metallic Cr and that the extra N_2O was consumed by the metallic Cr, instead of Cu. However, a blank test using $Cr_2O_3.3H_2O$, which was pretreated at 623K in H_2 for 1 hour, indicated that there was no N_2O decomposition on the pretreated Cr_2O_3 sample, ruling out that effect.

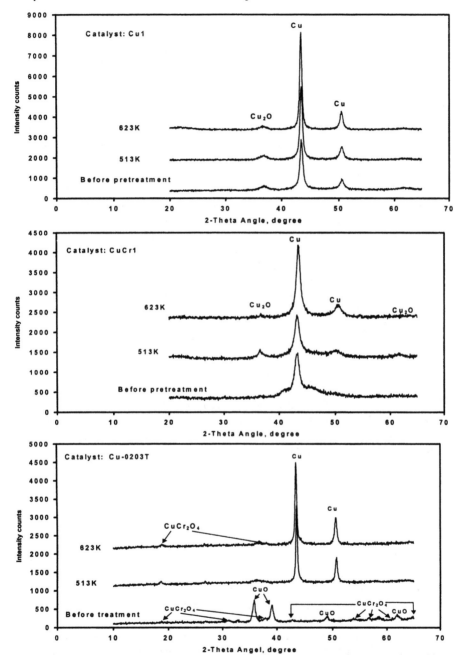

Figure 7 XRD profiles for the promoted (CuCr1) and unpromoted (Cu1) skeletal copper catalysts and commercial Cu-0203T catalyst that are pretreated in H_2 at the temperatures of 513K and 613K for 1 hour, respectively.

The XRD spectra of the three catalysts before and after pretreated at 513K and 623K (Figure 7) show the changes of Cu phase and crystallite size with thermal pretreatment temperature. The Cu-0203T catalyst originally consisted of CuO and $CuCr_2O_4$. After being treated in H_2 at 513 K for 1 hour, CuO was completely reduced to metallic Cu with a crystallite size of around 24 nm. Further increasing the pretreatment temperature to 623K narrowed the Cu (111) peak as a result of increasing the Cu crystallite size to 28 nm. No peak was observed for Cr_2O_3 or $CuCr_2O_4$ in the pretreated Cu-0203T samples, revealing that the chromium compounds in these catalysts existed as amorphous Cr_2O_3. The Cu1 catalyst initially had a Cu crystallite size of 13.7 nm. After treating at 513K, the Cu crystallite size grew to 16.7 nm, then to 19.9 nm at 623K. However, the Cu crystallite size of the CuCr1 catalyst, after being treated at 513K, only increased from 9.9 nm to 10.4 nm. Further increasing the temperature to 623K resulted in increasing the Cu crystallite size to 10.6 nm. As for the Cu-0203T catalyst, the Cr_2O_3 in the CuCr1 catalyst, after thermal pretreatment at 623K, remained in an amorphous status.

The results in Table 2 and Figure 7 also indicate that the chromia promoted skeletal Cu catalyst CuCr1 retains high BET and Cu surface areas even at temperatures up to 623K. The relatively high thermal stability of the very fine Cu crystallites makes the catalyst suitable for those reactions taking place at higher temperatures, possibly up to 623K.

Catalytic Activity and Stability of the Catalysts

Catalytic activities and stabilities of the promoted and unpromoted skeletal Cu catalysts for the gas-phase reactions of methanol dehydrogenation and methyl formate hydrogenolysis were determined. Comparison was also made with the commercial Cu-chromite catalyst.

Dehydrogenation of Methanol

The dehydrogenation of methanol to produce methyl formate can be expressed as:

$$2CH_3OH = CH_3OCHO + 2H_2 \tag{1}$$

The reaction was carried out at 101.3 kPa, 453K and GHSV c.a. $3.87 \times 10^4 h^{-1}$. The main products were methyl formate and hydrogen. Trace amounts of CO (< 1 mol%) and CO_2 (< 0.1 mol%) were also observed.

Figure 8 shows the activities and stabilities of different catalysts for the dehydrogenation of methanol. Methanol conversion over the Cu1 catalyst dramatically decreased from the initial 38% to 5% during the first 2.5 hours on-line and then maintained conversion of 5% for another 4.5 hours. Similar deactivation patterns were observed over a commercial Cu-ZnO/alumina (BASF

S3-85) catalyst [39] and over a Raney Cu catalyst [26]. The deactivation was explained as fouling of the active Cu sites by polymerized formaldehyde, an intermediate in the reaction [26]. Steam was found to be a very good agent for regeneration of the deactivated catalysts [39].

Relatively constant methanol conversions were obtained during the period of 7 hours for both the CuCr1 catalyst and the commercial Cu-0203T catalyst. This indicates that these catalysts have excellent stability for the methanol dehydrogenation reaction. The reason for the improved stability of the CuCr1 catalyst may be attributed by the presence of Cr_2O_3. Considering the Cu crystallite size difference between the Cu1 (16.7 nm) and CuCr1 (8.8 nm) catalysts together with the commercial Cu-0203T catalyst (24 nm), polymerization of formaldehyde, which causes the deactivation, seems to be independent of the copper crystallite size. Since both the CuCr1 and the Cu-0203T catalysts contain amorphous Cr_2O_3, this metal oxide may play a dominant role possibly by suppressing formaldehyde polymerization.

The results in Figure 8 show that both methanol conversion and the rate of formation of methyl formate over the promoted skeletal Cu catalyst (CuCr1) were higher than that over the Cu-0203T catalyst and that the stabilities of the two catalysts were almost the same.

Hydrogenolysis of Methyl Formate

The hydrogenolysis of methyl formate to methanol (reaction 2) is the reverse reaction of methanol dehydrogenation.

$$CH_3OCHO + 2H_2 = 2CH_3OH \qquad\qquad (2)$$

The mechanism obtained from isotopic labeling studies indicates that the reaction is involved with dissociation of methyl formate into two different intermediates, HCO- and $-OCH_3$, followed by the hydrogenation of the two fragments [40].

In this study, the three catalysts (Cu1, CuCr1 and Cu-0203T) were used for the hydrogenolysis of methyl formate at 101.3 kPa, 453K and GHSV: $1.17 \times 10^5 h^{-1}$ using a feedstock of H_2:CH_3OCHO at a ratio of 6:1. The products were mainly methanol with trace amounts (<0.1mol%) of CO and CO_2. A trace amount (<0.1 mol%) of water was only observed using the commercial Cu-0203T catalyst, and was thought to be formed during catalyst reduction.

Figure 9 shows the activities and stabilities of these catalysts for the reaction of methyl formate hydrogenolysis. Similar to the results for methanol dehydrogenation, about 80% of the initial activity of Cu1 catalyst was lost within 6.5 hours of which 70% was lost during the first 1.5hours. However, there was no significant activity decrease observed for the CuCr1 catalyst. The activity of CuCr1 catalyst after 1.5 hours reaction is more than 2 times higher than the Cu1 catalyst. It is also seen that the initial activity of the CuCr1 catalyst is much higher than the commercial Cu-0203T catalyst. As for the hydrogenolysis reaction, the improved activity and stability of CuCr1 catalyst can be explained by a process in which Cr_2O_3 on the catalyst surfaces suppresses the polymerization of the dissociated species on Cu active sites.

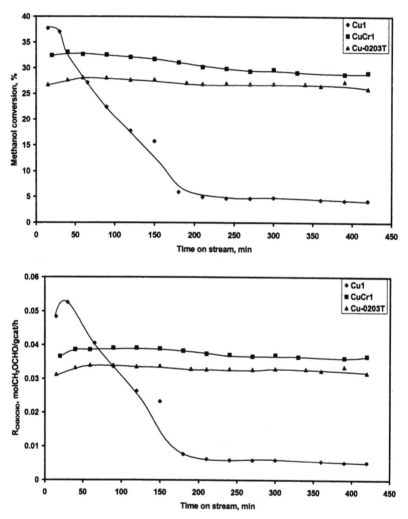

Figure 8 Comparison of the activity and stability of the promoted (CuCr1) skeletal Cu catalyst with the unpromoted (Cu1) skeletal Cu catalyst and the commercial Cu-0203T catalyst for the dehydrogenation of methanol to methyl formate at 101kPa, 453K and GHSV: c.a. $3.87 \times 10^4 h^{-1}$.

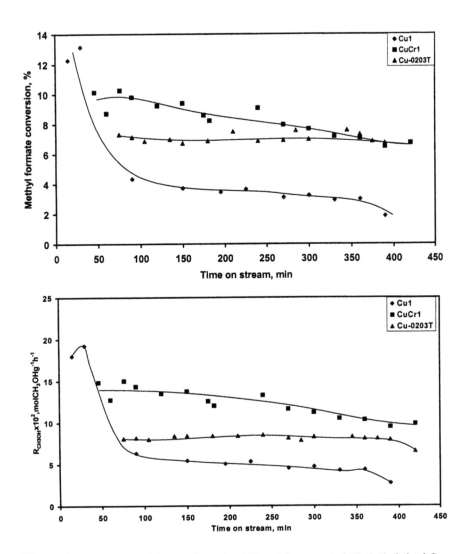

Figure 9 Comparison of the activity and stability of the promoted (CuCr1) skeletal Cu catalyst with the unpromoted (Cu1) skeletal Cu catalyst and the commercial Cu-0203T catalyst for the hydrogenolysis of methyl formate to methanol at 453K, 101kPa and GHSV: c. a. $1.17 \times 10^5 h^{-1}$.

The results from the hydrogenolysis experiments indicate that the Cr_2O_3 promoted skeletal Cu catalysts have great potential for application in industrial processes for ester hydrogenolysis.

CONCLUSIONS

1.   Chromia promoted skeletal Cu catalysts possess very high stability in caustic solutions. After being exposed to 6.1M NaOH at 323K for 170 hours, at 373K for 120 hours and at 423K for 100 hours respectively, the Cu crystallite sizes remain almost the same as the original, and the surface areas are maintained at very high levels (ie. BET areas are more than 40 m^2g^{-1}). However, the unpromoted skeletal Cu catalysts, under the same test conditions, suffer large losses in surface areas (from 25 m^2g^{-1} to less than 8 m^2g^{-1}) and the Cu crystallite sizes increase to more than two times that of the original.

2.   Chromia promoted skeletal Cu catalyst, after thermal pretreatment at 623K for 1 hour, has no loss of the Cu surface area, whilst that for an unpromoted skeletal Cu catalyst and a commercial Cu chromite catalyst suffer losses of Cu areas of 3 m^2g^{-1} and 4 m^2g^{-1} respectively due to Cu sintering.

3.   The activity and stability of skeletal Cu catalysts, after promotion with chromia, are improved significantly. No obvious deactivation is observed for a chromia promoted skeletal Cu catalyst for either dehydrogenation of methanol or hydrogenolysis of methyl formate during 7 hours on stream at 453K. However, an unpromoted skeletal Cu loses more than 85% its original activity during the same period for both reactions.

4.   The initial activities of a promoted skeletal Cu catalyst for both the reactions of methanol dehydrogenation and methyl formate hydrogenolysis are much higher than that of a commercial Cu-chromite (Harshaw Cu-0203T) catalyst and both catalysts show almost no deactivation due to surface fouling.

ACKNOWLEDGMENT

The support of this project by the Australian Research Council is gratefully acknowledged.

REFERENCES

1.   T. Turek, D.L. Trimm and N. W. Cant, The catalytic hydrogenolysis of esters to alcohols. *Catal. Rev.-Sci. Eng.*, **36**(4): 645-683, 1994.
2.   D.S. Brands, E.K. Poels, T.A. Krieger, O.V. Makarova, C. Weber, S. Veer and A. Bliek, The relation between reduction temperature and activity in copper catalysed ester hydrogenolysis and methanol synthesis. *Catal. Lett.*, **36**:175-182, 1996.

3.   D.S. Brands, E.K. Poels, A. Bliek, Ester hydrogenolysis over promoted Cu/SiO_2 catalysts. *Appl. Catal.* A: General, **184**: 279-289, 1999.

4.   G.H. Xu, Y. C. Li, Z. H. Li and H. J. Wang, Kinetics of the hydrogenation of diethyl oxalate to ethylene glycol. *Ind. Eng. Chem. Res.*, **34**:2371-2378, 1995.

5.   D. J. Thomas, Studies of the hydrogenolysis of dimethylsuccinate and diethyloxalate over copper based catalysts. Ph.D. Thesis, The University of New South Wales, Australia, 1990.

6.   T.S. Franczyk, Process to prepare amino carboxylic acid salts. U.S. patent No. 5,292,936, 1994.

7.   Urano, Yoshiaki, Kadono, Yudio, Goto and Takakiyo, Process for producing aminocarboxylic acid salts. U.S. Patent No. 5,220,055, 1993.

8.   O. Gomes, Jose, M. Ramon, Juan, S. Sanchez, D.D. Zori, Asuncion, Procedure for obtaining derivatives from acetic acid. U.S. patent No. 5,225,592, 1993.

9.   Goto, Takakiyo, Yokoyama, Hiromi, Nishibayashi, Hideyuki, Method for manufacture of amino-carboxylic acid salts. U.S. patent No. 4,782,183, 1988.

10. Takasaki Seiji, Takayashiki Kazuhiro, Imachi Yoshimi, Jpn. Kokai Tokkyo Koho (1997), 7 pp. CODEN: JKXXAF JP 09155195 A2 19970617 Heisei. CAN 127:67659 AN 1997:453955.

11. Virgil Jorge Gustavo, Ruiz Marta del Carmen, Eur. Pat. Appl. (2001), 13 pp. CODEN: EPXXDW EP 1067114 A1 20010110 CAN 134:71897 AN 2001:28608.

12. A. Mušič, J. Batista, J. Levec, Gas-phase catalytic dehydrogenation of methanol to formaldehyde over ZnO/SiO2 based catalysts, zeolites, and phosphates. *Appl. Catal.* A: General **165**: 115-131, 1997.

13. L. Domokos, T. Katona, Á. Molnár, A. Lovas, Amorphous alloy catalysis VIII. A new activation of an amorphous $Cu_{41}Zr_{59}$ alloy in the transformation of methyl alcohol to methyl formate. *Appl. Catal.* A: General **142**:151-158, 1996.

14. E.D. Guerreiro, O.F. Gorriz, G. Larson, L.A. Arrúa, Cu/SiO_2 catalysts for methanol to methyl formate dehydrogenation – A comparative study using different preparation techniques. *Appl. Catal.* A: General **204**:33-48, 2000.

15. E.D. Guerreiro, O.F. Gorriz, J. B. Rivarola, L.A. Arrúa, Characterization of Cu/SiO_2 catalysts prepared by ion exchange for methanol dehydrogenation. *Appl. Catal.* A: General **165**:259-271, 1997.

16. T. Matsuda, K. Yogo, C. Pantawong, E. Kikuchi. Catalytic properties of copper-exchanged clays for the dehydrogenation of methanol to methyl formate. *Appl. Catal.* A: General **126**:177-186, 1995.

17. J. W. Evans, N. W. Cant, D.L. Trimm and M.S. Wainwright, Hydrogenolysis of ethyl formate over copper-based catalysts, *Appl. Catal.*, **6**:355-362, 1983.

18. J. W. Evans, P. Casey, M.S. Wainwright, D. L. Trimm and N. W. Cant, Hydrogenolysis of alkyl formates over a copper chromite catalyst. *Appl. Catal.*, **7**:31-41, 1983.

19. D.M. Monti, M.S. Wainwright, D.L. Trimm and N.W. Cant, Kinetics of vapor-phase hydrogenolysis of methyl formate over copper on silica catalysts, *Ind. Eng. Chem. Prod. Res. Dev.*, **24**: 397-401, 1985.

20. J. W. Evans, M.S. Wainwright, N. W. Cant and D. L. Trimm, Structural and reactivity effects in the copper-catalyzed hydrogenolysis of aliphatic esters. *J. Catal.*, **88**:203-213, 1984.

21. M. S. Natal Sanitiago, m.A. Sánchez-Castillo, R.D. Cortright and J. A. Dumesic, Catalytic reduction of acetic acid, methyl acetate, and ethyl acetate over silica-supported copper. *J. Catal.*, **193**:16-28, 2000.

22. D. M. Monti, N.W. Cant, D. L. Trimm and M. S. Wainwright, Hydrogenolysis of methyl formate over copper on silica. I. Study of surface species by in situ infrared spectroscopy. *J. Catal.*, **100**:17-27, 1986.

23. M. A. Köhler, M.S. Wainwright, D.L. Trimm and N.W. Cant, Reaction kinetics and selectivity of dimethyl succinate hydrogenolysis over copper based catalysts. *Ind. Eng. Chem. Res.*, **26**(4):652-656, 1987.

24. T. Turek, D.L. Trimm, D. StC. Black, N. W. Cant, Hydrogenolysis of dimethyl succinate on copper-based catalysts. *Appl. Catal.* A: General, 116:137-150, 1994.

25. J. W. Evans, S. P. Tonner, M. S. Wainwright, D. L. Trimm and N. W. Cant, Evaluation of a 2 stage route to methanol. "Proc. of the 11th Australian Conference on Chemical Engineering", Brisbane, Australia, 1983, pp509.

26. S. P. Tonner, D. L. Trimm and M. S. Wainwright, Dehydrogenation of methanol to methyl formate over copper catalysts. *Ind. Eng. Chem. Prod. Res. Dev.*, **23** (3):384-388, 1984.

27. N.I. Onuoha and M.S. Wainwright, Kinetics of the hydrolysis of acrylonitrile to acrylamide over Raney[®] copper. *Chem. Eng. Commun.*, **29**:1-13, 1984.

28. J.C. Lee, D.L. Trimm, M.S. Wainwright, N.W. Cant, M.A. Kohler and N.I. Onuoha, Catalyst Deactivation 1987, B. Delmon and G.F. Froment eds., Stud. Surf. Sci. Catal., Elsevier Science Publishers B.V., Amsterdam, 34:235-243, 1987.

29. N.I. Onuoha and M.S. Wainwright, A.D. Tomsett and D.J. Young, Preparation and properties of Raney[®] copper foraminate catalysts. *J. Catal.*, **91**:25-35, 1985.

30. D.J. Thomas, D.L. Trimm, M.S. Wainwright and N.W.Cant, Modeling of the kinetics of the hydrogenolysis of dimethyl succinate over Raney copper, *Chem. Eng. Process.* **31**:241-245, 1992.

31. D.J. Thomas, M.R. Stammbach, N.W. Cant, M.S. Wainwright and D.L. Trimm, Hydrogenolysis of dimethyl succinate over Raney[®] copper catalysts. *Ind. Eng. Chem. Res.*, **29**:204-208, 1990.

32. M.A. Köhler, N.M. Cant, M.S. Wainwright and D.L. Trimm, The mechanism of the catalytic chemistry of ester hydrogenolysis on copper surfaces. *Proc. Int. Congr. Catal.*, **9**:1043-1050, 1988.

33. L. Ma and M.S. Wainwright, Development of skeletal copper-chromia catalysts I. Structure and activity promotion of chromia on skeletal copper catalysts for methanol synthesis. *Appl. Catal.* A: General, **187**:89-98, 1999.

34. L. Ma, D.L. Trimm and M.S. Wainwright, Structural and catalytic promotion of skeletal copper catalysts by zinc and chromium oxides. *Topics in Catal.*, **8**:271-277, 1999.

35. L. Ma, D.L. Trimm and M. S. Wainwright, Promoted skeletal copper catalysts for methanol synthesis. *Proc. of 12ᵗʰ International Symposium on Alcohol Fuels, Sept. 21-24*, Beijing, China, 1998, pp1-7.

36. J.W. Evans, M.S. Wainwright, A. Bridgewater, D.J. Young, On the determination of copper surface area by reaction with nitrous oxide. *Appl. Catal.* **7**:75-83, 1983.

37. A.J. Smith, L. Ma, T. Tran and M.S. Wainwright, Effect of chromate additive on the kinetics and mechanism of skeletal copper formation. *J. Appl. Electrochem.*, **30**:1097-1102, 2000.

38. L. Ma, A.J. Smith, T. Tran and M.S. Wainwright, Development of skeletal copper chromia catalysts II. Kinetics of leaching of aluminum and chromia deposition. *Chemical Engineering and Processing*, **40**:59-69, 2001.

39. C.J. Jiang, Studies of the production of hydrogen from methanol steam reforming at low temperatures. PhD Thesis, The University of New South Wales, Australia, 1992, pp168.

40. J.W. Evans, Studies of the copper catalysed hydrogenolysis of alkyl esters. PhD Thesis, The University of New South Wales, Australia, 1983, pp248.

19

Precious Metal Promoted Raney® Ni Catalysts

Stephen R. Schmidt
W.R. Grace & Co., Columbia, Maryland, U.S.A.

Setrak K. Tanielyan
Center for Applied Catalysis, Seton Hall University, South Orange, New Jersey, U.S.A.

ABSTRACT

A class of enhanced Raney® catalysts has been prepared using precious metal ("PM") dopants at efficient low levels (typically 0.25 wt% as PM). In aromatic nitro-to-amine reductions (exemplified here by p-nitrotoluene to p-toluidine in 95% EtOH), the modified catalysts show significantly greater initial activity than conventional Raney catalysts in batch reactions. The activity advantage is maintained through multi-batch recycling or extended continuous use. Increasing PM loading to the still modest 0.5-1.0% PM-on-Ni range lowers the threshold in H_2 pressure required for activity gains over unpromoted Ni.

X-ray photoelectron spectroscopy ("XPS") reveals that impregnation of Raney catalysts with basic-pH salts, e.g. ammine complex types containing Pd or Pt, rather than previously reported acid salts, modifies the placement of PM within the Ni catalyst's sponge structure. The PM placement is expressed as a surface/bulk ("S/B") ratio for the normalized dopant concentration. The resulting basic salt catalysts have lower S/B ratios and also generally perform better in activity and life tests than the acidic salt types.

INTRODUCTION

Raney® catalysts are used in a broad range of industrial hydrogenations. These include reductions of nitriles and dinitriles (e.g. for nylon intermediates), aldehydes (e.g. for sorbitol or alkane diols), olefins and alkynes (e.g. for monomer purification) and aromatic nitro compounds (e.g. for urethane intermediates).

Improving catalyst durability (delaying deactivation) is a worthy goal even in established processes. Among major industrial hydrogenation processes, nitroaromatic feedstocks are one particularly challenging type of chemical environment in which to sustain productivity. This partly stems from the nitro groups undergoing hydrogenolysis of N-O bonds, consuming 3 moles of H_2 per complete conversion of $-NO_2$ to $-NH_2$. The NO_2 groups and their subsequent reactive intermediates can deactivate (1) an initially zero-valent metal surface (e.g. Ni°) if this unusually high H_2 demand is not balanced by H_2 supply (the balance being known as 'hydrogen availability' (2)). Low hydrogen availability combined with amine-imine mixtures may also lead to higher order amines and coupling products (3), potential foulants of a catalyst surface. Therefore the nitroaromatic/Ni catalytic system is potentially self-poisoning. Although the desired reaction is relatively fast, in practice both initial conversion rate and catalyst life can depend on overcoming these inherent competing deactivation routes.

In a broader arena, base metal hydrogenation catalysts (e.g. Raney Ni) are often less durable than PM catalysts. The "noble" vs. "base" designation in this context can describe resistance to deactivation. In ranking net economic value across catalyst types, productivity/cost ratio is a key parameter. Less noble but less expensive Ni (e.g.) is evaluated against more noble, but more expensive Pd or Pt.

The primary objective for this study and related inventions (4) was to improve catalyst activity and durability over conventional Raney Ni. A guiding conceptual model proposed that a PM-doped Ni surface could be steadily supplied with chemisorbed hydrogen in a pressurized reactor. If Ni is thus maintained in hydrogen-rich form, sustainable higher activity should result (especially in reaction mixtures that deactivate hydrogen-depleted Ni surfaces). Achieving this affordably requires efficient use of expensive additives, so this was imposed as a constraint: the use of PM loadings much lower than the 2-5% typically dispersed on oxide or carbon supports. A secondary objective was to characterize the placement of PM on and within the Ni catalyst 'sponge' to yield a structural basis for observed correlations of activity with preparation method.

Compared to supported forms of Ni, Raney Ni is an especially convenient catalyst substrate for doping with promoter metals, in that its freshly-prepared surface holds adsorbed hydrogen. This hydrogen can be an *in situ* reducing agent for metal cations impregnated onto the Ni surface, maybe eliminating need for subsequent reduction under applied hydrogen pressure, unlike some supported forms of Pd, Pt or Ru.

Because typical Raney catalyst is immersed in a somewhat alkaline aqueous medium, the acid/base character of the chosen dopant metal salt is a key variable. Alkaline additives are expected to cause less abrupt deposition of dopant metal within the base metal sponge, i.e. gradual impregnation more controlled by reaction with the sponge metal surface. An additive less compatible with high pH solution may undergo premature reaction. Use of an acidic salt is expected to cause acid-base neutralization and thus faster dopant deposition, more controlled by pH swings than by redox with H-Ni. This might, at an anticipated extreme, even lead to dopant metal hydroxide precipitated as a distinct phase, rather than the intended outcome of metal being dispersed within, and bound to, the porous substrate.

PRIOR WORK

A common feature of much earlier work was use of strongly acidic precious metal chloride solutions (e.g. chloroplatinic acid (5)) nearly always applied to the Ni catalyst just before use. Both ageing and washing of the doped catalyst were generally avoided, rationalized as preventing 'burial'(6) of the Pt dopant at an inaccessible depth within the Ni catalyst structure. It is now known from our study that the most active and durable PM/Raney Ni catalysts can have a structure in which a greater fraction of the PM dopant is 'buried', i.e. more uniformly distributed within the Ni sponge, than via these acid salt precedents. Further, these new catalysts when used industrially demonstrate significant shelf life of at least several months, much like more standard Raney® catalysts.

Basic salt solutions formed *in situ* during catalyst activation (7) or formed just before or during impregnation (8) have also been reported, but with only minimal evidence of effect of preparation method on catalyst structure or performance.

EXPERIMENTAL

Catalyst Preparations and Characterizations

Ni substrate. All catalyst batches in Tables 1 and 2 were based on a W.R. Grace-produced powdered binary Ni-Al precursor alloy with 42 wt.% Ni. Those in Table 3 were based on a 41 wt.% Ni alloy. Alloy was gradually added to a 30 wt% NaOH solution under agitation, (molar ratio of NaOH to Al in the alloy of ~ 1.8 to 1). Temperature during alloy addition and a subsequent 4 hour digestion stage was maintained at ~80°C. This was the common origin of each modified ("doped") catalyst plus the baseline (Ni only) catalysts. The

Table 1 Catalyst Preparation and Characterization: Basic Salt Dopants

Sample	PM Input	Dopant (PM) Salt — Salt	PH	Pre-Doped Slurry pH	APS (u)	Bulk Analysis (ICP wt%) PM	Ni	Al	PM/Ni (at)	Surface Analysis PM/Ni (XPS)	S/B Ratio
1		None	-	-	24	-	-	-		-	-
2A	~1%	$(NH_3)_4PdCl_2 \cdot H_2O$	9.1	-	24	1.17	92.9	5.6	0.00695	0.17	24
2B	0.5%	$(NH_3)_4PdCl_2 \cdot H_2O$	9.1	11	28	0.53	93.9	5.3	0.00311	0.10	32
2C	0.5%	$(NH_3)_4PdCl_2 \cdot H_2O$	9.1	11	26	0.46	93.8	5.4	0.00270	-	-
2D	0.25%	$(NH_3)_4PdCl_2 \cdot H_2O$	9.1	11	26	0.25	94.0	5.5	0.00147	0.06	41
2E	0.25%	$(NH_3)_4PdCl_2 \cdot H_2O$	9.1	8	23	0.26	94.3	5.1	0.00152	0.08	53
2F	0.25%	$(NH_3)_4PdCl_2 \cdot H_2O$	9.1	9	21	0.28	93.8	5.7	0.00165	0.03	18
2G	0.25%	$(NH_3)_4PdCl_2 \cdot H_2O$	9.1	10	21	0.25	93.8	5.7	0.00147	0.04	27
2H	0.25%	$(NH_3)_4PdCl_2 \cdot H_2O$	9.1	11	26	0.22	92.9	6.6	0.00131	0.03	23
2I	0.25%	$(NH_3)_4PdCl_2 \cdot H_2O$	9.1	12	21	0.27	93.5	5.9	0.00159	0.05	31
2J	0.25%	$(NH_3)_4PtCl_2 \cdot H_2O$	8.1	11	23	0.27	94.3	5.1	0.00158	-	-
2K	0.125%	$(NH_3)_4PdCl_2 \cdot H_2O$	9.1	11	24	0.14	93.6	6.0	0.00083	0.02	24
3	0.25%	$(NH_3)_4PdCl_2 \cdot H_2O$	9.1	11	33	0.25	89.3	10.2	0.00155	0.08	52
4	0.25%	$(NH3)_4Pd(OH)_2 \cdot H_2O$ 13.7		11	20	0.22	93.5	6.0	0.00130	0.15	116

Table 2 Catalyst Preparation and Characterization: Acidic Salt Dopants

Sample	PM Input	Dopant (PM) Salt — Salt	PH	Pre-Doped Slurry pH	APS (u)	ICP Analysis (wt%) PM	Ni	Al	PM/Ni (at)	Surface Analysis PM/Ni (XPS)	S/B Ratio
5A	1%	$PdCl_2$	-0.2	9	27	0.92	93.8	5.0	0.00541	0.63	116
5B	0.5%	$PdCl_2$	-0.2	11	20	0.51	92.3	6.9	0.00305	0.27	89
5C	0.25%	$Na_2Pd(II)Cl_4$	4.0	11	21	0.29	93.8	5.6	0.00171	0.17	100
5D	0.25%	$Na_2Pd(IV)Cl_6$	2.0	11	24	0.28	93.5	5.9	0.00165	0.25	151
5E	0.25%	$Pd(NO_3)_2$	-0.7	11	21	0.28	93.7	5.7	0.00165	0.17	103
5F	0.25%	$PdCl_2$	-0.2	8	20	0.25	92.9	6.6	0.00149	0.18	121
5G	0.25%	$PdCl_2$	-0.2	9	22	0.26	93.9	5.5	0.00153	0.25	164
5H	0.25%	$PdCl_2$	-0.2	10	22	0.25	93.9	5.6	0.00147	0.17	116
5I	0.25%	$PdCl_2$	-0.2	11	28	0.27	93.3	6.1	0.00160	0.13	81
5J	0.25%	$PdCl_2$	-0.2	12	21	0.25	94.4	5.1	0.00146	0.13	89
5K	0.125%	$Pd\,Cl_2$	-0.2	11	24	0.124	95.0	4.6	0.00072	0.08	111

intermediate or 'crude' nickel catalyst was water washed to a selected pH of between about 8 and 11 before PM doping (see Table 1 for listing of variations). Ni-only baseline catalysts were washed to pH 9.

PM doping: The basic salts employed were mainly of the ammine complex type with chloride counter-ion, e.g. $(NH_3)_4PdCl_2.H_2O$ except for sample 4 which used the highly basic salt $(NH_3)_4Pd(OH)_2-H_2O$. The acidic salts for comparison were of the simple chloride type, e.g. $PdCl_2$ or partially neutralized type such as Na_2PdCl_4. The various doped catalysts were formed by adding to an agitated Ni catalyst slurry (~15% solids) a dopant salt equivalent to PM loading of between 0.1 and 1.0 wt.% of the slurry's nickel catalyst weight. A minimal amount of (typically) ~50 mL of water was used to pre-dissolve the dopant salt, which was then added all at once to (typically) ~500 mL of catalyst slurry. After 30 minutes of contact to ensure complete impregnation, agitation was stopped, and solids allowed to settle. Used doping solution was decanted and analyzed by ICP for precious metal content, resulting each time in immeasurably low levels, i.e. indicating 100% uptake of dopant by Ni. All the doped catalysts were decant-washed with water (at ~45° C) to a pH of about 9.

Tables 1–3 provide the data and description with respect to each of the series of samples produced. Besides parameters already described, this table also includes

- pH of dopant salt as ~3 wt.% solution
- pH of catalyst slurry before doping treatment; lower pH = further washing

Table 3 Catalyst Preparation and Characterization: New Series for Pressure and Dopant Type/Level Testing

Sample	Dopant (PM) Salt			Pre-Doped	
	PM Input	Salt	PH	Slurry pH	APS (u)
	-	None	-	-	16
7A	1%	$(NH_3)_4PdCl_2.H_2O$	9.1	11	16
7B	0.5%	$(NH_3)_4PdCl_2.H_2O$	9.1	11	16
7C	0.5%	$(NH_3)_4PtCl_2.H_2O$	8.1	11	15
7D	0.25%	$(NH_3)_4PdCl_2.H_2O$	9.1	11	16
7E	0.25%	$(NH_3)_4PdCl_2.H_2O$	9.1	9	15
7F	0.25%	$(NH_3)_4PtCl_2.H_2O$	8.1	11	16

Sample	Bulk Analysis (ICP wt%)				Surface Analysis	
	PM	Ni	Al	PM/Ni (at)	PM/Ni (XPS)	S/B Ratio
6	-	95.3	4.4	-	NA	
7A	0.91	94.1	4.7	0.00533	NA	
7B	0.47	94.5	4.7	0.00274	NA	
7C	0.54	95.0	4.2	0.00171	NA	
7D	0.26	95.6	3.8	0.00149	0.03	20
7E	0.24	95.4	4.1	0.00138	0.03	22
7F	0.22	95.0	4.5	0.00070	NA	

- Average particle size (mean diameter) = "APS" from Malvern light scattering method after 30 seconds ultrasound dispersion
- Bulk chemical analysis of doped catalyst by ICP
- Surface Analysis (XPS) of doped catalyst

Chemical Analysis: Total chemical assays of catalysts were determined by Inductive Coupled Plasma-Atomic Emission Spectroscopy analysis ("ICP") on samples dissolved in a mixture of $3HCl/1NHO_3$. The overall atomic ratio of precious metal to nickel was calculated from these "bulk" ICP results.

Surface Analysis: To learn the degree of penetration of the PM dopant within the porous Ni particles, X-ray Photoelectron Spectroscopy (XPS) was used to measure a complementary feature: average dopant concentration at the catalyst particle's "surface" (outer 'shell' of about 50 Å, the escape depth for photoelectrons in this context). For each measurement, about 0.5 g of water-wet catalyst was dried under flowing helium at a temperature of 130°C in a U-shaped Quantasorb(TM) sorptometry sample tube assembly. The dried sample was then transported under He in the closed tube assembly to a glove box serving as an antechamber to the XPS instrument, and there prepared further for analysis. In the glove box environment's Ar atmosphere, moisture content was no higher than 0.40 ppm and oxygen content was generally 0.00 ppm. The material was loaded as a thin layer onto double-sided tape mounted to a 1-inch diameter stainless steel stub and transferred into the XPS instrument *in vacuo* (10^{-6} torr). The final vacuum achieved in the analysis chamber was 10^{-8} to 10^{-9} torr.

Spectra were obtained using a PHI 5600 ESCA system from φ Physical Electronics using an aluminum x-ray source operating at 14.8 kV/25 mA energy and the detector positioned at 45° relative to the material surface being analyzed.

A 60 minute detailed scan on selected elements was performed with an energy resolution of 0.125 eV. Curve-fitting and atomic concentration determinations from the spectral data were done with software package (MULTIPAK v2.2a). Sensitivity factors for each element stored within the software were used in the atomic concentration calculations.

The Surface/Bulk ("S/B") ratio was calculated as follows (e.g., with Pd and Ni as dopant and base metal, respectively):

$$\text{S/B ratio} = \frac{\text{surface Pd/Ni}}{\text{bulk Pd/Ni}}$$

$$= \frac{[(\text{XPS Pd atomic concentration}/(\text{XPS Ni atomic concentration})]}{[(\text{ICP bulk \%Pd/atomic wt Pd})/(\text{ICP bulk \%Ni/atomic wt. Ni})]}$$

For a material in which the dopant was uniformly distributed through the Ni structure, the expected S/B value would be 1.00. Materials with relatively high

S/B values have a greater fraction of dopant concentrated near the outside of particles, while relatively low S/B indicates greater penetration of the dopant.

HYDROGENATION REACTOR TESTING

Outline:

a) batch activity and recycle stability testing for three representative catalysts at 140° C and 400 psig
b) batch activity testing at 125° C and a range of pressures and PM loadings
c) batch activity and recycle stability testing for selected catalysts from (b) at 125° C and 200 vs. 400 psig (for (b) and (c), variations among catalysts included PM type and loading level).
d) CSTR stability testing at 140° C/400 psig for selected 0.25% Pd catalysts

Batch Hydrogenation: *General methods for a-c.* For batch reaction testing the equipment used was an Autoclave Engineers Bench Top EZE Seal Reactor in The Center for Applied Catalysis at Seton Hall University. It is equipped with the "in house" customized pressure drop sensing module pictured in Figure 1. The module is divided into three sections corresponding to feed, high-pressure and low pressure respectively. The ports for gas delivery and vacuum are located in the feed section. The high-pressure section consists of a forward pressure regulator 4, varying volume ballast reservoir 5 and pressure transducer 3. The low-pressure section is in line with the reactor and its pressure is being monitored using a pressure transducer 3.

The hydrogenation reaction progress was monitored incrementally as gas uptake at constant reaction pressure, by measuring pressure drop in the calibrated ballast reservoir. The parameters of reaction time, pressure, temperature and ballast pressure were recorded at the rate of 12 points/minute during the first 10 minutes of reaction and then at increments of each 1 percent pressure drop in the ballast reservoir. The data obtained were plotted versus time and the reaction rates were calculated from the slope in the linear portion of the hydrogen uptake. At the completion of each reaction, aliquots of reaction solution were taken and analyzed by gas chromatography-mass spectrometry. The only materials identified were the starting 4-nitrotoluene, the 4-methylaniline product and a byproduct N-ethyl 4-methyl aniline.

a) Batch 140°/400 psig: Catalysts 1 (Ni), 2G (basic salt) and 5F (acidic salt) were tested. In each reaction, 65 mg of wet catalyst was transferred into the reactor, washed with 12 g of 95% ethanol/5% water (Pharmco Products) and then the reactor was sealed. The system was evacuated and filled with hydrogen followed by the addition via a gas-tight syringe of 28 g of an ethanol solution

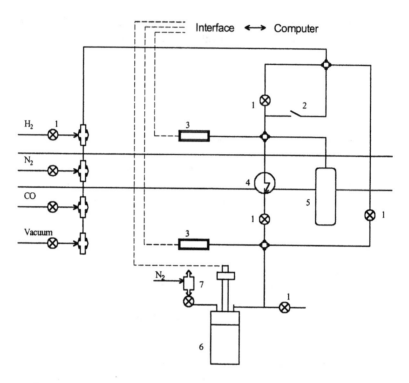

Figure 1 Component description for Automated Bench top EZE Seal Reactor: (1) two way shut-off valve, (2) DC actuated electromagnetic valve, (3) pressure transducer, (4) pressure regulator, (5) ballast tank, (6) autoclave, (7) – substrate injection reservoir.

containing 3.50 g of 4-nitrotoluene. The reactor was pressurized with hydrogen to the gauge pressure of 400 psi and then heated to a temperature 140°C. When the temperature/pressure parameters were reached, stirring (1300 rpm) and data acquisition were simultaneously initiated. After gas absorption ceased the reactor was cooled and the liquid phase extracted under pressure. For recycling, catalyst was left in the reactor and a new portion of feed was injected before continuing. Batch activity for each catalyst through five cycles is shown in Table 4.

b-c) Pressure and PM loading dependence: Using the catalysts of Table 3, the methods of section (e) were followed but using 1.00 g of p-NT in 25 g of solvent, at 125° C and at a series of pressures from 100 to 500 psig in 100 psi increments. Catalyst weight was reduced to 24 mg, which equates to a loading of 2.4%. These changes were all aimed at reducing absolute rates and increasing power of the testing to discriminate between catalysts.

Table 4 Catalyst Activity (mmol H2) in Batch Recycle
140°/400 psig/65 mg(1.86% loading)

Cycle	Cumulative mmol pNT/g	Sample 2G (basic Pd) min-1	(min-g)-1	Sample 5F (acidic Pd) min-1	(min-g)-1	Sample 1 (Ni) min-1	(min-g)-1
1	393	9.6	148	7.4	114	5.5	85
2	786	8.9	137	5.0	77	3.9	60
3	1179	5.0	77	2.9	45	2.6	40
4	1572	3.7	57	3.1	48	2.3	35
5	1965	3.2	49	3.2	49	2.2	34

Recycle stability: Selected catalysts from Table 3 were also recycled through 3 to 5 batch runs at the 200 psig and 400 psig pressure levels, as described in (e) above.

Results for (b) and (c) are summarized in Table 5a–5b.

Table 5a Batch Activity (mmol H2/min) w/Recycle at 125C, Indicated P
24 mg catalyst (2.4% loading); All catalysts are of basic salt type.

P(psig)	Cycle	7a 1 %Pd	7b 0.5%Pd	7c 0.5%Pt	7d 0.25%Pd	7f 0.25 Pt	6 Ni only
100	1	1.8	1.8		1.2		1.0
200	1	4.0	3.8	2.7	2.7	2.8	1.9
	2		2.5	2.3		2.4	2.0
	3		2.0	2.1		2.0	1.4
300	1	4.2	4.8		3.7		2.8
400	1	4.6	5.4		3.6	4.0	3.2
	2	3.5	4.4		3.4	3.2	2.5
	3	4.1	4.2		2.5	3.2	1.6
	4				2.6	2.3	1.9
	5				2.5	2.1	1.8
500	1	4.1	4.6		3.8		3.7

Table Vb Selectivity to PAT corresponding to batch runs of Table Va.

		Catalyst Number and Nominal Dopant Level					
		7a	7b	7c	7d	7f	6
P(psig)	Cycle	1 %Pd	0.5%Pd	0.5%Pt	0.25%Pd	0.25 Pt	Ni only
100	1						
200	1		91.1		91.4	91.9	94.8
	2						
	3						
300	1						
400	1	95.8	95.7		93.6	95.2	96.6
	2					95.6	
	3					94.7	
	4						
	5						
500	1						

d) CSTR stability testing: The reactor system used for continuous operation is at Seton Hall Univ./CAC and is depicted in Figure 2. In general the feed solution of 100 mg/mL in 95% EtOH was de-aerated and kept under N_2 in a reservoir bottle 1, ready for delivery by a Waters 510 HPLC pump 3.1 into the autoclave 7. Catalyst (65 or 36 mg as specified below), 20 mL of reactor-charging/purging carrier solvent, and 50 mL of feed solution were pressurized under hydrogen to 400 psig and heated to 140° C in the reactor. Stirring at 1300 rpm initiated the batch reaction that was used to reach essentially complete conversion (96% of theory based on hydrogen consumption). At that point inlet and outlet pumps 3.1 and 3.2 were started simultaneously. Level controller 10 interfaced to the controlling PC actuated outflow through valve 9. Samples were withdrawn every 30 minutes for GC analysis.

Net hydrogen consumption was determined by correcting for loss of hydrogen dissolved in product and as two-phase mixture through the reactor outlet. This was done by measuring the difference between 'hydrogen in' and apparent 'hydrogen out' at sustained 100% conversion under the operating conditions employed. As in the batch testing above, hydrogen consumption was used to determine rate, and GC analysis was used to confirm conversion and determine selectivity.

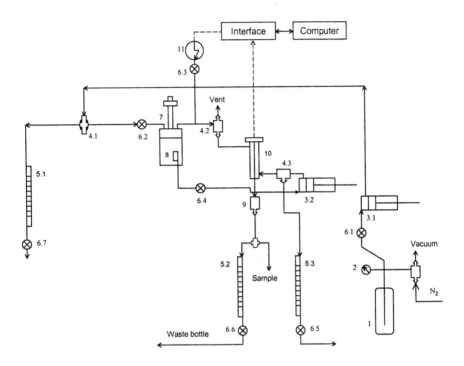

Figure 2 Component description for CSTRS: (1) – p-NT /ethanol storage bottle, (2) – pressure gauge, (3.1) – HPLC addition pump, (3.2) – extraction pump, (4.1-4.3) – three way valve, (5.1-5.3) – graduated cylinder, (6.1-6.7) – two way shut off valve, (7) – reactor, (8) – metal filter 0.5mkm, (9) DC actuated electromagnetic valve, (10) - level control unit, (11) – pressure regulator.

Desired feed flow rate was found iteratively, bracketing a range allowing convenient but meaningful discrimination between catalysts' levels of durability. Too high a flow rate (e.g. ~2 mL/min) deactivates catalysts prematurely by exceeding their ability to reduce p-NT, leading to rapid drop-off in % conversion. Too slow a flow rate (e.g. ~1 mL/min.) maintains long-term high conversions, delaying clear ranking of catalysts. Examples of these extremes are depicted in Figure 3 for Pd-doped catalyst 5F. An intermediate flow rate of 1.59 mL/min was then chosen. Combined with two different catalyst charges of 65 and then 36 mg, this allowed for fine-tuning the comparative test in terms of catalyst deactivation rate.

The three catalysts tested over 5 hours at these conditions were 5F (acidic salt 0.25%Pd), 1F (basic salt 0.25% Pd), and 1 (Ni baseline). Both their initial activities and indicators of deactivation (rates after extended runs times; also cumulative productivity) are listed in Table 4.

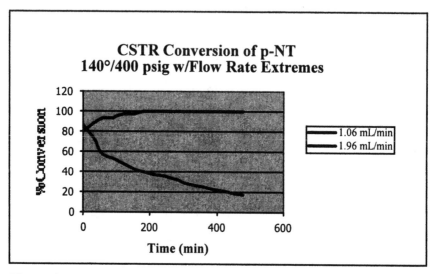

Figure 3

DISCUSSION

Structure: The S/B ratios of Tables 1–2 show that clearly separate regimes of dopant placement result from basic vs. acidic salt PM sources: usually S/B < 40 for moderately basic salts (ammine-PM-halide) and > 90 for most acidic ones. This seems to support our starting premise. We anticipated that a relatively small difference between salt solution pH and undoped catalyst slurry pH would yield lower S/B, i.e. less surface-enriched placement of the dopant within the catalyst sponge structure. (TEM comparisons attempting to corroborate this structural difference visually were somewhat inconclusive, due to the low levels of dopant present, and the resulting statistical obstacle to seeing enough 'islands' of PM against the 'sea' of Ni).

For extremely basic dopant salt (as used in catalyst 4) there is also evidence of instability when added to a mildly alkaline catalyst slurry. A high S/B of 116 resulted with this combination also, presumed due to pH 'shock' and hydrolysis. Thus there is suggestion of an optimum when both the catalyst slurry to be doped and the dopant solution are in roughly the pH 9-11 range. There is no clear or strong dependence of S/B on the pH of the Ni catalyst slurry prior to doping.

Although in every case observed the majority of the dopant is 'seen' within the 50 Å outer 'shell' (due to the electrochemical driving force for reducing the PM cations with H-Ni), the more extreme degree of surface-concentrated dopant for acid-salt preparations likens them more to 'eggshell' catalysts, structurally speaking. There is an implied slightly more uniform PM placement from basic salt preparations. It would be interesting to develop a method to selectively determine the dispersion of the PM. The S/B model predicts that PM dispersion would be greater for the basic salt type. In every case the measured surface area and CO chemisorption (9) of these catalysts fell into tight ranges typical of an undoped catalyst. The Pd or Pt observed in the XPS measurements (which also yielded the S/B results) was always 100% zero-valent. Thus any performance differences between types, as described below, appears unexplainable by oxidation states or total and metal (Ni + PM) surface areas.

Batch Activity: The initial activity and recycled activity results of Table 4 yield a ranking of 2G > 5F > 1, i.e. the low level doping with Pd makes a catalyst more active than the Ni substrate, and the basic salt type are initially more active, at these rather energetic test conditions. The absolute initial activities at 140°/400 psig were the highest observed overall in these studies, suggesting possible T-P synergy (3) but were not maintained through multiple cycles. The relative decline in activity across several cycles was actually fastest at this combination of relatively low catalyst:substrate loading (1.9%) with high temperature. Lower temperature and lower absolute amounts of catalyst and substrate were used in the remainder of the batch testing, at a slightly higher catalyst:substrate loading (2.4%). This was aimed at lower absolute rates but longer retention of activity.

The pressure- and PM loading-dependence matrix of Table 5a reveals several trends:

- For 0.25% PM loaded catalysts the maximum activity within this pressure range is not closely approached at below 300-400 psig
- For the higher PM loadings the threshold for highest activity is reached at lower pressure, between 200-300 psig
- 1% Pd loading adds no performance advantage over 0.5% Pd, possibly due to compromised Pd dispersion at the higher loading.
- Pt appears to be at least as effective a dopant as Pd at equal weight loadings, implying that the higher atomic wt. metal is superior on a per-atom basis.
- The most active PM-doped catalysts maintain an activity ratio advantage of roughly 1.5-2.5 over undoped Ni through 3-5 cycles. The advantage widens as Ni deactivates faster than PM/Ni

- The most efficiently-doped catalysts in this range (at 0.25% PM) may not immediately reveal an advantage over Ni but again the performance gap can increase over time.

The selectivity data corresponding to the above, listed in Table 5b, show possible evidence of worse hydrogen availability for the more active doped catalysts at 200 psig. This deficit (greater N-ethylation than for Ni) largely disappears at 400 psig where hydrogen availability would be expected to improve.

CSTR Testing: At the higher catalyst loading and thus lower weight space velocity, there is evidence of greater activity and durability for PM-doped catalysts taken as a group than for Ni. However, there is as anticipated little discrimination on those criteria between the two types of doped catalysts, due to the expected better hydrogen availability helping both types to perform adequately.

At the lower catalyst loading (36 mg) combined with flow rate still fixed as above, the effective weight space velocity nearly doubles, and the result is a clear discrimination among catalysts including acidic- vs. basic-salt-doped types. (At 180 minutes, conversion for the most active catalyst has dropped to ~88%, vs. 99% at the lower space velocity of Figure 3; the less active doped catalyst's conversion is at ~60%). Deactivation begins to overtake hydrogenation noticeably for both the Ni and acid-salt-doped catalysts such that they have steady state activities (Rs) of 1/3 to 1/2 that of the basic-salt-doped. Percent conversion and cumulative production fall off dramatically during the > 5 hrs duration of the experiments. As with the batch recycle testing at the same T and P, there is a performance difference correlated with the PM doping method.

FINAL OBSERVATIONS

Longer high-productivity catalyst life is expected for space velocities lower than those used in our CSTR comparative runs. That trend and the added constraint of safe heat removal explain why industrial reductions of nitroaromatics often use continuous reactors with modest feed flow rates and near 100% conversion in the reaction mixture at all times. (A contrasting, less desirable approach would use batch contact between full charges of nitro compound and catalyst).

Optimum selectivity was not the basis for the operating parameter decisions in this study, leaving potential for further improvement. Inherent selectivity differences between the dopant metals Pt and Pd in each reaction mixture are expected to impact this as well, although the primary function of the PM is believed to be indirect (maintaining the more prevalent active base metal sites in hydrogen-rich state).

Explaining the underlying causes for doping type-performance correlations is a goal of ongoing research. Speculated PM-to-Ni hydrogen spillover, bulk hydrogen storage, and reduced metal-hydrogen bond strengths (10) are areas for exploration.

There appears to be practical value in the general technique of doping sponge base metal catalysts with low levels of precious metals. Efficient, controlled-placement PM doping onto Raney metals can improve process economics, especially with recently observed volatile, high PM prices in effect. Sufficient PM loading onto Ni (and Cobalt, Cu, etc.), if combined with moderate applied hydrogen pressure can be an attractive alternative to highly-PM-loaded catalysts. Reductions and couplings of substrates other than nitro compounds are also anticipated applications; carbonyls, carbonyls plus amines (reductive alkylation), and C-C unsaturates are among them.

ACKNOWLEDGMENTS

The authors acknowledge the technical support of Robert Augustine, Yujing Gao and Jayesh Nair at Seton Hall University, and Linda Wandel, Douglas L. Smith, Ron Smaw and Jane Dowell at W.R. Grace.

REFERENCES

1) H. Debus, J.C. Jungers. *Bull. Soc. Chim. Fr.* 1959, 785 (1959).
2) P. Rylander in <u>Hydrogenation Methods</u>, Academic Press (1985) p. 31.
3) J. Wisniak, M. Klein. Reduction of nitrobenzene to aniline. *Ind. Eng. Chem. Prod. Res. Dev.* **23**, 44-50, (1984).
4) S.R. Schmidt. US Patent 6,309,758, (2001) and others pending.
5) E. Lieber, G.B.L. Smith. Reduction of Nitroguanidine.VI. Promoter action of platinic chloride on Raney Nickel catalyst *J.A.C.S.* **58**, 1417 (1936).
6) J.R. Reasenberg, E. Lieber. G.B.L Smith. The promoter effect of platinic chloride on Raney nickel. II. Effect of alkali on various groups. *J.A.C.S.* **61**, 384 (1939).
7) Brit. Pat. 1,119,512 (1968).
8) Levering,D.R.; Morritz,F.L.; Lieber,E. The promoter effect of platinic chloride on Raney nickel. I. General effects on type W-6 catalyst. *J.A.C.S.* **72**, 1190 (1950).
9) S.R. Schmidt. Surfaces of Raney® Catalysts, in Eds. Scaros and Prunier, <u>Catalysis of Organic Reactions</u>, M. Dekker (1995), pp. 45-59.
10) N. Khan, J.G. Chen. Low-Temperature HDS Activity of Ni/Pt(111) Bimetallic Surfaces in Univ. of Del./<u>CCST Annual Research Report</u> (2001).

20

Chemoselective Hydrogenation of Unsaturated Ketones in the Presence of Copper Catalysts

Nicoletta Ravasio, Rinaldo Psaro, and Federica Zaccheria
Centro CNR CSSCMTBSO and Dipartimento di Chimica I.M.A., Università di Milano, Milano, Italy

Sandro Recchia
Facoltà di Scienze CCFFMM, Università dell'Insubria, Como, Italy

In the fine chemicals industry, reduction of carbonyl groups mainly rely on the use of complex metal hydrides, sodium dihydrobis-(2-methoxyethoxy)-aluminate, commercialized as RedAl® or Vitride® being one of most used.

Catalytic hydrogenation of aliphatic ketones can be carried out in the presence of noble metals, usually Pt and Ru at 25-60°C and 1-5 atm H_2, or in the presence of Cu chromites and Ni catalysts. However, these last ones require about 50 atm of pressure and temperatures higher than 100°C [1]. Moreover, as the carbonyl group reduction ($\Delta H= -12$ kcal/mol) is thermodynamically unfavoured with respect to C-C double bond hydrogenation ($\Delta H=-30$ kcal/mol), problems of chemoselectivity arise in the hydrogenation of unsaturated ketones, and despite the potential for applications to large scale reduction of polyfunctional carbonyl-containing compounds, hydrogenation catalysts rigorously chemoselective for carbonyl reduction over alkene hydrogenation are still very rare. One of these is dimethylphenylphosphine stabilized copper(I) hydride [2].

During our studies on the use of supported copper catalysts in selective hydrogenation reactions, we found that these systems can be effectively applied to the chemoselective hydrogenation of α,β-unsaturated ketones giving quantitative yields in the corresponding saturated ketone also when another olefinic bond is present in the molecule [3].

Here we wish to report that chemoselective hydrogenation of the carbonyl group can be achieved in the presence of copper catalysts by using 2-propanol as hydrogen donor or as a solvent. 8% Cu/Silica catalysts prepared by a non-conventional technique and reduced at 270°C can be used. Molecular hydrogen addition is also possible, but for open chain substrates messy reactions are observed.

EXPERIMENTALS

Catalyst Preparation: Copper catalysts, all with a metal loading of 7-9 % in weight, were prepared by the chemisorption-hydrolysis method as already reported [3] by adding the support to a solution containing $[Cu(NH_3)_4]^{2+}$ and slowly diluting the slurry with water. The solids were separated by filtration, washed with water, dried overnight at 120°C and calcined in air at 350°C for 4 hours. The catalysts were reduced at 270°C with H_2 at atmospheric pressure, removing the water formed under reduced pressure, before the hydrogenation reaction.

Two kind of silica were used as the supports, namely a silica gel from Grace Davison (BET= 320 m^2/g, PV= 1.75 ml/g) and non porous Aerosil 380 from Degussa (BET= 380 m^2/g). The catalysts obtained showed the following features:
Cu/SiO$_2$ gel (hereafter referred to as Cu/SiO$_2$):
    BET= 263 m^2/g , PV= 0.78 ml/g, $S_{Cu(0)}$= 129 m^2/g_{Cu}
Cu/Aerosil : BET= 308 m^2/g, $S_{Cu(0)}$= 117 m^2/g_{Cu}

Catalytic Reaction: The substrates (100 µl) were dissolved in n-heptane or 2-propanol (8 ml) and the solution transferred under H_2 to a glass reaction vessel where the catalyst (100 mg) had been previously reduced. Reactions were carried out at 90°C and atmospheric pressure with magnetic stirring (700 rpm). Reaction mixtures were analyzed by GC-MS (HP 5971 series) and by GC (mesitylene as internal standard) using a DB-225 (30 x 0.25) capillary column for geranylacetone and SP-2560 (100 m) for 6-methyl-5-hepten-2-one and 5-hexen-2-one. After completion, the catalyst was filtered off, the solvent removed and the reaction mixture analyzed by ^{1}H-NMR (Bruker 300 MHz).

RESULTS

We have investigated the reactivity of three open chain unsaturated ketones, namely geranylacetone **1**, 6-methyl-5-hepten-2-one **2** and 5-hexen-2-one **3**.

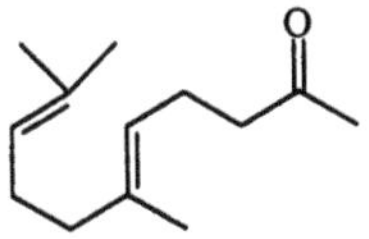

1	**2**	**3**

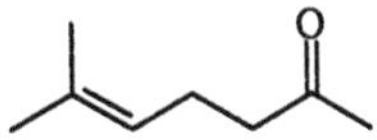

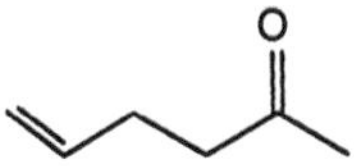

Selective hydrogenation of **1** was particularly challenging. Thus, when using Cu/silica gel a mess of products (up to 20) was obtained at total conversion after 2 hours. Among them we could identify beside the expected alcohol **4** (26%), bicyclic ethers **5** (6%), **6** and **7** (26%), α-,β-,γ-dihydroionones **8** and α-,β-,γ-dihydroionols **9**. Formation of these products can be accounted for by the presence of Brønsted and Lewis acid sites on the catalyst surface.

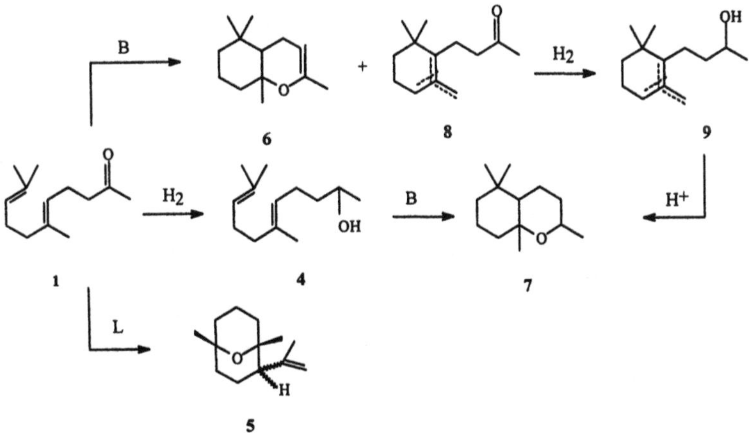

Scheme 1
(B= Bronsted acid sites, L= Lewis acid sites)

Thus, it is known that **1** gives ethers **5** as main products in the presence of Lewis acids such as BF₃-Et₂O, whereas Brønsted acids lead exclusively to chromenes **6** and dihydroionones **8** [4]. Products distribution over time (fig.1), that is simultaneous formation of **5**, **6** and **7** from the very beginning, suggests that Cu/SiO$_2$ under catalytic hydrogenation conditions promotes formation of ethers from the parent molecule **1** over Brønsted and Lewis acidic sites and from the hydrogenation product **4** over Brønsted ones (Scheme 1).

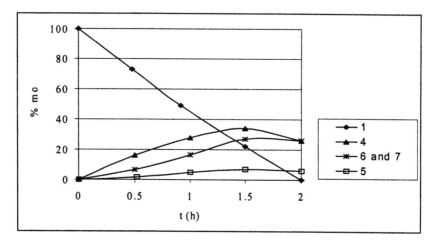

Figure 1 Reduction of geranylacetone **1** in the presence of Cu/SiO$_2$ (heptane, 1 bar H$_2$)

The presence of Brønsted and Lewis acid sites on supported copper catalysts has already been investigated by us. Brønsted sites present on the catalyst surface under catalytic hydrogenation conditions have been exploited to prepare tetrahydroedulans **7** starting from ionones in one step [5]. As far as Lewis sites responsible of epoxide rings opening [6, 7] or ene reactions [8] are concerned they are mainly due to the supports used. However, reaction of geranylacetone **1** in the presence of the bare silica gel gave, after 36 hours, only 5% of ethers **6**. Therefore, in the present case Lewis acid sites may be due to the presence of copper.

In order to suppress the acid catalyzed side reactions, we followed two strategies: to change the catalyst support and to change the solvent. Thus, we already experienced that the use of non porous silica as a support inhibits acidic reactions under hydrogenation conditions [3]. In the hydrogenation of **1**, substitution of silica gel as a support with a pyrogenic silica produced a very slow reaction requiring 48 hours to reach total conversion and giving again a mixture of products, although different from the previous one and probably all forming from the unsaturated alcohol **4** (Fig.2).

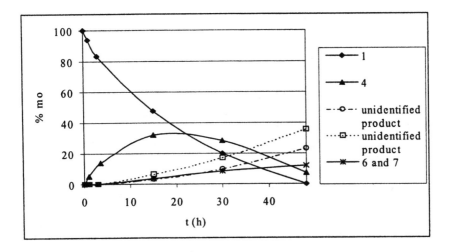

Figure 2 Reduction of geranylacetone **1** in the presence of Cu/Aerosil (heptane, 1 bar H$_2$)

On the other hand, the use of dioxane as solvent to poison the acidic sites, totally killed the activity of the catalyst, whereas 2-PrOH gave excellent results under both inert atmosphere and under molecular H$_2$, giving **4** with 95% yield.

Comparable yield in **4** can be obtained only through reduction of **1** with stoichiometric amounts of polymethylhydrosiloxane (PMHS) in the presence of catalytic amount of an active zinc compound [9]. However this process requires basic hydrolysis of the siloxane formed and a separation step.

Also the hydrogenation of 6-methyl-5-hepten-2-one **2** gave very low yield in the alcohol due to acidity of the catalyst. However in this case the reaction produced selectively 2,6,6-trimethyl pyran **10** with yields ranging from 78 to 87%.

This reaction represent a new case of bifunctional catalysis on the surface of a copper catalyst, that is the direct transformation of a ketone into an ether exploiting the hydrogenation activity of the catalyst giving the corresponding alcohol and the presence of acidic site activating the residual C=C double bond as a carbonium ion. As already reported for other substrates [4], this reaction takes place in one pot-one step avoiding formation of inorganic salts due to the use of mineral acids.

It is worth noting that acid catalyzed cyclization of alkenols is the most general method used to synthesize tetrahydropyranoids and tetrahydrofuranoids, valuable molecules in fragrance chemistry [10].

On the contrary, the use of copper supported on pyrogenic silica in the hydrogenation of **2** gave very good results allowing to reach a 95% yield in the unsaturated alcohol. The alcohol **4** thus obtained could be transformed in the tetrahydropiran derivative by allowing it to react separately on Cu/SiO$_2$ gel, but not improving the yield.

5-hexen-2-one **3** could not be hydrogenated to the corresponding unsaturated alcohol. With both catalysts and under both experimental conditions only the fully saturated alcohol was formed.

This is probably due to the steric availability of the unsubstituted double bond that makes its hydrogenation ready and competitive with the carbonyl group one. Isomerization of the C=C double bond towards the conjugate position cannot be excluded although 3-hexen-2-one was never identified in the reaction mixture. Looking at the products distribution over time, it is evident that by using Cu/SiO$_2$ in 2-PrOH and under N$_2$ (best conditions for ketone hydrogenation, Fig.3a) the unsaturated alcohol is the main product at low conversions, whereas with Cu/SiO$_2$ in n-heptane under H$_2$ (best conditions for C=C bond hydrogenation, Fig.3b) the saturated ketone is produced selectively with respect to the unsaturated alcohol. Over Cu/Aerosil no discrimination takes place (Fig.3c).

CONCLUSIONS

The support and reaction conditions can be finely tuned when using supported copper catalysts in order to obtain highly selective transformation. Best results obtained in this work are reported in Table 1.

In particular the use of Cu/SiO$_2$ in 2-propanol, under N$_2$ or under H$_2$, allows the chemoselective reduction of the carbonyl group also in polyfunctional molecules such as geranylacetone **1**. In hydrocarbon solvent the reaction can be addressed towards the one pot synthesis of valuable ethers.

It is worth remembering that α,β-unsaturated ketones gave the corresponding saturated ketones under both molecular H$_2$ addition and hydrogen transfer conditions [3].

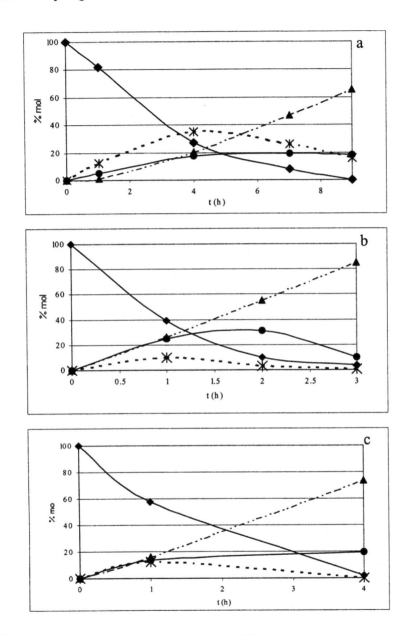

Figure 3 Reduction of 5-hexen-2-one **3** with three different catalytic systems:
(a) Cu/SiO$_2$, iPrOH, N$_2$; (b) Cu/SiO$_2$, heptane, H$_2$; (c) Cu/Aerosil, heptane, H$_2$
◆ 5-hexen-2-one; ▲ 2-hexanol; ● 2-hexanone; X unsaturated alcohol.

Table 1 Selective transformation of substrates **1**, **2** and **3** over supported copper catalysts.

	Substrate	Cond.	Catalyst	Product	Yield
1		2-PrOH, N_2, 83°C, 40 h	Cu/ SiO_2		95 %
2		2-PrOH, H_2, 83°C, 14 h	Cu/ SiO_2		97 %
3		2-PrOH, N_2, 83°C, 20 h	Cu/SiO_2		94 %
4		Heptane, H_2, 90°C, 10 h	Cu/Aerosil		95%
5		Heptane, H_2, 90°C, 3 h	Cu/SiO_2		87%
5		Heptane, H_2, 90°C, 1 h	Cu/SiO_2		85%
6		2-PrOH, N_2, 83°C, 8 h	Cu/SiO_2		65 %
7		Heptane, H_2, 90°C, 3 h	Cu/SiO_2		88 %
8		Heptane, H_2, 90°C, 3 h	Cu/Aerosil		73 %

REFERENCES

[1]   R. L. Augustine, *Heterogeneous Catalysis for the Synthetic Chemist*, Marcel Dekker, NY, 1995.

[2]   J-X. Chen, J.F. Daeuble, D.M. Brestensky, J.M. Stryker, *Tetrahedron* **56** (2000) 2153.

[3]   N. Ravasio, M. Antenori, M. Gargano, P. Mastrorilli, *Tetrahedron Letters* **37** (1996) 3529.

[4]   K.H. Schulte-Elte, G. Ohloff, *Helvetica Chimica Acta*, **58** (1975) 18.

[5]   N. Ravasio, V. Leo, F. Babudri, M. Gargano, *Tetrahedron Letters* **38** (1997) 7103.

[6]   N. Ravasio, M. Finiguerra, M. Gargano, *Catalysis of Organic Reactions*, F.Herkes ed., Marcel Dekker (NY) 1998, 513.

[7]   N. Ravasio, F. Zaccheria, R. Psaro, L. De Gioia, *Catalysis of Organic Reactions*, M.Ford ed., Marcel Dekker (NY)2000, p.601.

[8]   N. Ravasio, M. Antenori, F. Babudri, M. Gargano in *Heterogeneous Catalysis and Fine Chemicals IV,* H.U. Blaser, A. Baiker, R. Prins eds., *Studies in Surface Science and Catalysis*, **108** (1997) 625.

[9]   H. Mimoun, US Patent 6,245,952 to Firmenich SA(2001).

[10] P. Linares, S. Salido, J. Altarejos, M. Nogueras, A. Sanchez in "Flavour and Fragrance Chemistry" V. Lanzotti and O. Tagliatela-Scafati eds., Kluwer Ac. Publ. 2000, p.101.

21

Activated Base Metal Hollow Spheres: The Low Density Alternative for the Fixed Bed Hydrogenation of Organic Compounds

Daniel J. Ostgard, Monika Berweiler, Stefan Röder, and Peter Panster
Catalysts and Initiators Division, Degussa AG, Hanau, Germany

ABSTRACT

The new generation of fixed bed catalysts made from activated base metals (Metalyst^{TM}) is best described as an extremely durable collection of empty eggshells composed of high surface area activated metal. This technology offers an increase in volumetric activity at lower catalyst bulk densities without the presence of expensive, but not utilized, materials. In other words, the amount of metal in the reactor is reduced by as much as 40 to 77% leading to a more effective use of metal (1). In this paper, the effectiveness of this catalyst technology will be compared to others for the hydrogenation of various compounds (2, 3, 4).

INTRODUCTION

Powdered activated base metal catalysts (ABMC, also known as Sponge or Raney-type catalysts) have been successfully used for the hydrogenation of a variety of organic moeities. ABMC are used batchwise in the slurry phase where catalyst separation from the reaction mixture must be considered. Although considerable advances have been made in catalyst filtration (5), there is a level of desired product production where a continuous fixed bed process provides better economics (6). Moreover, the adjustment of various fixed bed reaction parameters (i.e., residence time (7)) can further improve catalyst selectivity and overall process flexibility.

Since the outer layer of a fixed bed catalyst is the most active due to internal mass transfer limitations (8), the use of activated base metal hollow spheres presents a far more effective use of catalytic material. This work discusses the effects of using alloys of different compositions, particle sizes, and textures on the stability, bulk density, and activity of the resulting catalytic hollow spheres. Catalytic hollow spheres ranging from 2 to 6.5 mm in size were produced resulting in crush strengths ranging from the acceptable 35 N to the outstanding 90 N, thus eliminating any fears of stability. The bulk densities of the different hollow spheres varied from 0.5 to 1.0 kg/l as compared to those of 1.65 to 2.2 kg/l for the current ABMC fixed bed catalysts, thereby representing a reduction of metal use by 40 to 77% on a volumetric basis. Another advantage of this system is that one can build up the catalytic shell with layers of different alloys allowing one to custom design the porosity and chemical composition of the catalyst with respect to shell depth (1). Examples of such layering techniques were examined in terms of the hollow spheres' physical characteristics and their ability to hydrogenate acetone in a trickle phase fixed bed reactor at 75°C and 5 bar.

The performance of the activated hollow spheres was compared to that of current fixed bed ABMC and supported base metal technology for the high-pressure hydrogenation of carbonyl (2), aromatic (3), nitrile (4), and other moeities (3). In each case, it was found that the activity of the catalytic hollow spheres were superior per unit of reactor volume and far superior when calculated per gram of metal in comparison to the current technology.

EXPERIMENTAL

The activated Metalyst^{TM} hollow spheres were produced according to the patent literature (1, 2, 3, 4, 9) by first spraying an aqueous polyvinyl alcohol containing suspension of the base metal-Al alloy and optionally a binder (e.g., Ni powder) onto a fluidized bed of styrofoam balls. This spraying can be performed in 1 or 2 steps depending on the need to either produce a thicker shell or to produce a catalyst with a designed layering of different alloys and/or metals. After impregnation, the coated styrofoam spheres were first dried and then calcined to burn out the styrofoam. The hollow spheres of alloy were then activated in a 20 wt.% caustic solution over 1.5 hours at 80°C. The activated catalyst was then stored under a mildly caustic aqueous solution (pH ~10.5) until use. Since the hollow spheres were preactivated, activation in the reactor before testing was not necessary. In one case, a batch of the Ni and Cr doped activated cobalt hollow spheres were treated with an aqueous LiOH solution to the point where the end catalyst contained ~0.2% LiOH. The crush strengths of the activated hollow spheres ranged from 35 to 90 N.

In accordance with the literature (1, 10, 11) the preparation of the shell activated Metalyst^{TM} tablets started with a homogeneous mixture of the desired alloy and shaping aids that was tableted, calcined, activated in caustic solution, washed, and stored under a mildly caustic aqueous solution (pH ~10.5) until use. Since Metalyst^{TM} is preactivated, activation in the reactor before testing was not necessary. The crush strength of Metalyst^{TM} tablets is approximately 300 N, thus eliminating catalyst attrition during reactor loading and the reaction itself.

Cobalt extrudates were made according to the patent literature (12) by extruding a mixture containing a 50 wt.%Co / 50 wt.%Al alloy and polymethylene copolymers at 190°C and a throughput of 10 kg/h with a double wave extruder. To decompose the polymethylene copolomers the extruded forms were then heated in an oven to 120°C followed by continual heating to 280°C within 90 minutes. Afterwards the temperature was ramped up to 800°C over 125 minutes and kept at this temperature for 140 minutes. The extrudates were then cooled, activated in a 20% caustic solution at 80°C over 120 minutes, washed, and stored under a mildly caustic aqueous solution (pH ~10.5) until use.

The rate of acetone hydrogenation was determined in the trickle phase with a 120-ml tubular reactor using 20 ml of either 3-mm activated tablets or hollow spheres. This hydrogenation was performed at 5 bar hydrogen pressure, 75°C, a hydrogen-to-acetone ratio of 5-to-1, and at LHSV of 4.0 h^{-1}. Acetone conversion and selectivities were determined by GC.

The rate of glucose hydrogenation was determined in the trickle phase with a 120-ml tubular reactor using 20 ml of either 4-mm activated tablets or hollow spheres. This hydrogenation was performed at 50 bar hydrogen pressure, 140°C, and at LHSV of 3.0 h^{-1} with a 40% aqueous glucose solution. Glucose conversion and selectivities were determined by HPLC.

The rate of butyronitrile hydrogenation was determined in the trickle phase with a 120-ml tubular reactor using 20 ml of either 3-mm activated tablets or hollow spheres. This hydrogenation was performed at 40 bar hydrogen pressure, 75°C, and at LHSV of 0.6 h^{-1} with a 20 wt.% butyronitrile in methanol solution. Butyronitrile conversion and selectivities were determined by GC.

Four different methods of adiponitrile hydrogenation were employed in this study. Method A used 40 ml of either 3 mm activated tablets or hollow spheres for the trickle phase hydrogenation of a 20 wt.% adiponitrile in methanol solution at 65 bar, 113°C, and at the LHSV values of 0.26 and 1.03 h^{-1}. Method B was carried out under the same conditions with the exception that 1.9 grams of NaOH were dissolved per liter of the 20wt.% adiponitrile in methanol reaction solution. Method C used 11.5 grams of catalyst placed in a basket located in the optimium mixing zone of a 0.5 liter autoclave that contained 30 grams adiponitrile and 180 grams methanol. Method C used the hydrogenation conditions of 150°C and 70 bar with the stirring rate of 1000 rpm for this slurry phase hydrogenation. Method D was the same as C except that enough liquid ammonia was added to reactor to

give an ammonia to adiponitrile molar ratio of 5. Adiponitrile conversion and selectivities for Methods A, B, C, and D were determined by GC.

The hydrogenation of a mixture of aromatic compounds found in petroleum
was carried out in a 0.5 liter autoclave stirred at 1000 rpm while containing 200
grams of the aromatic mixture and 11 grams of either 3 mm catalyst spheres or
extrudates placed in a basket located at the optimum mixing zone of the reactor at
35 bar and 150°C. During the experiment, the rate of hydrogen uptake was recorded. A commercially available 10% Ni on alumina catalyst was tested for this
application and prior to its use it was prereduced at 450°C. Activated Ni hollow
spheres were also used for this aromatic hydrogenation, however these catalysts
were not prereduced before use.

The rate of 1,4-dihydroxy-2-butyne hydrogenation was determined in the
trickle phase with a 120-ml tubular reactor using 40 ml of either 3-mm activated
tablets or hollow spheres. This hydrogenation was performed at 60 bar hydrogen
pressure, 135°C, and the LHSV values of 0.80 and 1.6 h^{-1} with a 50 wt.% 1,4-
dihydroxy-2-butene aqueous solution whose pH was adjusted to 7 with $NaHCO_3$.
1,4-dihydroxy-2-butyne hydrogenation conversion and selectivities were determined by GC.

RESULTS AND DISCUSSION

Catalyst Structure

The previous state of the art for fixed bed activated base metal catalysts was the
shell activation of tablets comprised of base metal aluminum alloy particles held
together with a small amount of metallic binder (7, 10, 11). The degree of activation of these tablets (see figure 1) can range from 10 to 70% with optimum values
ranging from 15 to 30% as determined by the application and its potential mass
transfer limitations. Although the shell-activated tablets have excellent catalytic
activity, selectivity, and mechanically stability over a long lifetime, they also have
rather high bulk densities.

As a solution to the high bulk density problem, we developed a catalyst
based on hollow spheres of activated base metal. As seen in figure 2, the activated
hollow spheres have an activation degree of 100% and all of the active metal is in
the "egg shell" region of the system, where mass transfer limitations are at a
minimum. The absence of unactivated alloy in the hollow spheres means that this
catalyst system will be more stable than the shell activated tablets under reaction
conditions that would react with the unactivated alloy. The leaching of the unactivated alloy in the tablets would lead to the degradation of their performance and
cause process problems due to the presence of the resulting Al-based compounds
in the product mixture. Since these catalyst systems are stored under an aqueous
caustic solution with a pH of ~10.5, the absence of unactivated alloy also means
that the activated hollow spheres will have a longer shelf life than the tablets.

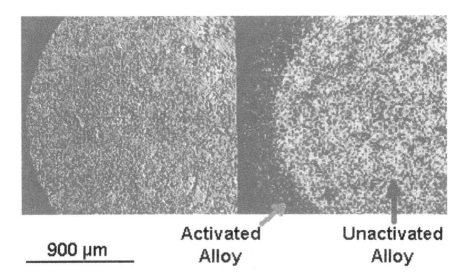

Figure 1 The distribution of Al before and after tablet activation (7).

Figure 2 A cross section of an activated Ni hollow sphere where the shell thickness ranges from 700 to 800 μm and the overall diameter of the sphere is 3 mm.

Table 1 Catalyst density, metal composition, and metal value per m^3.

Catalyst Type	Catalyst Form	% Base Metal[A]	%Al[A]	kg Catalyst per m^3 of Reactor	Metal Value per m^3 of Reactor[B]
Ni / Al	Tablets	60	40	2200	10690 Euros
Ni / Al	Hollow Spheres	90	10	700	4287 Euros
Potential reduction in metal value per m^3 for the hollow spheres					6403 Euros (60%)
Co / Al	Tablets	60	40	2200	33384 Euros
Co / Al	Hollow Spheres	90	10	700	14285 Euros
Potential reduction in metal value per m^3 for the hollow spheres					19099 Euros (57%)

[A] Probable base metal and Al contents. These values may change depending on the catalyst's use.
[B] Metal prices obtained in February of 2002 from the web sites of the LME and OMG.

Another benefit of the hollow spheres is the high porosity of their shell that allows reactants and products to easily flow in, through, and out of these spheres. While the hollow spheres are made by spraying an aqueous suspension of the alloy onto a template that will be burnt out later, the tablets are made by pressing together the particles of the alloy with shaping aids under a considerable amount of pressure. Thus, tableting results in a pore system that is tighter and more complex than that of the hollow spheres as readily shown by the comparison of figure 1 to figure 2. The lack of diffusion inhibiting material in the middle of the hollow spheres and the looseness of their pore system mean that reactions carried out with the hollow spheres are subjected to far less tortuosity in comparison to reactions performed with the tablets. In other words, the pathways to and from the active sites are freer for the hollow spheres leading to their higher activity on both a volumetric and weight basis along with improved selectivity, less catalyst fouling, and longer catalyst life.

The most noticeable advantages of the activated hollow spheres over the tablets are the drastic reduction in bulk density and the corresponding drop in metal costs. Table 1 displays these advantages in terms of catalyst density, metal composition, and metal value per m^3. For activated Ni the savings per m^3 is 6403 Euros leading to a 60% overall reduction of metal value. For activated Co the savings per m^3 is 19099 Euros leading to a 57% overall reduction of metal value.

The Reduction of Carbonyl Compounds with Activated Hollow Spheres

Acetone Hydrogenation

The effects of alloy preparation on catalyst performance and properties. The activated Ni hollow spheres used in the hydrogenation of acetone were prepared from 3 different Ni / Al alloys. Alloys A and B are two different variations

Table 2 The relative properties and acetone activities for activated hollow spheres made from alloys A, B, and C.

Property	Relative ranking of activated hollow spheres by alloy type
Mechanical Stability	$C \approx A > B$
Bulk Density	$C > A \approx B$
Acetone Activity	$B > A > C$ (see figure 3)
Shell Porosity	$B > C > A$ (see figure 4)

of rapidly quenched alloys and alloy C is one that was slowly solidified. The Ni / Al alloys used here were analyzed by energy dispersive X-ray analysis and found to have Ni_2Al_3, $NiAl_3$, and $Al-NiAl_3$ eutectic phases (7, 13, 14, 15). The amounts of the Ni_2Al_3 phase of these alloys were around 60%, the relative sizes of the Ni_2Al_3 phase domains were C >>> A > B, and the initial particle sizes of all these alloys were < 65µm. Table 2 displays the relative properties and performance of the activated Ni hollow spheres produced from each type of alloy.

As expected, the mechanical stability of the various hollow spheres roughly follows the trend of their bulk densities and the catalyst from alloy B with the most porous shell is the most active. However, the trend between activity and shell porosity for alloys C and A is somewhat unexpected. Figure 3 displays the acetone hydrogenation data on a per weight basis for the catalysts made from different alloys. Although the difference in activity between the catalysts made from alloys C and A is not that great, it is far more than the expected experimental error.

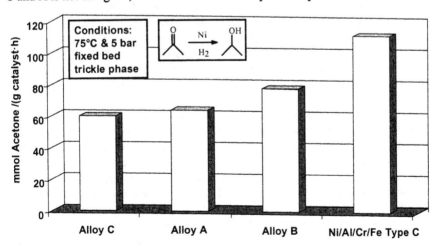

Figure 3 Acetone hydrogenation activity for the activated Ni hollow spheres made from A, B, C, and Ni/Al/Cr/Fe type C alloys. The particle size distributions of these alloys were < 65µm and the isopropanol selectivities for the above reactions were 99.9% or better.

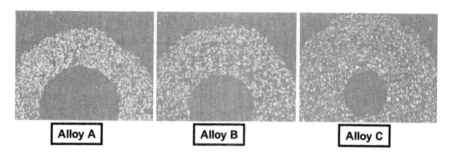

Figure 4 Cross-sections of activated Ni hollow spheres made from alloys A, B, and C. The shell thickness for each sphere ranged from 700 to 800 μm and the perceived differences in thickness are due to how close or far away the cross-sections were taken from the top of the spheres.

Since the activated hollow spheres made from alloy C have more shell porosity, one would think that this catalyst would be more active than that from alloy A. Moreover, it has been shown for alloys of the same particle size and containing ~60% of the Ni_2Al_3 phase that the smaller Ni_2Al_3 phase domains contain more Al after activation (14) and are consequently less active per mole of activated metal (7, 13, 14). Hence, activated hollow spheres from alloy A should also be less active per mole of activated metal in comparison to those from alloy C. Upon inspection of the cross-sections in figure 4, one notices that the very irregular particles of alloy C pack together to form mostly areas of very tight and complicated pores in combination with relatively few open gaps. These areas of tight porosity would be difficult to navigate for molecules in the trickle phase flowing at high throughputs, thereby leading to the lower than expected activity for the alloy C catalyst in comparison to the one from alloy A. In other words, this is a case where the type of shell porosity is more important than its amount. Hence, the overall activity of a catalyst is dependent on the balance between reactions conditions, the type of reaction, they type of alloy used, how porous the shell is, and the type of shell porosity.

Figure 3 also demonstrates that the addition of Cr and Fe promoters to alloy C practically doubles the activity of the catalyst for acetone hydrogenation. It has been shown in the literature (16) that the surface species of Cr on the activated catalyst exists as a low valent cation (Cr^{III}) that can coordinate with the unshared electrons from the oxygen of a carbonyl group resulting in the faster hydrogenation of this moiety. The surface species of Fe has also been proposed to be electropositive in nature so that it too can coordinate with a carbonyl group in a similar, albeit weaker, fashion as Cr^{III} for the enhance hydrogenation of this group (16). The incorporation of these promoters also increases the catalyst's surface area and stabilizes its residual Al content (17, 18), thereby improving its performance under harsh conditions such as hydrogenations in basic solutions.

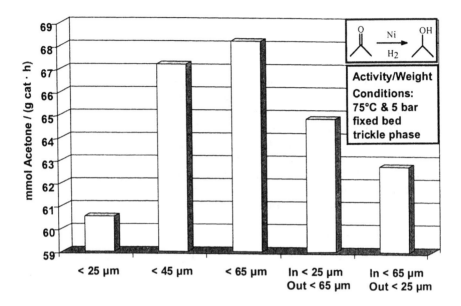

Figure 5 The effects of alloy particle size and the layering of different alloy particle sizes to the inner or outer part of the hollow spheres on the resulting catalyst's acetone hydrogenation activity. All the alloys used here were type C and the isopropanol selectivities for the above reactions were 99.9% or better.

Alloy particle size and layering effects on activated hollow spheres. As seen in figure 5, increasing the particle size of the alloy enhances the performance of its corresponding activated Ni hollow spheres for acetone hydrogenation. One also sees that the activity difference between the $< 65\mu m$ and $< 45\mu m$ alloy particle sizes is far less than that between the particle sizes of < 45 μm and < 25 μm indicating that the < 65 μm particle size may be close to optimal. Interestingly, the trend seen here between particle size and activity for the activated hollow spheres is opposite to what is observed for powder catalysts where decreasing the catalyst's particle size increases both its surface area and activity. The reason for this discrepancy could be in how the smaller alloy particles pack together to form a shell with a tighter and more tortuous pore system that could inhibit the free flow of reactants to and products from some of the catalyst's active sites. Hence, this is another case where the free flow of materials through the pore system of the shell is more important for the overall reaction rate than the catalyst's surface area. To test this theory, we made layered hollow spheres where the inner part of the shell consisted of one particle size and the outer part consisted of another. The first hollow sphere of this type had alloy particles < 25 μm in the inside and alloy particles < 65 μm on the outside, and although the resulting hollow spheres were more stable, they were about 5% less active than the pure < 65 μm activated hollow spheres. The second type of layered hollow spheres had the < 65 μm alloy in the inside and the < 25 μm alloy particles on the outside of the shell. The resulting catalyst from the second type of layered hollow spheres was about 8% less active than the hollow spheres made from only the < 65 μm particles and also less active than the first type of layered hollow spheres. This proves that the smaller particles are less active than the larger ones for these types of catalysts due to their resulting pore systems, and concentrating these particle on the outside of the spheres impacts the free flow of reactants over more of the catalyst leading to a lower activity. The general trend seen here is that as the particle size of the alloy making up the shell decreases the bulk density of the catalyst increases, the stability increases, and the activity decreases. While the layering technique is used here to make a point about the particle size effects, it could also be used to create other types of catalysts with different metals, materials, and/or alloys.

Glucose Hydrogenation

Figure 6 compares the glucose hydrogenation activity of the activated Metalyst[TM] tablets to those of the various activated Metalyst[TM] hollow spheres. In this

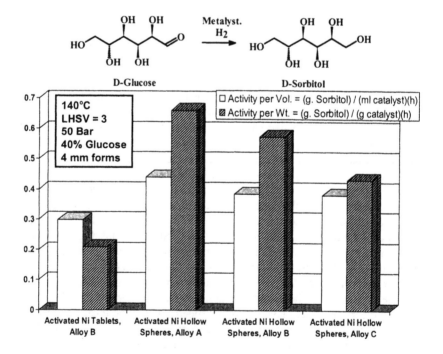

Figure 6 Glucose hydrogenaton data comparing the activated Metalyst[TM] tablets to the activated Metalyst[TM] hollow spheres. All the hollow sphere alloys were < 65μm.

example, one sees that all of the activated hollow spheres, regardless of alloy type, are more active than the tablets on both a per weight and a per volume basis. This could again be attributed to the improved flow of material to and from the active sites for this relatively viscous reaction solution. In contrast to the acetone hydrogenation data, the catalyst from alloy A is more active for the reduction of glucose than that from alloy B. As mentioned previously, the somewhat smaller Ni_2Al_3 phase domains of alloy B will result in an increased residual aluminum content in this catalyst leading to a lower activity (14). It has been shown before that, glucose hydrogenation is sensitive to the type of alloy where the more rapidly cooled alloys are less active per site than the slowly cooled ones. In the case of the tablets, alloy B was found to make the most active catalyst because it produced very porous tablets with far more, albeit less active, sites than a slowly cooled alloy (7, 13). In the case of the hollow spheres, the data suggests that the residual Al content of the catalyst is just as important as the shell's porosity for the two types of rapidly cooled alloys used here.

It is also worth mentioning that we were able to carry out these reactions at 50 bars of pressure in which is more characteristic of a slurry phase reaction with a powder catalyst (19) than the pressures of 150 to 400 bar as given in the literature for fixed bed sugar hydrogenations (7, 20, 21, 22). Such an improvement in the operating pressure of this reaction should lead to sizeable savings in its operation.

The Reduction of Nitriles with Activated Hollow Spheres

Butyronitrile Hydrogention

Figure 7 displays the relative butyronitrile hydrogenation activities on a per catalyst weight basis for the activated cobalt extrudates and four different generations of activated hollow spheres. Each successive generation of the cobalt hollow spheres had a higher shell porosity. As expected, all the hollow spheres perform better than the activated extrudates on a weight basis and the order of activity for the hollow spheres followed that of their shell porosity with the most porous being the most active. In figure 8 the same data is presented on a per volume of catalyst basis and the results, albeit more compressed, are still the same where the hollow spheres are more active than the extrudates and the activity of the hollow spheres depends on their shell's porosity. Once again, the improved flow of materials to the active sites for the hollow spheres have led to improve overall activity.

Another aspect to this reaction is its selectivity. As demonstrated in figure 9, the surface adsorbed butylimine can undergo a nucleophilic attack by butylamine to form 1-aminodibutylamine (23, 24) which can eliminate ammonia to give the corresponding Schiff base that can be reduced to dibutylamine (17, 18, 25, 26). This reaction is usually performed in the slurry phase where its selectivity is controlled by the addition of ammonia, KOH, NaOH, or other basic materials to the reaction mixture. While the addition of a base enhances selectivity, it can also

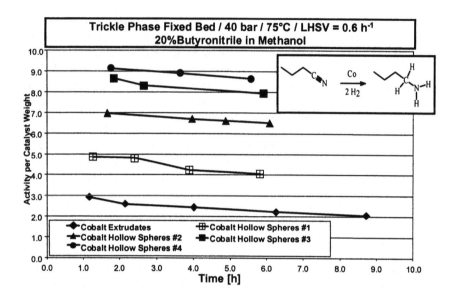

Figure 7 The comparison of buytronitrile hydrogenation activity on a weight basis for the activated cobalt hollow spheres to that of the activated cobalt extrudates. In each of these cases, primary amine selectivity was ~ 95%.

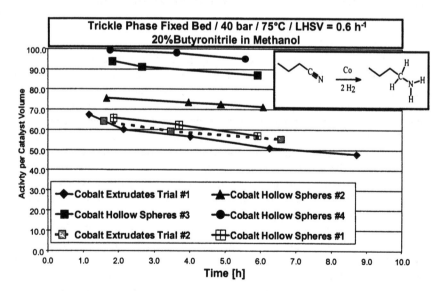

Figure 8 The comparison of buytronitrile hydrogenation activity on a volume basis for the activated cobalt hollow spheres to that of the activated cobalt extrudates. In each of these cases, primary amine selectivity was ~ 95%.

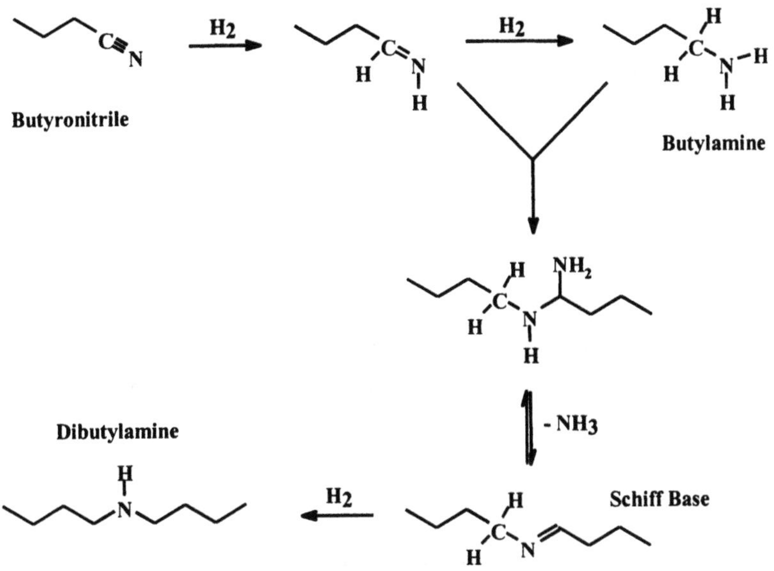

Figure 9 Reaction scheme for butyronitrile hydrogenation.

decrease catalyst lifetime via residual Al leaching. Hence, it would be ideal to enhance selectivity either without or with minimal base addition. The 1° amine selectivities for the hydrogenations shown in figures 7 and 8 were all ~ 95% and this high value is attributed to the very low contact time the butylamine has with the adsorbed imine during this fixed bed process. Thus, the utilization of a fixed bed process under the right conditions should offer the possibility to improve nitrile hydrogenation selectivity while eliminating or minimizing the addition of base.

Adiponitrile Hydrogenation

Table 3 shows the activity data for the trickle phase hydrogenation of adiponitrile (ADN) at the LHSV of 1.03 and 0.26. The Co hollow spheres were far more active than the Co tablets on both a weight and volume basis due to the increased porosity of the hollow spheres. Adding Cr and Ni dopants to the Co / Al alloy did not change the activity of the Co hollow spheres, however preadsorbing LiOH on the Cr and Ni doped Co hollow spheres did increase its activity at the LHSV h^{-1} of 1.03 by ~9%, while leaving the activity at the LHSV of 0.26 h^{-1} unchanged. The Cr and Fe doped Ni hollow spheres had the lowest activity of all the hollow spheres tested here and adding NaOH to the feed enhanced their activity by ~53% at the LHSV of 1.03 h^{-1} and by ~11% at the LHSV of 0.26 h^{-1}. This increase in activity for the above mentioned catalysts after the addition of base is at least partially due to the suppression of Schiff base formation (*vida infra*).

Table 3 The trickle phase adiponitrile hydrogenation activity data

Activated Catalyst	LHSV h^{-1}	% Conv.	Activity per g.[A]	Activity per ml[B]
Co hollow spheres	1.03	71.4	1.02	0.708
	0.26	99.9	0.39	0.247
Co Tablets	1.03	42	0.20	0.416
	0.26	86.5	0.10	0.216
Cr / Ni doped Co hollow spheres	1.03	71.8	0.85	0.711
	0.26	99.8	0.30	0.249
Cr / Ni doped Co hollow spheres with preadsorbed LiOH[C]	1.03	78.3	0.98	0.776
	0.26	99.9	0.32	0.249
Cr / Fe doped Ni hollow spheres	1.03	51.7	0.67	0.512
	0.26	88.3	0.29	0.220
Cr / Fe doped Ni hollow spheres with NaOH in feed[D]	1.03	79.1	1.02	0.783
	0.26	97.9	0.32	0.244

[A] Activity in units of [(mole Adiponitrile)/(gram catalyst· h)]
[B] Activity in units of [(mole Adiponitrile)/(ml catalyst· h)]
[C] This catalyst had LiOH preadsorbed on it to the level of ~0.2% before testing.
[D] This catalyst was tested according to method B where the feed contained 1.9 g NaOH per liter.

As seen in figure 10, the selectivity issues associated with ADN hydrogenation are somewhat more complicated than those for buytronitrile and under the wrong conditions, activated base metal catalysts are more likely to deactivate with dinitriles than with mononitriles (27). ADN has been shown to be reduced initially to aminocapronitrile (ACN) and then further to hexamethylenediamine (HMD) in a stepwise fashion (28, 29). ADN can react with excess base (see figure 10) to form a carbanion α to one of the nitrile groups that in turn can react with the other nitrile group via a Thorpe-Ziegler reaction to yield a cyclic imine that is quickly tautomerized to the enaminonitrile. This enaminonitrile strongly adsorbs on the catalyst's surface and hydrogenates slowly to the corresponding 1-amino-2-methylaminocyclopentane, thereby acting as an inhibitor for ADN hydrogenation (26). Since we avoided using heavily basic conditions in order to lengthen the catalyst's lifetime, we were also able to elude enaminonitrile formation and the problems associated with it. ADN can also form 1,2-diaminocyclohexane via a proposed adsorbed diimine species. Fortunately, this reaction is easily avoided by blocking the most active sites (26) and this can be done with the addition of base or, most probably, the formation of coke could suffice. Once formed, ACN could either be reduced to the desired HMD, it could react with HMD to form a linear Schiff Base (as previously described for buytronitrile), or it could react with itself to form a cyclic Schiff base (see figure 10). Regardless if the Schiff Base is cyclic

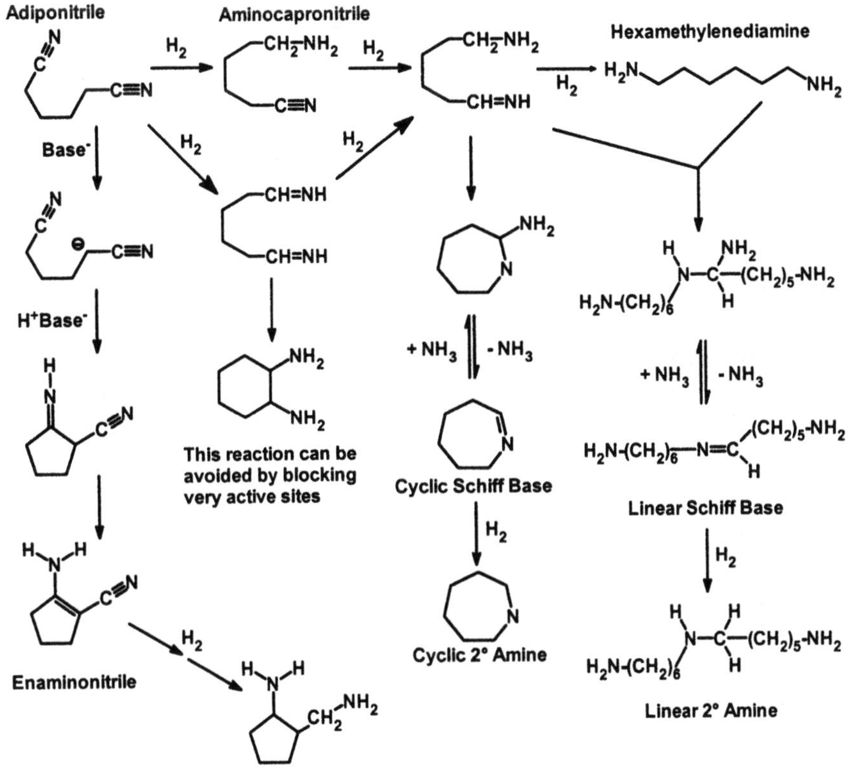

Figure 10 Reaction scheme for adiponitrile hydrogenation.

or linear, its presence will inhibit the formation of HMD due to its strong adsorption and slow hydrogenation rate (26). The order of the most common side products seen here was first the cyclic 2° amine, secondly the linear 2° amine and then the Schiff bases.

Table 4 displays the selectivity data for the trickle phase adiponitrile hydrogenation studies. Increasing the residence time of the reactants on the surface by lowering the LHSV resulted in higher HMD-to-ACN ratios, more 2° amines, higher linear-to-cyclic 2° amine ratios, and typically fewer Schiff bases. The shift to more saturated products with longer residence times should be expected due to increased hydrogen availability. The increase in secondary amine formation with lower LHSV is due to the corresponding increase in the contact time between the adsorbed imine and amines. It is interesting to note, that the linear-to-cyclic 2° amine ratio increase with lower LHSV could mean that the adsorbed imine has more of a chance to react with amines other its own amino group and/or it could also mean that the adsorbed cyclic Schiff base has more of a chance to form the linear variety via a transimination reaction (26).

Table 4 The trickle phase adiponitrile hydrogenation selectivity data

Activated Catalyst	LHSV h^{-1}	% HMD	% ACN	% 2° amines	% Rest	Ratio X[A]	Ratio Y[B]
Co hollow spheres	1.03	60.8	27.5	7.5	4.2	2.2	0.02
	0.26	79.7	0.1	17.5	2.3	796	0.33
Co Tablets	1.03	46.9	43.6	4.8	4.8	1.1	0
	0.26	71.0	13.6	11.7	3.7	5.2	0.35
Cr / Ni doped Co hollow spheres	1.03	43.6	34.8	12.9	9.0	1.3	0.01
	0.26	65.3	0.7	29.5	4.5	93.1	0.34
Cr / Ni doped Co hollow spheres with pread-sorbed LiOH[C]	1.03	69.1	27.3	1.8	1.8	2.5	0.00
	0.26	94.2	0.8	4.2	0.8	117.6	0.45
Cr / Fe doped Ni hollow spheres	1.03	32.3	29.4	25.7	12.6	1.1	0.18
	0.26	35.3	7.7	38.1	18.6	4.6	0.36
Cr / Fe doped Ni hollow spheres with NaOH in feed[D]	1.03	70.6	20.7	6.3	2.4	3.4	0.32
	0.26	89.0	3.5	5.7	1.8	25.7	0.44

[A] "X" is the ratio of hexamethylenediamine to aminocapronitrile.
[B] "Y" is the ratio of the linear 2° amine to the cyclic 2° amine.
[C] This catalyst had LiOH preadsorbed on it to the level of ~0.2% before testing.
[D] This catalyst was tested according to method B where the feed contained 1.9 g. NaOH per liter.

The Co hollow spheres without any doping agents and additives displayed the broadest range of HMD-to-ACN ratios between the two LHSV. At the LHSV of 0.26 h^{-1} it had the highest HMD-to-ACN ratio with a value of 796 and at the LHSV of 1.03 it dropped to 2.2. In contrast to the hollow spheres, the Co tablets were the most selective for ACN formation. One could try to explain this by saying that the active sites situated deeper in the tablets are more hydrogen deficient, however that would lead to an increase in other side products as well and that is not the case. Another reason may have something to do with the unactivated alloy in the tablets, but at this point, that would be speculative. Adding Cr and Ni dopants to the Co hollow spheres tended to decrease their HMD selectivity while substantially increasing the amount of 2° amines and other side products. Since Cr exists on the surface of the catalyst as an cation, it may coordinate with the un-shared electrons of the nitrile group or the nitrile group itself, thereby increasing its adsorption strength and allowing undesired reactions such as 2° amine formation to occur more readily. While the use of Cr as an doping agent makes the Co hollow spheres less selective, both it (17) and Ni (30) increase the catalyst's surface area and stabilize the residual Al so that the catalyst could better survive the use of base additives for selectivity improvement. Preadsorbing LiOH to the Cr and Ni doped Co hollow spheres improved their HMD selectivity to 94.2% with very few side products. The increase in selectivity is due to LiOH blocking the surface sites that cause side products and the increase in the reaction rate for the higher LHSV is due, at least partially, to the formation of fewer Schiff bases whose strong adsorption and slower hydrogenation rate inhibit ADN and ACN hydrogenation. The strong adsorption of LiOH on Co (24) makes it the perfect modifier for fixed bed hydrogenations on cobalt hollow spheres, because it does not always need to be present in the feed. Instead, LiOH would only need to be dosed in to pick up the selectivity when it starts to decline in order to replace preadsorbed LiOH that had been slowly washed off.

The Cr / Fe doped Ni hollow spheres displayed the worst selectivities of the group. According to the literature, the Cr and Fe of the Cr / Fe doped Ni exist as positively charged species on the surface (16) that can coordinate with unshaired electrons and electron rich groups. As was mentioned in the case of the Cr / Ni doped Co, the presence of these positively charged specie leads to the formation of side products due to the stronger adsorption strength of adiponitrile. However, the addition of Cr and Fe also increases the catalyst's surface area and stabilizes the residual Al so that NaOH or any other base could be used as an additive to increase selectivity with far less damage to the catalyst. As seen in table 4, adding NaOH to the feed of the Cr / Fe doped Ni catalyst greatly improved its HMD selectivity (70.6% at the LHSV h^{-1} of 1.03 and 89% at the LHSV of 0.26 h^{-1}), while drastically reducing the levels of secondary amines. Avoiding the formation of Schiff bases also resulted in higher activities for this catalyst when NaOH was used as a co-feed.

Table 5 displays the slurry phase adiponitrile hydrogenation data for the Co hollow spheres. As expected, the slurry phase data is slightly less selective than the fixed bed process at the lowest LHSV and the amount undesirable products, like the Schiff bases, are considerably higher. This increase in side products is from the much longer contact time amines have in the slurry phase with adsorbed imines. Adding ammonia to this reaction mixture increased the HMD selectivity to 89.4% and improved its activity by 22%. Ammonia increases the selectivity of this reaction by blocking nonselective sites and driving the equilibrium between the aminodialkylamine and the Schiff Base away from the Schiff Base. Avoidance of

Table 5 The slurry phase adiponitrile hydrogenation data

Activated Catalyst	Initial Activity[A]	% HMD	% ACN	% 2° amines	% Rest	Ratio Y[B]
Co Hollow Spheres	30.3	73.3	0.12	17.04	9.5	0.35
Co Hollow Spheres with NH$_3$[C]	37.0	89.4	0	4.48	5.7	0.41

[A] The initial activity in units of [(ml H$_2$) / (gram catalyst· min)]. Over the reaction time of 120 minutes, both catalysts came to an ADN conversion of 100% .

[B] "Y" is the ratio of the linear 2° amine to the cyclic 2° amine.

[C] This catalyst was tested in the presence of ammonia. The ammonia to ADN ratio was 5.

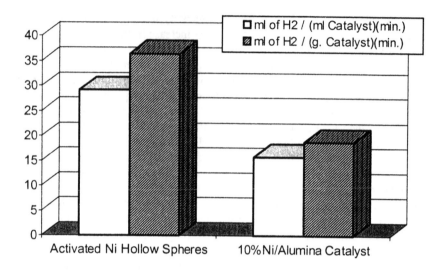

Figure 11 The hydrogenation data for the reduction of mixed aromatics in petroleum with activated Ni hollow spheres.

these strongly adsorbed and slowly hydrogenating side products such as the Schiff bases also results in an increase in activity.

The Reduction of Unsaturated C-C Bonds and Aromatics with Activated Hollow Spheres

The Hydrogenation of Mixed Aromatics in Petroleum

Figure 11 compares the activity of the activated Ni hollow spheres to that of a commercial $10\%Ni/Al_2O_3$ catalyst for the hydrogenation of mixed aromatics found in petroleum. The Ni hollow spheres exhibited approximately twice as much activity as the supported catalyst on both the weight and volume basis indicating their suitability for this application.

The Hydrogenation of 1,4-Dihydroxy-2-Butyne

As shown in Figure 12, the hydrogenation of 1,4-dihydroxy-2-butyne (ByD) occurs stepwise via formation of the 1,4-dihydroxy-2-butene (BeD) to the desired product 1,4-butanediol (BDO). Table 6 displays the trickle phase reaction data for the catalysts studied here at 135°C and 60 bar. The use of higher pressures and lower temperatures (e.g., 266 bar and 75°C (31)) would provide higher selectivities, however the conditions used here were chosen to make the reaction more demanding so that we could better differentiate between the various catalysts.

Table 6 The 1,4-dihydroxy-2-butyne hydrogenation

Activated Catalyst	LHSV h^{-1}	Act. Per g^A	Act. Per ml^B	% Con^C	% BDO	% t.BeD^D	% c.BeD^E	S^F	c/t^G
Ni – HS^H	0.8	0.92	0.80	99.5	70.6	3.6	8.7	5.74	2.42
	1.60	1.69	1.46	91.3	40.8	3.8	27.4	1.31	7.21
Mo doped Ni – HS	0.8	0.89	0.80	99.8	82.8	1.6	3.9	15.1	2.44
	1.6	1.72	1.55	92.8	51.7	2.8	20.5	2.22	7.32
Cr/Fe doped Ni – HS	0.8	0.92	0.75	94.1	59.3	2.0	20.7	2.61	10.4
	1.6	1.53	1.26	78.3	38.8	2.6	23.1	1.51	8.88
Ni Tablets	0.8	0.40	0.71	88.2	32.5	3.3	35.6	0.84	10.8
	1.6	0.57	1.01	62.8	18	1.5	29.5	0.58	19.7
Mo doped Ni Tablets	0.8	0.43	0.76	95.3	49.0	2.7	27.5	1.62	10.2
	1.6	0.64	1.14	71.4	20.7	1.3	27.5	0.72	21.2

^{A} Activity per weight in units of [(grams of 1,4-dihydroxy-2-butyne) / (gram catalyst· h)].
^{B} Activity per volume in units of [(grams of 1,4-dihydroxy-2-butyne) / (ml catalyst· h)].
^{C} The % of 1,4-dihydroxy-2-butyne conversion.
^{D} The % of 1,4-dihydroxy-trans-2-butene.
^{E} The % of 1,4-dihydroxy-cis-2-butene.
^{F} The ratio of [(1,4-butanediol) / (1,4-dihydroxy-trans-2-butene + 1,4-dihydroxy-cis-2-butene)]
^{G} The ratio of 1,4-dihydroxy-cis-2-butene to 1,4-dihydroxy-trans-2-butene
^{H} HS = Hollow Spheres

In comparison to the Ni tablets, the Ni hollow spheres were more active on both a weight and volume basis due to the easier flow of reactants through this system. In comparison to the Ni tablets, The hollow spheres were more selective-for BDO formation, fewer olefins were formed, and more of the *trans* olefin was generated. The increased BDO selectivity and lower olefin content could be contributed to the improved mass transfer of hydrogen through the hollow spheres. Since the hollow spheres are more selective than tablets, the increase in *trans* olefin content via the presence of acid/base sites on the hollow spheres has been ruled out. The higher *trans* content seems to mean that the half-hydrogenated state occurs more readily on the hollow spheres than with the tablets, however more studies are needed to explore this point in detail.

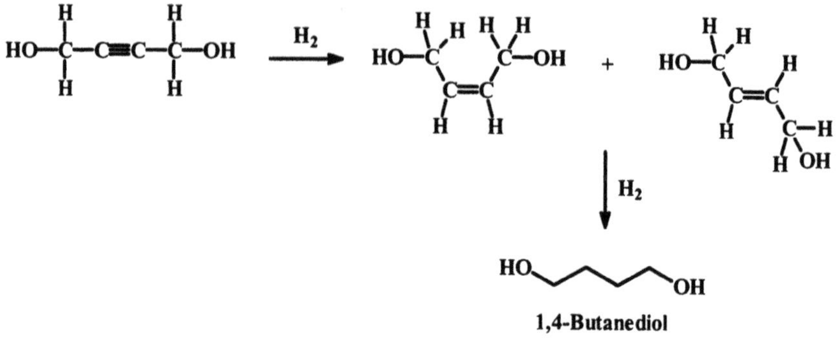

Figure 12 Reaction scheme for 1,4-dihydroxy-2-butyne hydrogenation.

The addition of Mo to the Ni hollow spheres and tablets greatly improves their BDO selectivity while slightly, if at all, increasing the activity. Another interesting aspect is that the addition of Mo does not effect the amount of *trans*-BeD. Hence it would seem that the interaction of the positively charged Mo surface species on ByD is strong enough to enhance BDO formation, but mild enough to leave other aspects of this reaction relatively unchanged. Although the use of Mo for this reaction has been mentioned in the literature (32), it had not been explained as to how it functions in this case. One possibility could be the coordination of one or both of the hydroxyl groups to surface Mo cations so that once the ByD is hydrogenated to the BeD, it does not go to far away before it rotates and readsorbs on the surface with its pi orbitals for the further hydrogenation to BDO.

As a comparison to the Mo doped catalysts, Cr / Fe doped Ni hollow spheres were also tested for this reaction. While the addition of Mo was beneficial, the use of Cr and Fe leads to a decrease in BDO selectivity in comparison to the catalyst without promoters. If it is the mild coordination of the hydroxyl groups to Mo that improves its BDO selectivity, then it would seem that the coordination of the hydroxyl groups to positively charged Cr and Fe surface species is too strong, thereby leading to an increase in side products.

CONCLUSIONS

Activated base metal hollow spheres are cost effective alternatives for the hydrogenation of organic compounds. In comparison to the older activated base metal catalyst technologies, the hollow spheres contain from 40 to 70% less metal per unit volume leading to savings of up to 60% in metal value per m^3 of reactor volume. The flow of reactants through the hollow spheres is far less tortuous in comparison to the older technologies meaning that the hollow spheres suffer fewer mass transfer limitations while enjoying enhanced activity, selectivity, and stability. The absence of unactivated alloy also means that this catalyst technology will have a longer shelf life in comparison to the "shell activated" technologies. The alloy type and particle size distribution also plays an important part in the performance of the activated hollow sphere. As the particle size distribution of the alloy becomes smaller, the resulting activated hollow spheres become mechanically more stable and catalytically less active with an increase in their bulk density. The shell itself can be designed by selectively building up different layers of various alloy types and/or metals so to as devise bimodal and/or multimetallic catalysts. It is important to note that both the type of porosity and the amount of shell porosity is important for the most effective use of these catalysts.

Activated hollow spheres have been found to be advantageous for the hydrogenation of carbonyl compounds, nitriles, aromatics, and unsaturated C-C bounds. In the case of carbonyl compounds, promoters (e.g., Mo and Cr) that exist as surface cations were found to be the most effective. In the case of nitriles, the use of promoters to stabilize the residual Al content of the catalyst so that it can be used with base modifiers was found to be the most useful combination. An example of this was the improved performance of the LiOH treated Cr / Ni promoted Co hollow spheres for the hydrogenation of adiponitrile to hexamethylenediamine. Some reactions were found to be more sensitive to the type of promoter they require. In the case of 1,4-dihydroxy-2-butyne, it was found that Mo worked satisfactory as a promoter while the Cr / Fe combination led to worse results. Nonetheless, for all of the reactions studied here it was found that the activate hollow spheres were more active than the activated tablets on both a volume and weight basis, thereby allowing increased flexibility in the use of promoters and other selectivity enhancing additives.

REFERENCES

1. D.J. Ostgard, P. Panster, C. Rehren, M. Berweiler, G. Stephani and L. Schneider, German Patent 19933450.1, 1999.
2. D.J. Ostgard, M. Berweiler, and S. Röder German Patent 10065029.5, 2000.
3. D.J. Ostgard, M. Berweiler, and S. Röder German Patent 10101646.8, 2001.
4. D.J. Ostgard, M. Berweiler, and S. Röder German Patent 10065031.7, 2000.
5. J.M. Lambert Jr., Catalysis of Organic Reactions, M. Scaros and M. Prunier editors, New York: Marcel Dekker Inc., 1994, pp 61-69.
6. J. Super, Catalysis of Organic Reactions, Michael Ford editor, New York: Marcel Dekker Inc., 2000, pp 35-49.
7. D.J. Ostgard, M. Berweiler, S. Röder, K. Möbus, A. Freund and P. Panster, Catalysis of Organic Reactions, Michael Ford editor, New York: Marcel Dekker Inc., 2000, pp 75-89.
8. C.H. Satterfield, Heterogeneous Catalysis in Industrial Practice, second edition, New York: McGraw-Hill, Inc., 1991, chapter 11.
9. D.J. Ostgard, M. Berweiler and S. Röder, German Patent 10101647.6, 2001.
10. S. Peter, R. Burmeister, D. Bertrand, H. Mösinger, H. Krause and K. Deller, European Patent 0648534 A1, 1994.

11. A. Freund, M. Berweiler, B. Bender and B. Kempf, German Patent 19721898 A1, 1998.

12. J. Sauer, T.Haas, B. Keller, A. Freund, W. Burkhardt, D. Michelchen and M. Berweiler, European Patent 0 880 996, 1998.

13. D.J. Ostgard, A. Freund, M. Berweiler, B. Bender, K. Möbus and Peter Panster, Chemie-Anlage+Verfahren, 9: 118-119, 1999.

14. S. Knies, M. Berweiler, P. Panster, H. E. Exner and D.J. Ostgard, Studies of Surface Science in Catalysis, A. Corma, F.V. Melo, S. Mendioroz, and J.L.G. Fierro editors, Amsterdam: Elsevier, 130: 2249-2254, 2000.

15. S. Knies, G. Miehe, M. Rettenmayr and D.J. Ostgard, Zeitschrift für Metallkunde, 92: 6: 596-599, 2001.

16. P. Gallezot, P.J. Cerino, B. Blanc, G. Flèche and P. Fuertes, J. Catal., 146:93-102, 1994.

17. C. DeBellefon and P. Fouilloux, Catal. Rev. - Sci. Eng, 36 (3):459-506, 1994.

18. S.N. Thomas-Pryor, T.A. Manz, Z. Liu, T.A. Koch, S.K. Sengupta and W.N. Delgass, Catalysis of Organic Reactions, Frank Herkes editor, New York: Marcel Dekker Inc., 1998, pp 195-206.

19. R. Albert, A. Strätz, and G. Vollheim, Chem.-Ing.-Tech., 52(7):582-587, 1980.

20. G. Darsow, US Patent 6020472, 2000.

21. H. Degelmann, J. Kowalczyk, M. Kunz and M. Schüttenhelm, German Patent 19701439 A1, 1998.

22. T. Mohr, E. Schwarz and P.-J. Mackert, German Patent 19929368 A1, 1999.

23. J.L. Dallons, A. Van Gysel, and G. Jannes, Catalysis of Organic Reactions, William Pascoe editor, New York: Marcel Dekker Inc., 1992, pp 93-104.

24. T.A. Johnson and D. P. Freyberger, Catalysis of Organic Reactions, Michael Ford editor, New York: Marcel Dekker Inc., 2000, pp 201-227.

25. J. Von Braun, G. Blessing and F. Zobel, Chem. Ber., 36:1988-2001, 1923.

26. P. Marion. M. Joucla, C. Taisne and J. Jenck, Heterogeneous Catalysis and Fine Chemicals III, M. Guisnet Editor, Amsterdam: Elsevier Science Publishers B.V., 1993, pp 291-298.

27. A.M. Allgeier and M.W. Duch, Catalysis of Organic Reactions, Michael Ford editor, New York: Marcel Dekker Inc., 2000, pp 229-239.

28. C. Mathieu, E. Dietrich, H. Delmas and J. Jenck, Chem. Eng. Sci., 47:2289-2294, 1992.

29. C. Joly-Vuillemin, D. Gavroy, G. Cordier, C. DeBellefon and H. Delmas, Chem. Eng. Sci., 49:4839-4849, 1994.

30. J.P. Orchard, A.D. Tomset, M.S. Wainwright and D.J. Young, J. Catal., 84: 189-199, 1983.

31. F.G. Low, French Patent 2029788, 1970.

32. W. DeThomas and E.V. Hort, US Patent 4153578, 1979.

22

Development of New Two-Step Technology for Polyols

Tapio Salmi, Päivi Mäki-Arvela, Johan Wärnå, Tiina-Kaisa Rantakylä, Valentina Serra-Holm, Janek Reinik, Kari Eränen
Åbo Akademi University, Åbo/Turku, Finland

INTRODUCTION

Diols and triols are key intermediates in the industrial production of lubricants, surface coatings and synthetic resins. The classical pathway in the diol synthesis is based on the use of alkali-catalyzed aldolization followed by the Cannizzaro reaction [1]. Equilibrium amounts of sodium formate ions (*i.e.* formic acid) are formed in the traditional reaction route. This is a serious drawback since formate has to be separated from the product mixture. The producer of diols and triols is forced to deal with a component which has a low market value.

An alternative synthesis route for diols and triols is based on the utilization of heterogeneous catalysis. The classical pathway in the diol catalysts is thus highly desirable. It has been discovered that it is possible to stop the reaction at the aldol stage by using mild bases such as tertiary amines as catalysts instead of NaOH [2]. Several solid catalysts for aldol condensation have been proposed in the literature, for instance modified hydrotalcites [3-7], mesoporous materials [8], and anion-exchange resins. Among the latter, the most promising catalysts are weakly basic anionic ion exchange resins, with tertiary amines as active groups [9-10]. The carbonyl group of the prepared aldol can be hydrogenated to diol or triol. Co-products are avoided as illustrated in the reaction scheme below, where R denotes an alkyl group.

Aldol condensation:

$$RCH_2C\underset{\diagdown H}{\overset{\diagup O}{}} \ + \ 2 \ \underset{H}{\overset{H}{}}C=O \ \xrightarrow{OH^-} \ RCC\underset{\diagdown H}{\overset{\diagup O}{}}$$

(with CH_2OH and CH_2OH substituents)

Diol hydrogenation (formation of triol):

$$R-\underset{CH_2OH}{\overset{CH_2OH}{C}}-C\underset{\diagdown H}{\overset{\diagup O}{}} \ + \ H_2 \ \rightleftharpoons \ R-\underset{CH_2OH}{\overset{CH_2OH}{C}}-CH_2OH$$

     According to the newest edition of a German textbook in industrial organic chemistry [11], the process based on heterogeneous catalysis is feasible when formaldehyde and isobutyraldehyde are used as starting molecules. In the case of linear aldehydes, elimination takes place as a side reaction in the presence of weak bases, which essentially complicates the process development. Elimination gives an unsaturated aldehyde, with propion- and butyraldehyde methyl- and ethylacrolein are obtained.

     The research carried out by our group has, however, shown that the production of trimethylolpropane and trimethylolethane in aqueous environment is possible over weakly basic anion exchange resin catalysts [12]. The system is chemically and physically complicated since several components and reactions appear and the system forms a liquid-liquid-solid dispersion in most cases.

EXPERIMENTAL SECTION

Aldolization Experiments and Formaldehyde Separation

Batchwise aldolization experiments were carried out in a 1.0 dm^3 jacketed glass reactor operating at atmospheric pressure. Formaldehyde, either as a freshly prepared aqueous solution of paraformaldehyde or as formalin (42 wt.-%), and n-butyraldehyde, the catalyst and the solvents, water and methanol, were loaded into the reactor at room temperature. Solid paraformaldehyde was dissolved in water by the addition of alkali (NaOH). Several anionic resin catalysts were screened [13].

     The amount of catalyst was 160 g, and the total liquid mass was about 800 g. The reactor was heated until the reaction temperature was attained (40-80°C). A careful temperature control was arranged with a thermostat and the temperature curve was stored on a PC. Glycol from the thermostate was used as a heating medium in the reactor jacket. A propeller agitator, set at 800 rpm, kept the liquid phase well mixed. The reactor was equipped with baffles to remove external mass transfer resistances.

A reflux condenser was placed on top of the reactor to prevent the escape of volatile compounds. The reflux condenser was cooled with recirculating glycol (-25°C) from a thermostat. In order to avoid the oxidation of the aldehydes, the reactor was maintained under nitrogen atmosphere, and an oil trap was installed at the outlet of the condenser. Nitrogen was bubbled through a three-way valve through which samples were withdrawn for chemical analysis. The first sample was taken at room temperature before switching on the heating system of the reactor. The reaction temperature was reached within 40-50 minutes from the beginning of the experiment.

Since formaldehyde is known to be a catalyst poison it was removed from the reagent mixture prior to hydrogenation by distillation. In order to prevent oligomerization of formaldehyde [14], water was gradually added to the mixture during distillation. The water feeding was adjusted in order to maintain a constant liquid volume in the distillation flask. Distillation was performed under atmospheric pressure and at 100 °C using a total batch volume of 400 ml (200 ml of aldolization product mixed with 200 ml of distilled water). As the first distilled droplets from the condenser were observed, the addition of extra water was commenced. In this way, the formaldehyde content of the solution was easily suppressed below 0.5 wt.%.

Hydrogenation Experiments and Hydrogen Solubility

The experiments were performed in a pressurized autoclave (Autoclave Engineers, 1 dm^3) equipped with a turbine impeller and a bubbling unit. The pressure, the temperature and the agitation velocity were controlled and registered during the experiment by a data acquisition program.

Fresh catalyst samples were activated prior to each kinetic experiment under dry conditions. The catalyst (5 g) was placed in the reactor and a silver ring packing was installed between the reactor cylinder and the lid. The catalyst was heated to the desired temperature at a rate of 5 °C/min under hydrogen flow (0.5 dm^3/min). The hydrogen flow (0.75 dm^3/min) was continued at elevated temperature for 1 h. After this activation, the gas flow was maintained constant, the heating jacket was removed and the reactor was cooled down.

The hydrogenation mixture was prepared by stirring 0.15 dm^3 formaldehyde separated aldol solution into 0.15 dm^3 methanol (>99.8 % J. T. Baker). The reaction mixture was saturated with hydrogen for 10-15 min in order to remove dissolved oxygen. The solution was injected into the reactor and temperature and pressure values were adjusted to the desired values. Liquid samples were withdrawn at pre-determined intervals and the duration of the experiments varied depending on process conditions.

The solubility of hydrogen in aldol and triol solutions at the experimental temperatures and pressures was measured with a gas chromatograph equipped with a packed column (Chromosorb 102) and a thermal conductivity detector (TCD) and two sample loops. Hydrogen pressure was applied, the temperature adjusted to the desired level and a HPLC pump was used to circulate the reactor contents (liquid phase containing dissolved hydrogen) between the sampling valve of the GC and the reactor. The quantitative analysis was calibrated with a standard H_2/N_2 gas mixture (996 ppm H_2 in N_2, AGA).

Chemical Analyses

Most of the analyses were carried out using a High Performance Liquid Chromatograph (HPLC) equipped with a LiChrosorb RP-18 column (5µm, 250x4mm) and diode array (DA) and refractive index (RI) detectors. Two different methods were applied to analyse the compounds: a derivatization method was used for the quantitative determination of aldehydes, aldols and unsaturated aldehydes, while a direct method was developed for measuring the contents of triols and organic acids. In the derivatization method [15], 2,4-dinitrophenyl-hydrazine (DNPH, Acros) was used as a reagent in acetonitrile (ACN) solution.

Catalyst Characterization

A commercial silica-supported nickel catalyst (Hoechst RCH 53/35, 69 wt.% Ni, 13 wt.% Cr, specific surface area 100 m^2/g) was used in experiments performed with butyr- and propionaldehyde in aqueous as well as methanolic solvent. The catalyst particles were cylindrical with a height of 5 mm and diameter of 6 mm. The particles were crushed and sieved to a fraction of 45-150 µm in order to avoid diffusional resistances inside the catalyst particles.

Before systematic kinetic experiments, the catalyst activation procedure was investigated by means of thermogravimetric analysis (TGA), X-ray photoelectron spectroscopy (XPS) and *in situ* hydrogenation experiments. An advantageous activation temperature was searched for. Three different temperatures were selected, *i.e.* 170°C, 300°C and 400°C. Since the critical sintering temperature of nickel is between 450-500°C, the highest temperature was limited to 400°C [16].

RESULTS AND DISCUSSION

Aldolization of Propion- and Butyraldehyde with Formaldehyde

The feasibility of the alternative process using heterogeneous catalysis was investigated by screening several anion-exchange resin catalysts [16] and optimizing the reaction conditions for the aldolization of butyraldehyde and propionaldehyde with formaldehyde; the phase distribution coefficients of the components in the liquid-liquid equilibria were determined and rate equations for the aldolization and elimination reactions were derived from molecular mechanisms.

The main products observed in the aldolization experiments were the corresponding aldol and alkyl acrolein. In fact, elimination of water always appears as a parallel reaction in the crossed aldolization as described in organic chemistry textbooks [17]. The α-hydroxymethyl-substituted aldehyde, appearing as a reaction intermediate, undergoes reversible dehydration, leading to the corresponding unsaturated aldehyde. The products D and W were detected by HPLC, whereas the amount of the intermediate, C, was evidently so low that it was outside the detection limit of HPLC.

$$RCH_2C\overset{O}{\underset{H}{<}} \ + \ \overset{H}{\underset{H}{>}}C=O \ \xrightarrow[\text{OH}^-]{R_3N^+} \ RCHC\overset{O}{\underset{H}{<}} \ \overset{CH_2OH}{|}$$

$$(A) \qquad\qquad (B) \qquad\qquad\qquad (C)$$

$$\overset{CH_2OH}{\underset{}{RCHC}}\overset{O}{\underset{H}{<}} \ + \ \overset{H}{\underset{H}{>}}C=O \ \longrightarrow \ \overset{CH_2OH}{\underset{CH_2OH}{RCC}}\overset{O}{\underset{H}{<}}$$

$$(D)$$

$$\overset{CH_2OH}{\underset{}{RCHC}}\overset{O}{\underset{H}{<}} \ \rightleftharpoons \ \overset{}{\underset{CH_2}{RCC}}\overset{O}{\underset{H}{<}} \ + \ H_2O$$

$$(X) \qquad\qquad (W)$$

Prior to the systematic kinetic experiments a screening of the mass transfer effect was carried out. A comparison of experiments performed with a catalyst particle mean diameter of 760-960 μm and of 250-500 μm showed quite similar results which indicate the absence of internal mass transfer resistances. Moreover, experiments performed with different agitation velocities (800 and 200 rpm) gave similar results, suggesting the absence of macro-mixing effects and external mass transfer resistances around the catalyst particles.

In addition to the main reactions, aldolization and elimination, we also observed side reactions in some of the experiments, such as acetalization of the reacting aldehydes. In fact, when methanol was used as solvent, it reacted with the aldehydes forming the corresponding acetals and hemiacetals. This reaction is highly undesirable, since it partially destroys the reagents. The change of the

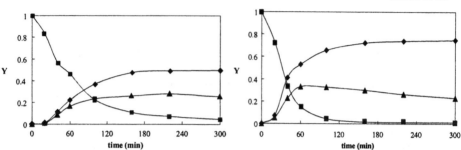

Figure 1 Aldolization experiments in methanolic (a) and aqueous (b) solvents: ( ) butyraldehyde; ($\blacktriangle$) ethylacrolein; ($\blacklozenge$) aldol. Conditions: formaldehyde-to-butyraldehyde molar ratio 4:1, temperature 70°C. ($Y_i = c_i/c_{0A}$; i = A, X, D)

methanol-water mixture to pure water was a breakthrough in the process development: acetalization was suppressed and a clearly more favourable product distribution was obtained. The results from experiments of aldolization of butyraldehyde carried out with methanol-water and pure water as the solvent are reported in Figure 1. In aqueous environment, two liquid phases were formed, so the lines in Figure 1 represent average concentrations in the emulsion. In the absence of methanol, significantly higher aldol yields were achieved, since acetalization was suppressed: the aldol-to-ethylacrolein molar ratio was about 3 in aqueous solvent and about 2 in methanolic solvent after five hours at 70°C. The initiation period seen in the product curves is caused by the temperature gradient at the initial stage of the experiment.

The influence of the water concentration in the reaction mixture was investigated under methanol-free conditions. This parameter considerably affects the product distribution: the aldol yield at the final stage of the experiment goes through a maximum and the corresponding yield of ethylacrolein through a minimum for a water-to-butyraldehyde molar ratio of 12:1. The aldol yields were 0.53 with a water-to-butyraldehyde molar ratio of 8:1, 0.76 with a water-to-butyraldehyde molar ratio of 12:1 and 0.62 with a water-to-butyraldehyde molar ratio of 16:1 at 70C° with a formaldehyde-to-butyraldehyde molar ratio of 4:1.

The effect of the reactant ratio on the product distribution was investigated and experiments were carried out with formaldehyde-to-butyraldehyde molar ratios varying from 2:1 to 8:1. The results from two experiments performed at 60°C with formaldehyde-to-butyraldehyde molar ratios of 2:1 and 4:1 are shown in Figure 2.

The reactant ratio had a considerable effect on the product distribution; a higher formaldehyde-to-butyraldehyde ratio increased the selectivity towards aldol. This observation is not surprising from a kinetic viewpoint, as water elimination from α-hydroxymethyl-substituted aldehyde (aldol 1) leading to the corresponding unsaturated aldehyde (ethylacrolein) is in competition with the reaction of the α-hydroxymethyl-substituted aldehyde with formaldehyde leading to aldol 2; this second reaction is favoured when an excess of formaldehyde is present.

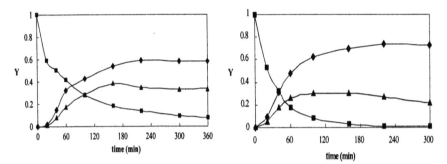

Figure 2 Aldolization experiments with a formaldehyde-to-butyraldehyde molar ratio of 2:1 (a) and 4:1 (b) in aqueous solvent at 60°C( ) butyraldehyde; (▲) ethylacrolein; (♦) aldol.

The temperature dependence of the product distribution was checked by plotting the concentrations of D and X at different temperatures but at fixed concentrations of B (formaldehyde). The results for butyraldehyde aldolization are displayed in Figure 3. The experiments showed that product distribution is essentially independent of temperature. Similar results were obtained for propionaldehyde aldolization.

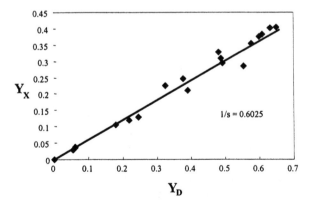

Figure 3 Yield of ethylacrolein versus yield of aldol with molar ratio of formaldehyde-to-butyraldehyde 3 to 1 in aqueous solvent at 50 – 70 °C.

Table 1 The values of the slopes of the c_D-c_X (aldol – ethylacrolein) curves for the aldolization of butyraldehyde in methanolic and aqueous environment

	methanolic solvent	aqueous solvent
formaldehyde:butyraldehyde 4:1	1.69	1.84
formaldehyde:butyraldehyde 2:1	1.16	1.48

Table 2 The values of the slopes of the c_D-c_X (aldol – methylacrolein) curves for the aldolization of propionaldehyde in methanolic and aqueous environment

	methanolic solvent	aqueous solvent
formaldehyde:propionaldehyde 4:1	5.01	9.35
formaldehyde:propionaldehyde 2:1	2.46	8.25

As can be observed in Tables 1 and 2, a clearly better product distribution with respect to aldol was obtained when operating in an aqueous environment. This was particularly visible for the aldolization of propionaldehyde, but the benefits were also clear in the case of butyraldehyde aldolization.

The overall reactions in aldolization can be summarized as follows:

$$A \; + \; 2B \; \longrightarrow \; D \qquad\qquad\qquad (I)$$

$$A \; + \; B \; \rightleftharpoons \; X \; + \; W \qquad\qquad (II)$$

where A = butyraldehyde or propionaldehyde, B = formaldehyde, D = aldol, X = methylacrolein or ethylacrolein, W = water.

Based on a detailed molecular reaction mechanism, the following rate equation were derived for aldolization and elimination, Serra-Holm et al. [18, 19].

$$r_I = k_1 c_A \left(\frac{K_B c_B}{1 + K_B c_B} \right)^2$$

$$r_{II} = k_2 \left(c_A \left(\frac{K_B c_B}{1 + K_B c_B} \right) - \frac{c_x}{K} \right)$$

The model was completed with the adsorption-oligomerization equilibria for formaldehyde, and the lumped kinetic parameters were estimated with non-linear regression. Generally, the model gave a very good agreement with experimental data obtained from aldolization of propion- and butyraldehyde.

Hydrogenation of TMP-aldol and TME-aldol to Trimethylolpropane (TMP) and Trimethylolethane (TME)

The second step of the process, the aldol hydrogenation, is in principle more straightforward, since no selectivity problems appeared with suitable metal catalysts. The challenge is to improve the reactivity of the bulky aldol molecule by optimizing the reaction conditions and to avoid catalyst poisoning caused by formaldehyde. Therefore, a procedure for the separation of formaldehyde was developed.

The thermogravimetric experiments showed that the catalyst reduced most efficiently at 400 °C. At 400 °C, the degree of reduction exceeded 10 %, whereas the degree of reduction was only 6 % at 170 °C.

X-ray photoelectron spectroscopy (XPS-ESCA) was used to analyse the chemical state of the catalyst surface. Two different activation temperatures were compared, 400 (sample 1) and 170°C (sample 2) for 1 hour under hydrogen flow. The samples were investigated closer within the energy domain where Ni appears in the spectra (855-862 eV). Two peaks indicate nickel; NiO and metallic Ni and that the NiO peak is much smaller for sample 1 than for sample 2.

The catalyst activation temperatures were also investigated by *in situ* hydrogenation experiments using samples reduced by hydrogen at 170°C and 400°C for 1 h. The actual hydrogenation experiment was performed at 80°C and 80 bar. A complete conversion was achieved in all experiments, but the hydrogenation velocity was most rapid over the catalyst activated at the highest temperature. Therefore, the activation procedure for systematic kinetic experiments was fixed at 400°C and 1 h hydrogen flow.

The hydrogen solubility was studied by using TMP-aldol and TMP-triol in aqueous mixtures. A correlation equation was obtained for solubility of hydrogen in aqueous solutions [20].

In the aldolisation step, pure water as a reaction solvent significantly improved the yield of aldol by suppressing the acetalization of butyraldehyde. The hydrogenation kinetics with pure water as a solvent were determined at pressures of 40-80 bar and temperatures of 60-90 °C. The triol selectivity was 100% in all experiments studied.

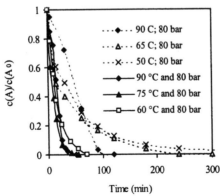

Figure 4 Hydrogenation rates of aldol with methanolic (dashed line) and non-methanolic (solid line) solvents.

Also in the experiments performed with aqueous solvent, an initial slowness of aldol hydrogenation caused by the presence of formaldehyde in the reaction mixture was observed as previously discovered in experiments carried out in methanolic solvent. However, the comparison of hydrogenation rates between methanolic and non-methanolic experiments revealed that the absence of methanol increased the total reaction velocity as illustrated in Figure 4.

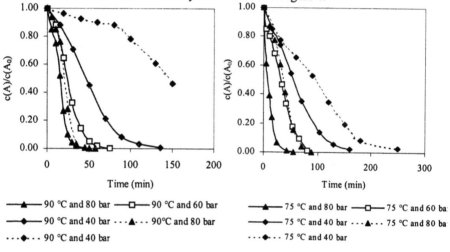

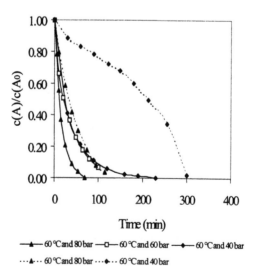

Figure 5   The normalized concentrations (c_A/c_{A0}) of TMP-aldol (solid) and TME-aldol (dashed) in the hydrogenation at different pressures and temperatures over NiCr-catalyst in aqueous solvent.

The hydrogenation rate of TME-aldol was found to be significantly lower than that of TMP-aldol when using low temperatures and pressures as illustrated in Figure 5. At the highest temperature and pressure studied (i.e. 90 °C and 80 bar), the hydrogenation rates were rather equal.

The superiority of pure water as a solvent was demonstrated even more clearly by an experiment performed with a high initial formaldehyde content; a complete conversion of TMP-aldol was achieved within 5 h, whereas the total reaction time for aldol hydrogenation was 10 h in methanolic solvent. Also the hydrogenation rate of formaldehyde was increased significantly by the selection of water as solvent as illustrated in Figure 6.

The initial hydrogenation rates of formaldehyde in aqueous and methanolic solvents were compared by fitting a pseudo-first order model with respect to the aldol to the kinetic data displayed in Figure 6. The ratio of the rates in aqueous and in methanolic solvent was 5.5. When the ratio of the rates was divided by the corresponding hydrogen solubilities it obtained the value 16, which reflects the huge difference in the intrinsic hydrogenation rates in water and methanol.

Since the kinetic experiments within the NiCr catalyst showed that formaldehyde retarded the hydrogenation kinetics of the aldol, especially at higher temperatures, pseudo-first order kinetics could not be applied for the aldol hydrogenation as illustrated in Figure 7. The logarithmic test plots are bent upwards, which indicate that the reaction order is less than one with respect to aldol.

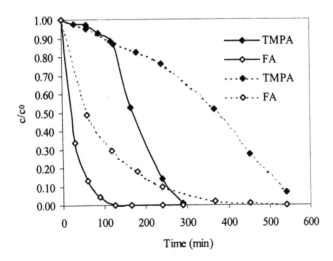

Figure 6 The hydrogenation of TMP-aldol and formaldehyde at 80 °C and 80 bar with a high initial FA-content x_{0FA}/x_{0A} = 0.7 using methanolic solvent (dashed) and pure water (solid) as solvents.

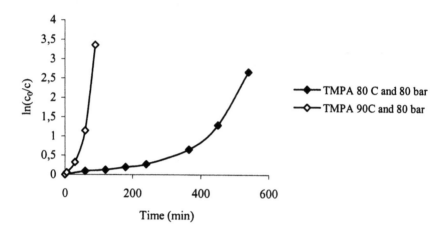

Figure 7 Semi-logarithmic plots for hydrogenation of TMP-aldol over the NiCr catalyst.

Some general principles are applied to the consideration of hydrogenation mechanisms. Aldol (TMP-aldol or TME-aldol) and hydrogen molecules are assumed to be adsorbed on the catalyst surface prior to the hydrogenation. The carbonyl group of aldol interacts with the catalyst surface and the adsorption of hydrogen is dissociative [21, 22]. However, transient kinetic studies indicate that hydrogen preserves its molecular identity during hydrogenation [23]. This leads to a mechanism, where the adsorbed aldol reacts with a pair of dissociatively adsorbed hydrogen giving the reaction product, triol, which is desorbed from the surface.

In the derivation of the rate equation, it is assumed that the surface reaction is irreversible and rate determining, whereas the adsorption steps of hydrogen and aldol are rapid enough for the quasi-equilibrium hypothesis to be applied. The desorption step of triol is assumed to be irreversible and very rapid $(c_{T*} \rightarrow 0)$.

Consequently, the hydrogenation rates are given by

$$r_j = k_j c_{i*} c_{H*}^2 \qquad (1)$$

where j denotes reaction index, $j =$ I or II and i denotes aldol (TMP- or TME-aldol) or formaldehyde, H is hydrogen and the asterisk (*) refers to the concentration of an adsorbed species.

Quasi-equilibrium approximations can be applied to the adsorption steps and a total balance gives a relation between the concentrations of the surface species:

$$\sum c_{i*} + c_{H*} + c_* = c_{Tot} \tag{2}$$

The rate equations for the hydrogenation reactions of aldol (I) and formaldehyde (II) can now be expressed as follows:

$$r_I = k_1 c_{A*} c_{H*}^2 = k_1 K_A K_H c_A c_H^2 c_*^3 \tag{3}$$

$$r_{II} = k_2 c_{FA*} c_{H*}^2 = k_2 K_{FA} K_H c_{FA} c_H^2 c_*^3 \tag{4}$$

By substituting quasi-equilibrium expressions into the site balance, the concentration of vacant sites is obtained and the final forms of the rate equations become:

$$r_I = \frac{k' c_A c_H}{(1 + K_A c_A + K_{FA} c_{FA} + \sqrt{K_H c_H})^3} \tag{5}$$

$$r_{II} = \frac{k'' c_{FA} c_H}{(1 + K_A c_A + K_{FA} c_{FA} + \sqrt{K_H c_H})^3} \tag{6}$$

where k' and k'' are lumped kinetic parameters: $k' = k_1 K_A K_H c_{Tot}^3$, $k'' = k_2 K_{FA} K_H c_{Tot}^3$. The component generation rates are obtained from

$$r_A = -r_I, \ r_{FA} = -r_{II}, \ r_T = r_I.$$

According to the experimental data, formaldehyde has a considerable retarding effect on the hydrogenation rate at high temperatures. Therefore, the rate equations (5) and (6) are modified to account for the inhibition. It is a well-known fact that adsorption enthalpies are influenced by the presence of existing adsorbates on the catalyst surface. The absolute value of the adsorption enthalpy decreases with increasing coverage. In the present case we assume the linear dependence of adsorption enthalpy on the surface coverage of formaldehyde.

The adsorption equilibrium constant is assumed to obey the law of van't Hoff. Provided that an analogous treatment can be applied on the adsorption of hydrogen, the lumped rate parameters become

$$k' = k_0' e^{(-E' - a'\Theta_{FA})/(RT)} \tag{7}$$

$$k'' = k_0'' e^{(-E'' - b'\Theta_{FA})/(RT)} \tag{8}$$

Eqs (7)-(8) clearly demonstrate, how the apparent rate constants diminish with increasing surface coverage of formaldehyde.

A semi-empirical modification is obtained by replacing the coverage of formaldehyde (Θ_{FA}) by its mole fraction in the liquid. Since the total concentration remains virtually constant we use $x_{FA}=c_{FA}/c_L$, where c_L is the total concentration of liquid. The final forms of the rate expressions thus become

$$r_I = \frac{k_0' e^{-(E' + \alpha' f)/RT} c_A c_H}{D^3}$$ (9)

$$r_{II} = \frac{k_0'' e^{-(E'' + \alpha'' f)/RT} c_{FA} c_H}{D^3}$$ (10)

where

$$D = 1 + K_A c_A + K_{FA} c_{FA} + \sqrt{K_H c_H} \quad \text{and} \quad f = \frac{c_{FA}}{c_L}.$$

The rate equations are combined with the mass balances of the components in order to determine the values of the kinetic parameters.

The mass balances for the organic components in the batch reactor are written in the following form:

$$\frac{dc_i}{dt} = r_i \varphi_B$$ (11)

where c_i is the concentration expressed in mol/kg and $\varphi_B = m_{cat}/m_L$ where m_L is the mass of the liquid. The stoichiometry gives the relation between hydrogenation rates (r_I and r_{II}) and the generation rates of the components (r_i).

The numerical parameter estimation was based on kinetic experiments out of which the temperatures and concentrations were recorded for the compounds. The kinetic and adsorption parameters were estimated by using the Simplex-Levenberg-Marquardt method, which minimizes the residual sum of squares between the estimated and the experimental concentrations with non-linear regression.

The estimated parameters were the adsorption constants, the lumped rate constants at the average temperature, and the apparent activation energies. The adsorption constants were assumed to be independent of temperature in order to suppress the number of adjustable parameters in regression.

Examples of the fit of the model to the data sets are provided in Figures 8 and 9. As can be seen from the figures, the proposed model describes the experimental data reasonably well.

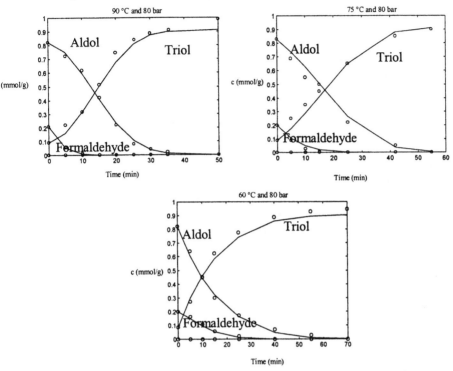

Figure 8 Examples of model fit. Measured (symbols) and estimated (curves) concentrations of TMP-aldol, formaldehyde and TMP-triol. The decreasing curves: TMP-aldol (upper), formaldehyde (lower). The increasing curve: TMP. Experiments with pure water over NiCr.

CONCLUSIONS

A new concept based on the use of solid catalysts for the production of two aldols, trimethylolpropane aldol and trimethylolethane aldol, was developed. The aldolization process is the first step for the production of the triols, trimethylolpropane and trimethylolethane. In the second step, these triols are obtained by catalytic hydrogenation of the aldols. The two-step technology was patented (FI Pat. Appl. 991519) [12].

Several weakly basic anion-exchange resin catalysts were screened in laboratory-scale and the results showed that a very good product distribution with respect to aldol can be obtained. Extensive kinetic studies demonstrated that weakly basic anion-exchange resin catalysts promoted both the aldolization and elimination processes, but the product distribution can be steered by selection of the solvent and the ratio of the reactants. A kinetic model based on the aldolization and elimination reaction mechanisms was developed and simplified. The model predictions were in good agreement with the experimental data, suggesting that separate treatment of the aqueous and organic phases can be replaced by the average concentrations and the distribution coefficients. The model will be useful in the scale-up of aldolization processes.

Hydrogenation experiments confirmed that the commercial catalysts studied, *i.e.* silica-supported NiCr was able to hydrogenate the aldol selectively to triol. The comparison of methanolic and water solvents revealed that the absence of methanol increased the total hydrogenation velocity, especially when the starting mixture contained high quantities of formaldehyde. The experimental data produced with NiCr were reasonably well described by the kinetics model, which account for competitive adsorption of reagents, surface reactions and the inhibitory effect of formaldehyde.

The developed two-step technology for the production of the triols is superior to the conventional technology, since the use of heterogeneous catalysts solves the catalyst separation problem and prevents the formation of a useless co-product, sodium formate.

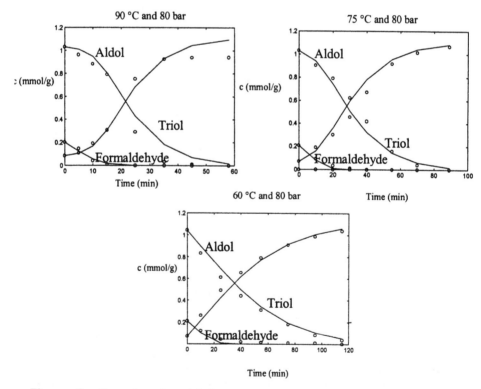

Figure 9   Examples of model fit. Measured (symbols) and estimated (curves) concentrations of TME-aldol, formaldehyde and TME-triol. The decreasing curves: TME-aldol (upper), formaldehyde (lower). The increasing curve: TME. Experiments with pure water over NiCr.

NOTATION

A, a', b'	parameters in rate equation
c_i	concentration (in mol/kg)
D	denominator in kinetic expression
E', E''	apparent activation energies
f	ratio of concentrations
k	rate constant
k', k_0', k'', k_0''	lumped rate parameters
K	equilibrium parameter
m	mass
n	amount of substance
P	pressure
r	rate
R	gas constant
s	slope
T	temperature
t	reaction time
x	mole fraction
Y	yield
α', α''	lumped adsorption parameters
θ	surface coverage
φ	catalyst mass-to-liquid mass

Subscripts and Superscripts

0	initial concentration
cat	catalyst
exp	experimental value
i	component index, general index
j	reaction index
L	liquid

REFERENCES

1. Kirk-Othmer, Encyclopedia of Chemical Technology, John Wiley and Sons, New York, 11, 1994.
2. Immel, H H Schwarz, O Weissel and H Krimm. Process for the preparation of trimethylolalkanes, US 4 122 290, Bayer AG.
3. M L Kantam, B M Choudary, Ch V Reddy. Aldol and Knoevenagel condensations catalysed by modified Mg-Al hydrotalcite: a solid base as catalyst useful in synthetic organic chemistry, Chem Commun 9: 1033-1034, 1998.
4. W T Reichle. Catalytic reactions by thermally activated synthetic anionic clay minerals, J Catal 94: 547-557, 1985.

5. K Koteswara Rao, M Gravelle, J V Valente, F Figueras. Activation of Mg-Al hydrotalcite catalysts for aldol condensation reaction. J Catal 173: 115-121, 1998.

6. A Guida, M H Lhouty, D Tichit, F Figueras P Geneste. Hydrotalcite as base catalysts. Kinetics of Claisen-Schmidt condensation, intramolecular condensation of acetylacetone and synthesis of chalcone. Appl Catal A General 164: 251-264, 1997.

7. E Dumitriu, V Hulea, C Chelaru, C Catrinescu, D Tichit, and R Durand. Influence of the acid-base properties of solid catalysts derived from hydrotalcite-like compounds on the condensation of formaldehyde and acetaldehyde. Appl Catal A General 178: 145-157, 1999.

8. B M Choudary, M Lakshmi Kantam, P Sreekanth, T Bandopadhyay, F Figueras and A Tuel. Knoevenagel and aldol condensations catalysed by a new diamino-functionalised mesoporous material. J Mol Catal A Chemical 142: 361-365, 1999.

9. M J Astle and J A Zaslowsky. Aldol Condensation. Ind Eng Chem 44: 12, 2869-2871, 1952.

10. G V Austerweil and R Pallaud. Les échangeurs d'anions dans les aldolisations, cétolisations et leurs réactions consécutives. Bull Soc Chim France: 678-680, 1953.

11. K Weissermel, and H-J Arpe. Polyhydric Alcohols. Industrial Organic Chemistry, 3rd ed. New York: VCH, 1997, p. 210.

12. T Salmi, V Serra-Holm, T-K Rantakylä, P Mäki-Arvela, L P Lindfors. Development of a clean production technology for triols. Green Chemistry: December, 283-287, 1999.

13. Serra-Holm, T Salmi, P Mäki-Arvela, E Paatero, L P Lindfors. Comparison of Activity and Selectivity of Weakly Basic Anion-Exchange Catalysts for the Aldolization of Butyraldehyde with Formaldehyde. Org Process Res& Devel 5: 368-375, 2001.

14. S Bezzi, A Iliceto. Sul sistema acqua-formaldeide- I: Costituzione delle soluzioni acquose di formaldeide. La Chimica e l'Industria 33: 212-217, 1951.

15. F Lipari, S J Swarin. Determination of formaldehyde and other aldehydes in automobile exhaust with an improved 2,4-dinitrofenylhydrazine method. J Chromatogr: 297-306, 1982.

16. P H Bolt, F H P M Habraken, J W Geus. On the Role of NiAl$_2$O$_4$ Intermediate Layer in the Sintering Behaviour of Ni/α-Al$_2$O$_3$. J Catal 151: 300-306, 1995.

17. J March, The Aldol Condensation. In: Advanced Organic Chemistry. 3rd ed. New York: John Wiley and Sons, 1985, p. 829.

18. V Serra-Holm, T Salmi, J Multamäki, J Reinik, P Mäki-Arvela, R Sjöholm, L P Lindfors. Aldolization of Butyraldehyde with Formaldehyde over a commercial anion-exchange resin – Kinetics and Selectivity Aspects. Appl Catal A General 198: 207-221, 2000.

19. V Serra-Holm, T Salmi, P Mäki-Arvela, L P Lindfors. Kinetics and product distribution of competing aldolization and elimination in a three.phase reactor. Catal Today 66: 419-426, 2001.

20. J Kuusisto. Measurement of Solubility of Hydrogen , Viscosity and Density in some Hydrogenation Processes. (in Swedish) M.Sc. Thesis, Åbo Akademi, 2000

21. P W Selwood. Adsorption and Collective Paramagnetism. Academic Press, 1962.

22. S Smeds, T Salmi, L P Lindfors, O Krause. Chemisorption and TPD studies of hydrogen on Ni/Al$_2$O$_3$. Appl Catal A General 144: 177-194, 1996.

23. C Mirodatos, J A Dalmon, G A Martin. Steady-State and Isotopic Transient Kinetics of Benzene Hydrogenation on Nickel Catalysts. J Catal 105: 405-415,1987.

23

Preparation and Characterization of High Activity and Low Palladium-Containing Debenzylation Catalysts

J. P. Chen, Deepak S. Thakur, Alan F. Wiese, Geoffrey T. White, and Charles R. Penquite
Engelhard Corporation, Beachwood, Ohio, U.S.A.

ABSTRACT

Activated carbon supported palladium catalysts have been widely used in fine organic chemical synthesis. Some of the typical applications are debenzylation, hydrogenation, reductive alkylation, reductive amination, etc. As an effective synthetic method, debenzylation has been used commercially in organic synthesis to deprotect various functional groups.

Literature review shows that high Pd-containing catalysts are favored for the debenzylation application. Due to the high demand of palladium metal, its price has been escalating over the last few years. It has therefore become necessary to design a new generation of high performance palladium catalysts with significantly lower Pd metal content while maintaining good debenzylation activity.

There are four types of widely used debenzylation reagents: benzyl ethers, benzylamines, benzyl carbamates and benzyl carbonates. Debenzylation reactions can be carried out either by using gaseous hydrogen or by catalytic transfer hydrogenation employing a palladium catalyst. In this work, we will report and discuss the results on the de-protection of 4-(benzyloxy) phenol, benzyl ether and dibenzylamine over our newly-developed high performance palladium catalysts using gaseous hydrogen.

The catalytic performance results demonstrate that our newly developed 3%Pd/CPS (carbon powder support) has activity equal to current commercially supplied 5%Pd/CPS catalysts. This lower metal content Pd catalyst not only maintains good activity, but also achieves high product selectivity.

The effects of palladium metal oxidation states, palladium metal dispersion and other properties on catalyst performance are discussed in this paper. Results on the effects of reaction conditions on the reaction rate, such as mixing speed, reaction temperature, solvent and feed impurity, are presented in this paper as well.

INTRODUCTION

Debenzylation has been widely used in fine chemical and pharmaceutical industries as one of the most powerful techniques in organic synthesis to protect various functional groups. Search results on US patents related to debenzylation reveal that about fourteen hundred patents were granted from 1976 to July 2001. As shown in Figure 1, about 200 patents were granted to debenzylation related inventions every five years from 1976 to 1990. The importance of the debenzylation applications has been increasing over the past decade, with 331 patents granted between 1991 and 1995, and 496 patents granted from 1996 to mid-2001. Thus, patents related to debenzylation have more than doubled during the last decade.

Literature review also shows that high Pd-containing catalysts are favored for debenzylation applications. Due to the high demand of palladium metal, its price has been escalating over the last few years. It has, therefore, become necessary to design a new generation of high performance palladium catalyst with significantly lower Pd metal content while maintaining good debenzylation activity and selectivity.

Rylander [1], Freifelder [2] and Seif et al. [3] have reported that an activated carbon supported palladium catalyst is the most commonly used catalyst for de-

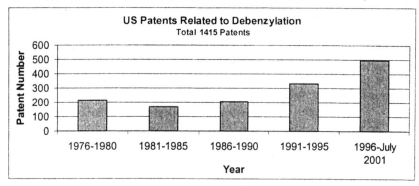

Figure 1 US patents related to debenzylation applications granted from 1976 to July, 2001.

benzylation applications because of higher activity and selectivity under mild reaction conditions. Using debenzylation not only can one improve the desired product selectivity, but also product yield and efficiency. Gao [4] has developed a more economical and efficient method of making optically pure isomers of formoterol by the reaction of an optically pure 4-benzyloxy-3-formamidostyrene oxide with an optically pure 4-methoxy-alpha-methyl-N-(phenylmethyl)-benzeneethanamine followed by debenzylation with a palladium catalyst. Reuschling [5] has successfully prepared highly *cis*-selective configuration of 4-substituted cyclohexylamines by reductive amination of 4-substituted cyclohexanol followed by debenzylation. This novel route has the advantage of high *cis*-selectivity and simplicity as compared with previous syntheses.

EXPERIMENTAL

Materials

Commercially supplied analytical grade substrates, benzyl ether, N-phenyl benzylamine and 4-(benzyloxy) phenol and the solvents (tetrahydrofuran, methanol, octanol, cyclohexane, acetone and methyl ethyl ketone) were used in this study.

Standard commercial catalyst samples were obtained directly from Engelhard. Catalyst samples were prepared via standard impregnation method. Palladium metal solution is introduced into a carbon powder slurry for absorption/impregnation. Catalyst preparation steps include absorption/impregnation, washing, filtration and drying. For pre-reduced catalysts, a reducing agent is introduced into the catalyst slurry before the filtration and washing steps.

Model Reactions Selection

To enable a relatively quick and easy evaluation of catalyst performance, the following three model reactions (Schemes A through C) were used in this catalyst development program.

Scheme A: Debenzylation of benzyl ether

Scheme B: Debenzylation of N-phenylbenzylamine

Scheme C: Debenzylation of 4-(Benzyloxy) phenol or hydroquinone monobenzyl ether

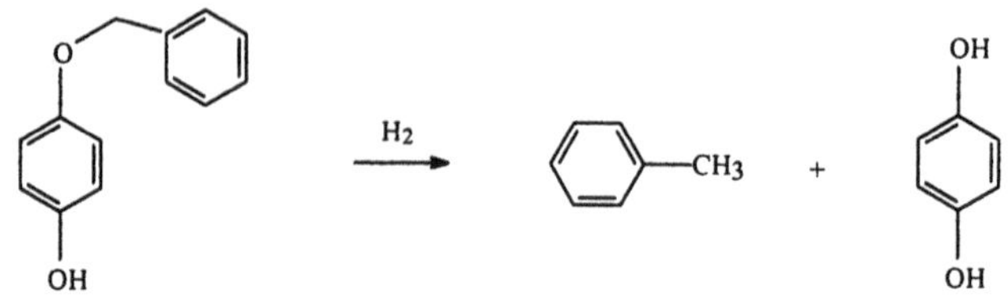

Evaluation Conditions

Figure 2 is a schematic representation of the experimental setup used for debenzylation reaction evaluation. The reactor is a 200 ml 2-neck flask. The reactor temperature is controlled by a constant temperature water bath temperature; Agitation rate is controlled by the rotation speed of the mixer (revolution per minute, rpm). Gas flow and pressure are maintained by Pressflow Model 1502 (Peteric Ltd.) control unit.

A fixed quantity (10 or 20 grams) substrate was dissolved in the desired solvent to a total volume of 100 ml. A catalyst loading level was selected at 3% or 5 wt% with respect to substrate. Reaction was carried out at an agitation rate of 2000 rpm (or other rate as specified in the text), a temperature range of 35 - 50°C and hydrogen pressure of 1.1 bar. Hydrogen uptake was continuously monitored and recorded during the reaction. GLC and GC/MS were used for product and feed composition analyses. Reaction rate constant calculations were based on a zero order reaction based upon hydrogen uptake within the first liter of hydrogen consumption. These were expressed in two different ways: Catalyst weight basis (moles/min./g cat), and Pd metal weight basis (moles/min./g Pd).

Catalyst Characterization

Catalyst particle size analysis was performed by using a Mastersizer S Analyzer (Malvern Instruments). Water is used to disperse the catalyst sample. Typical catalyst concentration is about 0.01 to 0.03%. In order to have improved catalyst dispersion in water, particle size distribution was measured after one minute ultrasound.

Catalyst filtration rate was measured by the time required for filtration of 350 ml of debenzylation product containing 10% 4-benzyloxy phenol in methanol and 1.2g of catalyst using a 55 mm diameter, #42 Whatman filter paper under 24" Hg vacuum.

CO pulse chemisorption measurements were carried out on a Micromeritics AutoChem 2910. Pd metal dispersion of catalyst was calculated from CO pulse chemisorption results. Before introducing CO, the catalyst samples were reduced at 75°C using 4% hydrogen in argon for 45 minutes. Catalyst samples were cooled to room temperature for CO chemisorption. Helium carrier gas flow rate was about 30 ml/min.

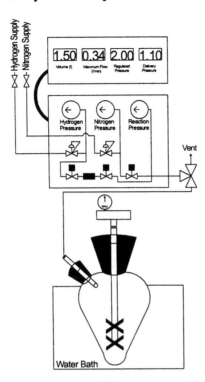

Figure 2 Schematic representation of the apparatus used for debenzylation reaction evaluation.

RESULTS AND DISCUSSION

Effects of Reaction Conditions

In this study, the effects of agitation speed and catalyst loading level on reaction rates were assessed at a constant hydrogen pressure of 1.1 bar. Debenzylation of both benzyl ether and N-phenylbenzylamine was carried out at 50°C, while that of 4-(benzyloxy) phenol was conducted at 35°C. Different debenzylation reac-tion models and catalysts will be characterized by distinct loading curves. To eliminate potential hydrogen diffusion limitation, one of the high activity catalysts, 5%Pd/CPS3, was selected for experimental parameters study.

Figures 3 and 4 show the effects of agitation speed and catalyst loading level on the reaction rate of benzyl ether debenzylation over 5%Pd/CPS3. The reaction rate remains nearly constant when the agitation speed exceeds 1500 rpm (Fig. 3) and the catalyst loading level is above 1.5 wt% (Fig. 4) under the given reaction conditions. To minimize hydrogen diffusion limitations, an agitation rate of 2000 rpm and a catalyst loading level of 1wt % were used for the benzyl ether study. Similar studies suggest using of 3% catalyst loading for both N-phenylbenzylamine and 4-(benzyloxy) phenol debenzylation studies.

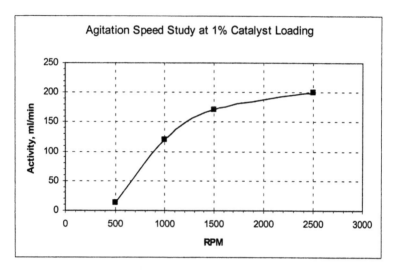

Figure 3 Effects of agitation speed on hydrogen uptake: 10% benzyl ether in THF solvent, 50°C, 1.1 bar hydrogen, 5%Pd/CPS3. The activity value is based on the time required for 1.0 liter hydrogen consumption.

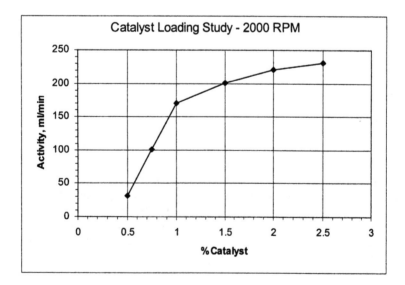

Figure 4 Effects of catalyst loading on activity: 10% benzyl ether in THF solvent, 50C°,1.1 bar hydrogen, 5%Pd/CPS3. The activity value is based on the time required for 1.0 liter hydrogen consumption.

Figure 5 depicts the hydrogen uptake curves of debenzylation of benzyl ether on 3Pd/CPS1 at 30, 40 and 50°C. Assuming zero order reaction, the rate constants at 30, 40 and 50°C are 4.7 X 10^{-3}, 9.2 X 10^{-3} and 1.58 X 10^{-2} moles/min./g catalyst respectively. Product analyses show that there is no selectivity change under these reaction temperatures, and hydrogenation of the benzene ring was not observed.

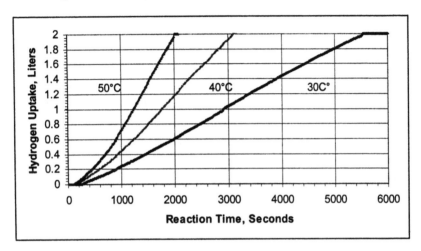

Figure 5 Hydrogen uptake curves for debenzylation of benzyl ether on 3%Pd/CPS1 at different temperatures. Other reaction conditions: 1.1 bar hydrogen pressure, 2000 rpm, 20 g benzyl ether diluted to 100 ml with THF, 0.333 g 3%Pd/CPS1 catalyst.

Newly Developed Catalysts vs. Current Commercial Catalysts

As indicated previously, our goal was to develop a new catalyst having lower palladium metal loading, while maintaining satisfactory activity and selectivity. The new catalyst has a Pd metal loading of 3wt%, which is 40% lower than the current commercial catalysts, 5%Pd/CPS2 and 5%Pd/CPS3. 5%Pd/CPS3 exhibits higher activity than 5%Pd/CPS2 for O-debenzylation.

The new 3%Pd/CPS4 catalyst has been tested for debenzylation of 4-(benzyloxy) phenol and N-phenylbenzylamine. Table 1 shows that under the same reaction conditions, this newly developed 3%Pd/CPS4 has approximately the same activity as the current commercially produced 5%Pd/CPS2 in both O- and N-debenzylation reactions on catalyst weight basis. Table 1 also shows that for 4-benzyloxy phenol debenzylation, the 3%Pd/CPS4 has lower activity than 5%Pd/CPS3 based on catalyst weight, however, these two catalysts have about the same activity based on palladium metal content. For N-phenyl benzylamine debenzylation, 3%Pd/CPS4 has the highest activity among these three catalysts.

Similar activities (based on catalyst weight) were similarly observed on 3%Pd/CPS4 and 5%Pd/CPS3 for 4-benzyloxy phenol when other solvents, e.g., n-octanol, THF, acetone, and acetic acid were employed.

Table 1 Comparison of Rate Constants Between New and Commercial Catalysts

| Catalyst | % Pd in cat | Loading Level, % | | Zero Order Rate Constant | |
		Catal. wt. Basis %	5% Pd basis	Moles/min/g cat. X 100	Moles/min/g Pd x 100
Debenzylation of 4-(benzyloxy) Phenol in methanol, 35°C					
5%Pd/CPS2	5	3	3	0.62	12.4
5%Pd/CPS3	5	3	3	1.03	20.5
3%Pd/CPS4	3	3	1.8	0.65	21.7
Debenzylation of N-Phenylbenzylamine in methanol, 50°C					
5%Pd/CPS2	5	3	3	1.53	30.5
5%Pd/CPS3	5	3	3	1.36	27.2
3%Pd/CPS4	3	3	1.8	1.61	53.6

Catalyst Particle Size vs. Filtration Rate

From a practical point of view, filterability of a powder catalyst in a slurry reaction is one of the critical properties. There are many factors that can impact the filtration rate, such as vacuum or pressure, temperature, slurry viscosity, catalyst particle size and shape, etc. Under given process conditions, catalyst particle size and its distribution play key roles on the filtration process. Smaller catalyst particle size and wider particle size distribution generally result in slower filtration rate. A narrower particle size distribution with proper particle size will provide a better filtration rate. Also, smaller particle size catalysts will enhance overall activity.

Figure 6 shows the percentage of the particles that are in the range from 0.05 to 1 micron for these three catalysts. 5%Pd/CPS3 has about 18% of the catalyst particles less than 1 micron. The other two catalysts, 5%Pd/CPS2 and 3%Pd/CPS4, have about 11~ 12% of their particles less than 1 micron diameter. In addition, they have about the same average particle size, 25 microns as shown in Table 2; however, 3%Pd/CPS4 has narrower particle size distribution than 5%Pd/CPS2 (Figure 7). Therefore, 3%Pd/CPS4 has the best filtration rate among these three catalysts.

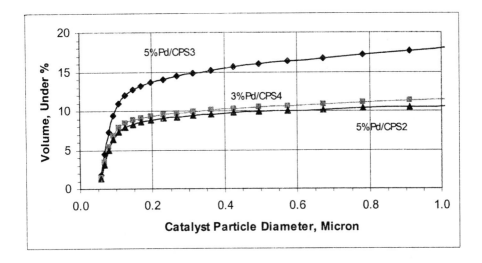

Figure 6 Catalyst particle size distribution of the Pd catalysts (Under % by volume is the percentage of particles of the sample less than the indicated diameter).

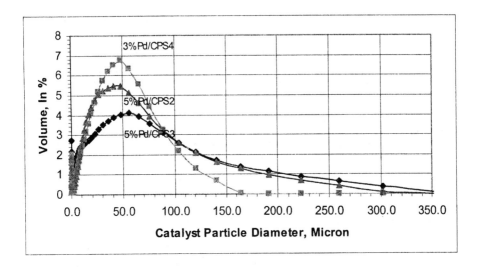

Figure 7 Catalyst particle size distribution of the Pd catalysts (In % by volume is the percentage of particles of the sample in this diameter).

Table 2 Catalyst Particle Size Distribution and Filtration Rate

Catalyst	Particle Size Distribution, microns			Filtration Time*
	D(volume,10%)	D(Volume,50%)	D(Volume, 90%)	
5%Pd/CPS2	0.57	24.8	87.8	7'11"
5%Pd/CPS3	0.10	16.5	93.7	10'33"
3%Pd/CPS4	0.3	25.5	69.6	5'40"

* Filtration time represented the time for filtration of 350 ml of debenzylation product of 10% 4-benzyloxy phenol in methanol and 1.2g of catalyst over 55 mm diameter #42 Whatman filter paper under 24" Hg vacuum.

Table 2 shows filtration rates and particle size distribution of the three catalysts discussed in the Table 1. The filtration rate of 5%Pd/CPS3 is the slowest among these three catalysts, filtration tome of 10'33" vs. 7'11" for 5%Pd/CPS2 and 5'40" for 3%Pd/CPS4. In a parallel experiments of filtration rate on CPSs, about the same filtration rates as their corresponding catalysts were observed. The slowest filtration rate of 5%Pd/CPS3 is due to its particle size distribution, i.e., 10% of this catalyst has 0.1 micron or smaller particle size.

Effect of Pre-activation on Catalyst Performance

It is of interest to evaluate the impact of catalyst pre-activation on performance. Catalyst manufacturers supply both pre-reduced and unreduced noble metal catalysts for a variety of applications. In order to assess the effect of Pd reduction, we evaluated unreduced and pre-reduced catalysts. Table 3 shows the reaction rate constants of debenzylation of 4-(benzyloxy) phenol on reduced and unreduced 3%Pd/CPS4 and 5%Pd/CPS4 under identical reaction conditions. These data clearly demonstrate the superior performance of unreduced catalysts over their pre-reduced counterparts. The reaction rate constants are significantly reduced when pre-reduced catalysts are used.

Analyses of palladium metal dispersion and CO chemisorption show that there is no clear relationship between reaction rate and palladium metal dispersion. There was a slight difference between metal dispersion of unreduced and pre-reduced 3%Pd/CPS4, whereas the metal dispersions of unreduced and pre-reduced 5%Pd/CPS4 are about the same. The reaction mechanisms for these two different type of catalysts are under further study in our laboratory.

Effects of Solvents on Activity and Selectivity

Like many other catalytic organic reactions, the debenzylation reaction is highly impacted by the organic solvents selected for the reaction. Solvents not only can change the catalytic reaction rate, but also can modify the reaction product distribution. Rylander [6] reviewed how to achieve selectivity improvement during hydrogenation with various solvents.

Table 3 Debenzylation of 4-(Benzyloxy) Phenol on Reduced and Unreduced Catalysts

Catalyst	Pd Metal Dispersion %	Pd Metal Surface Area m²/g cat.	CO Chemisorption ml/g cat.	Catalyst Reduction	Zero Order Rate Constant*	
					moles/min/ g cat x 100	Moles/min/ g Pd x 100
3%Pd/CPS4	34.7	4.63	2.19	No	0.64	21.32
3%Pd/CPS4	33.9	4.54	2.14	Yes	0.46	15.49
5%Pd/CPS4	30.4	6.78	3.21	No	1.12	22.31
5%Pd/CPS4	30.6	6.82	3.23	Yes	0.52	10.35

* Basis: 3% catalyst loading level, methanol solvent; 30°C, 2000 rpm

The combination of a good catalyst with a preferred solvent for the desired organic reaction can benefit the reaction system. Several solvents were used to study the effects of solvent on reaction rates and product selectivity on the newly-developed low palladium content catalyst. For N-phenylbenzylamine debenzylation, higher hydrogenation activity was obtained by employing methanol as a solvent. THF (tetrahydrofuran), and cyclohexane were found to be less effective. Studer and Blaser [7] studied the solvent effects on catalytic debenzylation of 4-chloro-N,N-dibenzyl aniline and found that the overall reaction rate was slower with the use of non-polar solvents.

Griffin, et. al. [8] studied the solvent (THF, water, acetone, MEK (methyl-ethyl ketone), MEK/water and water at pH 3) effects on hydrogenation activity of N-phenylbenzylamine. They claimed that there was no reaction between the ketone and the amine product. Aliphatic ketones were found to be the preferred solvents, with acetone being particularly effective.

In this study, three types of solvents, alcohol, ketones and alkane were used to study the effects of solvent polarity on hydrogenolysis activity. As shown in Table 4 the debenzylation (hydrogenation) reaction is favored by the use of high polarity (high dielectric constants, ε) solvents. Under the same reaction conditions, methanol (with the highest dielectric constant) is the best solvent among the three alcohol solvents for N-phenylbenzylamine (NPBA) conversion. Table 4 also shows that under the same reaction conditions, using non-polar solvent cyclohexane (dielectric constant ε is 2.02 at ambient temperature), results in low NPBA conversion, 27.1%.

Table 4   Effects of Solvents on N-Phenyl Benzylamine Debenzylation (NPBA) Reaction Activity and Selectivity on 3%Pd/CPS4

Solvent	ε^*	NPBA Conv. %	Product Distribution, %					Prod- ucts Total, %
			NPBA	Aniline	Toluene	Other Byproducts		
Methanol	33.0	95.0	5.3	44.9	48.8			93.1
Propanol	20.8	53.3	46.7	24.4	26.3			97.5
n-Octanol	10.3	31.9	68.1	13.3	14.7			96.1
Acetone	20.01	49.3	50.7	17.0	23.4	2.9 (1)	3.9 (2)	98.0
MEK	18.56	47.7	52.3	17.8	22.4	2.2 (3)	2.5 (4)	97.2
Cyclohexane	2.02	27.1	72.9	11.9	13.8			98.6

Reaction conditions: 3% catalyst loading with respect to NPBA, reaction time 3 hours, 50°C, 2000 rpm, 1.1 bar hydrogen
*ε Dielectric constants (or permittivity) at 293°K, from reference [9].
Acetone solvent: (1) = $C_9H_{11}N$ (2-methylindoline);
                                  (2) = $C_9H_{13}N$ (N-phenyl-isopropylamine);
MEK solvent: (3) = $C_{10}H_{13}N$;
                                  (4) = $C_{10}H_{15}N$.

Reasonable conversions of NPBA were observed when using ketones as solvents; however, product analysis showed increased byproducts formation with the use of ketones. When acetone is used as solvent, GC/MS analysis identified 2-methylindoline and N-phenylisopropylamine as the two major byproducts, which are produced via a secondary reaction between aniline (product), acetone and hydrogen (reductive amination). Similar byproducts were observed when using methyl ethyl ketone (MEK) as a solvent. In a parallel experiment, about 13% of aniline reacted with acetone under the same reaction conditions except in the absence of a catalyst. Most of the reaction product, about 11%, is non-hydrogenated product 2-methylindoline ($C_9H_{11}N$). These results clearly demonstrate that ketones are the least preferred solvents for N-debenzylation due to the secondary reaction between the desired product and the ketones.

Several Pd/CPS catalysts were also evaluated for O-debenzylation using 4-(benzyloxy) phenol substrate with methanol, n-octanol and THF as solvents. Substitution of n-octanol and THF for methanol resulted in lower hydrogenation activity.

Effect of Feedstock Impurities on Reaction Rate

The debenzylation reaction was found to be very sensitive to feedstock impurities. Figure 8 illustrates the hydrogen uptake of debenzylation reaction of 4-(benzyloxy) phenol from two different feedstock lots. As shown in Figure 8 that the hydrogen uptake was completed in 6000 seconds on a pure feed stock; however, under the same reaction conditions, only about 46% (0.55 liters) hydrogen uptake was completed on feedstock from a different lot. The GC-MS analyses show that the only difference between feedstocks is that the second feedstock contains about 0.5% 2-phenylmethyl 4-benzyloxy phenol, a triple ring compound.

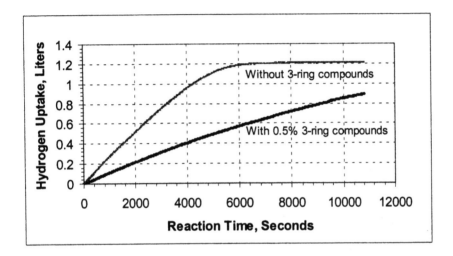

Figure 8 Effects of the feedstock impurity on the reaction rate for debenzylation of 4-(benzyloxy) phenol. Reaction conditions: methanol solvent, 35°C, 1.1 bar hydrogen, 2000 rpm, 3% catalyst loading (5%Pd/CPS3).

Based on hydrogen uptake curves, the reaction rate constant is 8.7×10^{-3} moles/g cat./min on the higher-purity feed stock, but only 3.3×10^{-3} moles/g cat./min on the lower-quality feedstock. Although only 0.5% 2-phenylmethyl, 4-(benzyloxy) phenol is contained in the feed stock, the reaction rate has been reduced by more than 50%.

One of the possible reasons for lower reaction rate on the second feedstock is an inhibition effect arising from the stronger adsorption of the impurity and its corresponding product onto the palladium metal surface. The poisoning action by this 3-ring compound and its hydrogenated counterpart is transient. Products analysis showed that the impurity was subsequently hydrogenated to its corresponding compounds: 2-phenylmethyl 1,4-benznediol and toluene (See the reaction scheme in Figure 9). It can be predicted from Figure 8 that the debenzylation reaction will take up more than 15,000 seconds to complete. This is more than double the reaction time for a pure feedstock, 6000 seconds. This is consistent with the reaction rate constant differences reported in the above discussion.

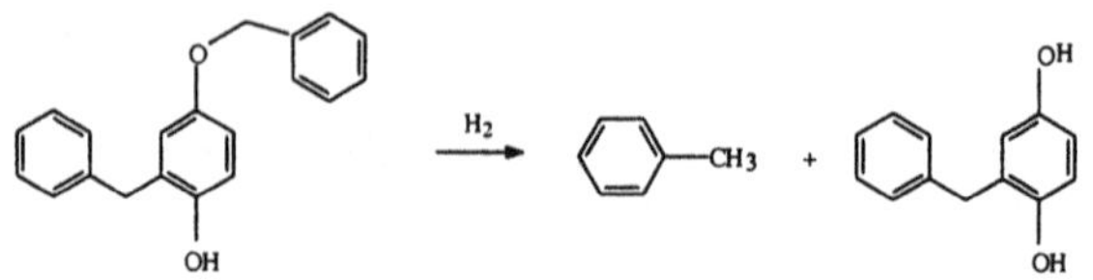

Figure 9 Hydrogenolysis reaction of 4-(benzyloxy) phenol impurity: 2-phenylmethyl, 4-benzyloxy phenol to 2-phenylmethyl 1,4-benznediol and toluene

CONCLUSIONS

The newly developed 3%Pd/CPS4 has equal or better activity than current commercial catalysts, 5%Pd/CPS2 and 5%Pd/CPS3 for a debenzylation application.

3%Pd/CPS4 has better filtration properties than 5%Pd/CPS2 and 5%Pd/CPS3 due to a narrower particle size distribution.

The unreduced Pd/CPS catalysts have higher debenzylation activity than their corresponding reduced counterparts.

Solvent type has a significant impact on debenzylation reaction rate and product selectivity. Use of polar solvents can result in better activity than non-polar or less polar solvents.

Ketones are not a good solvent for the -debenzylation reaction due to an undesirable secondary reaction between amine and ketones.

The debenzylation reaction is adversely affected by impurities in the feed stock and suggest a stronger adsorption of the contaminants on the active sites.

ACKNOWLEDGEMENT

The authors wish to thank Mr. Michael Baran, Ms. Eileen Davis and Mr. John Dormer for their support and contribution to this research effort.

REFERENCES

1. P. N. Rylander, Catalytic Hydrogenation over Platinum Metals, Academic Press, New York, 1967, p. 450.
2. M. Freifelder, "Practical Catalytic Hydrogenation, Technique and Applications", Wiley-Interscience, New York, 1971, p.398.
3. L. S. Seif, K. M. Partyka and J.E. Hengeveld, Selective Hydrogenolysis of Benzyl and Carbobenzylox Protecting Groups for Hydroxyl and Amino Functions, in Catalysis of Organic Reactions, ed. by Dale.W. Blackburn, Marcel Dekker, Inc. New York, 1990, p. 197.
4. Y. Gao, US Patent 6,268,533, Formoterol Process, July 31, 2001
5. D. Reuschling, Hoechst Schering Agrevo GMBH, WO 99/47487, Process for Preparing 4-substitute *cis*-cyclohexylamines, September 23, 1999

6. P. N. Rylander, "Use of Solvents to Achieve Improved Selectivities in Catalytic Hydrogenation, Chemical Catalyst News, Engelhard Corporation, October, 1989.

7. M. Studer and H-U, Blaster, J. Mol. Catal. A: Chemical, 112(1996), 437-445.

8. K.G. Griffin, S. Hawker and M.A. Bhatti, The Removal of Protecting Groups by Catalytic Hydrogenation, in Catalysis of Organic Reactions, ed. by Russell E. Malz, Jr., Marcel Dekker, New York, 1996, p.325.

9. C. Wohlfarth, "Permittivity (Dielectric Constant) of Liquids", in "CRC Handbook of Chemistry and Physics", David R. Linde, Editor-in-chief, 6-151, 82nd Edition, 2001-2002.

24

DuPHOS Rhodium(I) Catalysts: A Comparison of COD Versus NBD Precatalysts Under Conditions of Industrial Application

Christopher J. Cobley, Ian C. Lennon, Raymond McCague, James A. Ramsden, and Antonio Zanotti-Gerosa
Chirotech Technology Ltd., Cambridge, United Kingdom

ABSTRACT

We have examined the effectiveness of cyclooctadiene (COD) and norbornadiene (NBD) precatalysts of the type [Rh(DuPHOS)(diolefin)]BF_4 in catalytic asymmetric hydrogenation of prochiral olefins under conditions suitable for industrial application. The NBD precatalyst can give rise to the catalytically active species more rapidly than the corresponding COD complex, but as catalyst loadings were reduced to improve the process economics, the difference between the use of COD and NBD precatalysts became insignificant. For some substrates, identical reaction profiles were observed for either precatalyst, but an important determinant of reaction rate was the efficiency of hydrogen introduction to the mixture. In addition, the effectiveness of a catalyst loading of 50,000/1 for the hydrogenation of dimethyl itaconate with [Rh(DuPHOS)(COD)]BF_4, demonstrates that this precatalyst can be put to highly economic use.

INTRODUCTION

In a series of publications Heller et al (1,2) have examined the relative reactivity of rhodium complexes bearing cyclooctadiene (COD) and norbornadiene (NBD) ligands towards hydrogenation. Cationic complexes of rhodium bearing a chiral diphosphine ligand such as DuPHOS 1 are industrially important for the asymmetric hydrogenation of many prochiral olefins (3). The precatalysts are typically of the composition [Rh(DuPHOS)(diolefin)]BF$_4$, the active catalysts being generated by the removal of the diene ligand (e.g. cyclooctadiene or norbornadiene) *via* hydrogenation.

Recently, the asymmetric hydrogenation of two related substrates with precatalysts of the type [Rh(DuPHOS)(diolefin)]BF$_4$ has been investigated (4). It was demonstrated that the COD-containing complex required an induction time for the COD ligand to be removed from the precatalyst. This manifested itself as very low initial rate of hydrogen uptake, which increased as more catalyst became available. Conversely, for the NBD precatalyst there was no observable induction time and the hydrogenation reactions were complete in a fraction of the time required for the COD containing systems. Furthermore, at the end of the reactions they were able to show that in the case of the COD systems, approximately half the precatalyst introduced to the reaction system had not been converted to active catalyst.

Heller and Börner concluded *that the use of NBD precatalyst in the asymmetric hydrogenation has significant advantages over the application of the usually sold and applied COD complexes. By utilization of NBD precatalysts costs of ligands and catalysts can be significantly reduced.* They further asserted *that expensive DuPHOS ligands are wasted by more than 50%.*

Although we were initially surprised by these conclusions, a more detailed examination assured us that the conditions they had chosen for their studies; low hydrogen pressure, high dilution and high catalyst loading [molar substrate to catalyst ratio (S/C) of 100/1] were far removed from anything that may be regarded as economically viable for industrial application. Therefore, we decided to investigate the relative merits of NBD and COD precatalysts under conditions which, in our experience, more accurately reflect those found in the commercial utilization of asymmetric hydrogenation catalysts (5). In particular, we sought to investigate the use of lower catalyst loadings, since with a high value catalyst this is a critical parameter for process economy.

RESULTS AND DISCUSSION

Four substrates were chosen for our study, methyl-2-acetamidoacrylate 2, methyl acetamidocinnamate 3, dimethyl itaconate 4 and the Candoxatril precursor 5 (6) (Figure 1). The precatalysts [(*R,R*)-Me-DuPHOS Rh (NBD)]BF$_4$, (*R,R*)-7a, and [(*S,S*)-Et-DuPHOS Rh (NBD)]BF$_4$, (*S,S*)-7b, were prepared according to literature procedures (2). [(*S,S*)-Me-DuPHOS Rh (COD)]BF₄, (*S,S*)-**6a**, and [(*R,R*)-Et-DuPHOS Rh (COD)]BF₄, (*R,R*)-**6b** are commercially available (Figure 2).

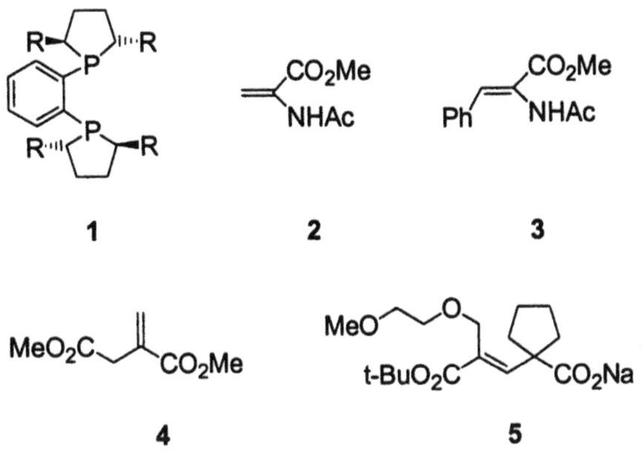

Figure 1 DuPHOS ligand and range of substrates chosen for the study.

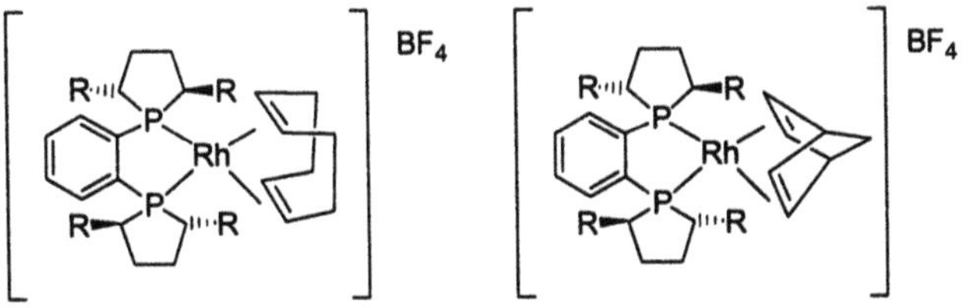

(*R,R*)-**6a** R=Me
(*R,R*)-**6b** R=Et

(*R,R*)-**7a** R=Me
(*R,R*)-**7b** R=Et

Figure 2 [(*R,R*)-DuPHOS Rh (COD)]BF₄ and [(*R,R*)-DuPHOS Rh (NBD)]BF₄ precatalyts.

Hydrogenation of methyl-2-acetamidoacrylate **2** with (*R,R*)-**7a**, at 1000/1 molar substrate/catalyst (S/C) ratio was very fast and reactions were complete in less than 10 minutes giving (*R*)-methyl *N*-acetyl alanine in > 99 % e.e. (3 bar H₂ pressure, room temperature). Under the same conditions, (*S,S*)-**6a**, gave (*S*)-methyl *N*-acetyl alanine in > 99 % e.e. Competition reactions were carried out in order to gain a measure of the relative productivity of the two catalysts. These experiments involved using an equimolar mixture of (*S,S*)-**6a** and (*R,R*)-**7a** in the same reaction. Since the precatalysts have opposite enantiomers of ligand, then the closer the overall productivity given by the NBD and COD precatalysts, the closer the product should be racemic. At a total S/C of 1,000/1, the (R)-isomer of product was obtained with 46 % e.e., reflecting a faster activation of the NBD precatalyst. However, at the more industrially acceptable substrate to catalyst loading of 10,000/1 the reaction was > 95 % complete after 60 mins and the product was obtained in only 15 % e.e. (R) (7).

Methyl acetamidocinnamate **3** was used in a more detailed study and allowed direct comparison with earlier reported work (2,4). The hydrogenation reactions were performed with a S/C = 2,000/1 of the Rh-Et-DuPHOS catalyst in methanol at 26 °C and 3 bar H_2 pressure. In the case of the COD containing complex (R,R)-**6b**, we observed an induction period qualitatively consistent with Heller's data (Figure 3). At the end of the induction period all of the COD precatalyst (R,R)-**6b** has generated the active catalyst. Thereafter, both reactions proceeded at the same rate giving products of > 99 % e.e. The difference in the overall reactions times (~5 min between (S,S)-**7b** and (R,R)-**6b**) is due to the induction time observed for the COD precatalyst (R,R)-**6b**. In a competition experiment with a 1:1 mixture of (R,R)-**6b** and (S,S)-**7b**, the product was obtained in 23 % e.e. (S)-isomer, which is produced by the catalyst generated from the NBD complex, (S,S)-**7b**. When the competition experiment was repeated at S/C = 5,000/1, the product was obtained after 50 min in 7 % e.e. (S), indicating that at lower catalyst loadings the two precatalysts tend towards equal effectiveness.

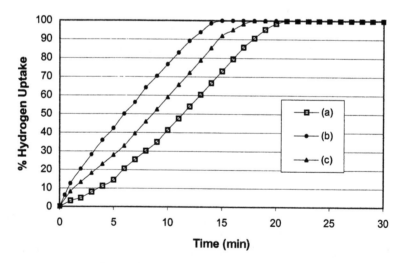

Figure 3 Hydrogenation of **3** comparing (a) (R,R)-**6b**, (b) (S,S)-**7b** and (c) 1:1 mixture of (R,R)-**6b**, and (S,S)-**7b**.

Dimethyl itaconate **4** is a very reactive substrate towards hydrogenation and complete conversion is achievable at low catalyst loadings. With a Rh-

MeDuPHOS catalyst at a S/C = 10,000/1 in methanol at 25 °C and 5 bar H_2 pressure, we observed no significant difference in the rates over the course of reaction between COD and NBD precatalysts (Figure 4). In both cases the reactions were complete in 50 mins with enantiomeric excesses in the range 97.4-98.0 %. Moreover, there was no evidence of any induction period and competition experiments with 1:1 (*S,S*)-**6a** and (*R,R*)-**7a** gave essentially racemic product (< 3 % e.e). Under these conditions, both the COD or NBD complexes are equally effective precatalysts. Further development was carried out on this process and a molar S/C = 50,000/1 was achieved for the hydrogenation of **4** using (*S,S*)-**6a**, this corresponds to a $^w/_w$ catalyst loading of 13,500/1. The reaction was complete in 10 h and the product was obtained in 97.5 % e.e. At this level of catalyst loading the cost of catalyst required to process 1 Kg of dimethyl itaconate is approximately 70 % of the cost of this very inexpensive substrate and < 1 % the value of the product (8). Using the reported loadings of S/C=100/1 (4), the catalyst cost would be approximately 350 times the cost of the substrate and more than 3 times the price of an existing catalogue supplier of the product.

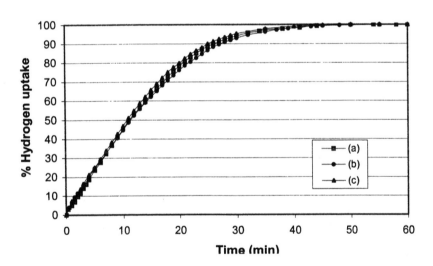

Figure 4 Hydrogenation of **4** comparing (a) (*S,S*)-**6a**, (b) (*R,R*)-**7a** and (c) 1:1 mixture of (*S,S*)-**6a**, and (*R,R*)-**7a**.

A series of ^1H and $^{31}P\{^1H\}$ NMR experiments shed some light on the reaction kinetics observed for the hydrogenation of methyl acetamidocinammic acid **3** and dimethyl itaconate **4**. Neither precatalyst (*S,S*)-**6b** or (*R,R*)-**7b** reacted with the substrates in the absence of hydrogen suggesting the possibility that neither precatalyst was activated by the substrate. When the NBD complex was treated with hydrogen in the absence of substrate the precatalyst was rapidly converted to the bis-solvate species **8** with the concomitant hydrogenation of the diolefin (Figure 5). These bis-solvate species are generally believed to be an active catalytic species. In the absence of hydrogen these species will form a stable complex **9** (Figure 6) with a substrate (9). These complexes react very rapidly with hydrogen to give the hydrogenation product. In a parallel experiment the COD complex reacted slowly with hydrogen to form the bis-solvate species.

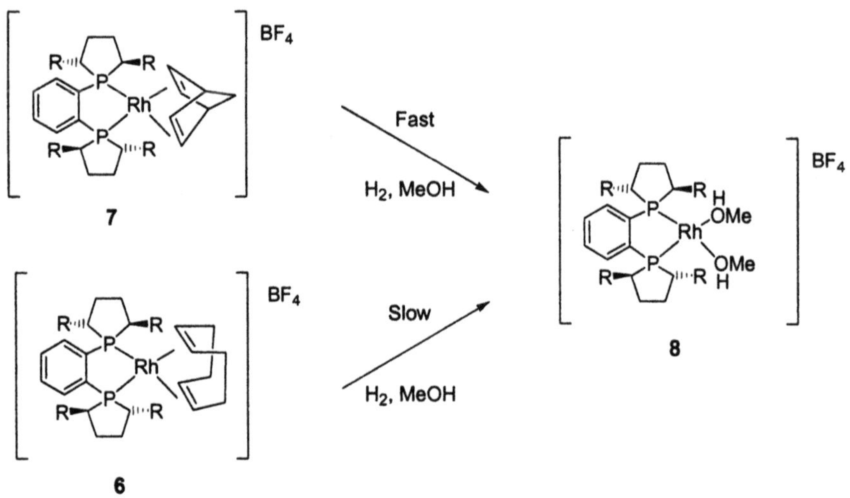

Figure 5 Reaction of NBD and COD precatalysts with H_2 in the absence of substrate.

When a mixture of NBD complex and 5 equivalents of the substrate was treated with approximately 2 equivalents of hydrogen two species were observed in the $^{31}P\{^1H\}$ NMR spectra: the unreacted NBD complex **7** and the substrate adduct **9**. Approximately 40 % of the substrate was converted to product. Using the COD precatalyst the only species observed in the $^{31}P\{^1H\}$ spectra was unreacted precatalyst **6** but there was a similar amount of hydrogenation product to that observed with the NBD complex. It is believed that in the first instance, in both experiments, a small amount of the precatalyst is hydrogenated into the active catalyst. In the case of the COD complex activation of the precatalyst is significantly slower than subsequent hydrogenation of the substrate; consequently any further hydrogen that dissolves in the reaction solvent is converted to product. The amount of active catalyst produced under these reaction conditions is below the detection limit of the NMR experiment. In the case of the NBD precursor the precatalyst is activated at a similar rate to the subsequent hydrogenation of the sub-

strate, consequently during the course of the reaction both active catalyst and hydrogenation product are produced. Once all the hydrogen has been consumed the
active catalyst reverts to its rest state, the substrate complex **9** (Figure 6).

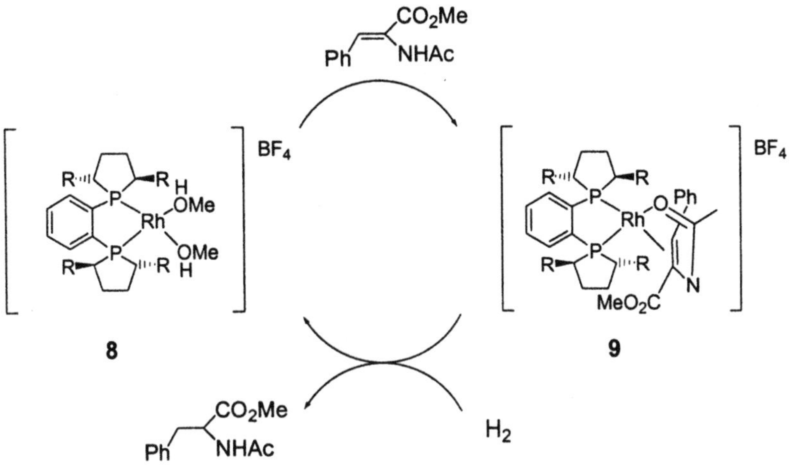

Figure 6 Catalytic cycle for the hydrogenation of methyl acetamidocinnammic acid.

It has been demonstrated in the NMR experiments and, more thoroughly in
the work of Heller (1), that the NBD complex is converted to active catalyst considerably more rapidly than in the case of the COD precatalyst. Yet, the competition reaction with dimethyl itaconate **4**, at low catalyst loading, gave essentially
racemic product. In both the NMR experiments and the larger scale hydrogenation, the rate of consumption of hydrogen by the substrate is dictated by the limitation of the slow rate of hydrogen dissolution into the reaction medium. These
reactions are therefore taking place in a regime where the overall rate is governed
by mass transfer of hydrogen rather than the intrinsic reactivity of the catalysts.
Part way through the reaction there will be both unconverted diolefin-rhodium
precatalyst and substrate-rhodium complex. With hydrogen limited, if the substrate-rhodium complex shows a higher affinity for hydrogen than the precatalyst,
then substrate will be converted to product and the precatalyst could be held up
unconverted. On the other hand if the precatalyst complex has a sufficient affinity
for the hydrogen, then that will have time to be converted to active catalyst regardless of whether the diolefin is COD or NBD. This may be the situation in the case
of the asymmetric hydrogenation of dimethyl itaconate **4**.

For our final example, we chose the Candoxatril precursor **5** as we had already developed a multi-kilogram hydrogenation process for this substrate (6). As

for dimethyl itaconate, no induction time was seen and both precatalysts, (*S,S*)-**6a** and (*R,R*)-**7a**, gave practically identical reaction profiles (S/C = 4,200/1, 5 bar H_2 pressure, MeOH, 26 °C) (Figure 3). The reactions were complete in 20 mins, providing product in > 98 % ee.

It is worth highlighting that under less efficient stirring conditions, the reaction took 90 min to go to completion. The hydrogenation reaction was conducted in a 600ml PARR hydrogenation vessel configured with two impellers. When the upper impeller is positioned just below the surface of the reaction solution, a more efficient transfer of hydrogen into solution is achieved (Figure 7, curves a and b) than when the impeller is placed deep within the bulk of the solution (Figure 7, curves c and d). This implies that for this type of hydrogenation, the reaction rate is dependent on the hydrogen mass transfer into solution (10).

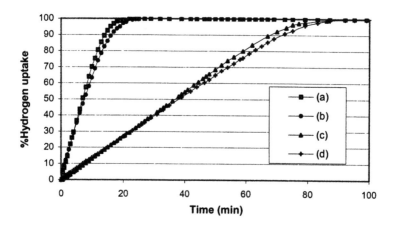

Figure 7 Hydrogenation of **5** comparing (a) (*S,S*)-**6a**, (b) (*R,R*)-**7a**, (c) (*S,S*)-**6a** with less efficient stirring and (d) (*R,R*)-**7a** with less efficient stirring.

Again we carried out competition reactions with a 1:1 mixture of (*S,S*)-**6a** and (*R,R*)-**7a** under the two stirring regimes. Not unexpectedly the reaction profiles were identical to the single catalyst profiles. In both cases the product had a small enantiomeric excess of 10-12 % of the (*R*) isomer produced by (*S,S*)-**6a**.

CATALYST IMMOBILIZATION

During recent years many processes (3,6) have been developed utilizing Rh-DuPHOS catalyzed hydrogenation. Substrate to catalyst ratios of up to 100,000/1 have been achieved for certain substrates (11) and reactions routinely run at S/C of

10,000-20,000/1, as outlined in this paper. An alternative approach to achieving economical use of precious metal catalysts is the immobilization of the homogenous precatalyst. Many attempts have been made in this direction with variable results being obtained (12). The Augustine immobilization method (13) has been successfully applied to the achiral precatalyst, [Rh(DiPFc)(COD)]BF$_4$ and was shown to be useful for a range of chemoselective hydrogenations (14,15). However, when [Rh(DuPHOS)(COD)]BF$_4$ precatalysts were immobilized and used for the asymmetric hydrogenation of methyl-2-acetamidoacrylate **2**, enantiomeric excess values of 92-95 % (variable over 4 runs) were obtained, using S/C of 100/1, even though four recycles were demonstrated (15). The homogenous reaction operates reproducibly at S/C of 10,000/1, providing product of >99 % ee. At the current state of development for immobilization technology, we conclude that for most fine chemical batch processes, using the highly active Rh-DuPHOS catalysts, it is better to optimize the catalyst loading for a single pass to obtain optimum economy. This is supported by the recently reported work on the manufacture of Metolachlor, where the authors state that the immobilized xyliphos catalysts showed lower activity and productivity. The decision was taken to use the homogenous iridium catalyst for multi-ton manufacture (16).

CONCLUSIONS

In conclusion, we have shown that when loadings of rhodium-DuPHOS precatalysts are reduced to levels conducive to economic industrial manufacture, the difference between the use of COD and NBD precatalysts becomes insignificant. Furthermore, with certain substrates and conditions we see no induction period with the COD precatalyst. This indicates that generally the more readily available COD precatalysts are appropriate for use under industrial conditions, at least in the case of rhodium-DuPHOS catalysts. The absence or presence of any induction period depending on the substrate or conditions we explain to be associated with effects of hydrogen availability at the catalyst and in particular by the rate of uptake of hydrogen by the COD precatalyst relative to that of the substrate-catalyst complex. Certainly, the hydrogen availability, as determined by the pressure vessel configuration, was critical to the overall hydrogenation rate. Finally, we have demonstrated that substrate to catalyst ratios of between 10,000 to 50,000/1 can be readily achieved for the hydrogenation of dimethyl itaconate **4** using [(R,R)-Me-DuPHOS Rh (COD)]BF$_4$ as the precatalyst, which negates the need for catalyst immobilization.

Experimental

In a typical experiment a 600 mL Parr hydrogenation vessel is charged with a MeOH solution of the substrate and pressurized with nitrogen to 10 bar under vigorous stirring. The system is allowed to equilibrate over 20 minutes before releasing the pressure. This procedure is repeated three times with nitrogen and three times with hydrogen. A solution of the catalyst in deoxygenated MeOH is introduced *via* syringe. The vessel is quickly purged a further three times with hydrogen (less than 20 seconds required) and pressurized to the required reaction pressure (gauge reading). The amount of hydrogen uptake is monitored at standard intervals and the pressure constantly maintained within 0.5 bar of the initial pressure reading. When no further hydrogen consumption is detected the pressure is released and samples taken for conversion and selectivity analysis (5). The curves of hydrogen uptake versus time are normalised to 100 % uptake.

Asymmetric Hydrogenation of Dimethyl Itaconate

The reaction was carried out in a 600 mL pressure vessel equipped with a mechanical stirrer, cooling/heating coil, injection port, temperature probe and fitted with a glass liner. Dimethyl itaconate (30.0 g, 0.190 mol) was added to the glass liner, followed by MeOH (273 mL). The vessel was assembled and pressurized to 10 bar with nitrogen. The reaction was stirred at the maximum rate (~750 rpm, stirrer assembly consisted of two propellers, one positioned just below the solvent surface and one near the bottom of the solution) for 30 minutes during which time the temperature was maintained at 20 °C (internal). The pressure was then released. In order to ensure an oxygen free atmosphere was attained, the vessel was recharged with nitrogen (10 bar) and after 10 minutes the vessel purged. This was repeated three times. After the final vent, the vessel was charged and purged three times with hydrogen (10 bar). [(S,S)-Me-DuPHOS Rh (COD)]BF$_4$ (S,S)-**6a**, (2.2 mg, 0.037 mmol, molar S/C 50,000/1) was added as a MeOH solution (10 mL, anhydrous and degassed by sparging with nitrogen prior to solution preparation) *via* syringe. The vessel was then charged with hydrogen (5 bar). The temperature was maintained at 20 °C throughout the course of the reaction and the hydrogen pressure readjusted to 5 bar after every 0.4 bar of hydrogen was consumed. After 10 h, no further hydrogen uptake was observed. The hydrogen pressure was released and the system flushed once with nitrogen (10 bar) prior to vessel disassembly. The reaction mixture was then analyzed by ^{1}H NMR and chiral GC for conversion and enantioselectivity. (complete conversion, 97.5 % e.e.). GC analysis: G-TA column, 40 °C to 130 °C at 5 °C/min, then to 170 °C at 15 °C/min, 21.8 min (S), 22.05 min (R).

ACKNOWLEDGEMENTS

We are grateful to Dr. Steve Challenger and Pfizer Ltd. for supporting this work through their supply of the Candoxatril precursor and useful discussion. We also thank Natasha Cheeseman of the *Chiro*Tech analytical team for her skilled technical assistance.

REFERENCES

1. H. J. Drexler, W. Baumann, A. Spannenberg, C. Fischer and D. Heller. *J. Organomet. Chem.*, **621**, 89 (2001).
2. D. Heller, S. Borns, W. Baumann and R. Selke, *Chem. Ber.*, **129**, 85 (1996).
3. M. J. Burk, *Acc. Chem. Res.*, **33**, 363 (2000).
4. A. Börner and D. Heller, *Tetrahedron Lett.*, **42**, 223 (2001).
5. C. J. Cobley, I. C. Lennon, R. McCague, J. A. Ramsden and A. Zanotti-Gerosa. *Tetrahedron Lett.*, **42**, 7481 (2001).
6. M. J. Burk, F. Bienewald, S. Challenger, A. Derrick and J. A. Ramsden, *J. Org. Chem.*, **64**, 3290 (1999).
7. Experimental conditions: MeOH, 3 bar H_2 initial pressure, room temperature. At S/C 1000/1: [Substrate]=1 M, [Rh]=1 x 10^{-3} M; at S/C 10000/1:[Substrate]=2 M, [Rh]=2 x 10^{-4} M.
8. The comparative costings are based on the catalogue prices for dimethyl itaconate (£50/500g, Lancaster) and (*R,R*)-**6a** (£92/100mg, STREM.), c.f. (+)-dimethyl (*R*)-methylsuccinate (£50/5g, Aldrich).
9. J. M. Brown and P. A. Chaloner, *J. Am. Chem. Soc.*, **102**, 3040 (1980).
10. Y. Sun, R. N. Landau, J. Wang, C. LeBlond and D. G. Blackmond. *J. Am. Chem. Soc.*, **118**, 1348 (1996).
11. Chirotech Technology Limited; unpublished results.
12. F. R. Hartley in Supported Metal Complexes. Dordrecht, (Reidel) 1985.
13. R. Augustine, S. Tanielyan, S. Anderson and H. Yang, *Chem. Commun.*, 1257 (1999).
14. M. J. Burk, A. Gerlach and D. Semmeril, *J. Org. Chem.*, **65**, 8933 (2000).
15. J. A. M Brandts, J. G. Donkervoort, C. Ansems, P. H. Berben, A. Gerlach and M. J. Burk in Catalysis of Organic Reactions. New York, (Marcal Dekker), 573-581 (2000).
16. H-U Blaser and F. Spindler. *Chimica Oggi/chemistry today*, July/August:17-20 (2001).

25

Enhanced Regioselectivity in Batch and Continuous Hydrogenation of 1-Phenyl-1,2-Propanedione over Modified Pt Catalysts

Esa Toukoniitty, Päivi Mäki-Arvela, Ville Nieminen, Matti Hotokka, Juha Päivärinta, Tapio Salmi, and Dmitry Yu. Murzin
Åbo Akademi University, Åbo/Turku, Finland

INTRODUCTION

In catalysis of organic reactions one of the most important requirements is the selectivity, broadly understood as chemo-, regio- and enantioselectivity [1]. The specific feature of many organic systems is that complex molecules possess several functional groups which adsorb differently on the catalyst surface and give a wide variety of products [2]. The selectivity towards a desired product can be controlled by the choice of catalyst, solvent and reaction conditions. The mode of adsorption is in some cases detrimental for the selectivity towards a particular product [2].

The importance of a specific adsorption mode is recognized especially in heterogeneous enantioselective catalysis, where intimate interactions of prochiral molecule with a modifier result in a specific enantioselectivity [3]. The role of the modifier is to steer the adsorption of the reagent in such a way that enantioselective hydrogenation is enabled. Heterogeneous enantioselective hydrogenation of α-keto esters over Pt catalysts modified with alkaloids has been studied intensively during recent years [4, 5, 6]. However, α-keto esters have just one reactive center and thus enantioselectivity aspects can be considered only. Diones, which possess two reactive carbonyl groups, can be hydrogenated over analogous catalytic system, and both enantio- and regioselectivity can be assessed. Very few publications exist concerning the enantioselective hydrogenation of diones [7, 8, 9]. The hydrogenation products, chiral α-hydroxyketones, are valuable building blocks in the asymmetric synthesis of biologically active compounds [10, 11].

In the hydrogenation of diones, hydroxyketones and diols are formed as products. Up to this moment, in the hydrogenation of diones, the main focus has been in the enantio- and diastereoselective product formation, while regioselectivity has received less attention. Regioselectivity in this context is understood as the preferable reduction of one of the two carbonyl groups. In the hydrogenation of asymmetric diones, e.g. 1-phenyl-1,2-propanedione and 2,3-hexanedione [12], the carbonyl groups have different chemical properties and the hydrogenation leads to the formation of two different regioisomers.

Keeping in mind the industrial applications of the products, the maximization of the yield of the desired product includes both the maximization of enantioselectivity and regioselectivity, and therefore the regioselectivity should not be overlooked. The present study mainly concerns the hydrogenation of 1-phenyl-1,2-propanedione, which gives possibilities to investigate regioselectivity due to its unsymmetrical structure.

Both batch and continuous reactor operation were studied. The latter was utilized, as although, fine chemicals are usually produced in batch-wise operating slurry reactors, continuous reactors offer several benefits, e.g. minimization of the time between batches, exclusion of catalyst separation and abrasion problems.

EXPERIMENTAL

Batch Reactor

1-Phenyl-1,2-propanedione (Aldrich, 99%) was hydrogenated in a pressurized reactor (Parr 4560, V=300 cm^3) in the absence of external and internal mass transfer limitation (verified experimentally). The reactor was equipped with an propeller type stirrer (four blades, propeller diameter 35 mm) operating at stirring rate of 1950 rpm. The hydrogen (AGA, 99.999%) pressure was 6.5 bar and temperature was 15 – 35°C. Pt/Al$_2$O$_3$ (Strem Chemicals, 78-1660) was used as a catalyst. The catalyst mass and liquid volume were 0.15 g and 150 cm^3, respectively The metal content was 5 wt.%, BET specific surface area 95 m^2 / g, the mean metal particle size 8.3 nm (XRD), dispersion 40% (H$_2$ chemisorption), the mean catalyst particle size 18.2 μm (Malvern). Catalysts were activated under hydrogen flow (100 cm^3 / min) for 2 h at 400°C prior to the reaction.

An in situ modification procedure was adopted: the deoxygenated solution, containing the solvent (ethyl acetate, FF Chemicals, 99.8%), the modifier (cinchonidine, Aldrich, 96%) and the substrate were injected into the reactor, where the activated catalyst was under hydrogen. The reaction commenced immediately. The modifier-to-catalyst mass ratio was 1:150 to 120:150. The initial concentrations of 1-phenyl-1,2-propanedione and cinchonidine were varied between 0.01-0.025 mol/dm^3 and $2.3 \times 10^{-5} - 2.7 \times 10^{-3}$ mol/dm^3, respectively.

Maximally 15 samples ($V_s = 1$ cm^3) were withdrawn from the reactor at different time intervals and analyzed with a gas chromatograph (GC) (Varian 3300) equipped with a chiral column (β-Dex 225; length 30 m, diameter 0.25 mm, film thickness 0.25 μm). Details of analytical procedure are presented in [9].

Fixed Bed Reactor

Hydrogenation of 1-phenyl-1,2-propanedione was carried out in a fixed-bed reactor (catalyst bed length 1.5 cm and 1.2 cm diameter) equipped with a back pressure regulator (Fairchild 10BP) at 25°C and 5 bar hydrogen in the absence of external mass transfer limitations (verified experimentally). A space time of 44 s (calculated based on the geometric bed volume and liquid flow rate) was used. The hydrogen flow was maintained constant with a mass flow controller (Brooks 5850E) and the liquid flow was controlled with a metering pump (Eldex CC-100-S-2CE).

In fixed-bed reactor technologies, an important factor is the pressure drop over the catalyst bed. In order to minimize the mass transfer effects, knitted silica fiber catalysts having small diffusion distance (< 10 μm) and low pressure drop, were applied in this study. 10 layers (0.4 g) of the catalyst was placed in the reactor between stainless steel nets and glass beads were used as inert packing material. The catalyst layer thickness was 0.15 cm resulting in catalyst bed length of 1.5 cm. The catalyst preparation and characterization procedures were described previously [13]. The main results from the catalyst characterization are summarized as follows: The Pt-loading was 5 wt.-%. The specific surface area (BET) of the support material was 59 m^{2}g^{-1}. The metal dispersion and the specific surface calculated per metal weight were 30.5% and 83.8 m^{2}g$^{-1}_{Pt}$ giving a mean Pt particle size of 3.3 nm.

Prior to the reaction, the catalyst was reduced *in situ* under flowing hydrogen at 400°C and 1 bar for 2 h. The liquid phase containing the modifier, the solvent (ethyl acetate), and the reactant, was degassed with hydrogen for 10 min before commencing the reaction. The experiments were carried out with co-current downward gas (60 cm^3 min^{-1}) and liquid flows (2.3 cm^3 min^{-1}). The concentrations of reactant and modifier were varied between 0.0075 - 0.05 mol dm^{-3} and $0 – 34 \times 10^{-5}$ mol dm^{-3}, respectively.

REACTION NETWORK

The main product of the hydrogenation of 1-phenyl-1,2-propanedione (**A**) was (*R*)-1-hydroxy-1-phenylpropanone (**B**, Figure 1), which is a key intermediate in the synthesis of L-ephedrine, pseudoephedrine, norephedrine, norpseudoepehdrine as well as adrenaline, amphetamine, methamphetamine, phenylpropanolamine and phenylamine [14]. The reaction scheme for the hydrogenation of **A** is displayed in Figure 1, where the reactant **A** is located in the center. The reactant has two carbonyl groups, which both can be hydrogenated under reaction conditions. In the absence of a catalyst modifier, hydrogenation of the phenyl-ring takes place. Therefore, the reaction network becomes more complex in the absence of the modifier and instead of eight products comprises of sixteen products. However, even in the absence of a modifier the yield of cyclohexyl derivatives remains low (<5%). Typical concentration vs. time plot is displayed in Figure 2 and it can be seen that the first hydrogenation step is rapid, whereas further hydrogenation to diols is somewhat slower.

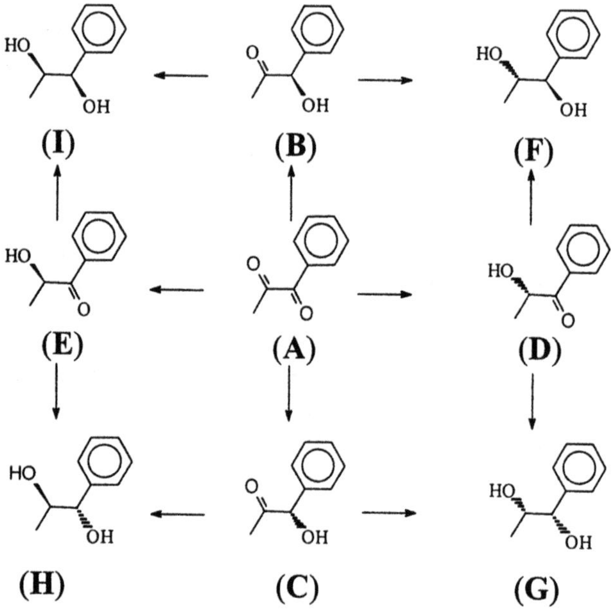

Figure 1 Reaction scheme of 1-phenyl-1,2-propanedione hydrogenation. A: 1-Phenyl-1,2-propanedione, **B**: (R)-1-Hydroxy-1-phenylpropanone, **C**: (S)-1-Hydroxy-1-phenylpropanone, **D**: (S)-2-Hydroxy-1-phenylpropanone, **E**: (R)-2-Hydroxy-1-phenylpropanone, **F**: (1R,2S)-1-Phenyl-1,2-propanediol, **G**: (1S,2S)-1-Phenyl-1,2-propanediol, **H**: (1S,2R)-1-Phenyl-1,2-propanediol I: (1R,2R)-1-Phenyl-1,2-propanediol.

ENANTIOSELECTIVITY

In the present study we are mainly concerned with regioselectivity, therefore enantioselectivity aspects of 1-phenyl, 1,2-propanedione will be mentioned only briefly. The enantiomeric excess (*ee*) of the main product **B** is defined as follows

$$ee = \frac{[\mathbf{B}]-[\mathbf{C}]}{[\mathbf{B}]+[\mathbf{C}]}$$

where $[\mathbf{B}]$ and $[\mathbf{C}]$ are the concentrations of (R)- and (S)-1-hydroxy-1-phenylpropanone, respectively.

In the absence of catalyst modifier, cinchonidine, no enantioselectivity was observed. The highest enantiomeric excesses (at 80% conversion of reactant) were about 60%, but varied as a function of the modifier concentration (Figure 3). As can be seen from Figure 3 the *ee* went through a maximum and the decreased with increasing modifier concentration. The *ee* increased at high reactant conversion (>90%) due to kinetic resolution, analogously to the regioselectivity (see below).

In continuous reactor the *ee* increased from 0% to the steady state level as the modifier coverage on the catalyst surface increased with increasing time-on-stream. The transient development of *ee* was proportional to the modifier feed. The steady state *ee* was dependent on the liquid phase concentration of the catalyst modifier and was independent on the space-time and reactant inlet concentration.

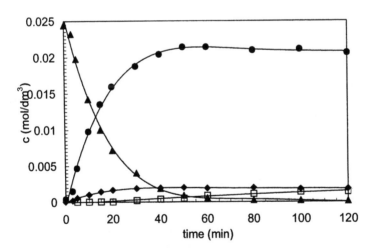

Figure 2 Hydrogenation of 1-phenyl-1,2-propanedione in ethyl acetate at 15°C. Catalyst: 5 wt.% Pt/Al$_2$O$_3$ modified *in situ* with (-)-cinchonidine. Symbols: (▲) 1-phenyl-1,2-propanedione (A), (●)1-hydroxy-1-phenylpropanone (B+C), (◆) 2-hydroxy-1-phenylpropanone (D+E), (□) 1-phenyl-1,2-propanediols (F+G+H+I).

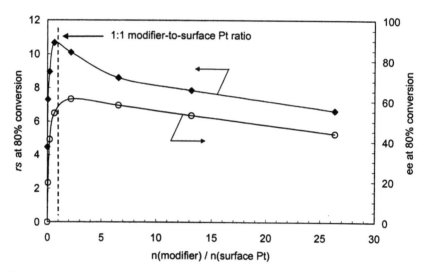

Figure 3 Regioselectivity (*rs*) and enantiomeric excess (*ee*) in the hydrogenation of 1-phenyl-1,2-propanedione in ethyl acetate at 15°C. Catalyst: 5wt.% Pt/Al$_2$O$_3$ modified *in situ* with different amounts of cinchonidine. Symbols: (◆) regioselectivity and (○) enantiomeric excess.

REGIOSELECTIVITY

The regioselectivity (*rs*) is defined as the ratio between the concentrations of 1-hydroxy-1-phenylpropanone (**1-OH**) and 2-hydroxy-1-phenylpropanone (**2-OH**) and can be expressed as follows.

$$rs = \frac{[B]+[C]}{[D]+[E]}$$

The regioselectivities obtained over Pt/Al$_2$O$_3$ catalyst in *batch reactor* were studied at different concentrations of reactant and modifier as well as at different temperatures.

Two distinct effects, namely initial *rs* and the increase of *rs* at high conversion levels of reactant due to kinetic resolution of **1-OH** and **2-OH**, should not be confused when elucidating the *rs* in the reaction system under consideration. The reaction is consecutive (dione -> hydroxyketone -> diol (Figure 1)) and the products giving rise to regioselectivity are the intermediates, **1-OH** and **2-OH**, which react further to diols. They react, however, with a much lower rate than the reactant hydrogenates to hydroxyketones. The *rs* is thus controlled at low reactant conversion by the production of hydroxyketones from the reactant and at high conversion (>90%) by their disappearance/consumption by further hydrogenation of **1-OH** and **2-OH** to diols. In the second hydrogenation step, **2-OH** reacts faster to diols than **1-OH** and the ratio of **1-OH** to **2-OH** changes and *rs* increases. The increase of *rs* at high reactant conversion levels due to the differences in the reaction rates of **1-OH** and **2-OH** is a clear case of kinetic resolution.

The experimental observations are summarized as follows: The reactant concentration had a very minor effect on the *rs*, the *rs* was about 10 as the concentration was varied between 0.01-0.025mol dm^{-3}.The effect of temperature on the regioselectivity was minor and *rs* decreased from 9.5 to 7.5 as the temperature was raised from 15 to 35°C. The most pronounced effect was observed at different concentrations of the modifier. In the absence of the modifier, the *rs* was around four and increased with increasing modifier concentration to a maximum (*rs*=10) around 1:1 surface Pt-to-modifier ratio, after which it gradually started to decrease (Figure 3). The most interesting observation is the over two-fold enhancement of *rs* due to the introduction of small amounts of modifier.

In the *continuous reactor*, over Pt/SiO$_2$ fiber catalyst, the regioselectivity had an initial transient period, after which a steady-state value, close to five, was attained (Figure 4). The transient development of *rs* was analogous to the transient development of *ee* in continuous reactor and could be correlated with the increasing modifier coverage on the catalyst surface. In Figure 4 the initial values of *rs* exceeding the steady-state *rs* level of about 5, are due to kinetic resolution, which increases *rs* whenever the reactant conversion was above 90%. In the absence of modifier the *rs* remained low, around two, and was increases only in the beginning

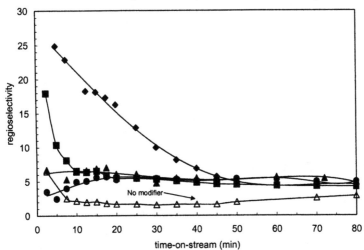

Figure 4 The effect of reactant inlet concentration on the regioselectivity. Inlet concentration of the modifier was $c_{OM} = 10 \times 10^{-5}$ mol dm^{-3} and space time (τ) of 44 s was used. Symbols: Concentration of A ($\blacklozenge$) 0.0075 M, ($\blacksquare$) 0.015 M, ($\blacktriangle$) 0.025 M, ($\bullet$) 0.05 and ($\triangle$) 0.025 M no modifier used.

of the reaction due to kinetic resolution (Fig. 4). It can be noticed that the SiO_2 –
supported catalyst gave systematically lower regioselectivities than the Pt/Al_2O_3
used in the batch reactor. However, an analogous, over two-fold *rs* enhancement
was observed over both of the catalysts. The SiO_2-supported catalyst gave lower
regioselectivities (*rs*=5) compared to the Al_2O_3-supported catalyst (*rs*=10) also in
the batch reactor, and therefore the differences in *rs* are not due to the choice of
reactor, but the choice of the catalyst.

The *rs* exhibits an analogous behavior to the enantiomeric excess as a function of the modifier concentration (Figure 3). Also the hydrogenation activity
shows a similar trend, but not as pronounced. These observations could call for an
explanation, which will assume that the regioselectivity is a consequence of similar interactions giving raise to enantiodifferentiation. However, the phenomena is
much more complex. The lack of enantioselectivity in the absence of catalyst
modifier is not in accordance with the *rs* value of four in the absence of the modifier and therefore contradicts with the direct connection between the *rs* and *ee*. A
further complication comes from the fact that the *rs* also depends on the structure
of the unsymmetrical dione. Previously the equal formation of both regioisomers
has been reported using 2,3-hexanedione [12] as a reactant, resulting in regioselectivity of 1, *i.e.* the lack of regioselectivity by the definition. Therefore, it is evident
that the *rs* is a rather complex phenomena, where the modifier, the reactant structure and the type of the catalyst play an important role.

The *rs* observed in the absence of the modifier, might be caused by the intrinsic differences in the reacting carbonyl groups of the reactant, *i.e.* differences
in the electronic properties of the carbonyl groups affect the reactivity and adsorption strength of the groups. The enhancement of regioselectivity on the other hand,
might be a consequence of the similar interactions, which induce also enantioselectivity.

In next section, the effect of the reactant structure on *rs* is evaluated based
on quantum chemical calculations using 1-phenyl-1,2-propanedion and 2,3-
hexanedione as model molecules in the calculations, after which the role of specific interactions responsible for the regioselectivity enhancement will be discussed.

THEORETICAL CALCULATIONS

The structure of the 1-phenyl-1,2-propanedione was optimized using the Hartree-
Fock (HF) approximation and second order Møller-Plesset perturbation theory
(MP2) with 6-31+G** [15, 16, 17, 18] basis set as well as with Density Functional
Theory (DFT). In the DFT calculations B3LYP [19] was utilized where Becke's
three parameter hybrid functional (B3) [20] is applied for the exchange part and
Lee-Yang-Parr(LYP) functional [21] for the correlation part. Gaussian98 [22] and
GAMESS [23] programs were used in the calculations. Mulliken charges, bond
orders and the geometry were evaluated at the optimized structure of the 1-phenyl-

1,2-propanedione. Numbering of the atoms and definition of the torsion angles are presented in the Figure 5. All the relevant torsion angles, bond distances, bond orders and Mulliken atomic charges at the equilibrium are reported in Tables 1-3. The calculated potential energy curve, which was evaluated with the aid of HF approximation, along the angle τ_2 is shown in Figure 6. In order to compare the results Mulliken charges, bond distances and bond orders of the optimized structure of the 2,3-hexanedione were evaluated. Numbering of the atoms of the 2,3-hexanedione is presented in Figure 5 and the essential results of the calculations are reported in Table 4.

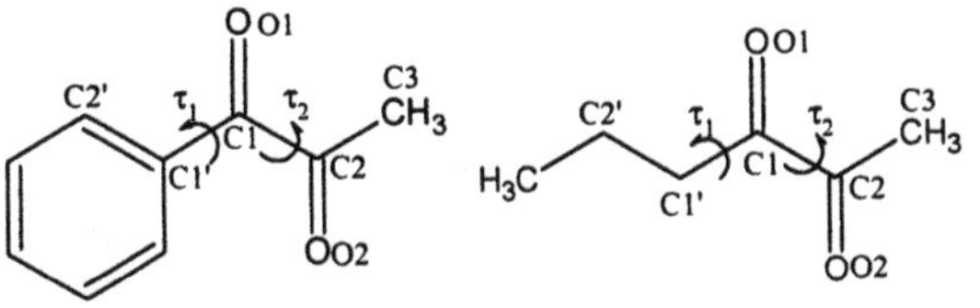

Figure 5 Plane structures of 1-phenyl-1,2-propanedione and 2,3-hexanedione, selectively numbered atoms and definition of torsion angles τ_1 and τ_2 (τ_1 = C2'-C1'-C1-O1, τ_2 = O1-C1-C2-O2).

There exist at least three factors which determine the potential energy surface (PES) over the dihedral angle O=C-C=O (*i.e.* torsion angle τ_2): (i) steric hinderance (Pauli repulsion), (ii) electrostatic interaction between the polar carbonyl groups and (iii) conjugation of the p-orbitals. Furthermore, at the optimized structure of 1-phenyl-1,2-propanedione, van der Waals interactions exist between oxygens and the *ortho*-hydrogens. The HF optimized structure of the 1-phenyl-1,2-propanedione has a torsion angle τ_2 = 136.6°, which implies that the carbonyl groups are not coplanar. This is in accordance with the experimental results obtained earlier for 1-phenyl-1,2-propanedione [24]. Moreover, this kind of behavior for O=C-C=O system has been reported previously in theoretical calculations for ethyl phenylglyoxylate [25].

Table 1 Selected bond distances in pm of optimized 1-phenyl-1,2-propanedione calculated using the HF, MP2 and B3LYB approximations, and bond orders using the HF approximation.

Bond	Bond distance / pm			Bond order
	B3LYB	**MP2**	**HF**	
C1'-C1	149	148	149	1.003
C1-C2	155	154	154	0.882
C2-C3	151	150	151	0.955
C1-O1	123	124	120	1.816
C2-O2	122	123	119	1.862

Table 2 Torsion angles τ_1 and τ_2 in degrees of optimized 1-phenyl-1,2-propanedione calculated using the HF, MP2 and B3LYB approximations.

	Torsion angle / degrees		
	B3LYP	**MP2**	**HF**
τ_1	-2.4	-13.3	-2.9
τ_2	154.0	138.2	136.6

Table 3 Mulliken charges for 1-phenyl-1,2-propanedione and 2,3-hexanedione calculated using the HF, MP2 and B3LYB approximations.

Atom	1-phenyl-1,2-propanedione			2,3-hexanedione
	B3LYP	**MP2**	**HF**	**HF**
C1'	0.04	0.51	0.67	-0.21
C1	0.34	-0.09	-0.18	0.30
C2	0.37	0.45	0.53	0.26
C3	-0.38	-0.46	-0.47	-0.37
O1	-0.46	-0.46	-0.44	-0.45
O2	-0.42	-0.46	-0.44	-0.46

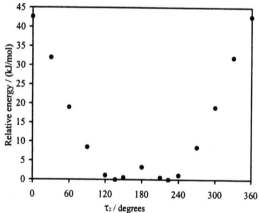

Figure 6 Potential energy curve of 1-phenyl-1,2-propanedione calculated over torsion angle τ_2. Energy of the optimized structure is set to zero.

The HF calculations indicate that the full rotation over torsion angle τ_2 is hindered because of the high energy barrier ($\Delta E = 42.6$ kJ mol^{-1}) at $\tau_2 = 0°$ (Figure 6). The energy barrier of the rotation over torsion angle τ_1 is also rather high ($\Delta E =$

24.4 kJ mol^{-1}) when $\tau_1 = 90°$. However, there may be low-energy paths on the τ_1-τ_2 surface. The torsion angle τ_1 at the optimized structure is close to 0° (-2.9°) meaning that the carbonyl group at position 1 is coplanar with the phenyl ring. This has been confirmed experimentally by Shen and Kolbjoern [24]. The planarity can be explained by the slight delocalization of the electrons between atoms C1'-C1-O1 which is indicated by several facts. The distance between atoms C1'-C1 is rather short (149 pm) and the bond order is quite high (1.003). In addition, at the optimized structure distances between the oxygens and *ortho*-hydrogens (265 and 246 pm) enable weak van der Waals interactions, which may stabilize this conformation. The Mulliken charge for carbon C1 (–0.18) indicates that the phenyl ring is attracting electrons from oxygen O1. For carbon C2, the Mulliken charge is appreciably higher (0.53). Notable is that the Mulliken charges for both oxygens are equal (-0.44). Difference in the C=O bond strengths can be seen in differences in bond orders and bond distances. The Bond order between atoms C1-O1 is lower than between atoms C2-O2 (1.816 and 1.862, respectively) and the bond distance between atoms C1-O1 is longer than between atoms C2-O2 (120 and 119 pm, respectively). Summarizing, all these results indicate that the C=O bond at position 1 is weaker than that at position 2. Notable is that the bond distance between carbons C1 and C2 is 154 pm and the bond order 0.882, which are characteristic to a single bond without any delocalization of the electrons nor partial double bond character. This has also been confirmed experimentally [24].

Table 4 Selected bond distances in pm of optimized 2,3-hexanedione calculated using HF optimization.

Bond	Bond distance / pm	Bond order
C1'-C1	151	0.952
C1-C2	154	0.873
C2-C3	150	0.959
C1-O1	119	1.862
C2-O2	119	1.854

The results of the calculations of the 1-phenyl-1,2-propanedione using MP2 are very similar to the results obtained from the HF calculations (see Table 1). Torsion angle τ_2 (= 138.2°) and Mulliken charges are similar to those in the HF optimized structure. Only torsion angle τ_1 (= -13.3°) differs slightly from the HF optimized structure.

The HF optimized structure of the 2,3-hexanedione is planar having dihedral angles of $\tau_1 = 0°$ and $\tau_2 = 180°$. Planar structure has also been observed for 2-hexanone [26] and experimentally for 2,3-butanedione [27]. The bond distance between carbons C1'-C1 (151 pm) is clearly longer in 2,3-hexanedione than in 1-

phenyl-1,2-propanedione. C=O bond distances are equal (119 pm). Moreover, the Mulliken charges of carbonyl carbons are almost equal (0.30 for C1 and 0.26 for C2) as are the C=O bond orders (for C1-O1 1.862 and for C2-O2 1.854). All these results indicate, that in 2,3-hexanedione C=O bond strengths are equal for both carbonyl groups.

Summarizing, these calculations provide an explanation of the reactant structure effects on the regioselectivity. A correlation between calculated values of the Mulliken charges, C=O bond distances and bond orders with regioselectivity was observed. The calculations indicated equal bond strengths for both carbonyl groups in 2,3-hexanedione, which could explains rs equal to 1, while weaker bond strength of the C=O bond in position 1 of 1-phenyl-1,2-propanedione favors the reduction of the C=O group in position 1 towards excess formation of 1-hydroxy-product and thus regioselectivity. An additional effect, which could contribute to the observed rs was the fact that 2,3-hexanedione adopted a planar conformation whereas 1-phenyl-1,2-propanedione was non-plan in the energy minimized conformation. This non-planarity might favor the adsorption of the C=O group 1 of 1-phenyl-1,2-propanedione whereas in 2,3-hexanedione the flat adsorption of the molecule would result in no discrimination between the two C=O groups and thereby to the observed lack of rs.

THE ROLE OF MODIFIER AND MODIFIER-REACTANT INTERACTIONS

The effect of reactant structure on rs was evaluated by the quantum chemical calculations. The difference between the reactive carbonyl groups was a probable reason for the differences in the reactivity and as a result regioselectivity was observed. However, also the differences in the energy minimized conformations of the reactants could contribute to the observed differences in rs.

The role of modifier in the enhancement of rs requires more explanation. First of all, the flat adsorption of the reactant on the catalyst surface would be responsible for the low $rs=4$ in the absence of the modifier at low concentrations of the reactant, when the surface is not too occupied and flat adsorption is facile. The flat adsorption of 1-phenyl-1,2-propanedione would also result in the production of cyclohexyl products via the phenyl ring hydrogenation.

At the reactant concentrations used, the rs remained virtually constant and no dependence on reactant concentration became observable. However concentration dependent changes in the adsorption mode were recently reported in selective hydrogenation of α,β unsaturated aldehyde [2]. It was supposed that the adsorption mode of cinnamaldehyde at high concentrations differs from that at lower concentrations, more precisely, cinnamaldehyde adsorbs perpendicular to the catalyst surface with the aromatic rings in parallel arrangements at high coverage. However, in our case the competitive adsorption of the modifier and the reactant complicates the situation.

An enhancement of the regioselectivity was observed when the catalyst modifier concentration was increased (Figure 3). The enhancement might be caused by 1) the modifier-reactant interaction in the liquid-phase, 2) the modifier-reactant interaction on the catalyst surface, 3) the coverage dependent adsorption modes of the reactant and modifier or a combination of factors 1-3.

The modifier-reactant interaction in the liquid-phase is the basis of the shielding effect model [4], which explains the rate enhancement and enantioselectivity in the case of ethylpyruvate hydrogenation over an analogous catalytic system as described in the present study. According to the model, the modifier-reactant complex formed in the liquid phase comes in contact with the catalyst surface and preferentially hydrogenates on one side of the reactant inducing enantioselectivity. There are, however, two factors which are not easily explained by the model. In our case, in the hydrogenation of dione, no significant rate acceleration was observed and therefore the maximum in the rs is not clearly defined, $i.e.$ one would expect the rs to increase with increasing amount of the modifier. Furthermore, the maximum in rs at a relatively high reactant-to-modifier ratio is not easily understood in the framework of the shielding effect model. The maximum in regioselectivity appears at reactant-to-modifier ratios of 370, $i.e.$ one reactant-modifier complex per 369 free reactant molecules in the liquid-phase. It is thus not convincing that the decline in the rs would take place at such high ratios and not closer to a 1:1 ratio. It would be acceptable when the reactant-modifier complex would react significantly faster than the reactant itself; however, no significant rate acceleration was observed in the presence of the modifier. The rate acceleration observed was maximally about 30% and appeared only at very low modifier concentrations, corresponding to reactant-to- modifier ratios of 370 or higher.

The modifier-reactant interaction on the catalyst surface is the basis of the model proposed by Baiker et $al.$ [28] for the rate enhancement and enantioselectivity in the case of ethylpyruvate hydrogenation over Pt catalysts modified by cinchonidine. The maximal in enantiomeric excess and rs were observed close to the 1:1 ratio of surface Pt-to-modifier (Figure 3) which supports the hypothesis that, the adsorption of the modifier on the catalyst surface would be the origin of the enantioselectivity and the enhancement in rs.

The dependence of the adsorption mode of cinchonidine on the coverage have been reported recently [29]. At a high coverage, the adsorption mode of the modifier changes and becomes unfavorable for the substrate-modifier interaction thus lowering the selectivity. This would imply that the rs and ee are strongly interrelated: after reaching the optimum modifier coverage, a further increase in the modifier concentration would result in a gradual decline in the ee and rs. The change of the adsorption mode would result in non-parallel adsorption of the quinoline moiety of cinchonidine. In such a tilted adsorption mode, the catalyst surface accommodates more modifier onto it and less dione and as a result a lower reaction rate would be also observed. This would result in a maxima in rs and ee as well as reaction rate, when the modifier concentration is increased. These observations are in accordance with current experimental results.

The coverage dependent adsorption mode of the modifier is coupled with the coverage dependent adsorption of the reactant as well. Over a non-modified catalyst, the reactant can readily adsorb in a planar mode, where both of the carbonyl groups and the phenyl ring are in the same catalyst plane. This assumption is supported by the experiments, which revealed that the phenyl ring of A is hydrogenated in the absence of the modifier. The formed cyclohexyl products can be taken as an indication of a planar adsorption mode of the A, *i.e.* in order to hydrogenate the phenyl ring, it has to be adsorbed parallel to the catalyst surface via p-bonding. As small amounts of the modifier are added, the phenyl ring hydrogenation is no longer feasible, indicated by the disappearance of the cyclohexyl products. This can be understood by the occupation of the surface due to the adsorption of the modifier in such a way that the planar adsorption of the modifier is no longer facile. This would also imply that the carbonyl group in the position 1 could be also adsorbed on tilted orientation. Therefore, it is evident that also the adsorption mode of the reactant depends on the modifier coverage on the catalyst surface.

To summarize, the enhancement in regioselectivity, due to the small amount of modifier added, was interrelated with the enantioselectivity (*es*) and can be most plausibly explained by similar interactions on the catalyst surface which are responsible for the enantiodifferentiation. The substrate-modifier interaction on the catalyst surface coupled with the coverage dependent adsorption modes of the modifier and reactant explain the enhancement of *rs* as well as *es* and their dependence on modifier concentration. In the light of presented data it is evident that one needs to incorporate the coverage dependent adsorption modes in the kinetic model for a correct description of the *rs* and *es*, otherwise the maximum in selectivities and selectivity dependence on modifier concentration cannot be described.

CONCLUSIONS

Hydrogenation of 1-phenyl-1,2 –propanedione was investigated in batch and continuous fixed bed reactors. The main product in the first hydrogenation step was 1-hydroxy-1-phenylpropanone; the ratio between 1-hydroxy-1-phenylpropanone and 2-hydroxy-1-phenylpropanone, defined as regioselectivity, was about 10 in the batch reactor and about 5 in the fixed bed reactor in the presence of the catalyst modifier, cinchonidine. In both reactors, over a two-fold enhancement of regioselectivity was observed as small amounts of catalyst modifier, cinchonidine, was added. In the continuous reactor, an initial transient period in *rs* was observed. The initial *rs* varied between 32 - 2 reaching a steady-state value of ca. 5. The amount of the modifier was the most important for the optimum *rs*, while temperature and reactant concentration had only a minor effect. The maximum in enantio- as well as regioselectivity was obtained close to the modifier-to surface Pt ratio of 1:1.

Quantum–chemical calculations provided an explanation for the regioselectivity in 1-phenyl-1,2-propanedione hydrogenation and also for the lack of regioselectivity in 2,3-hexanedione hydrogenation. The differences in the selectivity in 2,3-hexanedione hydrogenation. The differences in the electronic properties of the two reacting carbonyl groups as well as the differences in the energy minimized conformations were found to be a plausible explanation for the regioselectivity.

The enhancement of rs was strongly interrelated with the enantiomeric excess and was explained by the interactions between reactant and modifier on the catalyst surface. Interactions responsible for the enantiodifferentiation result also in the enhancement of regioselectivity. The coverage dependent adsorption modes of the modifier and reactant are responsible for the dependence of the rs and enantioselectivity on the modifier concentration.

REFERENCES

1. RL Augustine. Heterogeneous catalysis for the synthetic chemist. Marcel Dekker, 1995.
2. RJ Berger, EH Stitt, GB Marin, F Kapteijn, JA Moulijn. Chemical reaction kinetics in practice. CATTECH 5:30-60, 2001.
3. A Baiker. Progress in asymmetric heterogeneous catalysis: Design of novel chirally modified platinum metal catalysts. J Mol Catal A: Chem 115: 473-493, 1997.
4. JL Margitfalvi, E Tfirst. Enantioselective hydrogenation of α-keto esters over cinchona –Pt/Al$_2$O$_3$ catalyst. Molecular modeling of the substrate-modifier interaction. J Mol Catal A: Chem 139: 81-95, 1999.
5. T Bürgi, Z Zhou, N Kunzle, T Mallat, A Baiker. Enantioselective hydrogenation on chirally modified platinum: new insight into the adsorption mode of the modifier. J Catal 183: 505-408, 1999.
6. HU Blaser, HP Jalett, M Garland, M Studer, H Thies, A Wirth-Tijani. Kinetic studies of the enantioselective hydrogenation of ethyl pyruvate catalyzed by a cinchona modified Pt/Al$_2$O$_3$ catalyst. J Catal 173: 282-294, 1998.
7. E Toukoniitty, P Mäki-Arvela, J Wärnå, T Salmi. Modeling of the enantioselective hydrogenation of 1-phenyl-1,2-propanedione over Pt/Al$_2$O$_3$ catalyst. Catal Today 66: 411-417, 2001.
8. E Toukoniitty, P Mäki-Arvela, A Villela, A Kalantar Neyestanaki, T Salmi, R Leino, R Sjöholm, E Laine, J Väyrynen, J Ollonqvist, PJ Kooyman. The effect of oxygen and the reduction temperature of the Pt/Al$_2$O$_3$ catalyst in enantioselective hydrogenation of 1-phenyl-1,2-propanedione. Catal Today 60: 175-184, 2000.
9. E Toukoniitty, P Mäki-Arvela, M Kuzma, A Villela, A Kalantar Neyestanaki, T Salmi, R Sjöholm, R Leino, E Laine, DYu. Murzin. Enantioselective hydrogenation of 1-phenyl-1,2-propanedione, Journal of Catalysis, 2001, accepted.
10. W Adam, M Diaz, R Fell, C Saha-Möller. Kinetic Resolution of Racemic α-Hydroxy Ketones by Lipase-catalyzed Irreversible Transesterification. Tetrahedron Asymmetry 7: 2207-2210, 1996.
11. D Gala, D DiBenedetto, J Clark, B Murphy, D Schumacher, M Steinman. Preparations of Antifungal Sch 42427/SM 9164: Preparative Chromatographic Resolution, and Total Asymmetric Synthesis via Enzymatic Preparation of Chiral α-Hydroxy Arylketones. Tetrahedron Letters 37: 611-, 1996.

12. J Slipszenko, SP Griffiths, P Johnston, KE Simons, WAH Vermeer, PB Wells. Enantioselective hydrogenation.V.Hydrogenation of buta-2,3-dione and of 3-hydroxybutan-2-one catalysed by cinchona-modified platinum. J Catal 179: 267-276, 1998.

13. A Kalantar Neyestanaki, LE Lindfors. Catalytic Clean-up of Emissions from Small-scale Combustion of Biofuels. Fuel 77: 1727-1734, 1998.

14. VB Shukala, PR Kulkarni. L-Phenylacetylcarbinol (L-PAC): Biosynthesis and Industrial Applications. World Journal of Microbiology and Biotechnology 16: 499-506, 2000.

15. R Ditchfield, WJ Hehre, JA Pople. Self-consistent molecular-orbital methods. IX. Extended Gaussian-type basis for molecular-orbital studies of organic molecules. J Chem Phys 54: 724-728, 1971

16. WJ Hehre, R Ditchfield, JA Pople. Self-consistent molecular orbital methods. XII. Further extensions of Gaussian-type basis sets for use in molecular orbital studies of organic molecules. J Chem Phys 56: 2257-2261, 1972.

17. T Clark, J Chandrasekhar, GW Spitznagel, PR Schleyer. Efficient diffuse function-augmented basis sets for anion calculations. III. The 3-21 + G basis set for first-row elements, lithium to fluorine. J Comput Chem 4: 294-301, 1983.

18. PC Hariharan, JA Pople, Influence of polarization functions on MO hydrogenation energies. Theor Chim Acta 28: 213-22, 1973.

19. PJ Stephens, FJ Devlin, CS Ashvar, CF Chabalowski, MJ Frisch. Theoretical Calculation of Vibrational Circular Dichroism Spectra. Faraday Discuss 99: 103-119, 1994.

20. AD Becke. Density-functional thermochemistry. III. The role of exact exchange. J Chem Phys 98: 5648-5652, 1993.

21. C Lee, W Yang, RG Parr. Development of the Colle-Salvetti correlation-energy formula into a functional of the electron density. Physical Review B 37: 785-789, 1988.

22. Gaussian 98, Revision A.7, MJ Frisch, GW Trucks, HB Schlegel, GE Scuseria, MA Robb, JR Cheeseman, VG Zakrzewski, JA Montgomery, Jr., RE Stratmann, JC Burant, S Dapprich, JM Millam, AD Daniels, KN Kudin, MC Strain, O Farkas, J Tomasi, V Barone, M Cossi, R Cammi, B Mennucci, C Pomelli, C Adamo, S Clifford, J Ochterski, GA Petersson, PY Ayala, Q Cui, K Morokuma, DK Malick, AD Rabuck, K Raghavachari, JB Foresman, J Cioslowski, JV Ortiz, AG Baboul, BB Stefanov, G Liu, A Liashenko, P Piskorz, I Komaromi, R Gomperts, RL Martin, DJ Fox, T Keith, MA Al-Laham, CY Peng, A Nanayakkara, C Gonzalez, M Challacombe, PMW Gill, B Johnson, W Chen, MW Wong, JL Andres, C Gonzalez, M Head-Gordon, ES Replogle, JA Pople. Gaussian, Inc., Pittsburgh PA, 1998.

23. MW Schmidt, KK Baldridge, JA Boatz, ST Elbert, MS Gordon, JH Jensen, S Koseki, N Matsunaga, KA Nguyen, S Su, TLWindus, M Dupuis, JA Montgomery. General atomic and molecular electronic structure system. J Comput Chem 14: 1347-63, 1993.

24. Q Shen, H Kolbjoern. Molecular Structures and Conformations of Phenylglyoxal and 1-Phenyl-1,2-propanedione As Determined by Gas-Phase Electron Diffraction. J Phys Chem. 97: 985-988, 1993.

25. D Ferri, T Bürgi, A Baiker. Conformational isomerism of α-keto esters. A FTIR and *ab initio* study. J Chem Soc, Perkin Trans 2: 221-227, 2000.

26. C Alemán, E Navarro, J Puiggalí. Comparison between Diketones and Diamides: Effects of Carbonyl Groups on the Conformational Preferences of Small Aliphatic Segments. J Phys Chem 100: 16131-16136, 1996.
27. K Hagen, K Hedberg. Conformational analysis. IV. Molecular structure and composition of gaseous 2,3-butanedione as determined by electron diffraction. J Amer Chem Soc 95: 8266-8269, 1973.
28. T Bürgi, A Baiker. Model for Enantioselective Hydrogenation of α-keto ester over Chirally Modified Platinum Revisited: Influence of the α-keto ester Conformation. J Catal 194: 445-451, 2000.
29. D Ferri, T Bürgi, A Baiker. Chiral Modification of Platinum Catalysts by Cinchinidine Adsorption Studied by in situ ATR-IR Spectroscopy. Chem Commun: 1172-1173, 2001.

26

Asymmetric Catalysis Using Bisphosphite Ligands

Gregory T. Whiteker, John R. Briggs, James E. Babin, and Bruce A. Barner
The Dow Chemical Company, South Charleston, West Virginia, U.S.A.

ABSTRACT

Bisphosphite ligands were originally discovered at Union Carbide and have been extensively studied for olefin hydroformylation. Structural modification of bisphosphite ligands allows their use in asymmetric catalysis. These chiral bisphosphites lead to the highest combination of branched regioselectivity (b/l) and enantioselectivity (%*ee*) of any asymmetric hydroformylation catalyst reported to date. These catalysts have been applied to a highly efficient synthesis of *S*-Naproxen. In addition, the use of chiral bisphosphites in asymmetric olefin hydrocyanation is described. Details of the structural features responsible for high b/l and %*ee* in asymmetric hydroformylation and hydrocyanation are presented.

INTRODUCTION

Transition metal catalyzed olefin hydroformylation is an industrially important process for the production of oxygenated products.(1) Rhodium-catalyzed processes that employ phosphorus-based ligands allow increased control of rate and regioselectivity when compared to cobalt-catalyzed processes. The first generation of rhodium catalysts was based on triarylphosphine ligands. Second genera-

359

tion catalysts, based on bisphosphite ligands, were developed by Union Carbide in the 1980s.(2) Use of these ligands leads to higher regioselectivity, faster turnover rates and improved tolerance of functional groups. These features are especially beneficial when applying olefin hydroformylation to the synthesis of complex organic molecules.

We describe the extension of this class of bisphosphite catalysts to asymmetric hydroformylation and hydrocyanation of vinylarenes.(3) These enantioselective catalytic transformations are employed for the asymmetric synthesis of *S*-Naproxen, a widely used non-steroidal anti-inflammatory drug (NSAID). Factors which influence regioselectivity and enantioselectivity, as well as characterization of the catalyst resting states, are discussed.

EXPERIMENTAL

Asymmetric Hydroformylation

A catalyst solution consisting of $Rh_4(CO)_{12}$ (0.0109 g, 0.0582 mmole, 300 ppm), bisphosphite **1** (0.0950 g, 0.1179 mmole, 2:1 L/Rh ratio), styrene (6 g), and acetone (13.9 g) was prepared and charged to a 100 mL autoclave under nitrogen. The reactor was charged to 130 psi with 1:1 H_2/CO and the rate of reaction was monitored by measuring the rate of 5 psi pressure drops. After ten 5 psi pressure drops, the catalyst solution was removed from the reactor and analyzed by GC to determine regioisomeric ratio. A branched/normal isomeric ratio of 97:1 was observed. A small portion of the catalyst solution was oxidized with $KmnO_4$/$MgSO_4$ to convert the 2-phenylpropionaldehyde to 2-phenylpropionic acid. The acid product was analyzed by chiral GC on a Cyclodex B column to determine enantioselectivity. A ratio of 94:6 of *S*-2-phenylpropionic acid to *R*-2-phenylpropionic acid was observed.

Asymmetric Hydrocyanation. Caution: *Both HCN and acetone cyanohydrin are extremely dangerous. Use of a HCN monitor is advised.*

$Ni(COD)_2$ (22 mg, 0.077 mmol) and bisphosphite **1** (125 mg, 0.155 mmol) were combined in a 50 mL Schlenk flask in a nitrogen filled drybox. Deoxygenated THF (10.0 mL) was added via syringe, and the solution was stirred for 30 min. Acetone cyanohydrin (3.00 mL) and olefinic substrate (1-2 g) were added via syringe. The solution was then transferred via syringe to a Fisher-Porter bottle and heated to the appropriate reaction temperature. Styrene hydrocyanation reactions were analyzed by chiral capillary gas chromatography using an Astec Chiraldex G-TA column operating isothermally at 135 °C. Alternatively, solutions were analyzed by chiral stationary phase HPLC as follows: Solvent was removed from the reaction mixture by rotary evaporation. Catalyst was separated by filtration through silica gel (1:1 hexane-ether), and the filtrate was concentrated to an oil.

Enantiomeric excess was obtained by chiral HPLC using a Daicel Chiralcel OB column (styrene hydrocyanation: 99:1 hexane/2-propanol; 1.0 mL min^{-1}; 40°C. 6-methoxy-vinylnaphthalene hydrocyanation: 93:7 hexane/2-propanol; 0.5 mL min^{-1}; 40°C).

RESULTS AND DISCUSSION

Asymmetric Hydroformylation

The effects of bisphosphite ligand structure on regioselectivity and enantio-selectivity in asymmetric styrene hydroformylation are shown in Table 1. Catalytic reactions were preformed at ambient temperature and 130 psi CO/H$_2$. Hydroformylation regioselectivity was determined by GC of the product aldehydes. Enantioselectivity was determined by chiral GC after conversion to the carboxylic acid (eqn 1). The R,R-enantiomer of the bisphosphites in Figure 1 all produced the S-enantiomer of the product.

(eq. 1)

Bisphosphites which contain unsubstituted biphenols as end groups are required for high <u>linear</u> regioselectivity in α-olefin hydroformylation. However, for selective formation of <u>branched</u> aldehydes, end group biphenols with t-alkyl substituents in the 3,3'-positions produce the most selective catalysts. Comparison of bisphosphite $\underline{1}$ with bisphosphite $\underline{2}$, which lacks 3,3'-t-butyl substituents, reveals the large effect of this structural feature. Even ligands which contain 3,3'-iso-propyl substituents are ineffective (branched /linear = 7:1, 11 %ee). In addition, three-carbon bridged diols are significantly more effective in promoting asymmetric hydroformylation when compared with two-carbon and four-carbon bridged analogs. Similar structure-property relationships were also reported by van Leeuwen, $et\ al$.(4) Importantly, for all cases examined, enantioselectivity and regioselectivity were correlated. No ligands were found which were highly regioselective, but led to low enantioselectivity.

Table 1 Structure-property relationships for styrene hydroformylation. Conditions: 25 °C, 130 psi 1:1 CO/H$_2$, 2:1 Bisphosphite:Rh.

Ligand	Branched/linear	%ee
1	49:1	90%
2	5:1	14%
3	17:1	10%

The successful asymmetric hydroformylation of styrene using Rh/**1** catalyst suggested that this catalyst could be generally useful for the synthesis of optically active NSAIDs. Asymmetric hydroformylation of *p-iso*-butylstyrene using Rh/**1** proceeded to give the chiral aldehyde with b/l of 40:1 (eqn 2). Oxidation to the acid (Ibuprofen) revealed that the hydroformylation reaction occurred with 90 %ee. Likewise, asymmetric hydroformylation of 6-methoxynaphthalene using Rh/**1** proceeded with high branched regioselectivity (b/l = 90) and high enantioselectivity (87 %ee) (eqn 3) to *S*-Naproxen. Indeed, bisphosphite **1** is generally useful for asymmetric hydroformylation of vinylarenes.

Table 2 Structure-property relationships for styrene hydrocyanation. Conditions: 25 °C, 2:1 Bisphosphite:Ni.

Ligand	Branched/linear	%ee
1	2:1	65%
4	>200:1	0%
5	>200:1	13%
6	>200:1	40%

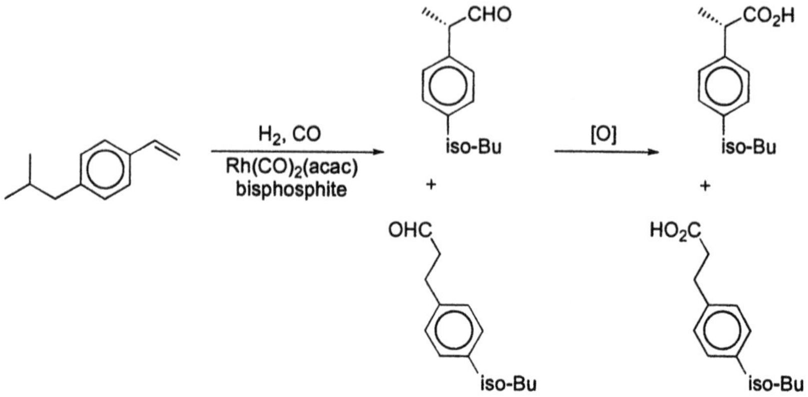

(eq. 2)

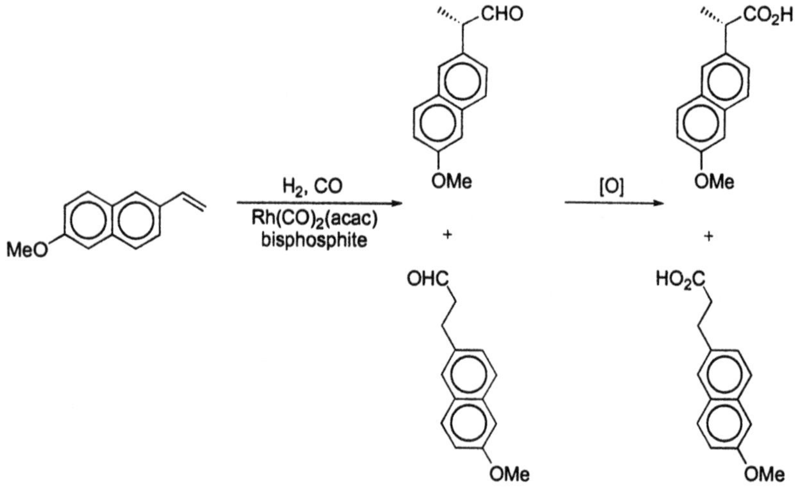

(eq. 3)

The optical purity of *S*-Naproxen produced by asymmetric hydroformylation with **1** can be increased by recrystallization. The solid aldehyde (87 %*ee*) from asymmetric hydroformylation of 6-methoxynaphthalene was recrystallized from acetone with an increase in enantiomeric purity to >98 %*ee*. In addition, the regiopurity of this aldehyde increased from 90:1 to >200:1 after one recrystallization.(5)

Asymmetric Hydrocyanation

Given the effectiveness of chiral bisphosphites in hydroformylation of vinylarenes, the use of these ligands in asymmetric hydrocyanation was investigated (eqn. 4). DuPont commercially employs homogeneous Ni catalysts based on triarylphosphite ligands in butadiene hydrocyanation.(6) Rajanbabu at DuPont reported that sugar-derived bisphosphinites form excellent catalysts for asymmetric vinylarene hydrocyanation.(7)   These catalysts regiospecifically produced the desired branched nitrile with high enantioselectivity.

(eq. 4)·

Hydrocyanation reactions were performed using catalysts prepared *in situ* from the reaction of chiral bisphosphites with Ni(COD)$_2$. Styrene hydrocyanation using bisphosphite **1** at 25 °C with acetone cyanohydrin as HCN source gave a mixture of regioisomeric nitriles with b/l = 2:1. Chiral HPLC revealed that the branched nitrile was formed in 60 %*ee*. A dramatic increase in regioselectivity was observed with ligand **4**, which lacks 3,3'-t-butyl substituents. This situation contrasts with that observed in asymmetric hydroformylation where 3,3'-t-alkyl substituents are required for high branched regioselectivity. Although the hydrocyanation regioselectivity is exceptionally high with bisphosphite **4**, the product nitrile is racemic. However, binaphthol-derived bisphosphites, **5** and **6**, exhibited very high branched regioselectivity with modest enantiomeric excesses.

Hydrocyanation of 6-methoxynaphthalene was preformed using Ni(COD)$_2$/**1**. The reaction proceeded with extremely high regioselectivity; no linear nitrile was observable by HPLC. Unfortunately, the enantioselectivity of this reaction was only 40%.

Catalyst Characterization

Spectral investigations of the catalytic resting states in both asymmetric hydroformylation and hydrocyanation using bisphosphite **1** were performed. Reaction of Rh(CO)$_2$(acac) with 1 equiv of **1** under 1:1 CO/H$_2$ led to clean formation of Rh(**1**)(CO)$_2$H. Infrared spectra of the reaction product exhibited bands at 2069, 2015 and 1984 cm^{-1}. Extensive characterization of this complex by NMR has been published by van Leeuwen.(4) Under 50 psi 1:1 CO/H$_2$ in the presence of styrene, identical bands were observed by *in situ* IR using the ReactIR system.

This observation demonstrates that the catalyst resting state under hydroformyla-tion conditions is Rh(1)(CO)$_2$H. In addition, these *in situ* infrared studies demon-strated that styrene hydroformylation using bisphosphite 1 is first order in both styrene and Rh concentrations.

The catalyst resting state in asymmetric hydrocyanation using bisphosphite 1 was studied by multinuclear NMR. Reaction of (COD)$_2$Ni with 1 equiv of 1 in 2:1 THF-d_8/styrene was monitored by ^{31}P{^{1}H} NMR. A large AB quartet was observed at δ164.1 (J$_{P-P}$ = 102 Hz) along with a much smaller AB quartet at δ165.0 (J$_{P-P}$ = 94 Hz) which was partially obscured by the larger AB pattern. Inte-gration of these ^{31}P resonances indicated a 94:6 ratio of these species which are assigned as two diastereomeric Ni(bisphosphite)(η^2-styrene) complexes. These species differ by which face of the prochiral styrene substrate is bound to the chi-ral P$_2$Ni fragment. Consistent with this formulation, ^{1}H NMR indicated the pres-ence of free 1,5-cyclooctadiene. These two diasteromeric P$_2$Ni(styrene) species were not observed in the absence of styrene, but were observable in hydrocyana-tion reaction mixtures indicating these P$_2$Ni(styrene) species are the catalyst rest-ing state. However, the diastereoselectivity of styrene binding in these P$_2$Ni(styrene) complexes is higher than the enantioselectivity observed in styrene hydrocyanation using bisphosphite 1. This observation indicates that the enantios-lectivity of the product nitrile is established by a step which occurs after binding of styrene.

CONCLUSIONS

Optically active bisphosphites are effective in mediating a variety of asymmetric homogeneous catalytic reactions. These catalysts are generally useful for asym-metric hydroformylation of vinylarenes, allowing efficient catalytic asymmetric synthesis of NSAIDs. These ligands are also effective for asymmetric hydrocya-nation, providing very high branched regioselectivities and, albeit modest, enan-tioselectivities.

ACKNOWLEDGMENTS

We gratefully acknowledge the technical assistance of Garland Johnson, Ott Huffman, Joe Kimble and Sue Chandler. The authors thank Arnold Harrison for providing NMR support. We thank the Dow Chemical Company for permission to publish this paper.

REFERENCES

1.   P.W.N.M. van Leeuwen, C. Claver, "Rhodium Catalyzed Hydroformylation", Kluwer Academic Press, Dordrecht (2000).

2. E. Billig, A.G. Abatjoglou, and D.R. Bryant, U.S. Patent No. 4,599,206, to Union Carbide (1986).

3. J.E. Babin and G.T. Whiteker, U.S. Patent No. 5,360,938, to Union Carbide (1992).

4. G.J.H. Buisman, E.J. Vos, P.C.J. Kamer, and P.W.N.M. van Leeuwen, *J. Chem. Soc. Dalton Trans.,*409 (1995).

5. B.A. Barner, J.R. Briggs, J.J. Kurland, and C.G. Moyers, Jr., U.S. Patent No. 5,430,194, to Union Carbide (1995).

6. K. Weissermel and H.-J. Arpe, "Industrial Organic Chemistry", VCH, Weinheim (1993).

7. T.V. RajanBabu and A.L. Casalnuovo, *J. Am. Chem. Soc.*, **114**, 6265 (1992).

27

Carbamate and Dicarbamate Syntheses from CO/O₂/Methanol/Amines

Pisanu Toochinda, Steven S. C. Chuang*, and Yawu Chi
The University of Akron, Akron, Ohio, U.S.A.

INTRODUCTION

Isocyanates and diisocyanates are important chemical intermediates for the pesticide, fertilizer, polymer, and pharmaceutical industries. The current process for isocyanate synthesis involves the use of highly toxic phosgene gas as follows:

$$Isocyanate\ synthesis\ from\ phosgene$$
$$RNH_2 + COCl_2 \rightarrow RNCO + HCl$$

where R is an alkyl or aryl group.

The phosgene toxicity, separation of hydrochloric acid from excess phosgene, and the use of chlorinated solvents as a reaction medium are the major drawbacks of this reaction process. Extensive studies have suggested that carbamate and dicarbamate can serve as environmentally benign precursors for the synthesis of isocyanate and diisocyanate (1-5). Figure 1 illustrates nonphosgene routes for the synthesis of two important diisocyanates [i.e., 4,4'-diphenylmethane diisocyanate (MDI) and toluene diisocyanate(TDI)] via carbonate and carbamate.

The key precursor for MDI synthesis is methyl-N-phenyl carbamate (MPC) which can be produced by 3 different oxidative carbonylation pathways

---

* *Corresponding author.*

as shown in Figure 1:
  i)   oxidative carbonylation of methanol (step 4) followed by the reaction of dimethylcarbonate (DMC) with aniline (step 5).
  (ii) the direct oxidative carbonylation of aniline with methanol (step 3), and
  (iii) oxidative carbonylation of aniline via urea (step 1 and 2),

  Various catalysts have been studied for these reaction processes. The major disadvantages of these reaction processes are (i) the need for an energy intensive separation step for catalyst recovery and (ii) limiting solubility of CO and O$_2$ in the liquid reactant medium. One of the objectives of our study is to develop a heterogeneous gas-solid catalytic reaction process for the synthesis of methyl-N-phenylcarbamate involving step 3 and dimethyl carbonate involving step 4 over Cu-based catalyst. This gas-solid process would eliminate the solubility limitation and catalyst separation step, thus enhancing the overall economics of the carbamate synthesis (6-8).

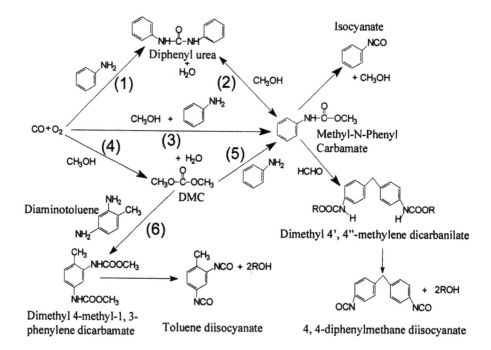

Figure 1 Non-phosgene routes for the synthesis of isocyanate, MDI, and TDI.

The second objective of this study is to determine the feasibility of using diphenyl phosphinic acid as a catalyst for synthesis of dimethyl 4-methyl-1, 3-phenylene dicarbamate (DMPD) from 2, 4-diaminotoluene and DMC (step 6) as follows:

$$CH_3\,C_6H_5\,(NH_2)_2 \; + \; (CH_3\,O)_2\,CO \rightarrow CH_3\,C_6H_5\,(NHCOOCH_3) + 2CH_3OH$$

Phosphinic acids are reported as an active catalyst the dicarbamate synthesis from aromatic diamines and diphenyl carbonate (9).

All the reaction processes were studied in an *in situ* infrared reactor cell. The product formation profiles were closely monitored. The results of this study show that gas-solid DMC and MPC synthesis processes are significantly more effective than the conventional gas-liquid synthesis processes. The results also show that diphenyl phosphinic acid is an effective catalysts for the synthesis of DMPD from DMC and 2, 4-diaminotoluene.

EXPERIMENTAL

Catalyst preparation: Al-MCM-41 was prepared by mixing solution A [tetramethyl amine hydroxide, Cab-O-Sil silica, and] with solution B [NaAlO$_2$, cetyltrimethylammonium chloride (CTMACl), NaOH, and H$_2$O]. The mixture was heated in an autoclave under autogenous pressure at 363 K for 48 hours. The resulting gel was calcined in a 30 cm³/min air flow at 823 K for 5 h to obtain MCM-41 (Si/Al ratio = 20). 5-wt% CuO/MCM-41 was prepared by incipient wetness impregnation of MCM-41 with a CuCl$_2$.2 H$_2$O (Sigma Chemical Co.) solution. The impregnated sample was dried in air at 298 K for 24 h and calcined at 823 K for 10 h to obtain CuO/MCM-41.

HZSM-5 was obtained by calcination of NH₄ZSM-5 (Zeolyst International Product, the ratio of Si/Al = 25) in 30 cm³/min air flow at 773 K for 24 h. 5-wt% CuCl$_2$/ZSM-5 was prepared by incipient wetness impregnation of HZSM-5 with CuCl$_2$.2H₂O (Sigma Chemical Co.). The impregnated sample was dried in air at 298 K for 24 h to obtain CuCl$_2$/HZSM-5. The ratio of the volume of salt solution to the weight of support was 1 cm³ to 1 g in the impregnation step during preparation of both catalysts.

In-situ Infrared study of DMC, MPC and DMPD syntheses: DMC, MPC and DMPD syntheses were studied in an *in situ* IR cell which was capable of operating up to 30 MPa. The solid catalyst sample was pressed into a self-support disk and placed between two CaF$_2$ rods in the transmission IR cell. Table 1 lists reaction conditions for these synthesis reactions. Liquid reactant mixture (0.2 cm³) was brought into the IR cell by injection into a known amount of gaseous reactants flowing into the reactor cell. For the DMPD synthesis, the liquid reactant mixture was brought into IR cell via helium flow as a carrier. The

Table 1. Reaction conditions of the in-situ infrared study of carbamate and dicarbamate syntheses.

Reaction	Catalyst/Promotor	T (K)	P (MPa)	Reactants	Reactant molar ratio (1)/(2)	CO/O_2	Time (min)	%Yield
I. MPC synthesis	5-wt% CuO/MCM-41 25 mg, NaI 7.2 mg	443	0.34	Aniline (1) MeOH (2)	1/42.7	10/1	7.35	20.9
II. DMC synthesis	5-wt% CuCl$_2$/HZSM-5 25 mg, NaI 7.2 mg	383	0.34	MeOH	-	10/1	2.00	8.1
III. DMPD synthesis	Diphenyl phosphinic acid 3 mg	403	0.1	DMT (1) DMC (2)	1/17.2	N/A	17.25	34.0

DMT = 2, 4-diaminotoluene, DMC = dimethyl carbonate.

IR spectra were collected during the entire course of reaction by a Nicolet MAGNA 550 Series II transmission infrared spectrometer. The background spectrum was taken upon injection of the reactant mixture. The background spectrum was subtracted from the spectra collected during *in situ* study to obtain the absorbance of the species in the reactor. The MPC yield is defined as the molar ratio of carbamate produced to aniline consumed. The DMC and DMPD yields were calculated by the same procedure with moles of methanol consumed and moles of DMC consumed, respectively.

RESULTS AND DISCUSSION

In-situ infrared (IR) study of MPC synthesis from the gas-solid oxidative carbonylation. Figure 2 (a) shows the IR spectra of the oxidative carbonylation of aniline and methanol over 5-wt% CuO/MCM-41 at 438 K and 0.34 MPa. The MPC's carbonyl bands at 1715 cm^{-1} emerged upon all the reactants entering the reactor. The mincrease in the intensity of the CO band during the first 0.37 min of the reaction is a result of filling CO/O$_2$ reactants into the reactor. The variation of MPC and aniline IR intensities corresponds to changes in their concentrations. The concentrations of aniline and MPC, obtained by converting their IR intensities with their calibration factors (i.e., extinction coefficients), are plotted along with time in Figure 2 (b). Increases in MPC concentration correspond to decreases in aniline concentration. The MPC concentration leveled off at 7.35 min. A 20.9 % yield of MPC was obtained. The initial rate, determined by the slope of MPC concentration vs. time profile, is 0.24 mol/(s. gmol catalyst) which is slightly lower than those reported for the reaction on Rh/C at 443 K and 4.8 MPa (10). However, our reaction condition (0.34 MPa, 443 K) is significantly milder than the reported condition on Rh/C (4.8 MPa, 443 K).

In-situ infrared (IR) study of DMC synthesis from the gas-solid oxidative carbonylation: Figure 3 shows the IR spectrum of DMC synthesis from the oxidative carbonylation of methanol over 5 wt% CuCl$_2$/HZSM-5. The IR intensity of DMC's carbonyl band at 1766 and 1789 cm^{-1} was leveled off after 2 min of reaction. A 8.1% yield of DMC was achieved. The key adsorbates observed are Cu+(CO)Cl at 2080 cm^{-1} and Cu$_2$(CO)$_4$Cl$_2$ at 2052 cm^{-1}.

In-situ infrared (IR) study of dimethyl 4-methyl-1, 3-phenylene dicarbamate (DMPD) from 2, 4-diaminotoluene and dimethyl carbonate (DMC). Figure 4 (a) shows the IR spectra of DMPD synthesis from the reaction of DMC with 2, 4-diaminotoluene using diphenyl phosphinic acid catalyst. The infrared spectra were plotted in the transmission mode. The carbonyl bands of DMPD at 1717 and 1734 cm^{-1} appeared upon the reaction temperature reaching 403 K. The intensity of these bands continued to grow with time while the IR bands of DMC at 1767 and 1780 cm^{-1}, 2, 4-diaminotoluene at 3228 cm-1, ammonium salt

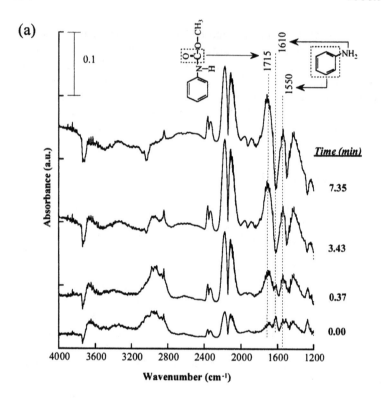

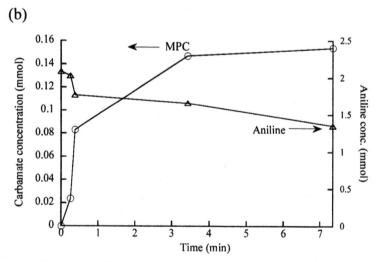

Figure 2 (a) *In situ* IR spectra of MPC synthesis at 0.34 MPa and 438 K over 5-wt% CuO/MCM-41 and (b) the concentration of MPC vs. time.

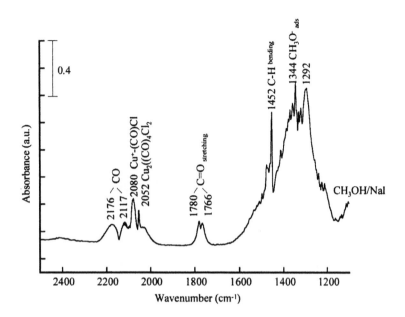

Figure 3 *In situ* IR spectrum of DMC synthesis at 0.34 MPa and 383 K over 5-wt% CuCl$_2$/HZSM-5.

intermediate species (Ph$_2$P(O)O$^-$H$_3$N(Tol)NH$_2$) at 1536 and 1558 cm^{-1}, and phosphinic ester intermediate (Ph$_2$P(O)OC(O)OCH$_3$) at 1287 cm^{-1} decreased. The IR assignment of the ammonium salt and phosphinic ester intermediates was based on the vibration frequency of amine salt and the phosphinic ester, respectively (11).

The variation of the intermediates' IR intensity and reactant/product concentrations was plotted with time in Figure 4 (b). The decreases in reactant (DMC) concentration and intermediate intensities along with the increase in DMPD oncentration and methanol intensity suggest the reactants were converted to ammonium salt and phosphinic ester intermediates which was further converted to DMPD and methanol. This reaction process may follow the following steps:

(i) Ph$_2$ P(O)OH + H$_2$N(CH$_3$C$_6$H$_3$)NH$_2$ $\longleftrightarrow$ Ph$_2$P(O)O$^-$ H$_3$N$^+$(CH$_3$C$_6$H$_3$)NH$_2$

(ii) Ph$_2$P(O)O$^-$ H$_3$N$^+$(CH$_3$C$_6$H$_3$)NH$_2$ + CH$_3$OC(O)OCH$_3$ $\rightarrow$
     Ph$_2$P(O)OC(O)OCH$_3$ + H$_2$N(CH$_3$C$_6$H$_3$)NH$_2$ + CH$_3$OH

(iii) Ph$_2$P(O)OC(O)OCH$_3$ + H$_2$N(CH$_3$C$_6$H$_3$)NH$_2$ $\rightarrow$
      H$_2$N(CH$_3$C$_6$H$_3$)NHC(O)OCH$_3$ + Ph$_2$ P(O)OH

where Ph = C$_6$H$_5$

(a)

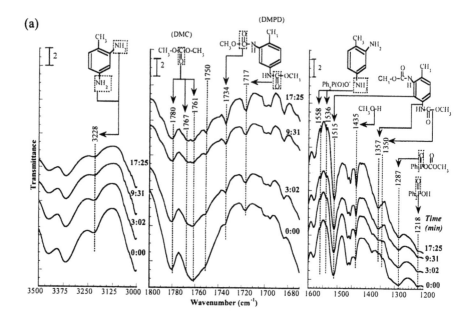

(b)

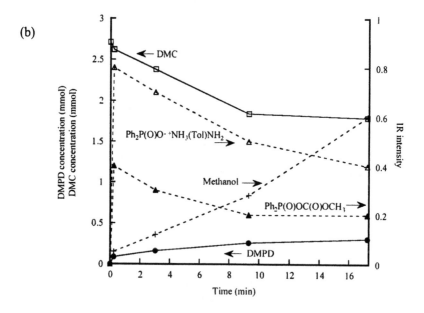

Figure 4 (a) *In situ* IR spectra of DMPD synthesis at 0.101 MPa and 403 K with dephenyl phosphinic acid catalyst. (b) The variation of methanol and intermediates' IR intensity and DMC/DMPD concentration vs. time.

The DMPD concentration approached a constant value at 17.35 min, and its yield was determined to be 34%. The initial rate of DMPD formation, determined from the slope of DMPD concentration vs. time profile, is 4.64×10^{-2} mol/(s.gmol catalyst) while the initial rate for diphenyl 4-methyl 1, 3- phenylene dicarbamate (DPPD) formation from diphenylcarbonate (DPC) and 2, 4- diaminotoluene is 2.95×10^{-4} mol/(s.gmol catalyst) (9). The significant difference in the rate may be due to the steric effect of a bulky phenyl group in DPPD compared to the methyl group in DMPD as shown below:

DMPD synthesis

$$CH_3C_6H_3(NH_2)_2 + (CH_3O)_2CO \rightarrow CH_3C_6H_3(NHCOOCH_3)_2 + 2CH_3OH$$

DPPD synthesis

$$CH_3C_6H_3(NH_2)_2 + (PhO)_2CO \rightarrow CH_3C_6H_3(NHCOOPh) + 2PhOH$$
where Ph = C_6H_5

Due to the difference in solubility of DMPD, DMC, 2,4- diaminotoluene, diphenyl phosphinic acid in H_2O and CCl_4, 2,4-diaminotoluene and DMC can be separated from the product/reactants/catalyst mixture by extraction with H_2O and CCl_4, respectively. However, separation of diphenyl phosphinic acid catalyst from DMPD remains a challenging problem. The immobilization of the diphenyl phosphinic acid catalyst on the MCM-41 support is currently underway, aiming at resolving catalyst recovery and product purification problems.

CONCLUSION

In situ IR study of methyl-N-phenyl carbamate and dimethyl carbonate syntheses over Cu-based catalysts shows that the gas-solid synthesis processes eliminate the solubility limitation, allowing the reaction to occur under milder condition. Diphenyl phosphinic acid was found to be an effective catalyst for the synthesis of DMPD from DMC and 2, 4-diaminotoluene using diphenyl phosphinic acid catalyst. The *in situ* IR study of the DMPD synthesis reaction revealed that ammonium salt (Ph₂P(O)O-+H₃N(Tol)NH₂) and phosphinic ester (Ph₂P(O)OC(O)OCH₃) may serve as reaction intermediates for the DMPD synthesis reaction. The heterogeneous gas-solid synthesis processes hold a great promise for replacement of conventional gas-liquid homogeneous synthesis processes.

ACKNOWLEDGEMENT

This work has been supported by the NSF Grant CTS 9816954 and the Ohio Board of Regents Grant R5538.

REFERENCES

1.   Weissermel, K., and Arpe, H. –J. in Industrial Organic Chemistry, 3rd ed., VCH: New York, 1997, pp 379-381.
2.   Lin, I. J. B., and Chang, C.-S., *J. Mol. Catal.* **73** (1992) 167.
3.   Fukuoka, S., and Chono, M., *J. Chem. Soc., Chem. Commun.* (1984), 399.
4.   Gupte, S. P., and Chaudhari, R. V., *J. of Catal.* **114** (1988) 246.
5.   Sato, Y. and Souma, Y., *Catalyst Surveys* **4** (2000) 65-74.
6.   Chuang, S. S. C., Toochinda, P., and Konduru, M., In *Green Chemistry*, Anastas, P. T., Heine, L. G., and Williamson T. C. (Eds)., Oxford University press, Oxford, (2001) 136-148.
7.   Chuang, S. S. C., Konduru, M., Chi , Y., and Toochinda, P., *Studies in Surface Science and Catalysis*, **130**, 737 (2000).
8.   Toochinda, P. and Chuang, S. S. C., *Environmental Science and Technology*.
9.   Aresta, M., Angela, D., and Quaranta, E., *Tetrahedron* **54** (1998) 14145- 14156.
10. Prasad, K. V. and Chaudhari, R. V., *J. Catal.* **145** (1994) 204.
11. Silverstein, R. M. and Webster, F. X. in Spectrometric Identification of Organic Compounds, 6th ed., Wiley: New York, 1997, pp 102-103, 142.

28

Supported Palladium Catalysts: A Highly Active Catalytic System for CC-Coupling Reactions

J. G. E. Krauter* and J. Pietsch
Degussa AG, Hanau, Germany

K. Köhler and R. G. Heidenreich
Technische Universität München, Garching, Germany

ABSTRACT

A variety of palladium on activated carbon catalysts differing in Pd dispersion, Pd distribution, Pd oxidation state and water content were tested in Heck reactions of aryl bromides with olefins. The optimization of catalyst and reaction conditions (temperature, solvent, base and Pd loading) allowed development of Pd/C catalysts with very high activity for Heck reactions of non-activated bromobenzene (TON ~ 18,000, TOF up to 9,000, Pd concentrations down to 0.005 mol%). The catalysts combine high activity and selectivity under ambient conditions (air and moisture), with easy separation and quantitative recovery of palladium. Determination of Pd in solution during and after the reaction and catalyst characterization before and after the reaction (TEM) indicate dissolution/re-precipitation of palladium during the reaction. The Pd concentration in solution is highest at the beginning of the reaction and is minimized (<< 1 ppm) at the end of the reaction. Pd leaching correlates significantly with the reaction parameters.

INTRODUCTION

The olefination of aryl halides (Heck reaction), one of the most important CC-coupling reactions in organic synthesis (1), is mostly catalyzed by palladium complexes in homogeneous solution. Important advantages of this reaction are the broad availability of aryl bromides and chlorides and the tolerance of the reaction for a wide variety of functional groups. Recently , the development of new highly active Pd complexes has allowed the activation and conversion of even aryl chlorides to a considerable extent (2–10). However, homogeneous catalysts suffer from the need for ligands (usually phosphine ligands) which may be difficult to handle (e.g. air sensitive basic phosphines); additionally, the ligands and the precious metal can be difficult to remove from the product.

Trying to overcome the issue of catalyst separation and Pd recovery, researcher have used immobilized Pd catalysts (polymer supported Pd catalysts (11)) and heterogeneous Pd catalysts (12–14) (Pd on activated carbon (15–17), Pd on metal oxides (18–21), and Pd on zeolites (22)). However, these systems do not reach the excellent activities that are observed with homogeneous catalysts. In addition supported Pd catalysts (mainly Pd/C) often give rise to unwanted dehalogenation of haloaromatics (15), convert only aryl bromides in some instances (18, 20, 21) (only one paper deals with aryl chlorides (22)), and often suffer from substantial Pd leaching (e.g. 14% of Pd is leached from Pd/C (23)). Since the use of supported Pd catalysts is motivated by the ease of catalyst separation, Pd leaching is the most important problem .

The latest work on Pd/C catalyzed Heck reaction has been published by Arai et al. (15). Whereas aryl iodides could be activated easily by the employed Pd/C catalysts, in case of aryl bromides low activities and dehalogenation was observed.

The present paper describes how Pd/C catalyst properties affect activity in the Pd/C catalyzed Heck reaction. Knowledge about this structure-activity relationship and optimization of reaction parameters for this heterogeneous system allowed development a Pd/activated carbon catalyst which exhibits the highest heterogeneous catalytic activity reported for the Heck reaction of aryl bromides with olefins. In addition, this paper describes the parameters that influence Pd leaching during the reaction and reveals that Pd dissolution / redeposition occurs. These findings make it possible to tune the Pd concentration in solution after the reaction down to 0.05 ppm Pd and open new insights into the mechanism of Pd/C catalyzed Heck reactions.

RESULTS

(1) Correlation of Catalyst Structure and Activity

Due to the lack of information about catalyst structure-activity relationships for Pd/C systems in CC-coupling reactions, we studied a variety of different Pd on activated carbon systems. In a first series of experiments, twelve Pd/C catalysts were tested in the Heck coupling of bromobenzene and styrene, as a model reaction for non-activated bromoarenes (Figure 1, R = H).

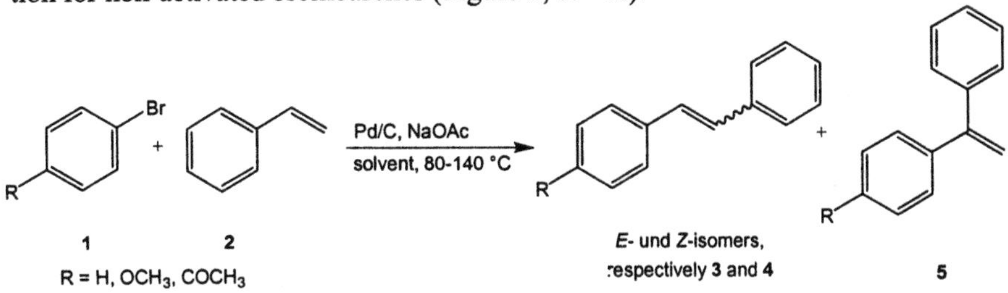

Figure 1   Heck coupling of bromoarenes with styrene.

All Pd on activated carbon catalysts contained 5 wt.% palladium and were characterized by a high (but varying) Pd dispersion (no thermal pretreatment, highest dispersion catalyst **3**). In addition to their Pd dispersion (average Pd crystallite size) the catalysts differed in the reduction degree of Pd, in their content of water (**D** for dry catalysts: ~ 5 wt.% water; **W** for wet catalysts: ~ 50–60 wt.% water) and in the Pd distribution (uniform impregnation, catalysts **3**, **5** and **6** or egg-shell catalysts, catalysts **1**, **2** and **4**). The reduction degree is defined as the amount of Pd^0 compared to the total amount of Pd (Pd^{II} and Pd^0) in the catalyst. Most of the palladium on the surface of catalysts **3–5** is Pd^{II} only, whereas catalysts **1**, **2** and **6** consist of Pd^{II} and mainly Pd^0 species.

Influence of Pd dispersion and reduction degree. The catalytic results summarized in Table 1 illustrate that all parameters investigated are of importance for the activity in the Heck reaction. All catalysts show conversions between 45 and 90 % of bromobenzene (after 20 hours) and high selectivities to *E*-stilbene (> 90 %); no dehalogenation of bromobenzene occurred in any experiment (Table 1).

A first tentative correlation gives the following requirements for high catalytic activity (Table 1): high Pd dispersion (catalyst **3**), low Pd reduction degree (catalysts **3–5**), high water content (> 50 %, i.e. wet catalysts; from W-series) and uniform Pd impregnation. In summary, catalyst **3W**, which combines all requirements listed above, is found to be the most active one.

Effect of thermal treatment. In order to support the correlation between Pd dispersion as well as reduction degree and activity, catalyst **3W** was thermally treated in forming gas ($N_2 + H_2$.) As expected, the dispersion of the Pd/C catalyst decreased with increasing temperature, which is illustrated by a decreasing amount of CO chemisorbed on the Pd surface (Table 2).

Table 1 Catalytic activities of twelve different Pd/C catalysts: Heck coupling of bromobenzene with styrene (Figure 1, R = H)[a].

entry	catalyst	conversion[b] [%]	yield 3[b] [%]	Pd-leaching [%][c]
1	1D	46	41	0.3
2	2D	61	54	1.9
3	3D	48	43	0.4
4	4D	51	45	0.6
5	5D	70	62	2.7
6	6D	60	55	1.7
7	1W	88	80	0.6
8	2W	59	54	3.0
9	3W	92	84	0.4
10	4W	82	75	1.9
11	5W	79	71	2.2
12	6W	78	70	2.2

[a] Reaction conditions: bromobenzene (10 mmol), styrene (15 mmol), NaOAc (12 mmol), 1.0 mol% Pd, DMAc (10 mL); T = 140 °C and t = 20 hours
[b] conversion of bromobenzene and yields of product 3 from GLC-analysis
[c] amount of palladium in solution/total amount of palladium in the reaction mixture

Table 2 Thermally treated catalysts in Heck coupling of bromobenzene with styrene (Figure 1, R = H)[a]

entry	pre-treatment temperature [°C]	pre-treatment atmosphere	CO_{ads} [mL/g cat]	conversion[b] [%]	yield 3[b] [%]	Pd-leaching [%][c]
1	3W	-	1.97	94	84	0.3
2	300	$N_2 + H_2$	1.79	58	52	0.7
3	500	$N_2 + H_2$	0.85	13	11	0.5
4	700	$N_2 + H_2$	0.16	4	3	0.9
5	3D	-	1.97	42	38	0.3

[a] reaction conditions: bromobenzene (10 mmol), styrene (15 mmol), NaOAc (12 mmol), 1.0 mol% Pd, DMAc (10 mL); T = 140 °C and t = 20 hours;
[b] conversion of bromobenzene and yields of product 3 from GLC-analysis;
[c] amount of palladium in solution/total amount of palladium in the reaction mixture.

The higher the thermal treatment temperature, the lower the dispersion became and consequently the lower the catalytic activity was in the Heck reaction (Table 2, entries 2–4.) Catalysts which were treated at 700 °C were found to have no catalytic activity at all (Table 2, entry 4.) This confirms the correlation between Pd dispersion and catalytic activity described above and in the recent literature

(20, 21.) The thermal treatment is accompanied by increased Pd reduction and loss of water, thereby also reducing the catalytic activity (Table 2, entry 5, added for comparison.) This illustrates the mutual interdependence of the parameters Pd dispersion, Pd oxidation state and water with respect to catalytic activity.

(2) Correlation of Reaction Parameters with Activity and Pd Leaching

After the selection of the best catalyst with optimized properties, the influence of the reaction parameters on activity and selectivity was studied in detail. We were also interested in the Pd concentration in solution after separation of the solid catalyst at the end of the reaction, which was expected to depend on these reaction parameters. Since the optimization of the reaction conditions included several parameters, in order to recognize interactions between parameters design of experiments (DoE) was used also and the results of 294 experiments were processed statistically (Statgraphics 3.1).

Each experiment was accompanied the determination of Pd in solution after hot filtration of the solid catalyst at the end of the reaction. Because simple Atomic Absorption Spectroscopy (AAS) was found to not be precise enough for the palladium analysis in this concentration range (detection limit too high.) ICP-OES and/or ICP-MS (Inductively Coupled Plasma - Optical Emission Spectroscopy or Inductively Coupled Plasma - Mass Spectrometry) were applied. To first approximation, the Pd leaching could not be correlated with the properties of the twelve different Pd/C catalysts described above ((1) Correlation of catalyst structure and activity.) There is, however, a strong correlation with the reaction parameters as described below.

Reaction temperature. The most active catalyst **3W** was used for these investigations. This Pd/C catalyst was tested at different reaction temperatures in the coupling of p-bromoacetophenone (Table 3, entries 1-4) and bromobenzene with styrene (Table 3, entries 5–7.) The catalytic runs were performed using only 0.1 mol% palladium (instead of 1.0 mol% relative to the bromoarene as in the catalytic experiments described before Table 1 and 2.)

As expected, activated bromoarenes like p-bromoacetophenone can be converted at lower temperatures than non-activated ones (Table 3, entries 2 and 7.) At reaction temperatures as low as 80 °C p-bromoacetophenone can be converted almost quantitatively. The catalytic activity in the coupling of the non-activated bromobenzene is highest at 140 °C (Table 3, entry 6.) The activity of the Pd/C catalyst is lower at 120 °C as expected, but also at higher temperatures, e.g. 160 °C (Table 3, entries 5 and 7.) The decrease of catalytic activity at 160 °C might result from the instability of the catalytically active Pd species (see below and (12, 16)) in solution at such a high temperature.

The palladium concentration in solution (leaching) also depends on the reaction temperature. This is illustrated for the coupling of bromobenzene and styrene with catalyst **3W** (Table 3, entries 5–7.) The most pronounced influence of the reaction temperature on the Pd concentration in solution is found in the Heck reaction of p-bromoacetophenone and styrene. At 60 °C no catalytic activity and almost no Pd (0.5 %) of the total Pd amount was found in solution (Table 3, entry 1.) Quantitative conversion of p-bromoacetophenone was obtained at 80 °C and higher temperatures (Table 3, entries 2–4.) Here, the Pd leaching increased significantly to 26 % of the total Pd content in the reactor at 80 °C, but surprisingly decreased again to approx. 1.0 % of the total palladium content at 100 °C and 120°C, respectively.

Table 3 Catalysis at different reaction temperatures: Heck coupling of p-bromoacetophenone or bromobenzene with styrene (Figure 1; R = COCH$_3$, H.)[a]

entry	R	catalyst	temperature [°C]	conversion[b] [%]	yield 3[b] [%]	Pd-leaching [%][c]	Pd in solution [ppm][d]
1	COCH$_3$	3W	60	6	6	0.5	0.3
2	COCH$_3$	3W	80	95	91	26.3	19.7
3	COCH$_3$	3W	100	100	95	1.1	0.8
4	COCH$_3$	3W	120	100	95	1.0	0.8
5	H	3W	120	79	71	1.1	0.9
6	H	3W	140	86	77	1.6	1.3
7	H	3W	160	80	73	0.7	0.5

[a] Reaction conditions: bromoarenes (10 mmol), styrene (15 mmol), NaOAc (12 mmol), 0.1 mol% Pd (3W), DMAc (10 mL); T = 60-160 °C and t = 20 hours
[b] Conversion of bromoarenes and yields of product 3 from GLC-analysis
[c] amount of palladium in solution/total amount of palladium in the reaction mixture;
[d] palladium in solution [μg]/[g] of solution.

Effect of the solvent. The influence of the solvent on catalytic activity and Pd leaching was studied in the coupling of p-bromoacetophenone and styrene with catalyst **3W** (Table 4.) The catalytic activity increased by switching from N,N-dimethylacetamide (DMAc) to N-methylpyrrolidone (NMP) (Table 4, entries 1 and 3.) In other solvents like toluene, acetonitrile, tetrahydrofuran (THF) or 1,4-dioxane, the catalytic activity at 80 °C was very low . For the reaction of bromobenzene with styrene in NMP quantitative conversion was found at 140 °C after 20 hours, while in DMAc only 86 % conversion was observed. This is consistent with the results found for p-bromoacetophenone.

The solvent also affects palladium leaching. For more precise determination, conditions favoring palladium leaching (T = 80 °C, as mentioned above) were applied in the coupling of p-bromoacetophenone and styrene (Table 4, entries 1–6). Both NMP and DMAc resulted in high Pd dissolution from the support: 22 % and 26% of the total Pd amount for NMP and DMAc, respectively. Very low Pd leaching (1.2–1.5 %) was found when toluene, THF or 1,4-dioxane were used.

This is accompanied by extremely low catalytic activity in these solvents. This is an indication that Pd in solution is the catalytically active species.

Table 4 Variation of the solvent: Heck coupling of p-bromoacetophenone with styrene (Figure 1, R = COCH₃.)[a]

entry	solvent	conversion[b] [%]	yield 3[b] [%]	Pd-leaching [%][c]	Pd in solution [ppm][d]
1	DMAc	94	90	26.0	19.5
2	toluene	0	0	1.4	1.1
3	NMP	100	95	22.0	16.4
4	acetonitrile	3	3	6.4	4.9
5	THF	3	3	1.5	1.1
6	1,4-dioxane	4	3	1.2	0.9

[a] reaction conditions: p-bromoacetophenone (10 mmol), styrene (15 mmol), NaOAc (12 mmol), 0.1 mol% Pd (**3W**); T = 80 °C and t = 20 hours; [b] conversion of p-bromoacetophenone and yields of product **3** from GLC-analysis; [c] amount of palladium in solution/total amount of palladium in the reaction mixture; [d] palladium in solution [µg]/[g] of solution.

(3) Activity under Optimized Reaction Conditions

Analysis and evaluation of the results discussed above lead to the following optimum reaction conditions for the Heck reaction catalyzed by Pd/C catalysts: 140 °C for non-activated bromoarenes (Table 3), NMP as solvent (Table 4) and sodium acetate as base. The high catalytic activity of the Pd/C catalysts under these conditions allowed reduction of decrease the typical palladium loading down to 0.01 mol%. Argon atmosphere helps to decrease the Pd leaching while keeping the high activity level (data not shown). The most active Pd/C catalyst **3W** should be used without thermal pretreatment (Table 1).

Under these optimized reaction conditions, non-activated bromoarenes like bromobenzene and even deactivated bromoarenes such as p-bromoanisole could be converted almost quantitatively within 2 hours (Table 5, entries 3 and 6.) The Pd content could be dramatically decreased down to 0.005 mol% palladium (1 µmol palladium for 20 mmol bromobenzene) in the coupling of bromobenzene and styrene (Table 5, entry 5.) This extremely high catalytic activity is reflected by turnover numbers (TON) up to 18,000 and turnover frequencies (TOF) up to 9,000 h⁻¹ for this reaction. Even with very low loadings of 0.005–0.01 mol% Pd all bromoarenes studied here could be coupled almost quantitatively within 2 to 6 hours (Table 5, entries 1, 3 and 9) with high selectivity (> 90%) for the E-products. Working in the presence of air (no purging with argon) bromobenzene can be converted almost quantitatively within 3 hours using a Pd loading of only 0.01 mol%. The TON and TOF for this reaction were 9,600 and 3,200 h⁻¹ respectively (Table 5, entry 4.)

Table 5 Optimized reaction conditions: Heck coupling of different bromoarenes with styrene (Figure 1)[a].

entry	R	catalyst [mol% Pd]	solvent	t [h]	conversion[b] [%]	yield 3[b] [%]	TON[c]	TOF[d] [h⁻¹]
1	COCH₃	0.01	NMP	4	96	90	9,500	2,375
2	H	0.01	NMP	1	73	65	7,100	7,100
3	H	0.01	NMP	2	97	87	9,500	4,750
4[e]	H	0.01	NMP	3	97	88	9,600	3,200
5[f]	H	0.005	NMP	2	90	82	18,000	9,000
6	OCH₃	0.10	NMP	2	92	80	900	450
7	OCH₃	0.02	NMP	2	72	63	3,500	1,750
8	OCH₃	0.02	DMAc	2	62	53	2,950	1,475
9	OCH₃	0.01	NMP	6	95	85	9,500	1,583

[a] Reaction conditions (except entry 5): bromoarene (10 mmol), styrene (15 mmol), NaOAc (12 mmol), 0.005–0.10 mol% Pd (**3W**), NMP or DMAc (10 mL); T = 140 °C
[b] Conversion of the bromoarenes and yields of the product 3 from GLC-analysis
[c] TON = moles of coupling products (all isomers) / moles of Pd
[d] TOF = moles of coupling products (all isomers)/moles of Pd per hour
[e] without exclusion of air and moisture (no purging with argon)
[f] bromobenzene (20 mmol), styrene (30 mmol), NaOAc (24 mmol), NMP (20 mL)

(4) Pd Leaching and Catalyst Evolution During the Reaction

Activity and leaching as a function of time. All determinations of the Pd concentration in solution described above were performed after hot filtration of the solid catalyst from the reaction mixture at the end of the reaction. However, in order to gain insight into how the catalytic reaction proceeds on a molecular level, if is important to look at the Pd concentration in solution during the reaction at a given time and correlate this with the corresponding catalytic activity.

Therefore, 25 samples of the reaction mixture were taken during the reaction of bromobenzene and styrene and analyzed by GLC as well as by ICP-OES or ICP-MS (to ensure higher accuracy in this experiment 1.0 mol% Pd was used.) Figure 2 shows that the palladium concentration in solution was highest at the beginning of the reaction. It dropped to 1 ppm when the reaction was finished. Over the reaction period the conversion rate decreased as the Pd concentration in solution decreased. The conversion-time behavior observed is expected for the Heck reaction independent on the Pd concentration in solution. The obvious analogy in conversion and Pd concentration could be understood for example by the influence of the concentration of starting material (potential ligands) on the Pd leaching.

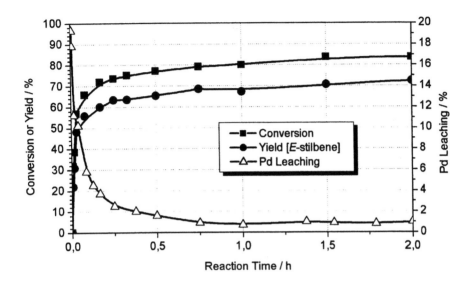

Figure 2 Investigation of activity and Pd leaching as a function of time: Heck coupling of bromobenzene with styrene; reaction conditions: bromobenzene (200 mmol), styrene (300 mmol), sodium acetate (240 mmol), 1.0 mol% Pd, catalyst **3W**, DMAc (200 mL); T = 140 °C; styrene was added last after catalyst, solvent, and bromo benzene had reached 140 °C.

Catalyst evolution. As stated above, the reaction is accompanied by a Pd dissolution/re-precipitation process (16.) This catalyst evolution is also reflected by Transmission Electron Microscopy (TEM) investigations of catalyst **3W** before and after the reaction (Figure 3.)

The fresh catalyst **3W** (Figure 3, top) clearly exhibits a higher Pd dispersion and more uniform distribution than the catalyst (Figure 2, bottom) after Heck coupling of bromobenzene and styrene under standard reaction conditions (DMAc, sodium acetate, 140 °C, 20 h.) The used catalyst was washed with methylene chloride and water before TEM investigations. The average Pd crystallite size is increased by one order of magnitude from 2.4 nm to 23 nm. In fact, the TEM figures indicate the agglomeration of small primary Pd crystallites into large aggregates (Figure 3.)

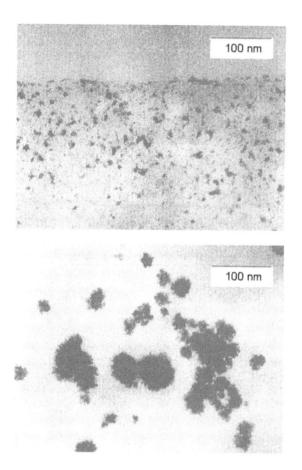

Figure 3 Transmission electron micrograph of catalyst **3W** before Heck coupling of bromobenzene with styrene and catalyst **3W** after the Heck reaction (bottom) (magnification 1:400,000). Reaction conditions: bromobenzene (10 mmol), styrene (15 mmol), sodium acetate (12 mol), 1.0 mol% Pd, DMAc (10 mL); T = 140 °C; reaction time 20 hours.

Multiple catalyst use. The dramatic decrease in Pd dispersion during the reaction should lead to a significant decrease in catalytic activity, when the catalyst is re-used in the same Heck reaction. This was in fact found in recycling studies with catalyst **3W** in Heck couplings of bromobenzene with styrene. The conversion after 6 hours with the recycled methylene chloride washed catalyst is roughly 50 % of fresh **3W**. A second recycling of the Pd/C catalyst decreases the reaction rate further and resulted in very low conversion (4 % after 6 hours). For higher or close to complete conversions reaction times of 20 hours or more are necessary. Re-used catalysts exhibit low activities comparable to other (fresh) heterogeneous systems in the literature (15). Easy and quantitative separation of the Pd/activated carbon catalyst allows, however, effective recovery of the noble metal.

Minimization of Pd leaching at the end of the reaction. Useful reaction procedures allow easy and practical control and minimization of Pd leaching at the end of the reaction. Pd re-deposition is achieved by (i) additional treatment at reaction or increased temperature, (ii) addition of reducing agents (sodium formate) and/or (iii) working under inert atmosphere (24). The Pd concentration in solution after the reaction can be reduced down to 0.05 ppm Pd in all investigated Heck reactions and thus allows practically complete separation of the noble metal.

Discussion and Conclusions. In the present work a large variety of Pd/C catalysts with different properties was studied as catalysts in Heck reactions of aryl bromides with olefins. The activity of the catalysts strongly depends on the Pd dispersion, the Pd oxidation state in the fresh catalyst, the water content (wet or dry catalysts) and the catalyst preparation conditions (impregnation method, pretreatment conditions). The effects are significant, i.e. Pd on the same activated carbon support is either found to be a nearly inactive catalyst or, in the other extreme, a catalyst with the highest activity ever reported up to now for heterogeneous systems for the conversion of aryl bromides (Table 6.)

Table 6 Comparison of activity and reaction conditions for heterogeneous and homogeneous catalysts in the Heck reaction of bromobenzene (Figure 1, R = H) with styrene (or methyl arcylate[a]) published in the literature very recently.

Catalyst	catalyst [mol% Pd][b]	T [°C]	t [h]	yield 3-5[%]	TON[c]	TOF[d] [h^{-1}]	reference
Pd/mesoporous silica	0.1	170	48	39	390	8	[18]
Polymer supported Pd carbenes	0.02	150	60	75	3,750	63	[11]
Pd/C[a]	>1	160	12	<20	≤20	2	[15]
3W (E 105 CA/W 5% Pd)	0.005	140	2	90	18,000	9,000	this paper
Pd/zeolites	0.05	140	0.5	96	1,920	3,840	[22]
Palladacycle	0.0001	130	72	29	290,000	4,028	[3]
Pd carbenes	5	184	1	100	20	20	[5]
Pd(OAc)$_2$	0.0009	130	96	85	94,400	984	[4]

[a] reaction of bromobenzene and methyl acrylate. High selectivity (up to 80%) to undesired benzene (dehalogenation)
[b] moles of Pd/moles of bromobenzene
[c] moles of coupling products (all isomers)/(moles of Pd)
[d] TOF = moles of coupling products (all isomers)/moles of Pd per hour.

The different Pd dispersion, catalyst preparation or pre-treatment conditions, respectively, explain well the differences in the catalytic activity of the various catalysts reported in the literature (15) in comparison to the high activity of catalyst **3W**. The influence of the Pd oxidation state (0 or II) can be interpreted in the context of Pd dispersion as well: each thermal treatment under reducing conditions to produce Pd(0), which is generally accepted as the active state (2), is connected with a decrease in Pd dispersion (Table 2.) Obviously, the *in situ* reduction of Pd(II) under Heck reaction conditions leads to the best catalysts with the highest dispersion.

The optimization of catalyst and reaction parameters allowed   development of a high performance catalyst for the reaction of bromoarenes with olefins whose activity exceeds that of any heterogeneous palladium catalysts reported in the literature by at least one order of magnitude (Table 6.) With the Pd/C catalysts used in this study no dehalogenation of bromoarenes occurred . This is in contrast to Arai et al. (15) where up to 80 % of benzene formation by dehalogenation was observed. Bromobenzene can be converted to 90 % within 2 hours and catalyst concentrations of only 0.005 mol% palladium. Turnover numbers (TON) up to 18,000 and turnover frequencies (TOF) up to 9,000 h^{-1} are obtained (Table 5, entry 5.)

The Pd concentration in solution is highest at the beginning of the reaction (see Figure 1), but is reduced to less than a few ppm after the reaction is finished. Under optimized reaction conditions the palladium concentration in solution can be minimized easily. The Pd concentration in solution correlates with the progress of the reaction, the nature of the starting materials and products, the temperature, the solvent, the base and the atmosphere (argon or air). These results do not limit the practical advantages of the heterogeneous system studied, because the Pd is re-precipitated onto the support at the end of the reaction and can thus be recovered almost quantitatively from the product mixture (<< 1 ppm Pd in the reaction mixture after filtration). This re-precipitation process can be controlled by careful choice of the reaction parameters (at the end of the reaction.) The experimental results, in particular the direct correlation of conversion and concentration of palladium in solution (Figure 1,) indicate a quasi-homogeneous reaction mechanism, i.e. Pd complexes or colloidal particles are active species in solution as has already been published earlier in the literature (13.) The activity at the beginning of the reaction seems to be extremely high before deactivation of the catalyst occurs. The Pd/C catalyst possibly represents a reservoir for the active palladium species in solution.

The catalysts combine all important requirements for practical application: high activity and selectivity without exclusion of air and moisture, extremely low Pd concentrations (0.005-0.1 mol%), easy and complete separation of Pd from the product mixture, easy recovery of palladium and commercial availability. The optimized Pd/C catalyst presented here can be regarded as an important step towards surprisingly simple systems supporting the needs of organic chemists in the laboratory and towards the commercial application of heterogeneous catalysts in C-C-coupling reactions.

REFERENCES

1. J Tsuji. Palladium Reagents and Catalysts. Chichester: Wiley 1995.
2. IP Beletskaya, AV Cheprakov. The Heck Reaction as a Sharpening Stone of Palladium Catalysis. Chem Rev 100:3009–3066, 2000.
3. WA Herrmann, C Broßmer, C-P Reisinger, TH Riermeier, K Öfele, M Beller. Palladacycles: Efficient New Catalysts for the Heck Vinylation of Aryl Halides. Chem Eur J 3:1357–1364, 1997.
4. MT Reetz, E Westermann, R Lohmer, G Lohmer. A Highly Active Phosphine-free Catalyst System for Heck Reactions of Aryl Bromides. Tetrahedron Lett 39:8449–8452, 1998.
5. E Peris, JA Loch, J Mata, RH Crabtree. A Pd complex of a tridentate pincer CNC biscarbene ligand as a robust homogeneous Heck catalyst. Chem Commun 201–203, 2001.
6. GY Li. The First Phosphine Oxide Ligand Precursors for Transition Metal Catalyzed Cross-Coupling Reactions: C-C, C-N, and C-S Bond Formation on Unactivated Aryl Chlorides. Angew Chem 113:1561–1564, 2001; Angew Chem Int Ed 40:1513–1516, 2001.
7. C Dai, GC Fu. The First General Method for Palladium-Catalyzed Negishi Cross-Coupling of Aryl and Vinyl Chlorides: Use of Commercially Available Pd(P(t-Bu)₃)₂ as a Catalyst. J Am Chem Soc 123:2719–2724, 2001.
8. T Rosner, A Pfaltz, DG Blackmond. Observation of Unusual Kinetics in Heck Reactions of Aryl Halides: The Role of Non-Steady-State Catalyst Concentration. J Am Chem Soc 123:4621–4622, 2001.
9. A Zapf, M Beller. Palladium Catalyst Systems for Cross-Coupling Reactions of Aryl Chlorides and Olefins. Chem Eur J 7:2908–2915, 2001.
10. M Beller, A Zapf, W Mägerlein. Efficient Synthesis of Fine Chemicals and Organic Building Blocks Applying Palladium-Catalyzed Coupling Reactions. Chem Eng Technol 24: 575–582, 2001.
11. J Schwarz, VPW Böhm, MG Gardiner, M Grosche, WA Herrmann, W Hieringer, G Raudaschl-Sieber. Polymer-supported Carbene Complexes of Palladium: Well-defined, Air-stable, Recyclable Catalysts for the Heck Reaction. Chem Eur J 6:1773–1780, 2000.
12. A Biffis, M Zecca, M Basato. Metallic Palladium in the Heck Reaction: Active Catalyst or Convenient Precursor? Eur J Inorg Chem 1131–1133, 2001.
13. A Biffis, M Zecca, M Basato. Palladium metal catalysts in Heck C-C coupling reactions. J Mol Catal A: Chem 173:249–274, 2001.
14. H-U Blaser, A Indolese, A Schnyder, H Steiner, M Studer. Supported palladium catalysts for fine chemical synthesis. J Mol Catal A: Chem 173:3–18, 2001.
15. F Zhao, BM Bhanage, M Shirai, M Arai. Heck Reactions of Iodobenzene and Methyl Acrylate with conventional Supported Palladium Catalysts in the Presence of Organic and/or Inorganic Bases without Ligands. Chem Eur J 6:843–848, 2000.
16. F Zhao, K Murakami, M Shirai, M Arai. Recyclable Homogeneous/Heterogeneous Catalytic Systems for Heck Reaction through Reversible Transfer of Palladium Species between Solvent and Support. J Catal 194:479–483, 2000.

17. F Zhao, M Shirai, M Arai. Heterogeneous catalyst system for Heck reaction using supported ethylene glycol phase Pd/TPPTS catalyst with inorganic base. J Mol Catal A: Chem 154:39–44, 2000.
18. CP Mehnert, DW Weaver, JY Ying. Heterogeneous Heck Catalysis with Palladium-Grafted Molecular Sieves. J Am Chem Soc 120:12289–12296, 1998.
19. S Iyer, VV Thakur. The novel use of Ni, Co, Cu, and Mn heterogeneous catalysts for the Heck reaction. J Mol Catal A: Chem 157:275–278, 2000.
20. M Wagner, K Köhler, L Djakovitch, S Weinkauf, V Hagen, M Muhler. Heck reactions catalyzed by oxide-supported palladium – structure-activity relationships. Topics in Catal 13:319–326, 2000.
21. K Köhler, M Wagner, L Djakovitch. Supported palladium as catalyst for carbon-carbon bond construction (Heck reaction) in organic synthesis. Catal Today 66:105–114, 2001.
22. L Djakovitch, K Köhler. Heck Reaction Catalyzed by Pd-Modified Zeolites. J Am Chem Soc 123:5990–5999, 2001.
23. A Eisenstadt, R Hasharon, Y Keren, K Motzkin. Process for the preparation of octyl methoxy cinnamate. (IMI Institute for Research), US patent 5,187,303.
24. RG Heidenreich, JGE Krauter, J Pietsch, K Köhler. J Mol Catal A: Chem in press.

29

Enantioselective Hydrogenation of Ethyl Pyruvate over Pt Colloids

J. L. Margitfalvi, E. Tálas, L. Yakhyaeva, E. Tfirst, I. Bertóti, and L. Tóth
Hungarian Academy of Sciences, Budapest, Hungary

ABSTRACT

Platinum nanocolloids with different average particle size were prepared using cinchonidine (CD) or cinchonine (CN) as stabilizing agents. These colloids were used as catalysts in enantioselective hydrogenation of ethyl pyruvate (EtPy) in various solutions. Two forms of the alkaloids were distinguished: (i) the stabilizing form ($(CD)_{st}$ or $(CN)_{st}$), and (ii) the excess form ($(CD)_{ex}$ or $(CN)_{ex}$), i.e., the amount of alkaloids added into the liquid phase. In acetic acid-methanol mixture high rates and ee values around 80 % were obtained. The ee values slightly decreased with the Pt particle size, however the activity decrease was very pronounced. This behavior was attributed to the increase of the unreduced forms of Pt and the presence of Cl ions in small Pt nanoclusters below 2.8 nm. The performance of this type of Pt nanocolloids was strongly affected by the concentration of both $(CD)_{ex}$ and acetic acid. Kinetic evidences were obtained with respect to the exchange between the stabilizing and the excess forms of the alkaloids. Results of molecular modeling showed that only the "shielded" substrate-modifier complex could be accommodated on the plain surfaces of these small Pt nanocolloids. It is considered that the results provided further support for our "shielding effect model" suggested earlier.

INTRODUCTION

The enantioselective hydrogenation of α-keto esters over Pt-cinchona alkaloids catalyst system is one of the most interesting and widely investigated heterogeneous asymmetric hydrogenation reactions [1-10]. This is a unique reaction as cinchona alkaloids, the most common chiral modifiers used bring about not only high enantioselectivities (ee > 95 %), but high rate acceleration (RA) [2]. All authors agree that in the enantio-differentiation step a 1:1 cinchona-substrate complex is involved. However, there is a strong distinction in the views on the issues: where and how the enantio-differentiation takes place. Those with heterogeneous catalytic background suggest that the above complex is formed on the platinum surface [3-7] (model I) (see Figure 1). In model I both the quinoline ring of the modifier and the substrate molecule (or its half-hydrogenated form) is adsorbed on the Pt surface. Contrary to that we had proposed that the formation of the substrate-modifier complex takes place in the liquid phase [8,9] and the formed shielded complex, as a *supramolecule,* is hydrogenated on the Pt sites [8-10] (model II) (see Figure 2).

Figure 1 Formation of the substrate-modifier complex formed on the Pt (111) surface [4]. Pt - gray , carbon - green , oxygen - red, hydrogen - white, nitrogen - dark blue*.

Figure 2 Accommodation of the shielded substrate-modifier complex formed in the liquid phase on the Pt (111) surface [8]. Pt - pink , carbon - green , oxygen - red, hydrogen - white, nitrogen - dark blue*.

---

* For color version, please see: www.orcs.org.

In model I the *open* conformer of the alkaloid is involved resulting in a site requirement around 1.1-1.2 nm (see Figure 1). Contrary to that in model II the *closed* conformer of the alkaloid forms a shielded substrate-modifier complex in the liquid phase. Due to the involvement of the closed conformer of the alkaloid the site requirement in the hydrogenation step is relatively small (see Figure 2).

In early nineties nice correlation was found between the particle size and the enantioselectivity, i.e., the larger the particle size of Pt the higher the enantioselectivity [5]. This correlation suggested that relatively large number of Pt sites has to be involved in the enantio-differentiation step. This view is reflected in model I. Contrary to that in the "shielding effect model" (model II), as it has been discussed above, there is no preference with respect to the size of the Pt nanoclusters, as in principle, the site requirement to hydrogenate the formed supramolecule (i.e., the substrate-modifier complex) is relatively small [8-10].

Recently it has been demonstrated that relatively high enantioselectivities can be obtained upon using platinum nanocolloids prepared in different ways [11-13]. These experiments clearly showed that high enantioselectivities can be obtained even on small platinum clusters (average particle size around 1.6 nm), where the accommodation of the 1:1 modifier-substrate surface complex (see Figure 1) is strongly hindered. Thus, these results created certain contraversion in the validity of the present models.

The formation of 1:1 modifier-substrate complex either on the Pt sites or in the liquid phase needs very characteristic chemical interactions in order to induce both rate acceleration and asymmetric induction. In this respect interactions, such as: van der Waals, ionic, covalent, donor acceptor, host-guest, $\pi-\pi$, etc. can be suggested. In model I the character of substrate-modifier interaction is monodentate with the involvement of a hydrogen bond between the quinuclidine nitrogen of cinchonidine and the proton of the half hydrogenated surface intermediate of the substrate [4,6,7]. Contrary to that in model II the character of the interaction is bidentate with the involvement of two different types of interactions: (i) interaction between the lone pair of the quinuclidine nitrogen and the carbonyl carbon of the keto group, and (ii) $\pi-\pi$ stacking between the large aromatic ring of the modifier and the conjugated double bond system of the substrate [8-10]. It is important to emphasize that the replacement of the quinoline ring of the alkaloid for phenyl or pyridyl resulted in complete loss of enantioselectivity [14]. Based on literature analogy we attributed this loss to repulsive $\pi-\pi$ stacking between the small phenyl ring and the conjugated double bond system of the substrate. It is worth for mentioning that in organic chemistry there are various persuasive data supporting the need of large aromatic ring system to induce "chemical shielding" responsible for enantio-differentiation [15]. In an other explanation the phenyl ring can not provide sufficient strength of adsorption to keep the 1:1 substrate-modifier complex on the platinum surface [14]. Unfortunately, in some of the surface models the character of chemical interactions in the 1:1 substrate-modifier complex was not even mentioned [12].

The aim of this paper is to clear the above-described contraversies. In this paper new results will be presented to demonstrate the peculiarities of this type of platinum nanocolloids in the enantioselective hydrogenation of ethyl pyruvate. Finally, upon using computer modeling it will be shown that the substrate-modifier complex in the shielded form can easily be accommodated on the small Pt nanoclusters resulting in enantio-differentiation.

EXPERIMENTAL

Preparation and Characterization of Platinum Nanocolloids

Pt nanocolloids were prepared according to the procedure described by Bönne-mann and co-workers [13]. The only difference is that cinchonidine (CD) and cin-chonine (CN) (Fluka) was used instead of 10,11-dihydrocinchonidine and the al-kaloid/Pt ratio was further increased. $PtCl_4$ (Fluka) was used without further puri-fication. The nanocolloids were analyzed for their C, N, and Pt content, while their particle size was determined by X-ray diffraction and TEM (CM20 Philips trans-mission electron microscope, 200kV). The TEM samples were prepared by peptiz-ing the colloid in ethanol and introduced onto a copper microgrid covered with carbon. UV/VIS spectroscopy was used for rapid test of particle size and degree of peptization of platinum nanocolloids determining the characteristic S values [16].

The X-ray photoelectron spectra were recorded by a Kratos XSAM 800 spectrometer using Mg $K\alpha_{1,2}$ radiation, with fixed analyser transmission (pass energy 40 eV). The spectra were referenced to the Au $4f_{7/2}$ line (of the Au spot deposited on some samples) set at binding energy B.E.=84.0 eV. For data acquisi-tion and processing the Kratos Vision 2000 program was applied. The Pt4f peaks were rather broad manifesting, in addition to the elemental state (Pt^0), two oxi-dized states of platinum. The ratio of these bonding states were determined by fitting the over-all peak envelopes to three doublet components with a separation of 3.33 eV between the $4f_{7/2}$ and $4f_{5/2}$ lines.

Catalytic Reactions

The hydrogenation of ethyl pyruvate (EtPy) was carried out in a glass lined 300 cm^3 SS autoclave equipped with an injection chamber for separate introduction of substrate. The Etpy was injected in t=0 minute by hydrogen pressure. Etpy (Fluka) was distilled under vacuum prior its use. The solvents (acetic acid, methanol, and toluene, (Reanal)) were used as received. Reaction conditions: temperature=12 °C, hydrogen pressure = 5 bar, [EtPy]$_o$ = 0.6 M, catalyst = 0.002-0.02g freeze-dried Pt nanocolloid. Stirring rate: 2000 rpm. The duration of catalytic experiments was 0.5-4 hours. Prior to the reaction the solution of the catalyst in the presence of added cinchona alkaloid was treated in ultrasonic bath for 90 minutes. GC analysis of reaction mixture was carried out on a modified cyclodextrine coated capillary column (Advanced Separation) resulting in complete separation of (R)- and (S)-ethyl lactate and EtPy. The optical yield was calculated as ee = ([R]-[S])/([R]+[S]). The kinetic treatment of the catalytic runs can be found elsewhere [8-10]. In this approach two first order rate constants, k_1 and k_2, were calculated describing the first 10-15 minute and the interval between 25 and 100 conversion, respectively.

Molecular Modeling and Monte Carlo Simulation

The molecular docking calculations were performed with some modules of the MSI: InsightII program package. We used the Discover (cvff forcefield, VA09A minimization method) and the Ampac/Mopac (AM1 semiempirical method) modules for optimizing the geometry of the complexes formed by the cinchonidine and the methyl pyruvate molecules. These calculations were performed for the whole systems from different initial positions of the examined complexes with *anti* conformations of the substrate.

The adsorption of the substrate-modifier complexes onto Pt (100) surface of the nanocluster was investigated using the Solids-Docking module of the Insight II package, which performs a Monte Carlo docking approach. The docking procedure is based on the random selection of positions and molecular orientations of the complexes above the surface followed by the minimization of the flexible guest complex to optimize its interaction with the fixed host surface.

RESULTS AND DISCUSSION

Preparation of Pt Nanocolloids

Influence of the alkaloid-Pt ratio

The synthesis of Pt nanocolloids was investigated using alkaloid/Pt ratios between 1:1 and 1:6. In the corresponding reference [13] the highest alkaloid/Pt ratio was 2. The characteristic feature of nanocolloids is given in Table 1. In addition of the use of cinchonidine (CD) cinchonine (CN) was also applied as a stabilizing agent.

As emerges from the data given in Table 1 the average particle size of the Pt nanocolloids was in the range of 1.8 – 2.8 nm. Typical particle size dependence is shown in Figure 1. The mean particle size of this nanocolloid is 1,9 nm. The particle size distribution is relatively narrow, the amount of particles larger than 2.8 nm is negligible. The slight asymmetry is due to the fact that particles below 0.9 were impossible to visualize. The particle size determined by X-ray diffraction was slightly smaller than that of determined by TEM. It is probably due to the fact that the nanocolloid has a relatively broad organic layer consisting of the alkaloid and acetic acid.

Data given in Table 1 indicate that the C/N ratio of the Pt colloids is not constant. In the series of colloids N° 1-4 it increases from 8.6 to 9.4. This fact indicates that upon increasing the alkaloid/PtCl$_4$ ratio the amount of excess acetic acid involved in the stabilization of the nanocolloid increases.

Table 1 Characteristic features of the nanocolloids prepared.

No	Alkaloid/Pt ratio[a]	Average particle size		Elemental analysis, w %				Pt^o/Pt_{total}
		TEM	XRD	Pt	C	N	Cl	
1	1	2.8	n.a.	40.6	32.7	3.8	2.2	0.65
2	2	2.4	n.a.	18.2	50.7	n.a.	3.0	0.58
3	4	2.1	1.8	17.9	49.1	5.1	4.2	0.26
4	6	1.9	n.a.	19.6	49.8	5.3	4.7	0.32
5[b]	6	n.a.	1.8	23.2	44.1	4.8	5.5	0.45

[a] The molar ratio of alkaloid/PtCl$_4$ during the reduction of PtCl$_4$. [b] cinchonine (CN) was used as stabilizing agent. The Pt^o/Pt_{total} ratio was determined from XPS results, n.a. = not available.

Valance state of platinum

In the original paper related to the preparation of this type of Pt nanocolloids it has been mentioned that the colloids are practically chlorine free (below 0.5%). However, as it is shown in Table 1 all colloids contain measurable amount of chlorine. The chlorine content depended also on the initial alkaloid/PtCl$_4$ ratio. This fact initiated the investigation of the valance state of platinum in these nanocolloids.

Results summarized in Table 1 indicate that the higher the alkaloid/Pt ratio during the preparation of nanocolloids the lower the Pt^o/Pt_{total} ratio. This fact has to be taken into account when the activity of the Pt nanocolloids will be compared.

The stabilization of the Pt in ionic form can be related to the formation of cinchonidine-PtCl$_4$ a complex in the liquid phase. This was evidenced by the difference in the circular dichroism spectra of CD measured in the presence and absence of PtCl$_4$ [17].

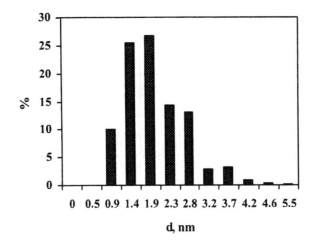

Figure 3 Typical particle size distribution of a Pt nanocolloid (sample N° 4 in Table 1, alkaloid/Pt ratio=6).

Catalytic Experiments

The influence of the type of nanocolloid

The activity and enantioselectivity measured on different Pt nanocolloids are summarized in Table 2. These results were obtained in the presence of added alkaloid (CD_{ex})). These results indicated that the performance of the Pt nanocolloids strongly depended on the mode of their preparation, namely on the alkaloid/Pt ratio used. The increase of the alkaloid/Pt ratio lead to a strong decrease of the activity and slight decrease of the enantioselectivity. The ee value obtained at low and full conversion was almost identical, i.e., in these experiments, within the experimental error, the ee appeared to be independent of conversion.

The change of CD to CN resulted in the preferential formation of (S) – product, it is reflected by the change of the sign in the ee value.

The influence of the alkaloid and acetic acid added to the reaction mixture

Based on the carbon content of the nanocolloids (see Table 1) it has been suggested that it should contain not only cinchonidine, but also acetic acid. For this reason the influence of the concentration of both acetic acid and cinchonidine on the performance of Pt nanocolloids has been investigated. These results are given in Tables 3 and 4, respectively.

Table 2 Enantioselective hydrogenation of ethyl pyruvate over different Pt nano-colloids.

N^o	Alkaloid/Pt ratio[a]	k_1, min^{-1}	k_2, min^{-1}	ee_{max}	ee_{end}
1	1	0.079	0.079	0.891	0.874
2	2	0.023	0.033	0.866	0.864
3	4	0.015	0.009	0.810	0.780
4	6	0.009	0.002	0.780	0.751
5^b	6	0.031	0.046	-0.562	-0.540

[a] The molar ratio of alkaloid/PtCl$_4$ during the reduction of PtCl$_4$. [b] cinchonine was used as stabilizing agent and the alkaloid added into the liquid phase, $[(CN)_{ex}]= 6.8 \times 10^{-3}$ M. Nanocolloid: 0.020 g (N^o1-5 in Table 1), $[(CD)_{ex}]=6.8 \times 10^{-3}$ M, $[Etpy]_o=0.6$ M, $p_{H2}=5$ bar, V=75 cm^3, solvent = CH$_3$COOH: methanol (5:1) in toluene, stirring rate =2000 rpm.

Table 3 Influence of acetic acid concentration on the hydrogenation of Etpy.

N^o	$[CH_3COOH]$, M	k_1, min^{-1}	k_2, min^{-1}	ee_{max}	ee_{end}
1	0.00	0.010	0.005	0.338	0.320
2	0.47	0.010	0.024	0.448	0.428
3	2.35	0.021	0.040	0.601	0.600
4	11.00	0.013	0.027	0.729	0.693

Nanocolloid: 0.020 g (N° 4 in Table 1), $[(CD)_{ex}]=6.8 \times 10^{-3}$ M, $[Etpy]_o=0.6$ M, $p_{H2}=5$ bar, V=75 cm^3, solvent = CH$_3$COOH in toluene, stirring rate =2000 rpm.

Table 4 Influence of excess cinchonidine, $(CD)_{ex}$ on the hydrogenation of Etpy.

N^o	$[(CD)_{ex,}]$ 10^{-3} M	k_1, min^{-1}	k_2, min^{-1}	ee_{max}	ee_{end}
1	no*	0.002	0.0005	0.240	0.196
2	no	0.011	0.029	0.499	0.477
3	0.7	0.011	0.028	0.509	0.509
4	3.4	0.015	0.031	0.591	0.577
5	6.8	0.018	0.034	0.662	0.600

Nanocolloid: 0.020 g (N° 4 in Table 1), $[Etpy]_o=0.6$ M, $p_{H2}=5$ bar, V=75 cm^3, solvent = 2.35 M CH$_3$COOH in toluene, stirring rate =2000 rpm, * - solvent = pure toluene.

Results presented in Tables 3 and 4 strongly indicate that, although the nanocolloids prepared are active in the enantioselective hydrogenation of Etpy both their activities and enantioselectivities are strongly influenced by the concentration of both cinchonidine and acetic acid added to the reaction mixture.

The increase of $(CD)_{ex}$ from zero to 6.8 x 10^{-3} M resulted in only minor changes in the values of first order rate constants, k_1 and k_2, however the changes in the enantioselectivity values were very pronounced. This fact indicates that the amount of alkaloid involved in the stabilization of the Pt nanocolloids is not sufficient to induce high enantioselectivity. We suggest that extra amount of alkaloid is required to form the 1:1 modifier-substrate complex.

The influence of the concentration of acetic acid appeared to be even more complex. If the concentration of acetic acid is not sufficient in this case even in the presence of added CD both the ee values and the rates are extremely low. In the experiment without added CD and acetic acid (see exp. N° 1 in Table 4) the reaction rate was slow that only 20 % conversion was achieved after 4 hours of reaction. From these results it can be concluded that the presence of acetic acid is crucial in this catalytic system.

Exchange between the stabilizing and excess forms of the alkaloid

The results has been obtained so far indicates that two forms of the alkaloids can be distinguished: (i) the stabilizing form $((CD)_{st}$ or $(CN)_{st})$, and (ii) the excess form $((CD)_{ex}$ or $(CN)_{ex})$, i.e., the amount of alkaloids added into the liquid phase. The Pt colloid prepared upon using cinchonine $(Pt_{CN}$, see catalyst N°5 in Table 1) was used to investigate the possible exchange between the two forms of the alkaloid. These results are presented in Figure 4. In the first experiment the Pt_{CN} colloid was used and the concentration of $(CN)_{ex}$ was 6.8 10^{-3} M. In this experiment the ee was independent of the conversion and leveled of at ee = - 0.6.

In the second experiment instead of $(CN)_{ex}$ $(CD)_{ex}$ was used and its concentration was also 6.8 10^{-3}M. In this experiment the initial ee value was in the same range as in the previous experiment, however, above 40 % conversion the ee value increased and at 100 % conversion the final ee value was close to zero. This result indicates that there is an exchange between the two forms of the alkaloid. The initial ee values (ee = -0.6) show that at low conversions the initial $(CN)_{st}$ form is involved in the events controlling the asymmetric induction. As the reaction proceeded the $(CN)_{st}$ form was exchanged by $(CD)_{ex}$ resulting in an increase in the ee values. The final ee value (ee = 0) indicates that the above exchange is almost quantitative.

In the third experiment the $(CD)_{ex}$ was added in a calculated amount prior to the treatment with ultrasound. In this experiment the ee value was constant but was opposite in sign, i.e., during the ultrasonic treatment full exchange between the two forms of the alkaloid took place (see Figure 4). The phenomenon appeared to be fully reproducible. Further studies reviled that this exchange is strongly influenced by the concentration of acetic acid, too [18].

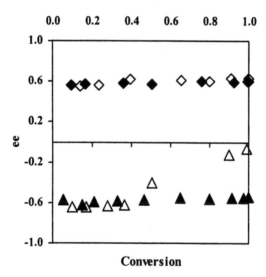

Figure 4 The ee – conversion dependencies obtained in the presence of Pt$_{CN}$ varying the character of excess alkaloid ((CN)$_{ex}$. or (CD)$_{ex}$). ▲ – experiment in the presence of (CN)$_{ex}$; △ - experiment in the presence of (CD)$_{ex}$ without treatment with ultrasound, ◆,◇ - experiment in the presence of (CD)$_{ex}$ after treatment with ultrasound. The concentration of excess alkaloids is 6.8 x 10^{-3} M.

Computer Modeling

In our earlier studies [8-10] the Monte Carlo method has been applied to simulate the adsorption of the modifier-substrate into the Pt(111) surface. The results of these simulations indicated that the above complex, as a supramolecule, maintains its entity after chemisorption (see Figure 2). Slight differences were observed in the orientation of the quinoline ring towards the Pt(111) surface, i.e., orientations either parallel or slightly bent to the Pt(111) surface were obtained (8-10). However, the "shielding effect" provided by the quinoline ring was maintained in each case.

In this work two hypothetical Pt clusters, with particle size of 1.6 and 3.6 nm were build. The nanoclusters strongly resemble the nanoclusters presented in ref 12. Monte Carlo simulation was carried out in such a way that the centre of mass of the supramolecule was placed 5-8 nm above the center of the Pt(100) surface of the nanocluster and the surface-docking program was initiated. The Pt (100) surface of the smaller nanocluster was 3 x 3 in size. In the larger colloid the above size increased to 5 x 5. It has to be emphasized that none of these sites can accommodate the substrate-modifier complex shown in Figure 1.

A   B

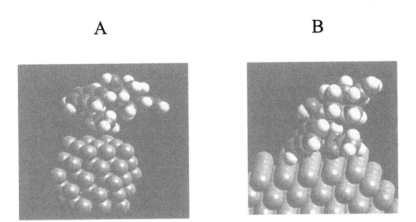

Figure 5
A. Monte Carlo simulation of the adsorption of the modifier-substrate complex onto the Pt(100) surface of a 1.6 nm Pt nanocolloid.
B. Monte Carlo simulation of the adsorption of the modifier-substrate complex onto the Pt(100) surface of a 3.6 nm Pt nanocolloid.
Colours: Pt - gray , carbon - green , oxygen - red, hydrogen - white, nitrogen - dark blue[*].

The results of the simulation are shown in Figures 5A and 5B. These results clearly show that the shielded modifier-substrate complex can be accommodated onto the small platinum nanocluster similar to its accommodation at large Pt surfaces [8-10]. After docking both the "shielding effect" provided by the quinoline ring and the "directionality" between the quinuclidine nitrogen and the carbon atom of the keto group are maintained. The N-C distance (between the quinuclidine nitrogen and the carbon atom of the keto group) is 0.361 and 0.346 nm, what is close two the value calculated for the same surface complex formed over large flat Pt (111) surface (N-C = 0.371 nm).

CONCLUSIONS

In summary, in the presence of acetic acid and added amount of alkaloid the Pt nanocolloids prepared by using cinchona alkaloids appeared to be highly active and selective in the enantioselective hydrogenation of ethyl pyruvate. We consider that in this type of Pt nanocolloids both the quinuclidine nitrogen and the quinoline ring is involved in the stabilization of the colloidal form, however the presence of acetic acid seems to be also very crucial in this process. The prepared Pt nanocol-

---

[*] For color version, please see: www.orcs.org.

loids contained also chloride anions and Pt cations and their amount increased upon increasing the alkaloid/PtCl₄ ratio in the preparation step. The presence of these ionic forms strongly decrease the activity of this catalytic system, however the enantioselectivity decreased only slightly. In the presence of acetic acid the stabilizing alkaloid can be replaced by the alkaloid added into the solution, i.e., there is an equilibrium between the stabilized and excess forms of alkaloid. Upon using Monte Carlo simulation the accommodation of the shielded substrate-modifier complex has been demonstrated. These new results can be considered as further proves with respect to our "shielding effect model" suggested earlier [8-10].

ACKNOWLEDGEMENT

Thanks to Dr. I Sajó for the performance of XRD measurements.

REFERENCES

1. Y. Orito, S. Imai and S. Niwa, *J. Chem. Soc. Japan*, 1118 (1979).
2. H.U. Blaser et al., Sud. Surf. Sci. Catal.,**41**, 153 (1988).
3. H.U. Blaser and M. Müller, *Stud. Surf. Sci. Catal.*, **59**, 73 (1991).
4. I.M. Sutherland, A. Ibbotson, R.B. Moyes, and P.B. Wells, *J.Catal.*, **125**, 77 (1990).
5. H.U. Blaser, H.P. Jalett, D.M. Monti, A. Baiker, and J.T. Wehrli, *Stud. Surf. Sci. Catal.*, **67**, 141 (1991).
6. H.U. Blaser, H.P. Jalett, M. Müller, and M. Studer, *Catal. Today*, **37**, 441 (1997).
7. H.U. Blaser, H.P. Jalett, M. Garland, M. Studer, H. Thies, A. Wirth-Tijani, *J.Catal.* **173**, 282 (1998).
8. J.L. Margitfalvi, M. Hegedűs and E. Tfirst, *Stud. Surf. Sci. Catal.* (11th International Congress on Catalysis) **101**, 241 (1996).
9. J.L. Margitfalvi, M. Hegedűs, E. Tfirst, *Tetrahedron: Asymmetry*, **7**, 571 (1996).
10. J.L Margitfalvi, E. Tálas, E. Tfirst , C.V Kumar, A Gergely, *Applied Catalysis A: General*, **191**, 177 (2000).
11. X. Zuo, H. Liu, M. Liu, *Tetrahedron Letters*, **39**, 1941 (1998).
12. X. Zuo, H. Liu, D. Guo, X.Y., *Tetrahedron*, **55**, 7787 (1999).
13. H. Bönnemann, G.A. Braun, *Angew. Chem. Int. Ed. Engl.*, **35**, 1992 (1996).
14. M. Schürch, T. Heinz, R. Aeschimann, T. Mallat, A. Pfaltz, A. Baiker, *J. Catal.*, **173**, 187 (1998); K.E. Simons, G. Wang, T. Heinz, T. Giger, T. Mallat, A. Pfaltz, A. Baiker, *Tetrahedron Asymmetry*, **6**, 505 (1995).
15. U. Maitra, P. Mathivanan, *Tetrahedron: Asymmetry*, **5**, 1171 (1994).
16. D.G. Duff, P.P. Edwards, B.F.G. Johnson, *J. Phys. Chem.* **99**, 15934 (1995).
17. J. L. Margitfalvi, F. Zsila, E. Tálas, to be published.
18. J.L. Margitfalvi, E. Tálas, to be published.

30

Palladium-Catalyzed Amination of Fluorohaloarenes: A Chemoselective Synthesis of Fluoroquinolone Antibacterials

John R. Sowa, Jr.,[*†] Joseph P. Simeone,[††] Stuart Hayden,[†‡§] Robert Brinkman,[†] Jason Kodah,[†§] Jodie Brice,[†,††] and Amanda Villoresi[†,††]**
Seton Hall University, South Orange, and Merck & Co., Inc., Rahway, New Jersey, U.S.A.

ABSTRACT

The palladium-catalyzed amination reaction is developed for the preparation of fluoroanilines from fluorohaloarenes (82 – 98 % yields) and the synthesis of fluoroquinolone antibacterials including Norfloxacin (58 – 75 % yields). For both sets of substrates, the fluoro-substituent remains intact, thus, the chemoselectivity is quantitative.

---

[*] *Corresponding author*, e-mail: sowajohn@shu.edu, FAX: 973-761-9772.
[†] Department of Chemistry and Biochemistry, Seton Hall University.
[‡] Merck & Co., Inc.
[§] *Current affiliation*: The R. W. Johnson Pharmaceutical Research Institute, Raritan, New Jersey, U.S.A.
[**] *Current affiliation*: Albany Molecular Research, Inc., Albany, New York, U.S.A.
[††] Undergraduate research participant.

INTRODUCTION

Currently there is considerable interest in fluoroquinolone antibacterials as these pharmaceuticals are used for the treatment of infections resulting from exposure to *Bacillus anthracis* (anthrax).[1] In particular, Ciprofloxacin is specifically recommended; however, utilization of other fluoroquinolones such as Levaquin and Tequin is under investigation.[2] The first fluoroquinolone to establish clinical usage was Norfloxacin (1).[3,4] This pharmaceutical also serves as an excellent example of the typical synthesis of fluoroquinolones which are almost invariably prepared via nucleophilic addition (S_NAr) of an amine to a fluorochloroquinolone (2) as the final step (Scheme 1).[3,5] However, it is known that this step may produce 10 – 25 % of an undesired fluoro-substituted byproduct (3).[6,7]

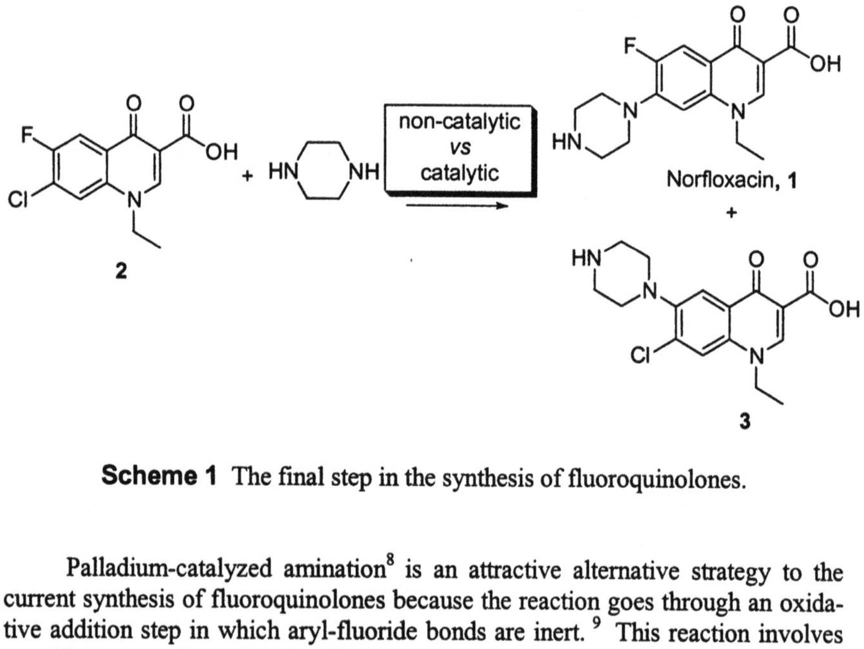

Scheme 1 The final step in the synthesis of fluoroquinolones.

Palladium-catalyzed amination[8] is an attractive alternative strategy to the current synthesis of fluoroquinolones because the reaction goes through an oxidative addition step in which aryl-fluoride bonds are inert.[9] This reaction involves an efficient coupling of aryl halides and amines (eq 1) and is successful for aryl halides with a broad range of functionalities. In this paper, we report model studies of the palladium-catalyzed amination of fluorohaloarenes to produce fluoroanilines. Indeed, the model studies demonstrate that the catalytic reaction is chemoselective.[9] Moreover, in this paper, we demonstrate that the palladium-catalyzed amination methodology can be extended toward chemoselective syntheses of fluoroquinolones.

RESULTS AND DISCUSSION

Our initial studies were directed toward the coupling of simple fluorohaloarenes and amines as model systems (eq 1). Preliminary attempts using the $Pd_2dba_3/P(o\text{-}tol)_3$/NaOtBu and Pd_2dba_3/dppf/NaOtBu catalyst systems were unsuccessful.[10] However, the use of the *in situ* prepared catalyst system,[11] $Pd_2(dba)_3$/binap/NaOtBu, led to excellent yields (> 87 %) for the reaction of several fluorohalobenzenes and N-methylpiperazine (4-6, Table 1).

X = I, Br, Cl(slowest)

$$\text{F-arene} + \text{NHRR'} \xrightarrow[\substack{\text{toluene or neat} \\ 75 - 115\,°C \\ 5 - 42\,h}]{\substack{Pd_2(dba)_3 \\ \text{Binap} \\ \text{NaOtBu}}} \text{product} \quad (1)$$

82 - 98 %

Furthermore, in each of these reactions < 1 % fluoro-substitution was observed by GC-MS, thus, selectivity was completely reversed relative to a hypothetical S_NAr reaction.[12] Iodo- and bromo-substituents demonstrated greater reactivity as these reactions went to completion in 5-7 h at 110 °C. However, substitution of the chloro-substituent was more difficult and required heating the reaction mixture in an excess of neat N-methylpiperazine for 42 h at 115 °C. Nevertheless, for chloroarenes, this is the highest yield reported for the Pd_2dba_3/binap/NaOtBu system, which suggests that the electron withdrawing fluoro-substituent activates the more labile C-Cl bond toward oxidative addition.[13]

A greater array of amine substrates has also been explored for this reaction utilizing the 4-fluorobromobenzene as the fluoroarene. The results (Table 1) show excellent yields (85 – 93 %) for the primary (7) and secondary (8, 9) amines. Although no reaction was observed with aniline, 93 % yield was obtained with N-methylaniline (8). Derivatives prepared from the cyclic amines, morpholine (10) and piperidine (11), were also obtained in excellent yields of 82 – 98 %. A preliminary attempt at a disubstitution reaction using two equiv of piperazine with one equiv of 4-bromofluorobenzene gave the disubstituted product (12) in only 35 % yield. However, it is likely that this yield could be improved using an excess of 4-bromofluorobenzene.

This model study demonstrates the excellent chemoselectivity that the palladium-catalyzed amination reaction provides due to the inability of the catalyst to activate the C-F bond. Furthermore, since nucleophilic substitution is not a viable pathway for the synthesis of these non-functionalized substrates, this procedure represents the most general route yet available to simple fluoroanilines.

Table 1 Results for Pd-Catalyzed Aminations of Fluorohaloarenes.[a]

Arene	Product	Temp (°C)	Time (h)	Yield
$o\text{-BrC}_6\text{H}_4\text{F}$	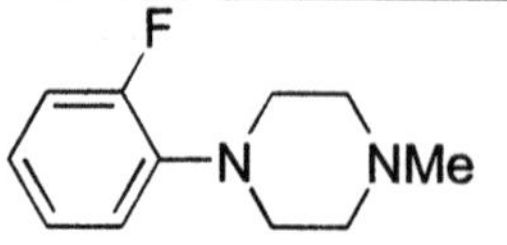 ![F-phenyl-N-piperazine-NMe] (4)	80	5	89 %
$m\text{-BrC}_6\text{H}_4\text{F}$	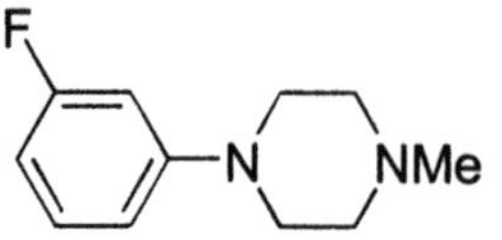 ![F-phenyl-N-piperazine-NMe] (5)	90	7	97 %
$p\text{-BrC}_6\text{H}_4\text{F}$	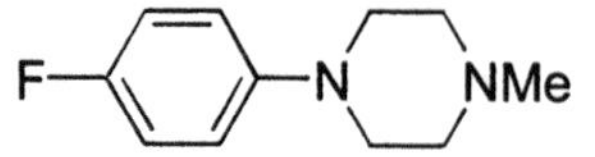 F—⟨ ⟩—N‿NMe (6)	75	5	98 %
$o\text{-IC}_6\text{H}_4\text{F}$	4	85	5	87 %
$o\text{-ClC}_6\text{H}_4\text{F}^{\text{b}}$	4	115	42	91 %
$o\text{-ClC}_6\text{H}_4\text{F}^{\text{b,c}}$	4	115	>50	0 %
$p\text{-BrC}_6\text{H}_4\text{F}$	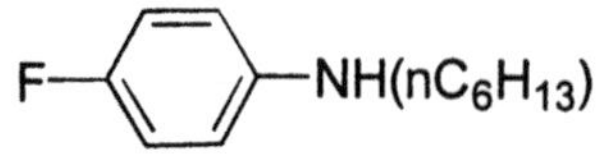 F—⟨ ⟩—NH(nC$_6$H$_{13}$) (7)	85	6	85 %
$p\text{-BrC}_6\text{H}_4\text{F}$	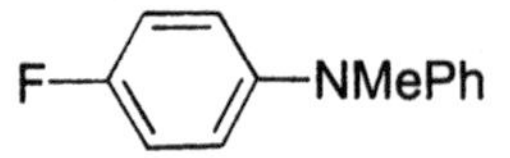 F—⟨ ⟩—NMePh (8)	85	5	93 %
$p\text{-BrC}_6\text{H}_4\text{F}$	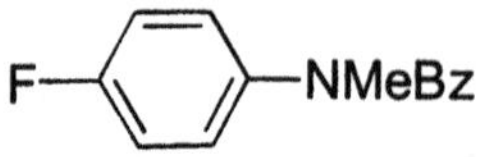 F—⟨ ⟩—NMeBz (9)	85	8	86 %
$p\text{-BrC}_6\text{H}_4\text{F}$	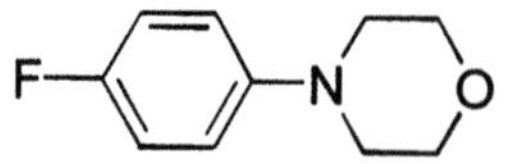 F—⟨ ⟩—N‿O (10)	85	5	97 %
$p\text{-BrC}_6\text{H}_4\text{F}$	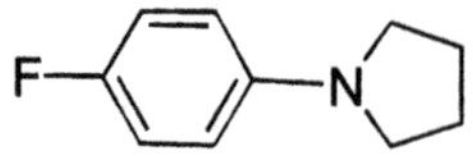 F—⟨ ⟩—N‿ (11)	85	5	82 %
$p\text{-BrC}_6\text{H}_4\text{F}$	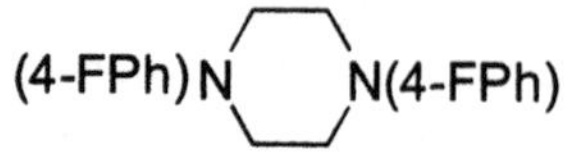 (4-FPh)N‿N(4-FPh) (12)	85	6	35 %

[a]Unless otherwise stated the following reaction conditions were used: toluene solvent, Pd$_2$(dba)$_3$ (2 mol %), binap (1.5 mol %), NaOtBu (2 equiv) and amine (1 – 2 equiv).
[b]Reaction was performed neat with 5 equiv of amine.
[c]No catalyst was used for this reaction.

Synthesis of the Fluoroquinolones

The synthesis of fluoroquinoline antibacterials almost invariably involves substitution of the chlorofluoroquinolone with an amine as the final step (Scheme 1).[3,5] Thus, the above model studies indicate excellent potential for the palladium-catalyzed amination reaction to succeed. However, initial attempts to couple the chlorofluoroquinolone derivative **2** with piperazine using the Pd$_2$dba$_3$/binap catalyst system and NaOtBu in toluene solvent resulted only in the recovery of unreacted starting material. Changing to more polar solvents (DMSO, DMF) or the addition of iodide salts (in an attempt to generate the iodo derivative) had no effect. It was believed that the insolubility of the carboxylic acid **2** played a role in its failure to react and that the ethyl ester would be a more productive substrate. Conveniently, the ethyl ester of **2** is an intermediate in the standard synthesis of Norfloxacin, thus, the synthesis of **13** was readily accomplished (eq 2).[3]

$$(2)$$

With this compound in hand, the original conditions were repeated. However the nucleophilic character of the NaOtBu base resulted in a product from a non-catalytic attack of the t-butoxide anion at the fluorine and subsequent hydrolysis of the t-butyl ether during workup to give phenol **14** (eq 3).

$$(3)$$

The first signs of the desired product **15** were obtained in 29% yield by switching to the non-nucleophilic base, cesium carbonate after 48 hours in refluxing toluene.[14] Further optimization was achieved by switching to the higher boiling DMF and increasing the amount of piperazine to five equivalents. These modifications gave yields that ranged from 58 - 66 % after three hours at reflux for the piperazine (**15a**) and 72 - 75% for the N-methylpiperazine (**15b**) analogs (eq 4).

(4)

13

15a R=H, 58-66 %
15b R=Me, 72-75 %

Detailed analysis of the reaction mixtures revealed that the reaction occurs with quantitative chemoselectivity as no fluoro-substituted products are observed. However, further analysis of the synthesis of **15a** revealed 22 % of the known hydrodechlorinated product (**17**).[15] Although the reducing agent has not been identified, reductions in palladium-catalyzed aminations have been observed where the amine acts as the reducing agent via a β-hydrogen elimination pathway.[16]

17

Although this byproduct is removed by chromatography in the work-up, efforts to eliminate its formation during the reaction are in progress. Unfortunately, preliminary attempts to improve yield and eliminate **17** with ligands such as $P(tBu)_2$(2-biphenyl), $P(Cy)_2$(2-(2'-dimethylamino)biphenyl) and $P(tBu_3)$ have thus far been unsuccessful.[8] When the non-catalytic reaction was run under the same conditions, the desired **15a** formed in 24 h which is much longer than the reaction with catalyst present. However, the formation of the undesired fluoro-substituted product was inconclusive. Nevertheless, the substantially shorter reaction time (3 h) when catalyst is present suggests that a catalytic reaction is indeed occurring.

The syntheses of the fluoroquinolones were completed by subjecting the corresponding esters to basic hydrolysis.[3] Treatment of **15a** and **15b** with 2N sodium hydroxide solution for two hours at reflux resulted in near quantitative yields of Norfloxacin (**1**)[3] and the N-methylpiperazine derivative **16**.[3]

Another strategy employed to enhance the chemoselectivity of fluoroquinolone syntheses has been the use of boron reagents to increase the electrophilicity of the C-7 position by chelating to the β-ketoacid but ~ 5% side-product (**3**) formation is still observed.[6b] Our study shows that the palladium-catalyzed pathway completely eliminates the formation of this by-product.[7] Although palladium-catalyzed amination has been extensive developed for simple as well as moderately complex substrates, its compatibility with fluoroquinolones is noteworthy because of the complex functionality this substrate possesses.[10] Efforts are underway to optimize this reaction toward higher yields and lower reaction temperatures.

Experimental Section

Synthesis of the Fluoroanilines

The following general procedure was followed. For each compound, reaction temperatures, times, yields and procedural changes are listed in Table 1. Compounds **4 – 6**,[9] **7**,[17a] **9**,[17b] **10**,[17c] **11**,[17d] **12**[17e] are previously reported.

A 50 mL, two-necked round-bottom flask was equipped with a magnetic stirbar, condenser and a thermometer. Under a flow of nitrogen, the flask was charged with a fluorohalobenzene (3.0 mmol), an amine (1.1 equiv for primary amines or 2.1 equiv for secondary amines),[18] NaOtBu (830 mg, 8.5 mmol), (R) or (S)-binap (63 mg, 1.0 mmol), and Pd$_2$(dba)$_3$ (30 mg, 0.3 mmol). Toluene (air-free, anhydrous) was added via syringe and the reaction mixture was heated to 75 - 85 °C. The reaction mixture was stirred under an atmosphere of nitrogen and monitored periodically by thin layer chromatography (TLC) using silica gel plates (fluorescent indicator) with an eluent of 10 % methanol in methylene chloride.

After 5 - 7 h, the reaction mixture was cooled to room temperature. The reaction mixture was diluted with methylene chloride (20 mL), and vacuum filtered (in air). The filtrate was carefully concentrated by evaporation of the solvent on a rotary evaporator. (It was necessary to apply the vacuum gradually and to use a cool water bath as the fluoroaniline product is very volatile.) The crude product was purified by column chromatography on silica gel. The product was eluted with a solvent mixture of either 3 % methanol in methylene chloride or 50 % ethyl acetate in hexanes as determined by TLC analysis. The isolated fractions were carefully concentrated on a rotary evaporator to remove most of the solvent. The remaining traces of solvent were removed under a vacuum pressure of 10 mm Hg at 25 °C.

Synthesis of Fluoroquinolines

Catalytic synthesis of **1-ethyl-6-fluoro-1,4-dihydro-4-oxo-7-(1-piperazinyl)-quinoline-3-carboxylic acid ethyl ester (Norfloxacin ethyl ester) (15a)**. To a dry 15 mL recovery flask equipped with a reflux condenser was added **13** (150 mg, 0.5 mmole),[3] (R)-binap (10 mg, 0.017 mmole), cesium carbonate (360 mg, 0.950 mmole), piperazine (215 mg, 2.5 mmole), Pd$_2$(dba)$_3$ (10 mg, 0.010 mmole), and 1.5 mL of DMF. The reaction vessel was purged with nitrogen and heated to reflux. After three hours, the reaction mixture was cooled to room temperature, and the solvent was removed under reduced pressure. The residue was purified by thin-layer chromatography on a 500 micron silica gel preparative plate with an elution mixture of chloroform:methanol:water:ammonia (80:20:2:0.2) to give 101 mg of **15a** as an off-white solid in 58% yield and, separately, 29 mg of **17** as a light brown solid in 22% yield.[15] Using this procedure, the yield of **15a** ranged as high as 66 %. Compound **15a** was saponified in refluxing 2 N NaOH and neutralized to the carboxylic acid with 2 N HCl to give 1 in 96 % yield.[3] The ^{1}H NMR spectrum of 1 was identical to an authentic sample purchased from Sigma.

Catalytic synthesis of 1-ethyl-6-fluoro-1,4-dihydro-4-oxo-7-(1-methyl-piperazinyl)-quinoline-3-carboxylic acid ethyl ester (15b). To a dry 15 mL recovery flask equipped with a reflux condenser was added **13** (150 mg, 0.5mmole),[3] (R)-binap (10 mg, 0.017 mmole), cesium carbonate (360 mg, 0.95 mmole), N-methylpiperazine (122 µL, 1.1 mmole), $Pd_2(dba)_3$ (10 mg, 0.01 mmole), and 1.5 mL of DMF. The reaction vessel was purged with nitrogen and heated to reflux. After three hours, the reaction mixture was cooled to room temperature and the solvent was removed under reduced pressure. The residue was purified by thin-layer chromatography on a 500 micron silica gel preparative plate with an elution mixture of methylene chloride:methanol (9:1) to give 131 mg of **15b** as an off-white solid in 72% yield. Using this procedure, the yields ranged as high as 75 %. Compound **15b** was saponified in refluxing 2 N NaOH and neutralized to the carboxylic acid with 2 N HCl to give **16** in 96 % yield.[3]

ACKNOWLEDGMENTS

This research was, in part, supported by a grant from the Labeled Compound Synthesis Department of Merck Research Laboratories. We also acknowledge support from the Clare Booth Luce Fund at Seton Hall University for providing summer research internships A. Villoresi and J. Brice. We are grateful for Prof. N. Snow for use of his GS/MS.

REFERENCES

1. a) D. J. Kelly, J. D. Chulay, P. Mikesell, A. M. Friedlander, *J. Infect. Dis.*, **166**, 1184 (1992) b) C. M. Terriff, A. M. Tee, *Am. J. Health-Syst. Pharm.*, **58**, 233 (2001).
2. E. Silverman, "Bioterrorism anxieties may spell end of red tape." The Sunday Star-Ledger: Newark, NJ, Oct. 28, 2001, p. 11.
3. H. Koga, A. Itoh, S. Murayama, S. Suzue, T. Irikura, *J. Med. Chem.*, **23**, 1358 (1980).
4. *Quinolone Antimicrobial Agents, 2nd ed.*; Hooper, D. C.; Wolfson, J. S., Eds. American Society for Microbiology: Washington, DC, 1993; Ch. 1, 3.
5. Chu, D. T. W. In *Organofluorine Compounds in Medicinal Chemistry and Biomedical Applications*, Filler, R. et al. Eds. Fluoroquinolone carboxylic acids and antibacterial drugs. Elsevier: New York, 1993.
6. a) U. R. Kalkote, V. T. Sathe, R. K. Kharul, S. P. Chavan, T. Ravindranathan, *T. Tetrahedron Lett.*, **37**, 6785 (1996). b) I. Hermecz, L. Vasvari-Debreczy, B. Podanyi, G. Kereszturi, M. Balogh, A. Horvath, P. Varkonyi, *Heterocycles*, **48**, 1111 (1998). c) M. Preiss, *Eur. Patent Appl.* 0,461,501,A1 Jan 6. 1991.

7. Although the chemoselectivity is not discussed, the following patents claim > 90 % yield of fluoroquinolone products using protected piperazines: S. W. Park, Y. S. Kim, J. H. Jin *US Patent* 5051505, 1991; V. Scherrer-Pangka *CA Patent* 1326239, 1994.

8. For reviews see: a) J. F. Hartwig, *Angew. Chem. Int. Ed. Engl.*, **37**, 2046 (1998). b) J. P. Wolfe, S. Wagaw, J.-F. Marcoux, S. L. Buchwald, *Acc. Chem. Res.*, **31**, 805 (1998). c) J. F. Hartwig, *Synlett*, 329 (1997).

9. For a preliminary communication of this work see: S. Hayden, J. R. Sowa, Jr. In *Catalysis of Organic Reactions*, Herkes, F. E., Ed. Synthesis of fluoroaniline derivatives by selective palladium-catalyzed coupling of N-methylpiperazine and fluorhaloarenes. Marcel Dekker: New York, 1998; pp. 627-632.

10. a) M. S. Driver, J. F. Hartwig, *J. Am. Chem. Soc.*, **119**, 8232 (1997). b) S. Wagaw, R. A. Rennels, S. L. Buchwald, *J. Am. Chem. Soc.*, **119**, 8451 (1997).

11. J. P. Wolfe, S. Wagaw, S. L. Buchwald, *J. Am. Chem. Soc.*, **118**, 7215 (1996).

12. A. Paine, *J. Am. Chem. Soc.*, **109**, 1496 (1987).

13. There are a number of catalytic systems that are useful for aryl chloride substrates and would likely enable this reaction to be performed under milder conditions. See ref. 6.

14. J. P. Wolfe, S. L. Buchwald, *J. Org. Chem.*, **65**, 1144 (2000).

15. T.-L. Tsou, S.-T. Tang, J.-R. Wu, JY.-W. Hung, Y.-T. Liu, *Eur. J. Med. Chem.*, **34**, 255 (1999).

16. a) B. C. Hamann, J. F. Hartwig, *J. Am. Chem. Soc.*, **120**, 12706 (1998). b) B. C. Hamann, J. F. Hartwig, *J. Am. Chem. Soc.*, **120**, 7369-7370 (1998). c) B. C. Hamann, J. F. Hartwig, *J. Am. Chem. Soc.*, **120**, 3694 (1998). d) J. F. Hartwig, S. Richards, D. Baranano, F. Paul, *J. Am. Chem. Soc.*, **118**, 3626 (1996).

17. a) S. K. Srivastava, P. M. S. Chauhan, A. P. Bhaduri, *Synth. Commun.*, **29**, 2085 (1999). (b) B. L. Fox, R. J. Doll, *J. Org. Chem.*, **38**, 1136 – 1140 (1973). (c) S. M. Gasper, C. Devadoss, G. B. Schuster, *J. Am. Chem. Soc.*, **117**, 5206 (1995). (d) G. H. Kerr, O. Meth-Cohen, E. B. Muttock, H. Suschitzky, *J. Chem. Soc., Perkin Trans. 1*, 1614 (1974). (e) E. R. Bissell, *J. Heterocycl. Chem.*, **14**, 535 (1977).

18. The choice of 1.1 equiv for primary amines and 2.1 equiv for secondary amines is empirically based on the following reference: J. P. Wolfe, S. L. Buchwald, S. L. *J. Org. Chem.*, **61**, 1133 (1996).

31

Catalytic Hydrogenation of Cinnamaldehyde: The Effects of Support and Supported Metal on Activity and Selectivity

Nabin K. Nag
Engelhard Corporation, Beachwood, Ohio, U.S.A.

ABSTRACT

Hydrogenation of cinnamaldehyde has been studied in slurry-phase using an auto-clave, under 100 psig of hydrogen, and at 100°C. The catalysts used are alumina-supported Pt, Pd, Ru and a combination Pt and Ru. It has been found that the ac-tivity and the selectivity for hydrogenation to various products depend on several factors including the nature of the support alumina, the active metal constituting the catalyst, and the method of reduction used to fix the metals on the support.

Electron Spectroscopy for Chemical Analysis (ESCA) studies on some se-lected catalysts indicate that higher concentration of Pt on the support surface is more conducive to the selective hydrogenation of the terminal C=O bond of cin-namaldehyde to the corresponding unsaturated alcohol.

INTRODUCTION

Catalytic hydrogenation of α,β-unsaturated aldehydes like cinnamaldehyde is of interest from both fundamental and commercial points of view. On one hand, the effect of different catalysts on the selectivity for various products can throw some light on the catalytic functionality *vis-a-vis* product selectivity, and on the other, new catalysts capable of generating desired selectivity patterns may be designed from the gathered information. From the industrial point of view, there is a need for selective hydrogenation of cinnamaldehyde to the corresponding unsaturated alcohol (1).

Marinelli et al. (2) have discussed the thermodynamic constraint in selectively hydrogenating the C=O group in cinnamaldehyde in presence of the C=C group, stressing the fact that thermodynamics favors the C=C bond hydrogenation. As a result, they rightly point out that replacing one metal by another will have no substantial effect in selectively hydrogenating the carbonyl group. One of the ways to achieve high C=O hydrogenation selectivity, then, is to manipulate the kinetics of the process. This may be achieved by taking advantage of electronic and steric effects (stemming from some unique properties of some metals or combinations thereof) and also by choosing suitable supports (2, 3). Tuley and Adams (4) were the first to show the beneficial effect of iron chloride (as a co-catalyst) on Pt for increased selectivity for the C=O group in the liquid phase hydrogenation of cinnamaldehyde. Various other workers (5-22) have since confirmed the general phenomenon of enhancement in the C=O bond hydrogenation selectivity by metals like Pt and Ru with other metals like Sn and Fe acting as promoters. Poltarzewski et al. (5) reported that tin oxide, when incorporated into a Pt/Nylon catalyst, increases the C=O bond hydrogenation selectivity. They explained this augmented selectivity as an effect of Lewis acidity imparted to the active metal Pt by Sn. Similar electronic effects on C=O bond hydrogenation selectivity have been found by others using Pt-Fe (6); Pt-Sn (7a); Pt-Ge (7b); and Ru-M (8) where M=Sn,Fe,Zn,Ge and Sb. In the catalysts used by the above workers, enhanced C=O bond hydrogenation selectivity has been attributed to electron transfer from the promoting metals (M) to Pt or Ru. A similar effect of Lewis acidity, imparted by Sn^{x+} to Ru, in Ru-Sn catalysts, has also been reported (9).

In addition to the electronic effects, as noted above, steric effects have been found to play a crucial role in directing the kinetics in favor of C=O bond hydrogenation in conjugated aldehydes. For instance, Blackmond et al. (10) found with Ru, Rh and Pt catalysts, supported on alkali metal-substituted Y zeolites, that steric hindrance restricts the penetration of the whole molecule into the zeolite pores. As a result, the centrally located C=C bond is at least partly prevented from getting adsorbed on the metal atoms located inside the cages. However, the adsorption of the molecule through the terminal C=O group is still possible, and hence the observed higher selectivity for the unsaturated alcohol. These authors (10) have also found that increased electron density around the caged active metal, brought about by changing the exchanged alkali metal in the zeolite, is a contributing factor for higher selectivity for the unsaturated alcohol. Enhanced C=O bond hydrogenation selectivity may also be achieved by substituting a hydrogen atom of one of the carbons associated with the C=C bond by a methyl group. This type of effect has been reported by Beccat et al (11) on a Pt(111) face, and also on a $Pt_{80}Fe_{20}$ (111) single crystal face.

As regards the support effect, Vannice and Sen (12) found that Pt on titania, especially when reduced at high temperatures, is more selective towards the hydrogenation of a C=O bond in presence of a conjugated C=C bond (like in crotonaldehyde) than Pt on silica and alumina. Special sites generated at the metal-support interface (an SMSI effect) of Pt-titania are attributed to the enhanced selectivity for the unsaturated alcohol. A similar effect was reported by Wijsmeier et al. (13) on titania-supported Ru in comparison with silica-supported Ru.

From the above review it is clear that higher C=O bond hydrogenation selectivity can be achieved by: 1) adding suitable promoters to the catalyst, 2) taking advantage of steric effects arising due to appropriate substituents in the molecules, and 3) choosing a suitable catalyst support that provides the necessary electronic and/or steric effects necessary for controlling the selectivity of the products.

The purpose of the present work has been to study the effect of various parameters that control the activity and selectivity patterns during the catalytic hydrogenation of cinnamaldehyde. The parameters studied are: support type, active metal (or combination of metals) and catalyst preparation procedure. The last parameter has not been studied well, as is evident from the above literature survey.

EXPERIMENTAL

Catalyst Preparation

Support. Two different commercial alumina powders, designated as Al-A and Al-B, obtained from two different manufacturers (who used different manufacturing processes) processes, were used as the catalyst supports. The alumina powders were calcined in air to generate gamma (in Al-A) and mixed gamma-delta phases (in Al-B) before metals loading. Al-A had higher pore volume (about 2.2 ml/g in the meso and macro regions) and lower bulk density (about 0.35 g/ml), and Al-B had lower pore volume (about 0.55 ml/g in the meso region only) and higher bulk density (about 0.7 g/ml). The surface areas of both the aluminas were around 200 m^2/g.

Metal Loading and Reduction: A slurry phase adsorption technique was used to impregnate the alumina powders with soluble salts of Pt, Ru and Pd. The adsorbed metals were then reduced chemically at elevated temperatures. Two different methods of reduction (methods –1 and –2) using two different chemicals were used. The temperatures of reduction remained the same. Additionally, a

3%Pt/Al-B catalyst was made by reducing by gaseous hydrogen (4% in Ar) at 300°C a catalyst precursor that was obtained by incipiently wetting Al-B by chloroplatinic acid solution. The bimetallic Pt-Ru catalysts were prepared on Al-B by co-impregnation of the metal salts followed by the chemical reduction (method-1) at the same temperatures that were used for the monometallic catalysts, as described briefly above.

Catalyst Activity Testing

A one-liter autoclave (Autoclave Engineers) was used to conduct the hydrogenation reaction. After 7.5 g of cinnamaldehyde (Aldrich) was dissolved in 375 ml methanol and charged into the reactor, 1 g of dry catalyst was introduced. The air inside the autoclave was purged with nitrogen, followed by three times flushing with hydrogen. After this, the autoclave was pressurized with hydrogen to 30 psig, and the heater and the impeller (at 1500 rpm) were switched on. When the temperature reached 100°C, the pressure (generally about 60 psig) was raised to 100 psig by hydrogen. This point was arbitrarily taken as the 'zero' time. During the reaction, the temperature was maintainedat100±1°C. Samples were taken for analysis at different time intervals during the run that continued up to 6 h.

Product Analysis

A 30 m long capillary column packed with DB-1, maintained at 125°C, was used to separate and quantify the products. Peak areas were calibrated against standard solutions of the reactant and the products.

ESCA

A Surface Science machine (model SSX-100) using AlKα x-rays (energy: 1486 eV) was used for the investigation. Any differential charging arising due to the insulating effect of the catalyst support was compensated by applying an electron flood gun (2 eV). The binding energies were corrected by taking the C_{1s} (originating from adventitious carbon in the samples) peak (BE: 284.6) as the standard. Scofield sensitivity factors were used for the quantitative analyses with the intensities of the peaks.

RESULTS

In this work, the activity of the catalysts is reported as the reciprocal of the time (hour) taken for 50% conversion of cinnamaldehyde. The selectivity for a particular product is defined as:

$$\text{Selectivity} = \frac{(\text{ number of moles of a product }) \times 100}{\text{total number of moles of reactant converted}}.$$

Typical products distribution, and selectivity profiles are shown in Figures 1 and 2, respectively.

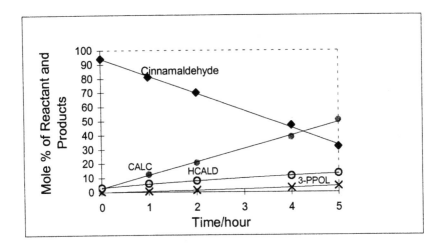

Figure 1 Reactant and products distribution as a function of reaction time.

Monometallic Catalysts

Most of the catalysts studied are Pt/alumina; however, studies were also made with one Pd/Al and one Ru/Al in order to get an idea about the effect of different metals on the performance of the catalyst.

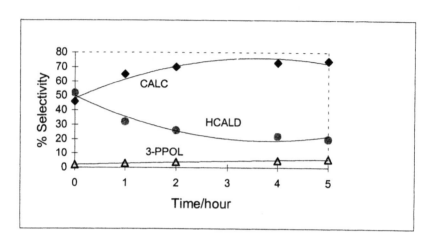

Figure 2 Selectivities for various products as a function of reaction time.

The reaction network (10), showing different hydrogenation products of cinnamaldehyde hydrogenation, is shown below:

Cinnamaldehyde
(CAL)

Cinnamyl Alcohol
(CALC)

(Isomerization)

Hydrocinnamaldehyde
(HCAL)

3-Phenyl Propanol
(3-PPOL)

Pt/Al-A and Pt/Al-B: The results of cinnamaldehyde hydrogenation on various monometallic catalysts, based on Al-A and Al- B, are given in Tables 1and 2 respectively.

It is observed in Table 1 that although the selectivity pattern obtained with different catalysts based on Pt (and using reduction method 1) remains virtually the same, the activity increases as a function of Pt loading. However, when reduction method 2, instead of 1, is used to make the catalyst, both activity and selectivity are changed. For example, the 1.5%Pt/Al-A catalyst, reduced by method-2, shows much higher activity as compared with the similar catalyst reduced by method-1 (compare row 1 with row 4 in Table 1). But this gain in activity comes at a substantial loss in selectivity for C=O hydrogenation.

Table 1 Activity and Selectivity Data for Al-A-based Catalysts

Catalyst	Activity	Selectivity for CALC	Selectivity for HCALD	Selectivity for 3-PPOL
1. 1.5% Pt[1]	0.25	60	34	6
2. 3.0% Pt[1]	0.40	57	36	7
3. 5.0% Pt[1]	0.50	63	33	4
4. 1.5% Pt[2]	0.36	34	54	10
5. 5.0% Pd[1]	>1	<1	78	21
6. 1.5% Ru[1]	0.13	51	35	9

Note: Superscripts [1] and [2] indicate reduction methods 1 and 2 respectively. In cases when the sum total of selectivities is less than 100%, the balance is accounted for by some unidentified minor low- molecular-weight by-products.

Pd in 5% Pd/Al-A (Table 1), on the other hand, is far more active than its Pt counterpart, although the former has very high selectivity for the C=C bond hydrogenation and virtually none for the C=O bond. It appears that on Pd/Al-A the reaction occurs almost exclusively through the route: CAL--> HCAL--> 3-PPOL (see the reaction network above).

Ru in Ru/Al-A (See Table 1) shows much lower activity than its Pt counterparts in spite of the fact that 1.5% Ru contains almost twice as much metal atoms as 1.5% Pt. However, the selectivity patterns of the Pt and Ru catalysts are almost indistinguishable.

Table 2 Activity and Selectivity Data for 3% Pt/Al-B Catalysts.

Catalyst	Activity	Selectivity for CALC	Selectivity for HCALD	Selectivity for 3-PPOL
1. Red. Meth-1	0.26	69	27	4
2. Red. Meth-2	0.17	32	47	5
3. Red. By H_2	3.8	4	3	3

Note: Catalysts 2 and 3 in the above table produced 16% and 90% undesirable bi-products (not identified), respectively.

Profound effects of the method of reduction of the Pt/Al-B-type catalysts on activity and selectivity are demonstrated by the results given in Table 2. The catalyst reduced by method-1 is 1.5 times more active, and two times more selective for the hydrogenation of the C=O bond, as compared with method 2. Recall that an opposite trend in activity is observed for catalysts based on Al-A (Table 1), although the selectivity trends remain the same in both series, based on Al-A and Al-B. This is clearly a support effect. The hydrogen-reduced catalyst stands alone in its activity and selectivity pattern. As observed in table 2, this catalyst has very high activity but almost no selectivity for the hydrogenation of either C=C or C=O. The products, on analysis, were found to contain some condensation and some low molecular-weight compounds that were not fully identified or quantified.

Bimetallic Catalysts on Al-A

In Table 3 the results of cinnamaldehyde hydrogenation on two Pt-Ru bimetallic catalysts are given. A widely varying activity among these bimetallic catalysts is observed; however, the selectivity pattern shows much less sensitivity towards catalyst composition than the activity trend. The significance of these results will be discussed in more detail in the following section.

Table 3 Activity and Selectivity Data on bimetallic /Al-A Catalysts (Reduction Method 1)

Catalyst	Activity	Selectivity for CALC	Selectivity for HCAL	Selectivity for 3-PPOL
1. 1.5% Pt + 0.75% Ru (Pt:Ru=1.04)	0.26	58	28	4
1a. 1.5% Pt/Al-A	0.25	60	34	10
1b. 1.0% Ru/Al-A	0.13	51	35	9
2. 0.75% Pt + 1.5% Ru (Pt:Ru= 0.26)	1.25	65	27	8

Note: 1a and 1b are copied from Table 1 for the sake of easy comparison.

DISCUSSION

Monometallic Catalysts

Effect of Metal Loading. From the activity and selectivity data (Table-1) of the Pt catalysts based on Al-A, it is observed that the activity of the catalysts increases considerably as a function of Pt loading insofar as the metal is reduced by the same method (in the present case, reduction method 1). However, the selectivity for various products remains virtually unchanged even though the activity increases by a factor of 2, when the Pt loading increases from 1.5 to 5.0%. It appears, therefore, that the intrinsic activity of the sites remains unchanged, but their surface concentration increases with increasing metal content, resulting in increasingly higher overall activity of the catalysts as a function of metal loading. The attributes responsible for the activity and selectivity of the catalysts (stemming from different coordination environment, and/or the exposure of different faces of the metal crystallites), are altered or modified, however, when the reduction procedure is changed. This is reflected by substantially higher activity of the catalyst and almost a total reversal of the selectivity pattern for the hydrogenation of the C=C and C=O bonds (compare rows 1 and 4 in Table 1) when reduction method 1 is replaced by 2.

Effect of Catalyst Support: There is a considerable effect of the pore structure and the source of the alumina used as support on the activity and selectivity of the catalysts. Thus, by comparing row 2 in Table 1 with row 1in Table 2 it is observed that Al-A generates a catalyst with considerably higher activity, and relatively lower selectivity for C=O bond hydrogenation, as compared with Al-B. It is important to note that Al-A has higher pore volume (and bimodal pore size distribution with substantial amount of macro pores) as compared with Al-B, while the surface areas remain almost the same.

Effect of Reduction Method: As indicated above, the method of reduction also has a profound effect on the activity and selectivity of the catalysts. The pertinent data are given in Table 2. The highest activity is obtained by reducing the catalyst by hydrogen. As seen in Table 2, this catalyst is radically different from the other two that are obtained by *in situ* chemical reduction. While these two catalysts have relatively lower activity, they are very selective for the hydrogenation of the C=O and C=C bonds. On the contrary, the hydrogen-reduced catalyst has very little selectivity for the three products of interest.

Focusing attention on the other two (chemical) reduction methods, it is observed from rows 1 and 2 of Table 2 that reduction method 1 generates more active as well as more C=O bond hydrogenation selective catalyst, as compared with method 2. In case of Al-A-based catalysts a similar effect of the reduction method was found on the activity (see rows 1 and 4 in Table 1) of the catalysts.

However, the effect on the selectivity pattern was reversed. This confirms the profound effect of the reduction method used to fix the metals on the support on the activity and selectivity of the Pt/Al catalysts.

Catalyst Characterization by Electron Spectroscopy for Chemical Analysis (ESCA): In order to learn more on the various effects discussed above, ESCA was applied to characterize some selected catalysts. Specifically, the purpose was to see if any discerning features of surface properties, as revealed by ESCA, would help explain the differences in catalytic activity and selectivity caused by the various factors discussed above. The ESCA results are summarized in Table 4.

Table 4 ESCA Data for Some Selected Catalysts, all Containing 3% Pt

Catalyst	100x(Pt:Al)	BE, Pt4f $_{7/2}$	FWHM, eV
1. Al-A, Red. Meth 1	0.65	72.25	1.70
2. Al-B, Red Meth 1	6.0	70.85	2.02
3. Al-B, Red Meth 2	0.40	71.30	1.72
4. Al-B, Red by H$_2$	0.45	71.90	1.80

The data indeed show some discriminating features. Thus, by comparing the ESCA-generated Pt:Al ratio (this ratio reflects metal concentration on the catalyst surface) in the catalysts, it is observed that the Al-B-based catalyst (row 2, Table 4) is highly 'edge coated' by Pt. In other words, in this catalyst Pt has much higher (by about an order of magnitude) surface concentration as compared with the Al-A-based catalyst (row 1, Table 4). Both catalysts were prepared and reduced by the same method. It is this high surface concentration of the active metal that contributes substantially to the C=O bond hydrogenation selectivity.

This is explained by the fact that when the metal is mostly concentrated near the pore mouth (that is, in an 'edge coated' situation), the terminally located C=O group of the reacting cinnamaldehyde molecule gains easy access to the metal, leaving the moiety containing the aromatic ring and the C=C bond outside the pores. This helps in getting the C=O group selectively hydrogenated to the unsaturated alcohol. This situation is relatively more favored than when the whole molecule has to penetrate the pores for the double bond to be activated by adsorption on the metal atoms located deep inside the support pore structure. As mentioned earlier, Blackmond et al. (10) found that the saturation of the C=O bond in the hydrogenation of cinnamaldehyde is facilitated on Y-zeolite-supported catalysts containing metals like Pt. They explained this using the model that the activation of the C=O bond is favored by its location at the opposite end of the bulky aromatic ring. This configuration allows the O atom of the C=O group an easy access to the active centers near the pore mouths, leaving the aromatic ring outside the pore. By the same token, the activation of the C=C bond is statistically less favored because this being located at the center of the molecule it is necessary for the whole molecule to be inside the pores to reach an active center. But this penetration is hindered to some extent by the steric restriction encountered by the molecule at the pore mouths. With the present catalysts, high degree of edge coating means that the activation of the C=O bond by active sites located near the pore mouth is statistically more favored than the activation of the C=C bond by the sites located inside the pores of the 'edge-coated' catalyst. There is further evidence to support the above argument. It has been found (23) that under similar experimental conditions the C=O bond in cinnamaldehyde is more selectively hydrogenated than the C=C bond on highly edge-coated Pt/carbon powder catalysts, as compared with highly dispersed Pt the same supports.

The other significant difference in the surface properties of the catalysts supported on Al-A and Al-B (compare 1 and 2 in Table 4) is that the surface Pt on the latter is substantially more reduced, as reflected by the lower binding energy (BE) of the $4f_{7/2}$ electrons. Additionally, the full width at half-maximum (FWHM) of the $Pt4f_{7/2}$ peak in the latter is considerably higher. This higher FWHM may partly be due to higher carrier-catalyst interaction (most likely an electronic interaction effect) (24). Therefore, the Pt species on the Al-B indeed show some unique surface properties that may be attributed, at least partially, to its relatively higher selectivity for the C=O bond hydrogenation.

Focusing attention now to the three Al-B-based catalysts (Table-4, rows 2-4), obtained by three different methods of reduction, the BE data show that the Pt in the most active, but most non-selective catalyst (that is, catalyst 4 in Table 4) is the least reduced of the three. The methods by which Pt was incorporated into the support and fixed (by reduction) were completely different from the method applied to the other three. This, plus the relatively low temperature of hydrogen reduction (300°C), may be the reason the Pt in this catalyst is the least reduced. These may also be the reasons why this catalyst behaved so differently than the others. No clear-cut trend between performance and surface properties of the catalysts can be established by the ESCA data. However, it appears, at least qualitatively, that the degrees of reduction and high surface concentration of Pt are the two factors that have the most influence on the activity and selectivity of the catalysts. This is one aspect that calls for further systematic investigation.

Other Metals (Pd and Ru): The activity and selectivity data relative to two other metals, Ru and Pd, show some notable features. Comparing rows 5 and 6 in Table1 one observes that the Pd-based catalyst is extremely active, but this activity is almost exclusively directed for the hydrogenation of the C=C bond with nearly total indifference to the C=O bond. This is a very important observation, because this catalyst gives a means of selectively hydrogenating a C=C bond in the presence of a conjugated C=O bond. Ru behaves differently: It is observed from Table-1 that Ru gives the least active catalyst of the three metals studied, although its selectivity pattern is quite comparable to that of the Pt-based catalysts. From these data it may be concluded that the active sites on the Ru-based catalyst are similar in nature to those of the Pt-based catalysts. However, from its lower activity it appears that the sites on Ru catalyst have either lower intrinsic activity or have lower surface concentration or both. On the other hand, the Pd-based catalyst, judged by its high activity and its reverse selectivity pattern, seems to have highly active sites that are different in nature than those of the Ru- and Pt-based catalysts. Further probing into this aspect is not possible with the present data.

Bimetallic Catalyst

It is well known that the catalytic activity and selectivity of a metal for a certain reaction may be substantially altered by the presence of a second metal in the system (25). In the present work, although Pt has been the metal of primary interest, two experiments were done with the bimetallic Pt-Ru/Al-A system in order to get an idea about the effect of a second metal on the activity and selectivity of Pt. From Table 3 it is observed that an equimolar amount of Ru, when combined with 1.5% Pt, does not affect the activity and selectivity of the Pt-alone catalyst. When the amount of Ru is doubled (see item 2 in Table 3), however, the activity increases by a factor of about 5, while the selectivity for various products remains virtually unchanged. This dramatic increase in activity is probably due to the formation of new, highly active sites at the bimetallic interface. The selectivity patterns of the two bimetallic catalysts are very similar to that of the Pt-alone catalyst. This implies that the new sites on the bimetallic and those on Pt-alone catalysts are of same nature as far as the selectivity is concerned.

CONCLUSIONS

From the results of the present investigation the following conclusions are made:

1. The activity and selectivity for various products of cinnamaldehyde hydrogenation on Pt/Al catalysts are dependent on the physical properties and the make of the support alumina.

2. The method of reduction applied during the catalyst preparation step has profound influence on the activity and selectivity of the catalysts.

3. For the Pt/Al catalysts, the activity increases as a function of metal loading without much affecting the selectivity trend. This indicates that the nature of the active sites remains practically the same, but the increased activity is due to an increased number of active sites on the catalyst surface

4. Pd/Al is very active, but it is less selective toward the hydrogenation of the C=O bond. On the other hand Pt/Al, although less active than Pd/Al, is much more selective toward the hydrogenation of the C=O bond.

5. Ru/Al is the least active catalyst; however, its selectivity pattern is comparable to that of the Pt/Al catalysts.

6. For the bimetallic Pt-Ru /Al system, Ru shows no effect on the activity and selectivity of Pt at lower Ru concentration. However, at high Ru concentration, the activity of Pt-Ru is considerably increased without substantially affecting the selectivity pattern. New, highly active sites generated at the Pt-Ru interface are probably responsible for the augmented activity.

7. Higher metal concentration on the support surface (that is, near the pore mouths of the support, resulting in a highly edge-coated situation), and more reduced Pt on the surface are more conducive to the selective hydrogenation of the C=O bond.

REFERENCES

1.   T. Birchem, C.M.Pradier, Y. Bertier, and G. Cordier, *J.Catal.*, **146**, 503(1994)
2.   Marinelli, T.B.L.W., Nabuurs, S., and Ponec, V., *J.Catal.* **151**, 431 (1995).
3.   Englisch, M., Jentys, A., Lercher, J. A., *J. Catal.*, **166**, 25 (1997).
4.   Tuley, W.F., and Adams, R., *J. Am. Chem. Soc.*, **47**, 3061 (1925).
5.   Poltarzewski,S.,   Galvagno,   R.,   Pietropaolo,   R.,   and   Staiti,   P., *J.Catal.*,**102**,190(1986).
5a. P.Gallezot, A. Giroir-Fendler, and D. Rihards, in Catalysis of Organic Reactions, ed. W.E.Pascoe, p.1, Marcel Dekker,Inc, 1992.
6.   Goupil, P., Foillous, P., and Maurel, R, React. *Kinet. Catal. Lett.*,**35**, 185 (1987).
7.   (a) Galvagno, S., Poltarzewski, S., Donato, A., Neri, G., and Pietropaolo, R., *J. Mol.Catal.* **35**, 365 (1986); (b) *J.Chem. Soc., Chem. Comm.*, 1729 (1986).
8.   Coq, B., Figueras, F., Moreau, C., Moreau, P., and Warawadkar, M., *Catal. Lett.*, **22**, 189 (1993).
9.   Deshpande, V.M.,Patterson, W.R., and Narasimhan, C.S., *J.Catal.*, **121**, 165-182 (1990).
10. Blackmond, D., Oukaci, R., Blanc, B., and Gallezot, P., *J.Catal.*, **131**, 401 (1991).

11. Beccat, P., Bertolini, J.C., Gauthier, Y., Massardier, J., and Ruiz, P., *J. Catal.*, **126**, 451 (1990).
12. Vannice, M.A., and Sen, B., *J.Catal.*, **115**, 65 (1989).
13. Wismeijer, A.A., Kieboom, Ap.P.J., and van Bekkum,H., React. *Kinet. Catal. Lett.*, **29**, 311 (1985)., *Appl. Catal.*, 25, 181 (1986).
14. Van Mechelen, C.,and Jungers, J.C., *Bull. Soc. Chim. Belg.* **59**, 597 (1950).
15. Jenek, J., and Germain, J.E.,*J.Catal.*, **65**, 141 (1980)
16. Noller, H., and Lin, W.M., *J.Catal.*, **85**, 25 (1984).
17. Nagase, Y., Hattori, H., and Tanabe, K., *Chem. Lett.*, 1615 (1983).
18. Richard, D. Ockelford, J., Giroir-Fendler, A, and Gallezot, P., *Catal. Lett.*, 3, 53 (1989).
19. Waghari, A., Oukaci, R., and Blackmond, D., *Catal. Lett.*, 14, 115 (1992).
20. Tronconi, E., Crisafulli, C., Galvagno, S., Donato, A., Neri, G., and Pietropaolo, R., *Ind. Eng. Chem., Res.*, **29**, 1766 (1990).
21. Gallezot, P., Giroir-Fendler, A, and Richards, D., in Catalysis of Organic Reactions, ed. W.E.Pascoe, p.1, Marcel Dekker,Inc. N.Y. 1992
22. Wismeijer, A.A., Kieboom, Ap.P.J., and van Bekkum,H., React. Kinet. Catal. Lett., **29**, 311 (1985)., *Appl. Catal.*, 25, 181 (1986).
23. Nag, N.K., Unpublished data.
24. Nag, N.K., *J.Phys. Chem.*, **91**, 2324 (1987).
25. Sinfelt. J., Bimetallic Catalysts, John Wiley and Sons, New York, 1983.

32

Novel Synthesis of 2-Alkylhexahydropyrimidines from Nitriles and 1,3-Diamines

Frank E. Herkes
The DuPont Company, Wilmington, Delaware, U.S.A.

ABSTRACT

Dytek ®EP (1,3-diaminopentane) is a DuPont commercial diamine produced by the cyanobutylation of ammonia with *cis*-2-petenenitrile followed by batch hydrogenation of the aminonitrile, 3-aminopentanenitrile, with Raney® Ni or Co catalysts. During the reduction step, we observed the formation of 2-methyl- and 2-ethyl-4-ethylhexahydropyrimidines as co-products. These hexahydropyrimidines appear to form by catalytic dehydrogenation of 1,3-diaminopentane followed by addition of 1,3-diaminopentane or alkylation of 3-aminopentanenitrile with 1,3-diaminopentane and subsequent thermal rearrangement of the intermediate enamine to primary imines that are then converted to the hexahydropyrimidines. The postulation of a third route where acetonitrile and propionitrile are intermediates, led to a new synthetic route to 2-alkylhexahydropyrimidines by reduction of nitriles in the presence of a 1,3-diamine. Use of an aminoalcohol, 3-amino-1-propanol with acetonitrile provided a new route to 2-methyl-tetrahydroxazine.

INTRODUCTION

In 1985 DuPont commercialized a process for the synthesis of 1,3-diaminopentane (1) from a co-product produced in their adiponitrile process. The commercial product, Dytek® EP is a colorless low-odor branched diamine that is oxidatively stable to color formation. Unlike other alkyl diamines, 1 has pronounced amine differential reactivity. The pK_a's of 9.5 and 8.3 provide differences in basicity that are desirable in polyurethane and epoxy curing and condensation polymerizations. This basicity difference is undoubtedly due to steric hindrance around the 3-amino group. A comparison of the reactivity of the two amino groups in 1 vs. other alkyl and cycloalky diamines with non-equivalent amino groups is 30:1, 4:1 and 2.5:1 for the single Michael addition of 1, isophoronediamine and 2-methylpentamethylenediamine, respectively to acrylonitrile at 25 °C. 1 is prepared by catalytic reduction of 3-aminopentanenitrile (2) employing Raney® Ni or Co slurry catalyst. The aminonitrile is produced by cyanobutylation of ammonia with cis-2-pentenenitrile (3), a co-product formed during the Ni catalyzed hydrocyanation of butadiene (1).

Cyanobutylation of ammonia with 3 occurs in high yield and conversion employing aqueous ammonia at 90-110 °C and autogeneous pressure. The initial product mixture is two-phase at the start and quickly becomes a single phase at ~ 40% conversion to 2. Water at a minimum level is necessary for high yield and conversion during this equilibrium addition reaction. Competitive with this addition is isomerization of 3 to t-2-pentenenitrile and t-3-pentenenitrile. After removal of water and unreacted pentenenitriles from the amination product, 2 is refined and subsequently batch hydrogenated using a Raney® Ni or Co catalyst containing NaOH promoter. The latter is present to reduce secondary amine formation. Prior synthesis of 1 and 2 was performed by metal catalyzed addition of ammonia to 3 followed by catalytic hydrogenation of 2 in methanol to 1 in low yield (2). 1 has also been prepared by condensation of 1-penten-3-one with hydrazine in ethanol followed by hydrogenation in liquid ammonia using Raney Ni® catalyst (3).

During the hydrogenation, we observed the formation of two hexahydropyrimidines along with isomeric diamine coupling products. Since one of these hexahydropyrimidines, 4-ethyl-2-methylhexahydropyrimidine (4), has a boiling point close to that of 1 in the refining step, we sought to understand the origin of its formation and determine how to reduce or eliminate its presence.

EXPERIMENTAL

Batch hydrogenations of **2** with Raney® Ni and Co catalysts in the presence of NaOH were performed in a 300 mL stainless steel Autoclave Engineers mag-nedrive packless autoclave equipped with a thermocouple, internal cooling coils, sample dip tube containing a stainless steel 5 micron Mott filter and Dispersimix turbine type draft tube agitator containing a rotating impeller. Hydrogen uptake kinetics were followed by the pressure drop in a 1 L hydrogen reservoir feeding the autoclave and transmitted to a Yokogawa HR1300 recorder. Hydrogen uptake data collected every minute throughout the run was monitored both graphically and electronically, and fed into a data file for post-analysis.

Product analysis of hydrogenation product was done using a Hewlett Pack-ard 5890 gas chromatograph GC), equipped with a DB 1701 (5% crosslinked phenyl-methyl-silicone) megabore column (30 m long, 0.33 ID, 0.25 um film thickness) and a flame ionization detector. The temperature program was 60 °C for 2 min + 8 °/min to 230 °C for 15 min. The column flow rate was 1.5 cc/min helium and split vent flow rate of 60 cc/min helium. The injector and detector temperatures were 250 °C and 265 °C, respectively. Product and by-product iden-tification was done by GC/MS analysis using chemical ionization techniques.

The activated Raney® metal slurry catalysts (W. R. Grace Co.) employed in this study were primarily Ni and Co. Ra 2800 contains only Ni and Ra 2400 con-tains nickel along with chrome and iron promoters. Ra 2724 contains cobalt pro-moted with chrome and iron.

Synthesis of 1,3-Pentanediamine

One hundred grams (99% purity, 1.01 moles) **2** and 5 g of wet Raney® Ni cata-lyst (W. R. Grace, Ra 2400) were charged to a 300-mL stainless steel autoclave. After closing, the reactor was purged 3x with hydrogen. The temperature was raised to 110 °C under 50 psig hydrogen with slow stirring. At reaction tempera-ture, the pressure was raised to 750 psig with hydrogen and maximum (~ 1200 rpm) stirring commenced. A 500 ml stainless steel ballast cylinder was inserted in series with the hydrogen feed to allow measurement of hydrogen uptake at con-stant pressure. After 2 h, the mixture was cooled . GC analysis of the filtered product showed 98% conversion of **2** and 90.4% selectivity to **1**. Further analysis showed a 3.3% and 1.8% selectivity to **6** and the hexahydropyridines **4** and **5**, respectively.

Synthesis of 2,4-Diethylhexahydropyrimidine

In a typical batch run, 70 g (99% purity, 0.66 mole) **1**, 80 g (1.45 moles) propioni-trile and 3 g of wet Raney® Ni catalyst (W. R. Grace, Ra 2800) were charged to a 300-mL stainless steel autoclave. After closing, the reactor was purged 3x with hydrogen. The temperature was raised to 90 °C under 50 psig hydrogen with slow stirring. At reaction temperature, the pressure was raised to 650 psig with hydrogen and maximum (~ 1200 rpm) stirring commenced. A 500 ml stainless steel ballast cylinder was inserted in series with the hydrogen feed to allow measurement of hydrogen uptake by propionitrile at constant pressure. The pressure in the ballast was set to 1000 psig and repressured when the gauge dropped to 650 psig.

After four hours of heating and stirring, the mixture was cooled to rt and allowed to stand 16 hr at 50 psig. GC analysis indicated a 72% conversion of **1** and 96% selectivity to **5**. After replacing the reactor atmosphere with nitrogen, the product was filtered away from the catalyst. Fractional distillation under reduced pressure on a 26 in. Teflon® spinning band column yielded 27 g of clear product **5** with > 99% purity having a bp of 84 °C at 22 psig.

Synthesis of 4-Ethyl-2-methylhexahydropyrimidine

Sixty grams of **1** (96.7% purity, 0.569 mole), 60 g acetonitrile (1.46 moles) and 1 g of wet Raney® Ni (Ra 2800) were charged to a 300 mL stainless steel autoclave reactor. The mixture was stirred and heated at 90 °C and 500 psig hydrogen pressure for 6 hr. Capillary GC analysis of the cooled product indicated a 36% conversion of **1** and 97.4% selectivity to **4**. The reactor was recharged with 19.6g additional acetonitrile and run at 90 °C and 500 psig for 21 hr. Re-analysis indicated a 92.1% conversion of **1** and 98% selectivity to **4**. Ethyl amine was the only low boiling impurity. Distillation in a Teflon spinning band column gave pure **4** with a bp of 71 °C at 20 psig. The structure of **4** was confirmed by comparison with a mass spectrum of **4** obtained by condensation of **1** with acetaldehyde at 50 °C.

Results and Discussion

The formation of **4** and 2,4-diethylhexahydropyrimidine (**5**) during the hydrogenation of **2** with Raney® Ni was totally unexpected and new for this type of 1,3-aminonitrile hydrogenation. Their molar concentration in all instances was approximately 1:1. Confirmatory identification was made by independent synthesis of **4** and **5** by condensing **1** with acetaldehyde and propanal (4). Secondary amines, produced from amine coupling, are also by-products formed during the hydrogenation of **2**. As expected, the amine coupling product, N-(3-aminopentyl)-1,3-pentanediamine (**6**) was the major isomer observed.

In attempting to understand the origin of **4** and **5** during the reduction of **1**, runs were performed under different reaction conditions to determine what reaction parameter or parameters promote or minimize their formation. Most of the early work on the process development synthesis of **1** employing Raney® Ni displayed lower yield due to **4**, **5** and **6** production during the hydrogenation. Combined hexahydropyrimidine yield losses of 1-3% were typically observed with Raney® Ni at hydrogenation temperatures of 100-120 °C. When Raney® Co was employed under similar conditions, yield losses to **4** and **5** were substantially reduced.

On the surface it would appear that the mechanism of **4** and **5** formation arises by generation of equal molar amounts of acetonitrile and propionitrile from **1** by a metal catalyzed *retro Thorpe-Ziegler Reaction* (5) to generate equal moles of the two alkyl nitriles. These, in turn, are *in-situ* hydrogenated to their respective alkylideneamine intermediate (**8**) which in turn adds **1** to form the N-(aminoalkyl)-1,3-pentanediamine (**9**) intermediate. The latter rapidly loses ammonia and cyclizes by addition of the free primary amine to the N-alkylidene-1,3-pentanediamine (**10**). The postulated formation of acetonitrile and propionitrile as intermediates suggests that the **1** is dehydrogenated in successive steps to **2** and then to the intermediate 3-iminopentanenitrile (**7**) which then rearranges to pro-

duce the alkyl nitriles. However, no detection of the alkyl nitriles, however, was observed in the reaction mixture or cold trap during control experiments of **1** with catalyst.

A second and more plausible mechanism is the reductive alkylation of **2** by **1** to produce a 185 molecular weight intermediate **12**, that can cyclize to **13** and thermally rearrange to the two imines, **14** and **15** (Thomas A. Johnson and Richard V. C. Carr, Air Products and Chemicals, Inc., private communication, 2002).

Unlike the hydrogenation of **2** to **6**, the conversion of **12** to **13** and **13** to **14** appear to be thermal processes and quite slow under our typical hydrogenation temperatures range (90-120 °C). **14** cyclizes directly to **4** and **1** adds to **15** with loss of ammonia to produce **5**. This mechanism also predicts formation of equal molar amounts of **4** and **5**.

In hydrogenation experiments of **2** with low catalyst loading of Ra Ni and using 100% excess ammonia to prevent secondary amine formation, **12** was observed to build-up as an intermediate and then decrease with time to produce **6**. It is believed that **11** may be the stable intermediate in solution because of the high concentration of ammonia, but then thermally decomposes to **12** in the GC or GC/MS injection port. The intermediate **12** hydrogenated slowly to **6**.

In separate experiments we observed the formation of **4** and **5** by simply heating **1** at 140 °C with Raney® Ni at 1 atm. Subsequent experiments at 140-150 °C with Raney® Ni and Co catalysts resulted in conversion of **1** to **4** and **5** in a combined selectivity of > 80%. The observation that **2** was produced in these experiments lent earlier support for **2** as a reaction intermediate for the first mechanism. A control experiment of **2** alone with Ra 2800 under similar conditions, however, failed to show any conversion to **4** and **5**. Catalytic dehydrogenation of **1** to the primary imine followed by addition of **1** also produces **11** on route to **4** and **5**. Table 1 summarizes the effect of temperature, catalyst type and reaction time on hexahydropyrimidine production from 1 and Raney® Ni. A typical example from this table shows the heating a stirred mixture of 1 and 1.6% catalyst loading of Ra 2800 for 48 h at 150 °C produced 4 and 5 in 72.2% combined selectivity at 44% conversion of 1.

This transformation of 1,3-diamines to hexahydropyrimidines appears to be indigenous to Raney® Ni, Co and ruthenium catalysts but not to Pd. With a 5% Ru/C catalyst, a 5% conversion of 1 occurred after 5 h producing 4 and 5 in 95% combined selectivity. When NaOH was added to the Raney® catalyst, it greatly slowed the reaction to 4 and 5 supporting the metal catalysis of this transformation by reducing its adsorption ability for 1.

Table 1 Effect of temperature, catalyst loading and time on formation of 4, 5 and 2 from 1 employing Raney® Ni, Co and Ru catalysts at autogeneous pressure.

Catalyst	Time hr	% cat. loading	T, °C	% 1 conv.	% 4 sel.	% 5 sel.	% 2 sel.
Ra 2800	47	0.13	160	5.4	37.6	43.6	18.9
Ra 2400	47	0.13	160	9.2	28.8	33.0	38.2
Ra 2400	22	1.3	160	33.7	36.3	42.8	14.1
Ra 2400	48	1.6	150	15.1	39.6	42.5	6.1
Ra 2800	48*	1.6	150	44.0	35.0	38.2	5.6
Ra 2724	6	2.0	180	32.0	7.9	12.9	nd
5% Ru/C	24	6.0	140	5.0	49.8	45.2	5.1

* 2 wt% water added

A Cr promoted (Ra 2400) Raney® Ni catalyst was twice as active compared to an unpromoted Ni catalyst and addition of 2% water greatly increased the conversion of 1 perhaps by increasing the adsorption of 1 on the catalyst surface. Caustic addition (e.g., 400 ppm NaOH) to this mixture almost completely inhibited 4 and 5 formation suggesting decreased adsorption of 1 on the catalyst surface. The kinetic rate of formation of 4 and 5 at 180 °C, as expected, are similar (Figure 1). Some reduction of 5 is seen with time suggesting possible ring opening at 180 °C. A 7.6-fold excess of 6 was also observed confirming the intermediacy of 11 and 12 in the formation of 4, 5 and 6.

Reduction of 2-methyl-3-aminobutyronitrile (17) in methanol at 90 °C and 800 psig with a Raney® Ni catalyst produced the expected 4-methyl-1,3-diaminobutane (18) along with small amounts of by-product 2,4,5-trimethylhexahydropyrimidine (19) and 4,5-dimethyl-2-ethylhexahydropyrimidine (20) by the latter same mechanism described above. Two observable isomers (e.g., *cis/trans*) of (19) and (20) were seen in yields of 2.6% and 3.2%, respectively. The starting nitrile, 17 consisting of two diastereomers was synthesized in 13% yield from aqueous ammonia and *t*-2-methyl-2-butenenitrile (16) at 150 °C for 12 hr.

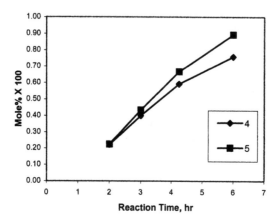

Figure 1 Formation of **4** and **5** with time at 180 °C and autogeneous pressure

Hexahydropyrimidine Synthesis

The first hypothesis of an *in-situ* reduction of acetonitrile and propionitrile in the presence of **1** to **4** and **5** suggested that one might be able to use this approach as a new synthetic route to this class of heterocyclics. Hexahydropyrimidines are conventionally prepared by condensation of aldehydes or ketones with 1,3-diamines (4). Water is a by-product in these reactions and must be removed either to favor the imine or enamine equilibrium or for product purification. Generally, the condensation is acid or base catalyzed and run in solvents (6). In some cases

the open-chain tautomer is produced as a co-product where the 2 position is un-substituted. In the above chemistry only volatile ammonia is produced and no open-chain isomers are observed. The high selectivity to the hexahydropyri-midines and easy product isolation makes this a desirable synthetic route. A num-ber of nitriles were hydrogenated in the presence of **1** at 90 °C with Raney® Ni or supported Ni catalysts to produce the respective hexahydropyrimidines (21) in high yield (Table 2). A 2-4 molar excess of alkylnitrile was employed in most of the reductions. Hydrogenation of a 3.1 molar excess of acetonitrile to 1 at 90 °C and 650 psig employing Ra 2800 for 7 hr produced 4 in 92% yield. 3-Methylamino-1-aminopentane (R = C_2H_5, R^1, R^2 = CH_3) produced 2-methyl-3-methylamino-4-ethylhexahydropyrimidine (21b) in 92% yield with Raney® Ni and acetonitrile. A 33% yield of 2-(4-cyanobutyl)-4-ethylhexahydropyrimidine (21d) was produced when adiponitrile was employed as the nitrile substrate and 1 as the diamine.

Table 2 Synthesis of hexahydropyrimidines from batch reduction of nitriles in the presence of 1 at 90 °C and 650 psig.

R	R^1	R^2	Id.	Catalyst	Molar ex's amine	Time hr	% 1 conv.	% sel. 21
GH_5	H	C_2H_5	5	Ra 2800	2.2	20	72	96
C_2H_5	H	i-C_4H_9	21a	Ra 2800	2.4	6	64	97
C_2H_5	H	CH_3	4	Ra 2800	3.4	6	92	98
C_2H_5	H	CH_3	4	Ra 2800	3.1	7	98	94
C_2H_5	H	CH_3	4	Ni/Al_2O_3	1.7	6	43	94
C_2H_5	H	CH_3	4	Ra 2724	1.7	5	14	92
C_2H_5	H	CH_3	4	Ra 2800	1.7	6	85	98
C_2H_5	H	CH_3	4	Ra 2800	4.8	7	96	83
C_2H_5	CH_3	CH_3	21b	Ra 2800	3.1	7	98	94
H	H	CH_3	21c	Ra 2800	3.9	10*	98	88
C_2H_5	H	CH_3	4	Ra 2724	3.9	10*	36	86
C_2H_5	H	ADN	21d	Ra 2800	2.2	10	56	59

70°C and 800 psig

Tetrahydroxazine Synthesis

With the success of reducing nitriles in the presence of a 1,3-diamine to make hexahydropyrimidines, we turned our attention to the use of 3-hydroxy-1-aminopropane (22) as a surrogate for the 1,3-diamine. Hydrogenation of acetontrile with Raney® Ni in the presence of 22 in a 1:1 mole ratio at 95 °C and 500 psig produced tetrahydro-2-methyl-1,3-oxazine (23) in 90% selectivity and 70% conversion of 22 by intramolecular cyclization of the hydroxy to the intermediate enamine. This synthesis is quite different from other syntheses (7) which involve Pd or Pt homogeneous complexes for the intramolecular hydroamination of aminopropyl vinyl ether to give 23.

Hydrogenation with Raney®Co

Having understood the origin of hexahydropyrimidine formation from 1, attention was turned on finding ways to reduce or eliminate its formation. Use of Raney® Co at lower catalyst loading and NaOH addition greatly reduced 4 and 5 formation during the reduction of 2. Use of Raney® Co (8) instead of Raney Ni at 2 wt% loading and 90-125 °C with 500-1000 ppm NaOH resulted in < 0.1% combined selectivity of 4 and 5 (Table 3) and > 90% yield to 1. Without the base using a Raney® Co catalyst, the combined selectivity is an order of magnitude higher than with base present.

Table 3 Effect of temperature, catalyst and NaOH concentration on 4 + 5 formation in the batch reduction of 2 at 800 psig

Catalyst	% cat. loading	Temp °C	NaOH ppm	% 1 sel.	% 4 + 5 sel.	% 6 sel.
Ra 2400	2	125	500	81.6	0.40	12.9
Ra 2724	2	125	500	82.0	0.01	13.4
Ra 2400	2	100	500	86.9	1.3	1.5
Ra 2724	2	100	500	96.9	0.35	1.4
Ra 2400	5	100	500	83.8	0.78	7.6
Ra 2724	5	100	500	92.1	0.06	6.6
Ra 2724	2	90	1000	99.5	0.10	0.14
Ra 2724	2	90	0	94.6	0.99	1.1
Ra 2724	5	90	1000	98.5	0.16	0.41

CONCLUSIONS

Batch hydrogenation of 2 with Raney® Ni produces significant amounts of 4 and 5 in addition to the desired product 1. Use of Raney® Co catalyst with NaOH greatly reduces the formation of **4** and **5** and displays > 90% yield to **1**. The by-products, **4** and **5** are produced by *in situ* addition of **1** to the imino group of 3-aminopentylideneamine during reduction of **2** or by catalytic dehydrogenation of **1**. The adduct, **11** loses ammonia to form the enamine, **12**. In the hydrogenation step, **12** is converted mainly to the triamine, **6** or cyclizes followed by thermal rearrangement to **4** and propylideneamine, **15**. The latter adds **1** with loss of ammonia to produce **5**. This route to produce by-product hexahydropyrimidines appears to be new for 1,3-aminonitriles and their concentration can be enhanced or mitigated by the catalyst choice and reaction conditions. This finding provided the basis for a new high yield synthetic route to 2-alkylhexahydropyrimidines from nitrile and 1,3- diamines with minimal by-products. This hydrogenation approach also provided a new synthetic entry to 2-alkyl-tetrahydro-1,3-oxazines in high selectivity from 1,3-aminoalcohols.

REFERENCES

1. C. A. Tolman R. J. McKinney, W. C. Seidel, J. D. Druliner and W.R. Sevens. Homogeneous Nickel-Catalyzed Hydrocyanation. In: D. D. Eley, H. Pines and P. B. Weisz ed, Advances in Catalysis, volume 33, 1985, pp 2-45.
2. E. W. Kluger, Tien-Kuei Su, T. J. Thompson. US Pat. 4,211,725 to Milliken Research Corp. (1980); E. W. Kluger, Tien-Kuei Su, T. J. Thompson. US Pat 4,260,556 to Milliken Research Corp. (1981).
3. Francis L. Scott. US Pat 3,119,872 to Pennsalt Chemical Corp., 1967.
4. (a) J. L. Reibsomer and G. H. Morey. J Org Chem 15: 245-8, 1950; (b). R. F. Evans, Aust J Chem 20: 1634-61, 1967.
5. (a) H. Baron, F. G. P. Remfry and J. F. Thorpe. J Chem Soc 85: 1726, 1904. (b) J. P. Schaefer and J. J. Bloomfield, Org. Reactions 15:1, 1967; (c) Advances in Organic Chemistry, E. Taylor and A. McKillop. "The Chemistry of Cyclic enaminonitriles and o-aminonitriles", NY, Interscience, p1 1970.
6. M. Hajek, K. Wagner, W. Uerdingen and W. Wellner. US Pat 4,404,379 to Bayer 1983.
7. (a) M. P. Shaver, C. M. Vogels, A. I. Wallbank, T. L. Hennigar, K. Biradh, M. J. Zaworotko and S. A. Westcot. Can J Chem 78: 568-576, 2000; (b) C. M. Vogels, P. G. Hayes, M. P. Shaver and S. A. Westcott. Chem Commun 1: 51-52, 2000.
8. F. E. Herkes and J. L. Snyder. US Pat 5,898,085 to DuPont 1999.

33

Steric and Electronic Effects in the Hydrogenation of N-Containing Heterocyclic Compounds

M. Campanati, S. Franceschini, and A. Vaccari[*]
Dipartimento di Chimica Industriale e dei Materiali, INSTM, Udr di Bologna, Bologna, Italy

O. Piccolo
Chemi SpA, Cinisello Balsamo (MI), Italy

ABSTRACT

The liquid-phase hydrogenation of alkylquinolines, 2-chloro and 2-hydroxyquinoline, and alkylindoles under mild reaction conditions (T = 373 K, PH_2 2.0 MPa) was investigated using a commercial Rh/Al_2O_3 (Rh 5 wt.%) catalyst and compared with the behaviour of quinoline and indole. Unlike quinoline, for which only a partial hydrogenation occurred, the alkylquinolines showed the formation of fully hydrogenated products, due to the presence of an alkyl side chain that avoided any deactivation effect. The nature and position of the substituent had a significant role. On the contrary, for the alkylindoles a negative effect related to the presence of electron-donating substituents on the carbocyclic ring was observed, with catalytic performances worse than those detected with the indole.

---

[*] *E-mail*: vacange@ms.fci.unibo.it

INTRODUCTION

The hydrogenation of N-containing heterocyclic compounds, such as indole, quinoline and their derivatives, are of wide industrial interest, being widely employed in the production of intermediates and fine chemical as well as in petrochemical processes [1-3]. Heterogeneous catalysts containing Pd, Ni, Rh, Pt, Cu etc. have been investigated for the liquid-phase hydrogenation of quinoline (Q) to decahydroquinoline (DHQ) [4]. These studies showed the hydrogenation occurred in two steps (Fig. 1A). First, 1,2,3,4-tetrahydroquinoline (py-THQ) and small amounts of 5,6,7,8-tetrahydroquinoline (bz-THQ) are formed and second, Py-THQ and bz-THQ are hydrogenated to DHQ. The first step is conducted under mild temperature and H_2 pressure (T = 373 K, PH_2 = 0.1-7.0 MPa) using an alcohol as solvent. On the contrary, the formation of DHQ required drastic reaction conditions: an acid as solvent, temperature up to 533 K and PH_2 ranging from 11.0 to 21.0 MPa. The aim of this work was to investigate the effect of electron-donating or electron-withdrawing substituents (located either on the heterocyclic or carbocyclic ring) in the liquid phase hydrogenation of quinoline and its derivatives under mild reaction conditions. The study has been also extended to indole (I) and its alkylderivatives (Fig. 1 B), in order to compare the behaviours of these two classes of N-containing heterocyclic compounds.

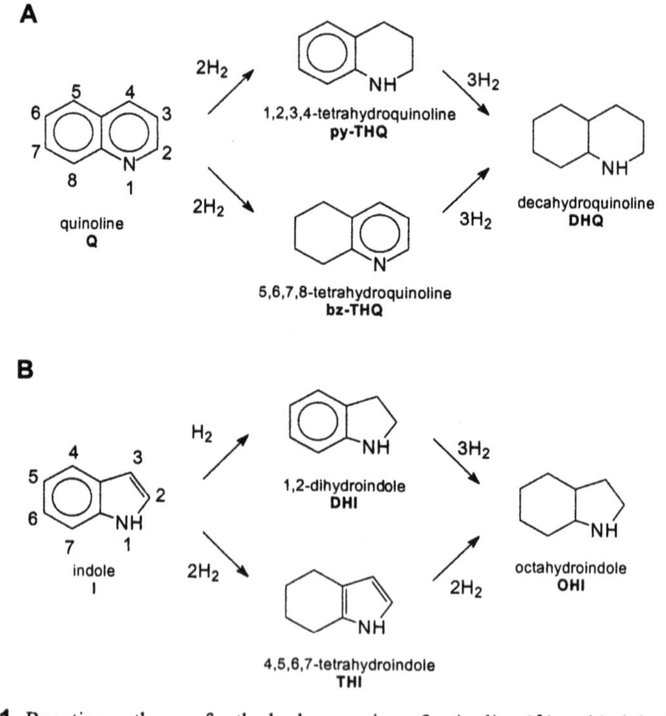

Figure 1 Reaction pathways for the hydrogenation of quinoline (**A**) and indole (**B**).

EXPERIMENTAL

The commercial 5 wt.% Rh supported on Al_2O_3 (Rh/Al_2O_3) catalyst was supplied by Engelhard and used as received, while the 2 wt.% Rh-containing pillared lay-ered clay catalyst (Rh-PLC) was prepared by ion exchange of a Mg/Al/Ce pillared montmorillonite and reduced under H_2 flow at 673 K for 2 h [5]. High resolution TEM micrographs were taken with a JEM 3010 (JEOL), operating at 300 kV and using a 5 µL dried droplet of a catalyst suspension in i-propanol onto a wholly amorphous carbon film, coating a 200 mesh copper grid (TAAB Laboratories Equipment). Quinoline (Q), 2-methylquinoline (2mQ), 3-methylquinoline (3mQ), 4-methylquinoline (4mQ), 6-methylquinoline (6-mQ), 8-methylquinoline (8-mQ), 2-chloroquinoline (2ClQ), 2-hydroquinoline (2OHQ), indole, 2-methylindole (2mI), 2,3-dimethylindole (2,3dmI), 7-ethylindole (7eI), n-heptane and i-propanol were purchased from Aldrich Chemicals (purity   98.0 %) and used as received. The tests were carried out using a 300-mL stainless steel Parr reactor, equipped with magnetic stirrer and digital oven temperature controller. The typical condi-tions for the tests were: 0.45 g of 5 wt.% Rh/Al_2O_3 catalyst, 0.13 moles of sub-strate in 135.0 mL of i-propanol, PH_2 = 2.0 MPa and T = 373 K. Quantitative analyses were carried out using a Perkin Elmer Autosystem XL gas chromato-graph, equipped with a PE-5 column (30 m x 0.25 mm, film thickness 0.25 µm) and a FID detector, while qualitative analyses were carried out using a GC-MS Hewlett-Packard GCD 1800A system, equipped with an HP-5 column (30 m x 0.25 mm, film thickness 0.25 µm) and a mass spectrometer detector. The yields were calculated, as moles of hydrogenated product formed divided by the moles of quinolines or indoles fed and expressed as percent.

RESULTS AND DISCUSSION

Under mild reaction conditions (T = 373 K, PH_2 = 2.0 MPa, 5 wt% Rh/Al_2O_3) quinoline was only partially hydrogenated to py- or bz-THQ (Fig. 2), since the complete hydrogenation to DHQ was hampered by catalyst deactivation due to the adsorption of a reaction intermediate(s) on the active sites (probably one or more dihydroquinolines formed in the step Q → py-THQ) [6].

No formation of DHQ was observed with increasing either the temperature from 373 to 473 K or the H2 pressure from 2.0 to 4.0 MPa. Also the addition of excess CH_3COOH did not modify activity and selectivity, while the addition of a small amount of a 3 M NaOH solution completely inhibited the Q conversion [6]. Thus, a Brønsted acid media does not affect the adsorption equilibrium of the de-activating intermediate(s) on the active sites of the catalysts, while a Brønsted base interacts with them too strongly.

On the contrary, the addition of an appropriate Lewis base may compete with the reaction intermediate(s) for the adsorption, favouring the complete hy-drogenation of Q to DHQ. This Lewis base has to be stable in reaction conditions and without any catalytic reactivity in the aromatic substitution of Q. A test carried out feeding a reaction mixture containing also *N,N*-diisopropylethylamine (Di-PEA) (molar ratio Q/DiPEA = 1.0), i.e. a strong base [7] which has sufficient steric hindrance to avoid too strong adsorption on the active sites, gave rise to a yield in DHQ of 17% ca. [6], confirming the hypothesis that a Lewis base may partially avoid the poisoning of the catalyst active sites.

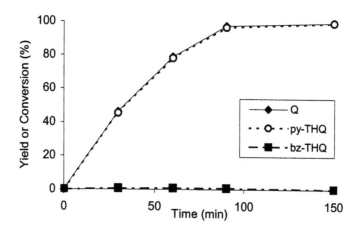

Figure 2 Catalytic data for the hydrogenation of quinoline as a function of the reaction time [T = 373 K; PH_2 = 2.0 MPa; 0.45 g of 5 wt.% Rh/Al_2O_3; 0.13 moles of quinoline in 135.0 mL of i-propanol].

An alternative way to smooth the catalyst deactivation due to the irreversible adsorption of reaction intermediate(s) is to reduce the metal particle size. The Rh-PLC catalyst exhibited Rh crystallites smaller than those of the Rh/Al_2O_3 catalyst ($\leq$ 2 nm instead of 4 nm) (Fig. 3), without significant increases in the size of the metal particles after reaction for both catalysts [5].

A 373K the Rh-PLC catalyst gave only partial hydrogenation of Q to py-THQ (Fig. 4), although with a significantly lower conversion than with $Rh-Al_2O_3$ catalyst (referred to the same Rh-content), attributable to its microporous structure and steric hindrance of Q. [5]. However, when the temperature was increased to 473 K, the Rh-PLC catalyst, unlike the commercial Rh/Al_2O_3 catalyst, gave rise to a good yield in DHQ (Fig. 4), underlining the key role played by the metal particle size in the hydrogenation of other quinoline ring systems, as for example in the case of cinchonidine, with a corresponding loss of enantioselectivity [5,9].

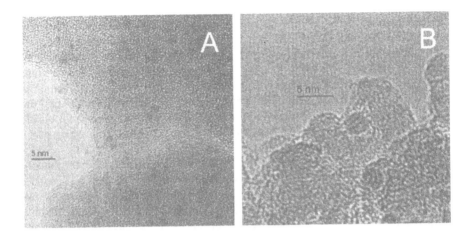

Figure 3 High resolution micrographs of the 2 wt.% Rh-PLC (A) and 5wt. % Rh/Al₂O₃ (B) catalysts before the catalytic tests (magnification 400,000x).

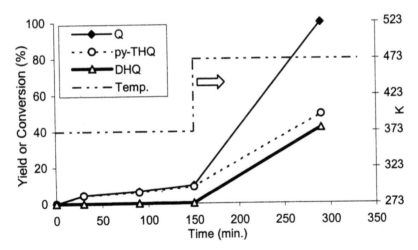

Figure 4 Catalytic data for the hydrogenation of quinoline as a function of the reaction time and temperature [PH₂ = 2.0 MPa; 1.125 g of 2 wt.% Rh-PLC; 0.13 moles of quinoline in 135.0 mL of i-propanol].

Unlike that reported for quinoline, using the same conditions the hydrogenation of methylquinolines resulted in fully hydrogenated products (Fig. 5). It is noteworthy that the higher conversion detected in the presence of an electron-donating group is apparently in contrast with the current scientific literature. Analogous to naphthalene [9,10], the steric effect of the methyl group should affect negatively the hydrogenation of the aromatic rings. In fact, the strength of substrate adsorption on the active sites decreases in presence of an alkyl substituent and the hydrogenation of an electron-rich ring is less favoured than that of an electron-poor ring. On the contrary, in the present case the steric hindrance of the methyl group may hamper the irreversible adsorption of intermediate(s) on the active sites, responsible for the catalyst deactivation, allowing the formation of the fully hydrogenated products.

The electron-donating properties affected also the intermediate distribution: when the methyl group was located on the heterocyclic ring of the quinoline both the possible partially hydrogenated intermediates were formed, because the N-containing ring is electronically stabilized. On the contrary, when the methyl group was located on the carbocyclic ring, the only intermediates formed were the 1,2,3,4-tetrahydromethylquinolines since the carbocyclic ring is more stable to hydrogenation than the heterocyclic ring. Furthermore, the following comprehensive scale of activity has been found:

2mQ   8mQ   6mQ   4mQ   3mQ

showing that higher yields in decahydro-products were obtained when the methyl group was close to the nitrogen atom. This latter result may be due to a reduction of the interaction between the lone pair of electrons on nitrogen of the substrate and the active sites of the catalyst, attributable to the steric hindrance of the methyl group. On the basis of the above results it is possible to hypothesize that the lone pairs of electrons on nitrogen plays a relevant role in the catalyst deactivation observed in the hydrogenation of quinoline [6].

It must be pointed out that opposite results were obtained when an electron-withdrawing substituent was present: for example, the 2-chloroquinoline (Cl-Q) gave rise to a very low yield in 1,2,3,4-tetrahydro-2-chloroquinoline (<1.5%), with formation mainly by dehalogenation of Q (20.5%), py-THQ (73.3%) and alkylation products (4.7 %). On the other hand, feeding the 2-hydroxyquinoline (OH-Q) a fully different reaction pathway was observed (Fig. 6), with formation mainly of the corresponding hydrogenated lactam.

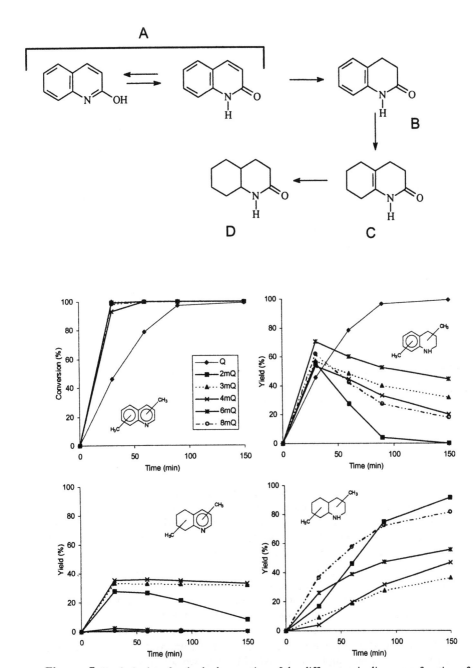

Figure 5 Catalytic data for the hydrogenation of the different quinolines as a function of the reaction time [T = 373 K; PH_2 = 2.0 MPa; 0.45g of 5 wt.% Rh/Al_2O_3; 0.13 moles of quinoline or methylquinoline in 135.0 mL of i-propanol].

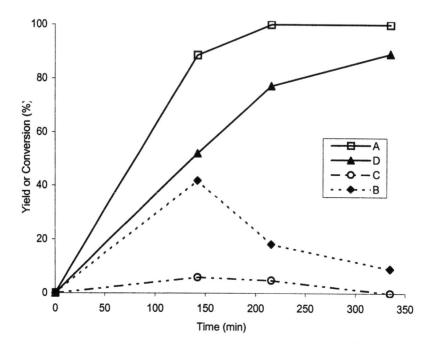

Figure 6 Catalytic data and proposed reaction pathway for the hydrogenation of 2-hydroxyquinoline (OH-Q) as a function of the reaction time [T = 373 K; PH_2 = 2.0 MPa; 0.45g of 5 wt.% Rh/Al_2O_3; 0.13 moles of 2-hydroxyquinoline in 135.0 mL of i-propanol].

Unlike quinoline, the hydrogenation of indole (I) did not show any catalyst deactivation, forming mainly the fully hydrogenated octahydroindole (OHI). The main reaction intermediate was 2,3-dihydroindole (DHI) with only traces of the intermediates hydrogenated on the carbocyclic ring (esahydroindole or EHI). However, under the same condition used in the quinoline hydrogenation (T = 373 K, 2.0 MPa, and i-propanol as solvent) but using indole, cracking products were formed in high amounts (Fig. 7). It is noteworthy that the solvent employed affected the amount and nature of the cracking products obtained: for example, with n-heptane almost three times more by-products formed than using i-propanol or t-butanol. Furthermore, the former solvent formed mainly a cracking product tentatively identified on the basis of the GC-MS spectrum as N- (2-aminoethyl)-OHI instead of the N-ethylaniline obtained with other two solvents.

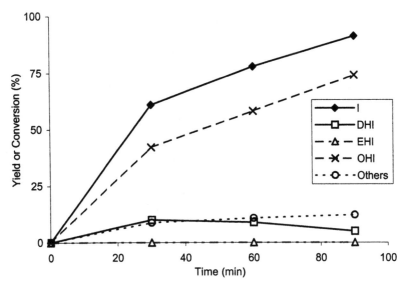

Figure 7 Catalytic data for the hydrogenation of indole (I) as a function of the reaction time [T = 373 K; PH_2 = 2.0 MPa; 0.45g of 5 wt.% Rh/Al$_2$O$_3$; 0.13 moles of indole in 135.0 mL of i-propanol].

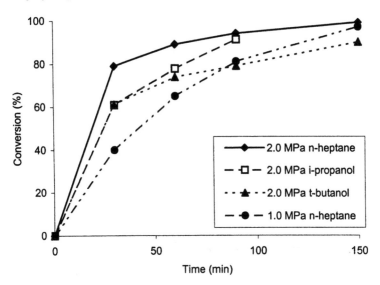

Figure 8 Catalytic data for the hydrogenation of indole (I) as a function of the solvent and pressure [T = 373 K; 0.45g of 5 wt.% Rh/Al$_2$O$_3$; 0.13 moles of indole in 135.0 mL of solvent].

To significantly reduce the extent of hydrogenolysis (never observed in the hydrogenation of quinoline or methylquinolines) the H_2 pressure was reduced to 1.0 Mpa and n-heptane instead of polar i-propanol was used as solvent. With these conditions complete conversion of indole was always achieved (Fig. 8). Under these conditions the hydrogenation of 2mI and 7eI showed lower conversion values than that of indole (Fig. 9), again with the products hydrogenated in position 2 and 3 as the main reaction intermediates. Furthermore, the hydrogenation of the intermediates was more difficult when the alkyl chain was located on the carbocyclic ring (2.3-dihydro-7-ethylindole), according to the higher stability of alkyl-substituted benzene in the hydrogenation. This effect increased with the number of substituents, as observed for the 2,3dmI. Furthermore, the hydrogenation of the 2,3dmI gave rise to a different reaction intermediate (4,5,6,7-tetrahydroindole, see Fig. 10), because of the stability of the pyrrolic substituted ring that decreased in comparison to those obtained with I, 2mI and 7eI..

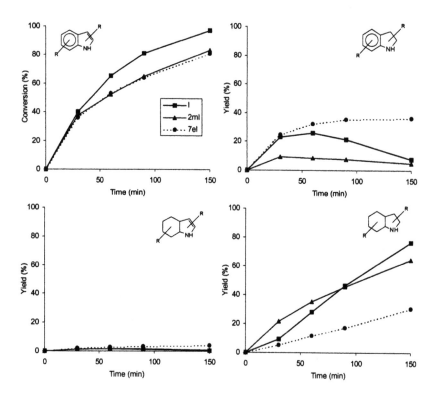

Figure 9 Catalytic data for the hydrogenation of the different indoles as a function of the reaction time [T = 373 K; PH_2 = 1.0 MPa. 0.45 g of 5 wt.% Rh/Al_2O_3; 0.13 moles of indole or alkylindole in 135.0 mL of n-heptane].

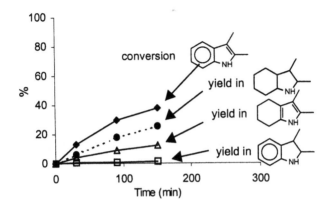

Figure 10 Catalytic data for the hydrogenation of 2,3-dimethylindole as a function of the reaction time [T = 373 K; PH$_2$ = 1.0 MPa; 0.45 g of 5 wt.% Rh/Al$_2$O$_3$; 0.13 moles of 2,3-dimethylindole in 135.0 ml of n-heptane].

CONCLUSIONS

Steric and electronic effects play different roles in the hydrogenation of N-containing heterocyclic compounds. With the quinolines, the presence of an electron-donating group surprisingly increases the activity, allowing one to obtain the fully hydrogenated products. This effect depends on the position of the substituent, suggesting a role of the lone pair of electrons on nitrogen in the catalyst deactivation during quinoline hydrogenation. On the contrary, the indoles showed regular trends, with lower conversion values for the alkylindoles than for indole. This deactivation increased with the number and position of alkyl chains.

ACKNOWLEDGMENTS

This work was performed in the frame of the Program Agreement between the CNR (Rome, Italy) and the MURST (Rome, Italy) (Law 95/95). Thanks are due to Engelhard for providing the commercial 5 wt.% Rh/Al$_2$O$_3$ catalyst.

REFERENCES

[1] G. Perot. The Reactions Involved in Hydrodenitrogenation. Catal. Today **10**: 447-472, 1991.
[2] M. Freidelder. Hydrogenation of Pyridine and Quinolines. Adv. Catal. **14**: 203-253, 1963.
[3] J. E. Shaw, P.R. Stapp. Regiospecific Hydrogenation of Quinolines and Indoles in the

Heterocyclic Ring. J. Heteroc. Chem. **24**: 1477-1483, 1987.

[4]   C. W. Curtis, D.R. Cahela. Hydrodenitrogenation of Quinoline and Coal using Precipitated Transition Metal Sulfides. Energy & Fuels **3**: 168-174, 1989.

[5]   M. Campanati, M. Casagrande, I. Fagiolino, M. Lenarda, L. Storaro, M. Battagliarin, A. Vaccari. Mild Hydrogenation of Quinoline. 2. A Novel Rh-containing Pillared Layered Clay Catalyst. J. Mol. Catal. A Chemical in press.

[6]   M. Campanati, A. Vaccari, O. Piccolo. Mild Hydrogenation of Quinoline. 1. Role of Reaction Parameters. J. Mol. Catal. A Chemical **17**: 287-292, 2002.

[7]   L.A. Carpino, A. El-Faham, Effect of Tertiary Bases on O-Benzotriazolyluronium Salt-Induced Peptide Segment Coupling, J. Org. Chem. **59**: 695- 698, 1994.

[8]   V. Morawsky, U. Pruβe, L. Witte, K.-D. Vorlop. Transformation of Cinchonidine during the Enantioselective Hydrogenation of Ethyl Pyruvate to Ethyl Lactate. Catal. Commun. **1**: 15-20, 2000.

[9]   A. W. Weitamp. Stereochemistry and Mechanism of Hydrogenation of Naphtalenes on Transition Metal Catalysts and Conformational Analysis of the Products. Adv. Catal. **18**: 1-110, 1968.

[10] TH. J. Nieuwstad, P. Klapwijk, H. Van Bekkum. Hydrogenation of Alkyl-substituted Naphthalenes over Palladium. J. Catal. **29**: 404-411, 1973

34

The Mechanism of Ethanol Amination over Nickel Catalysts

S. D. Jackson[1,4], J. R. Jones[2], A. P. Sharratt[5], L. F. Gladden[3], R. J. Cross[1], and G. Webb[1]

[1] *Department of Chemistry, The University of Glasgow, Glasgow, Scotland.*
[2] *Department of Chemistry, University of Surrey, Guildford, United Kingdom*
[3] *Department of Chemical Engineering, University of Cambridge, Cambridge, United Kingdom*
[4] *R&T Group, Synetix, Billingham, Cleveland, United Kingdom*
[5] *Ineos Fluor Ltd, Cheshire, United Kingdom*

ABSTRACT

The amination of ethanol over a 10% Ni/silica catalyst was investigated using [^{2}H], [^{3}H], [^{13}C]-tracers. The kinetics of the reaction was also studied. Analysis of the results suggested that the accepted mechanism did not explain the observed behaviour. Other previously proposed mechanisms were also found to be unable to describe the results. A new mechanism is proposed.

INTRODUCTION

The amination reaction, forming primary, secondary, and tertiary amines, can be between alcohols, aldehydes, or ketones, and ammonia and hydrogen. Sabatier and Mailhe in 1909 [1] first reported the synthesis of amines from alcohols. Over the course of the 20th century the importance of these compounds has increased and amines are key intermediates and final products within the fine chemical, agrochemical, and pharmaceutical industries. The amination of alcohols to form amines has been the subject of a number of mechanistic studies [2-8] proposing a variety of mechanisms. In this study we have investigated the

453

amination of ethanol over Ni/silica catalysts and have found that, using a combination of kinetic and tracer studies, the current mechanism cannot account for all the processes occurring. Therefore a new mechanism that rationalises results is proposed.

EXPERIMENTAL

A 10% Ni/silica catalyst was used throughout this study. The catalyst was prepared by aqueous impregnation. Sufficient acidic nickel nitrate solution was added to a silica support (ICI, S.A. 475 m^2g^{-1}) to produce a weight loading of 10 % w/w nickel. The excess water was removed by evaporation and the sample dried at 353 K overnight. The catalyst was then calcined in air at 673 K.

A recirculating reactor system was used throughout. The catalyst (typically 0.2 g) was reduced in situ by heating to 673 K in a flow of hydrogen (25 cm^3min^{-1}) and holding at this temperature for 3 h. The catalyst was then cooled in flowing hydrogen. A reaction mixture of ammonia (1.23 – 2.45 mmol.) and ethanol (0.18 – 1.8 mmol.) in hydrogen were recirculated (30 cm^3min^{-1}) through a by-pass for 0.5 h to ensure complete mixing then the flow was directed over the catalyst bed. The reaction temperature was varied between 383 K and 453 K. At intervals 1 cm^3 samples were withdrawn and analysed by GC and NMR.

The ^2H, ^{13}C, and ^{15}N labelled materials (Merck Sharp & Domme) were all used as received. The following isotopically labelled species were prepared, [^3H]-ammonia, [1-^3H]-ethanol, and [G-^2H]-ethanol, purified and used along with [^3H]-hydrogen [9].

RESULTS

The catalyst (0.2 g) was reduced and subjected to a reaction mixture ($C_2H_5OH:NH_3:H_2$, 1:4:15 molar ratio with ethanol at 0.63 mmol., total pressure of 1.5 atm.) at 383 K a typical product distribution is shown in Figure 1. The temperature was varied and activation energies determined. The activation energy for ethylamine formation over Ni/silica was determined at 33±5 kJ.mol^{-1}, while the activation energy for ethanol conversion was 85±10 kJ.mol^{-1}.

An amination reaction was performed using d_6-ethanol. A kinetic isotope effect with $\eta = 6$ was observed. Analysis of the ethylamine revealed that the major product was $CD_3CH_2NH_2$.

When an amination reaction was performed using [1-^3H]-ethanol, the [^3H]-hydrogen in the OH group was rapidly distributed between the ammonia and hydrogen gas. Using [^3H]-H_2 (tritium) it was possible to trace the incorporation of hydrogen into ethanol and ethylamine. Therefore a reaction was performed using [^3H]-H_2 as the hydrogen source in the reaction mixture. Analysis of the ethanol and ethylamine by NMR revealed that no tritium was incorporated into ethanol in any position other than the –OH group. Tritium was immediately observed in the ethylamine. Table 1 shows the extent of incorporation with time.

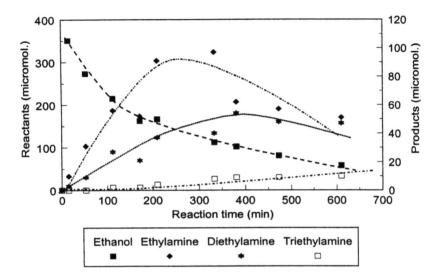

Figure 1 Reaction profile.

Table 1 Incorporation of Tritium into Ethylamine Product.

Time (min)	53	112	172	210	333	381
Conversion (%)	70	80		>90		
Methyl	No	No	No	No	Yes	Yes
Methylene	Yes	Yes	Yes	Yes	Yes	Yes

Amination using [1-^{13}C]-ethanol was investigated over the 10% Ni/silica catalyst to determine whether or not the carbon atoms were exchanged during the amination process. In Figure 2 the top spectrum is a natural abundance DEPT (distortionless enhancement by polarisation transfer) 135 ^{13}C NMR spectrum of the products of an amination reaction. The DEPT 135 pulse sequence produces a spectrum where the methylene signals are 180° out of phase relative to the methyl signals. Hence we can identify the methyl signal at 20 ppm and the methylene signal at 50 ppm. The bottom spectrum shows the DEPT 45 ^{13}C NMR spectrum of the products of an amination reaction using [1-^{13}C]-ethanol. In this case the pulse sequence does not change the phase. Clearly the label is only present in the methylene group.

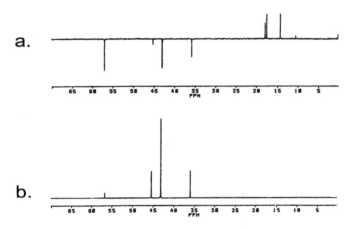

Figure 2 ^{13}C NMR spectra of amination products. a. DEPT 135 spectra following an amination reaction using natural abundance, b. DEPT 45 spectra following an amination using [1-^{13}C]-ethanol.

The effects of changing the hydrogen and ammonia concentrations were also studied (Table 2).

Table 2 Effect of Varying Hydrogen and Ammonia Concentration on Rate[a].

	Ratio of concentrations (molar)			
	NH$_3$	C$_2$H$_5$OH	H$_2$	Rate
Ni/silica[b]	4	1	15	11.0
	4	1	0.7	11.3
Ni/silica[c]	4	1	15	2.4
	6	1	15	4.9

[a] Rate in µmol.g^{-1}.min^{-1}.
[b] Conditions : temperature 413 K
[c] Conditions : temperature 383 K

The reaction of ethanol with the catalyst with and without hydrogen was investigated. On passing ethanol in a helium carrier over the catalyst at 413 K, four gas-phase products were detected ethanal, hydrogen, water and ethene/ethane (these were not separated on the GC column). When ethanol was passed over the catalyst at 413 K in a hydrogen carrier only water and ethene/ethane were detected.

DISCUSSION

There is no universally agreed mechanism for the amination of alcohols. Numerous mechanisms have been proposed [2-8] but many have been developed to interpret specific results that relate to the system under study, e.g. none of the mechanisms specifically relate to amination over a nickel catalyst. This is not an unusual occurrence but the widely differing types of catalysts used for amination makes the generation of a general mechanism more difficult. Most of the detailed studies have been performed over copper [6] or iron [4] catalysts. However our results do allow us to assess the general applicability of these mechanisms to the amination over nickel catalysts.

When we examine the reaction of ethanol over the catalyst in the absence of hydrogen we see four products, hydrogen, water, ethanal, and ethene/ethane. These are formed by dehydrogenation and dehydration.

$$CH_3CH_2OH \Leftrightarrow CH_3CHO + H_2$$
$$CH_3CH_2OH \Leftrightarrow CH_2CH_2 + H_2O$$

As the catalyst is a good hydrogenation catalyst we can expect that a portion of the ethene may be hydrogenated to ethane. When ethanol is passed over the catalyst using a hydrogen carrier we find that the dehydrogenation reaction is completely suppressed. This is not surprising given that $K_{eq} = 5x10^{-3}$. Therefore in the presence of hydrogen, dehydration is the predominant reaction. A similar result was found by Roberts and co-workers [10] when examining ethanol adsorption over Pt(111). In this study the transformation from ethanol to ethanal occurred at 295 K but above 350 K the adsorbed species lost oxygen and formed a hydrocarbon on the surface [10]. These results would favour the mechanism proposed by Popov [5, 8] where ethene was the proposed intermediate.

Popov Mechanism.

$$NH_3(g) \Leftrightarrow NH_3(a)$$
$$C_2H_5OH(g) \Leftrightarrow C_2H_5OH(a)$$
$$C_2H_5OH(a) \rightarrow C_2H_4(a) + H_2O$$
$$C_2H_4(a) + NH_3(a) \rightarrow C_2H_5NH_2(a)$$
$$CH_3CH_2NH_2(a) \Leftrightarrow CH_3CH_2NH_2(g)$$

However retention of the molecular asymmetry of the reactant, intermediates, and product was demonstrated by amination using [1-^{13}C]-ethanol. This confirms that the mechanism proposed by Popov [5, 8] cannot be correct as ethene is symmetrical and some scrambling of the carbon atoms would be expected. These results however also suggest that ethanal is unlikely to be a favoured intermediate as the amination reaction is run in an excess of hydrogen.

The most widely accepted mechanism is that proposed by Schwegler and Adkins [2]. It has as the rate determining step the production of adsorbed

Schwegler-Adkins Mechanism

$H_2(g) \Leftrightarrow 2H(a)$
$CH_3CH_2OH(g) \Leftrightarrow CH_3CH_2OH(a)$
$CH_3CH_2OH(a) \rightarrow CH_3CHO(a) + 2H(a)$
$NH_3(g) \Leftrightarrow NH_3(a)$
$CH_3CHO(a) + NH_3(a) \rightarrow CH_3CH=NH(a) + H_2O(a)$
$CH_3CH=NH(a) + 2H(a) \Leftrightarrow CH_3CH_2NH_2(a)$
$CH_3CH_2NH_2(a) \Leftrightarrow CH_3CH_2NH_2(g)$
$H_2O(a) \Leftrightarrow H_2O(g)$

ethanal by loss of hydrogen from adsorbed ethanol. Many of the tracer results are in keeping with this mechanism, however there are some significant discrepancies.

The studies by Baiker [6], Fridman [3], and others [4], all found an isotope effect when the methylene hydrogen was labelled. However it is worth noting that isotope effect was larger in our study. The value obtained by Baiker [11] for n-octanol amination over a copper catalyst 1.9, whereas over a Ni/silica catalyst the kinetic isotope effect was measured at 6. Our tracer results are in agreement that the rate determining step involves the loss of a methylene hydrogen from the kinetic isotope effect observed in the amination of d_6-ethanol. The picture is further complicated however in that the effect of varying the hydrogen and ammonia concentrations also indicates that they are involved in the rate determining step (rds). When the hydrogen concentration was reduced the reaction rate increased very slightly. This would be in agreement with the loss of a methylene hydrogen as the rds as lowering the hydrogen pressure would favour the dehydrogenation reaction. The increase of rate seen when the ammonia concentration was increased indicates that ammonia is also involved in the rds. The involvement of ammonia in the rds is not within the scope of the Schwegler-Adkins [2] mechanism.

We are left therefore with a situation that none of the mechanisms proposed can accurately describe the results obtained. In developing a new mechanism that can accommodate the results it is important to use as much of previous mechanisms as possible. As we have shown above, much of what has been proposed is in agreement with the results. However there has been much greater understanding of molecular adsorption/desorption since the publication of the Schwegler-Adkins mechanism. The adsorption of alcohols on Ni(111) [12] and our own exchange results show that the O-H bond is the easiest to break in this system. Therefore the initial adsorption under reaction temperatures will result in the formation of an ethoxy species on the surface. Similarly the adsorption of ammonia has been shown to be highly favoured when small quantities of adsorbed oxygen are present on the surface [13] leading to the formation of imide. In our system it is likely that there is a sub-monolayer presence of oxygen [14] and hence such a reaction would be favoured. Therefore both imide and ethoxy can be expected to be major surface species. However we have shown that in the presence of hydrogen ethanol will also dehydrate to give a hydrocarbon fragment. This leads us back to the mechanism proposed by Popov [5, 8]. However from the tracer results molecular asymmetry is maintained therefore an ethyl fragment is proposed rather than ethene. A new mechanism is set out below that takes account of all the results.

$$H_2(g) \Leftrightarrow 2H(a) \tag{1}$$
$$NH_3(g) \Leftrightarrow NH_3(a) \tag{2}$$
$$NH_3(a) + O(a) \Leftrightarrow NH_2(a) + OH(a) \tag{3}$$
$$NH_2(a) + O(a) \Leftrightarrow NH(a) + OH(a) \tag{4}$$
$$CH_3CH_2OH(g) \Leftrightarrow CH_3CH_2O(a) + H(a) \tag{5}$$
$$CH_3CH_2O(a) + H(a) \rightarrow CH_3CH_2(a) + OH(a) \tag{6}$$
$$\mathit{CH_3CH_2(a) + NH(a) \rightarrow CH_3CHNH(a) + H(a)} \tag{7}$$
$$CH_3CHNH(a) + H(a) \Leftrightarrow CH_3CHNH_2(a) \tag{8}$$
$$CH_3CHNH_2(a) + H(a) \Leftrightarrow CH_3CH_2NH_2(g) \tag{9}$$
$$CH_3CH_2NH_2(a) \Leftrightarrow CH_3CH_2NH_2(g) \tag{10}$$
$$OH(a) + H(a) \rightarrow H_2O(g) \tag{11}$$

The rate determining step is highlighted in italics (equation (7)). This step will have give a positive rate effect when ammonia concentration is increased and hydrogen concentration decreased and will have a kinetic isotope effect on the loss of the methylene hydrogen. The reaction of ethoxy with an imide group

$$CH_3CH_2O(a) + NH(a) \rightarrow CH_3CHNH(a) + OH(a) \tag{12}$$

could also be a potential rds however on balance we believe that the results and literature favour equation (7).

ACKNOWLEDGEMENTS

The authors would like to thank Dr. R. Griffiths of ICI for the preparation of the silica.

REFERENCES

1. P. Sabatier and A. Mailhe, *Compt. Rend.*, **148**, 898 (1909).
2. E. D. Schwegler and H. Adkins, *J. Am. Chem. Soc.*, **61**, 3499 (1939).

3. R. A. Fridman, E. I. Bogolepova, R. M. Smirnova, Yu. B. Kryokov, G. A. Zhokova, and A. N. Bashkirov, *Neftekhimiya*, **12**, 91 (1972).
4. G. A. Kliger, L. F. Lazutina, R. A. Fridman, Y. B. Kryukow, A. N. Bashkirov, Y. S. Snagovskii, and R. M. Smirnova, *Kinet. Katal.*, **16**, 660 (1975).
5. M. V. Klyuev and M. L. Khidekel, Russ. *Chem. Revs.*, **49**, 28 (1980).
6. A. Baiker and J. Kijenski, *Catal. Rev. Sci. Eng.*, **27**, 653 (1985).
7. T. Mallat and A. Baiker, in "Handbook of Heterogeneous Catalysis", G.Ertl, H.Knozinger, J.Weitkamp, eds., Chpt.4.7, VCH, Weinheim, 1997.
8. M. A. Popov, Thesis, Institute of General Chemistry of the USSR Academy of Sciences, Moscow, 1952.
9. A. P. Sharratt, Ph.D. Thesis, University of Surrey, 1991.
10. M. K. Rajumon, M. W. Roberts, F. Wang, and P. B. Wells, *J. Chem. Soc. Faraday Trans.*, **94** (1998) 3699.
11. A. Baiker, W. Caprez and W.L. Holstein, *Ind. Eng. Chem., Prod. Res. Dev.*, **22**, 217 (1983).
12. R. Raval, L. J. Shorthouse and A. J. Roberts, *Surf. Sci.*, **480**, 37 (2001).
13. A. F. Carley, P. R. Davies, and M. W. Roberts, *Chem. Commun.*, 1793 (1998).
14. S. D. Jackson, J. Willis, G. J. Kelly, G. D. McLellan, G. Webb, S. Mather, R. B. Moyes, S. Simpson, P. B. Wells, and R.Whyman *PCCP*, **1**, 2573 (1999).

35

Continuous Heterogeneous Catalytic Hydrogenation of Organic Compounds in Supercritical CO₂

Venu Arunajatesan and Bala Subramaniam
University of Kansas, Lawrence, Kansas, U.S.A.

Keith W. Hutchenson and Frank E. Herkes
The DuPont Company, Wilmington, Delaware, U.S.A.

INTRODUCTION

Heterogeneously catalyzed hydrogenation is typically carried out either in slurry reactors or fixed-bed adiabatic reactors in series. Slurry reactors are often operated in multiphase mode involving the sparging of hydrogen through a slurry of the liquid reactant and finely powdered catalyst particles. The liquid phase in slurry reactors serves to absorb the heat generated by the exothermic hydrogenation reactions. However, the low solubility of hydrogen in the solvent leads to severe gas-liquid mass transfer limitations. The other drawbacks of the conventional process include the possibility of runaway conditions with adiabatic reactors in series, large reactor volumes for slurry reactors, and costs associated with separating the product and the catalyst from the product mixture.

The use of supercritical fluids as reaction media has received increased attention in recent years. The pressure-tunable physical and transport properties of a supercritical fluid may be exploited to find an optimal reaction medium characterized by liquid-like density and heat capacity, yet significantly better (more gas-like) transport properties. The interest in SCF-based processes is evidenced by several recent reviews involving homogenous (1,2), immobilized (3), and heterogeneous catalysis (4,5) that have appeared in recent years including books devoted to this topic (6,7).

In the case of solid-catalyzed hydrogenation, solubilizing the reactants (organic substrate and hydrogen) in a single, environmentally benign solvent such as supercritical carbon dioxide ($scCO_2$) eliminates gas-liquid mass transfer limitations. This leads to enhanced reaction rates. The ensuing reduction in reactor volumes, and therefore the holdup of hazardous reactants, improves the inherent process safety. Furthermore, due to its liquid-like heat capacity, $scCO_2$-based hydrogenation offers the possibility of performing the reactions in a fixed-bed reactor with effective temperature control. In contrast to slurry phase operation, there is minimal catalyst entrainment in the reactor effluent stream. Further, the product(s) may be separated from CO_2 by depressurization of the reactor effluent stream. Examples of fixed-bed hydrogenations in *sc* media include Fischer-Tropsch synthesis (8), hydrogenation of cyclohexene (9), hydrogenation of functional groups (10,11), enantioselective hydrogenation (12) and hydrogenation of oils (13).

We have systematically investigated the hydrogenation of organic compounds in $scCO_2$ taking into account the phase behavior of the reaction mixture, the temperature control in the reactor during operation, and possible deactivation of the noble metal hydrogenation catalysts due to CO formation via the reverse water gas shift reaction between CO_2 and H_2. The goal is to develop a fundamental understanding of the underlying physicochemical processes that should aid in rationally addressing process feasibility and reactor design issues. As model systems, we investigated the exothermic hydrogenation of cyclohexene over Pd/C catalyst and the selective hydrogenation of toluene to methylcyclohexenes over Ru/Al$_2$O$_3$ with CO_2 as the reaction medium.

EXPERIMENTAL

The hydrogenation experiments were carried out in a fixed-bed reactor as described in our earlier paper (9). The schematic of the experimental apparatus is shown in Figure 1. Additional details and any changes made to the experimental apparatus or analytical procedure are detailed below. As earlier, approximately 1 g of the catalyst was loaded in the fixed-bed reactor (inner diameter = 12 mm, outer diameter = 19 mm, bed length = 25 mm) reduced *in situ* at 30 °C (for 2% Pd/C) or 200 °C (for 1% Ru/Al$_2$O$_3$) in flowing hydrogen (50 standard cm^3/min) for 2 h. The catalyst bed temperature in the present study was measured at 0, 3, 6, 12, and 25 mm along the axis of the reactor using a profile thermocouple. The reactor temperature and pressure were maintained at desired set points with heaters and a micrometering valve, respectively, which were interfaced with a Camile® Data Acquisition and Control System.

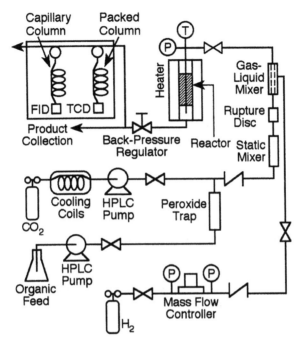

Figure 1 Schematic of the experimental unit for performing continuous catalytic hydrogenation in *sc*CO₂ (Reprinted from ref. 9 with permission from Elsevier Science).

The cyclohexene hydrogenation was conducted with equimolar amounts of cyclohexene and hydrogen (0.24 gmol/h) in 90% CO₂ (4.32 gmol/h) at 70 °C and 138 bar. The selective hydrogenation of toluene to methylcyclohexene was investigated using a feed composition of 2:1 molar ratio of hydrogen to toluene (>99%, Fisher Scientific) in 90% CO₂ at 60 °C and 138 bar. The toluene flowrates (0.11 gmol/h) used in a typical experiment provided a toluene weight hourly space velocity (WHSV) of 10 h⁻¹ and an overall WHSV of ~145 h⁻¹. For experiments using higher toluene WHSV, the flowrates were increased proportionally.

The effluent analysis was conducted using an HP 5890 GC equipped with a TCD and an FID for the analysis of permanent gases and hydrocarbons. The details of the GC setup are described elsewhere (9). In order to improve the ability to obtain good hydrogen balance, the TCD was calibrated for hydrogen using a vacuum technique described by Snavely (14). The use of argon as a carrier gas provided a sensitive and linear response of the TCD for the expected range of hydrogen concentrations. The effluent sample from the reactor was analyzed after allowing the pressure in the sample loop to equilibrate to atmospheric pressure.

The FTIR experiments were conducted in the transmission mode using a Nicolet Protégé 460 FTIR spectrophotometer equipped with a high pressure transmission cell (HPL-TC-13-1, Harrick Scientific). The cell, equipped with

ZnSe windows, is capable of operating up to 330 bar and 260 °C. The internal volume of the cell was 0.25 cm³. Approximately 3 mg of the alumina supported Pd or Ru catalyst (in wafer form) were used as the catalyst. The 50 micron thick wafer (8 mm dia) was prepared by pressing the finely ground catalyst powders in a hydraulic press (Model M, Carver Inc.) at 1000 bar for 30 seconds. The catalyst wafer was loaded in the cell and reduced *in situ* as in the fixed-bed reactor. The wafer was then heated to 70 °C and exposed to 5 mol% hydrogen in CO_2 at 138 bar. The H_2/CO mixture was transferred to the cell from a 250 cm³ high pressure sample bomb. The IR spectrum was then collected at predetermined time intervals.

The phase behavior of the feed and the reacting mixture at the reactor operating conditions was observed experimentally using a well-mixed fixed volume view cell equipped with sapphire windows (20 cm³, 400 bar MAWP) as described earlier (9). The temperature and pressure at which a homogenous mixture was formed was determined. Based on these experimental measurements, we chose to operate the reactor at 70 °C and 138 bar for cyclohexene hydrogenation, and at 60 °C and 138 bar for the selective hydrogenation of toluene.

The total peroxide content in the cyclohexene feed was measured using the iodometric titration procedure (ASTM D 3703-92). Chemical identification of the peroxides was done through GC/MS analysis as described elsewhere (9).

RESULTS & DISCUSSION

Cyclohexene Hydrogenation

The exothermic hydrogenation of cyclohexene to cyclohexane (ΔH = -118 kJ/mol) has been studied as a model reaction by several authors (15-18). Hanika et al (15,16) have extensively studied the temperature effects of this reaction in the gas and liquid phase in a trickle bed reactor. They observed that as the "hot spot" temperature in the bed increased from 70 °C to 170 °C, the selectivity toward cyclohexane dropped from 100% to 15% in favor of the dehydrogenation product benzene. Hitzler et al. (10) have studied the continuous fixed-bed hydrogenation of cyclohexene in $scCO_2$ over Pd/Deloxan catalysts. Large axial temperature gradients (on the order of 100°C) in the reactor were reported (10). Such temperature gradients could limit the productivity due to either equilibrium limitations or the loss in selectivity.

The heat capacity of supercritical fluids is liquid-like at near critical conditions but is lower farther away from the critical point (19). Hence, by operating the reactor at near critical conditions, it should be possible to significantly minimize the "hot spot" temperatures in an adiabatic reactor. On the other hand, operating the reactor in the gas phase could lead to a dramatic adiabatic temperature rise in the bed (9).

Axial Temperature Profile in Adiabatic Fixed-bed Reactor

Figure 2 shows the axial temperature profile in an adiabatic reactor during the startup. A schematic of the reactor with the position of the thermocouple (dots) along the axis of the bed is shown below the transient temperature profiles. As seen from the figure, there was a rapid rise in the axial bed temperature in the initial 15 minutes of the reaction that stabilized ~25 minutes after the startup. The bed temperature was constant for the reminder of the 24 or 48 h experiment while a stable catalyst activity was observed. The sigmoidal-shaped steady-state temperature profile is typical of what one would observe in an exothermic fixed-bed reactor, characterized by an exponential increase in temperature closer to the reactor inlet (wherein a majority of the conversion occurs) followed by a flattening of the temperature profile due to reactant depletion. The maximum adiabatic temperature in the bed was observed to be 16 °C above the feed temperature. This is significantly less than the 167 °C predicted for total conversion of cyclohexene to cyclohexane if gas-like CO_2 is used (9) or the 52 °C observed during gas phase operation wherein the cyclohexene conversion is 34% (15). Clearly, in addition to significantly lowering the "hot spot" temperature in the bed, the use of the near-critical reaction medium can render the process "parametrically insensitive" to temperature leading to safer operation of the reactor.

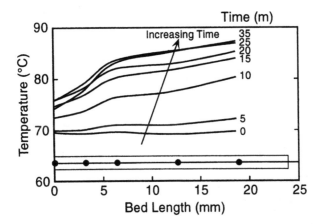

Figure 2 Temporal axial temperature profile during hydrogenation of cyclohexene over 2% Pd/C (T_0 = 70 °C, P = 134-138 bar, H_2/cyclohexene = 1:1, CO_2 = 90%, Olefin WHSV = 20 h⁻¹).

Figure 3 shows the effect of feed temperature on the axial temperature profile. In all these cases, the cyclohexene conversion is ~80% with total selectivity toward cyclohexene. The "hot spot" temperature in the bed increases proportionally to the feed temperature and does not show any runaway tendencies. This is due to the liquid-like heat capacity provided by the near-critical reaction medium. The effect of pressure on the axial temperature profile during cyclohexene hydrogenation was also studied (Figure 4). At the various operating pressures, the cyclohexene conversion is nearly identical (77-79%). As the process pressure was increased from 124 bar to 138 bar, the "hot spot" temperature decreased from 24 °C to 12 °C. This is attributed to the increased heat capacity of the reaction medium at the higher pressures.

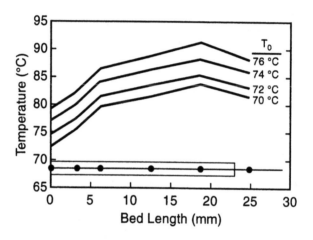

Figure 3 Effect of feed temperature on the steady-state axial temperature profile during hydrogenation of cyclohexene over 2% Pd/C (T_0 = 70-76 °C, P = 138 bar, H_2/cyclohexene = 1:1, CO_2 = 90%, Olefin WHSV = 20 h^{-1}).

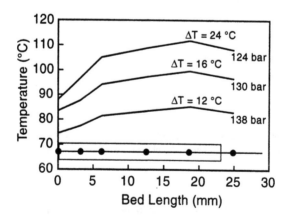

Figure 4 Effect of process pressure on the steady-state axial temperature profile during hydrogenation of cyclohexene over 2% Pd/C (T_0 = 70 °C, P = 124-138 bar, H_2/cyclohexene = 1:1, CO_2 = 90%, OWHSV = 20 h^{-1}).

Effect of Organic Peroxides

The cyclohexene feed contained roughly 180 ppm peroxide as received. Organic peroxides, known to cause catalyst deactivation (9,20), were reduced to <6 ppm before the reaction and were further reduced using the in-line alumina trap. If the organic peroxides in the feed are not reduced to <6 ppm, this could lead to significant deactivation of the catalyst (9).

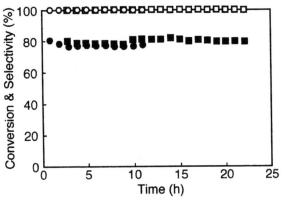

Figure 5 Repeatability of the steady conversion (■,●) and selectivity(□,○) over (2% Pd/C) on the removal of organic peroxides in the feed (T_0 = 70 °C, P = 138 bar, H_2/cyclohexene = 1:1, CO_2 = 90%, Olefin WHSV = 20 h^{-1}).

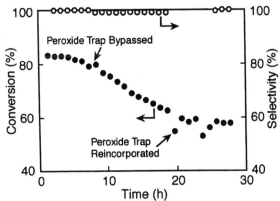

Figure 6 Effect of removal of organic peroxides in the feed on the catalyst (2% Pd/C) activity (T_0 = 70 °C, P = 138 bar, H_2/cyclohexene = 1:1, CO_2 = 90%, Olefin WHSV = 20 h^{-1}) (Reprinted from Ref. 9 with permission from Elsevier Science).

Figure 5 shows the stable and repeatable activity obtained on the mitigation of these peroxides using the in-line alumina trap. To conclusively demonstrate the detrimental effect of organic peroxides on catalyst activity, an experiment was performed in which the alumina trap was bypassed in the middle of a run to expose the catalyst to the feed peroxides, and then reincorporated after a few hours. As seen in Figure 6, nearly constant catalyst activity was observed when the peroxides in the feed were removed by the alumina trap. On the other hand, when the alumina trap was bypassed after ~8 h to allow the organic peroxides to reach the catalyst bed, a clear deactivating trend (2%/h) was observed. Reincorporating the alumina trap into the cyclohexene feed line after nearly 12 h prevented further deactivation of the catalyst. This stable activity was observed until the run was ended approximately 8 h after reincorporating the alumina trap.

Toluene Hydrogenation

Selective hydrogenation of organic compounds is one of the most important reactions in the chemical and pharmaceutical industries. Selectivity in any reaction can be classified as Type I, II or III according to the nature of reaction involved (21). *Type I selectivity* occurs between two different products formed in parallel reactions that differ by at least one reactant. *Type II selectivity* involves selectivity of one product over another in parallel reactions involving the same reactants. And, *Type III selectivity* involves the selectivity between products in a series reaction. Both Type I and III selectivities are affected by mass transport rates (22).

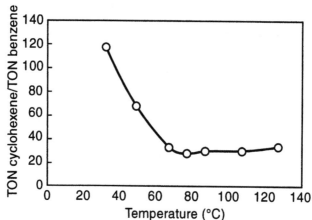

Figure 7 Relative rates of benzene and cyclohexene hydrogenation over Ru/SiO$_2$ catalyst [adapted from ref. 16].

Selective hydrogenation of aromatic compounds to cycloalkenes has gained significant import owing to the versatility of cycloalkenes as intermediates in a variety of industries. However, the hydrogenation of an aromatic compound tends toward cycloalkane, the ultimate hydrogenation product, as it is thermodynamically favored. The selective hydrogenation of benzene or toluene to the corresponding cycloalkene and cycloalkane is considered to involve both Type II and

III selectivity (21). In order to enhance the selectivity toward the intermediate, the cycloalkene should be removed from the catalyst pores before any further reaction can occur. For this reason, a reaction medium that provides enhanced mass transfer rates for removal of intermediate products from the catalyst pores is preferable. In addition, a reaction medium with enhanced heat capacity is also desirable, since the exothermicity of the reaction could lead to lowering of selectivity at higher temperatures as shown by Figure 7 (16).

The hydrogenation of toluene has been studied as a model reaction for the hydrogenation of aromatics (23-30). Although several catalysts have been used for this reaction, ruthenium has been determined to be the most selective catalyst for the partial hydrogenation reaction. While negligible selectivity toward the intermediate has been observed in the gas or liquid phase in the absence of any modifiers, 20% selectivity has been reported with modifiers (31).

Due to the significant reduction in heat and mass transfer limitations achievable using supercritical reaction medium, it is possible to increase the selectivity toward the intermediate in a series reaction. Bochniak and Subramaniam (8) observed that the reduction in pore diffusion limitations during Fisher-Tropsch synthesis using supercritical n-hexane as the reaction medium led to an increased production of the intermediate α-olefins. In the present study, the hydrogenation of toluene will be studied employing $scCO_2$ as the reaction medium to investigate if the selectivity toward methylcyclohexene can be enhanced.

Conversion & Selectivity Studies

The as-received toluene contained <5 ppm peroxides. The lower peroxide content is to be expected since aromatics have lower tendency than alkenes to form organic peroxides. However, the feed was treated on-line to further mitigate these peroxides. The activity and selectivity of 1% Ru/Al_2O_3 were determined in a fixed bed reactor at supercritical conditions (60 °C, 138 bar) using 2:1 molar ratio of hydrogen to toluene and 90% CO_2 as the solvent. Both 1-methyl-cylohex-1-ene and 1-methyl-cyclohex-3-ene were formed during the reaction. The methylcyclohexene reported here is the sum of these two isomers. As shown in Figure 8, >10% selectivity toward methylcyclohexenes was observed at the operating conditions in the absence of any modifier. The maximum temperature rise observed in the bed was ~18 °C. A gradual decrease in the activity of the catalyst (1%/h) was observed with time-on-stream. A similar deactivation during selective hydrogenation of toluene was attributed to the formation of alkylation products leading to the fouling of the active site (26).

Carbon and hydrogen balance closures were achieved during the experiment. The typical mass balance during the experiment is shown in Figure 9. The effects of various parameters such as temperature, pressure, H_2/Toluene ratio, and space velocity on the selectivity of the Ru/Al_2O_3 at supercritical condition were investigated. The selectivity was more significantly affected by space velocity than any other variable tested.

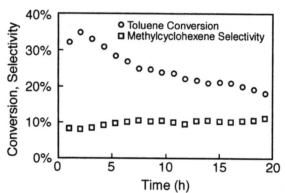

Figure 8 Activity and selectivity during hydrogenation of toluene over 1% Ru/Al$_2$O$_3$ (T$_0$ = 70 °C, P = 138 bar, H$_2$/toluene = 2:1, CO$_2$ = 90%, Toluene WHSV = 10 h^{-1}).

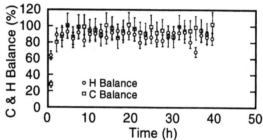

Figure 9 Carbon and hydrogen balance during the selective hydrogenation of toluene to methylcyclohexene over 1% Ru/Al$_2$O$_3$ (T$_0$ = 70 °C, P = 138 bar, H$_2$/toluene = 2:1, CO$_2$ = 90%, Toluene WHSV = 10 h^{-1}).

Effect of Space Velocity

The effect of space velocity on the selectivity of the catalyst was tested by proportionally increasing the flowrate through the catalyst bed in the fixed-bed reactor. The selectivity in each case was noted at 15 h. Doubling the toluene weight hourly space velocity from 10 h^{-1} to 20 h^{-1} nearly doubled the methyl cyclohexene selectivity from 12% to about 22%. Increasing the space velocity further to 30 h^{-1}, improved the selectivity further by a small amount as shown in Figure 10. These trends are to be expected: at higher space-velocities, the decreased residence time leads to lower conversion and enhances intermediate product selectivity. The improved selectivity in the supercritical reaction medium is attributed to the higher diffusivity compared to the liquid phase and better "hot spot" temperature control compared to the gas phase. Our results show that with further improvements in catalyst design, it should be possible to rationally exploit scCO$_2$ reaction medium to enhance intermediate product selectivity in partial hydrogenation reactions.

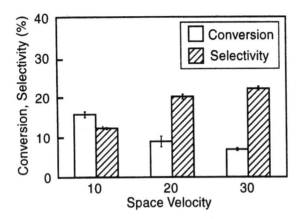

Figure 10 Effect of space velocity on the selectivity toward methylcyclohexene during the selective hydrogenation of toluene over 1% Ru/Al_2O_3 ($T_0 = 70$ °C, P = 138 bar, H_2/toluene = 2:1, $CO_2 = 90\%$, Toluene WHSV = 10-30 h^{-1}).

In situ FTIR Studies

Minder *et al.* (32) investigated the batch hydrogenation of ethyl pyruvate to (R)-ethyl lactate over a modified Pt/Al_2O_3 catalyst in supercritical reaction media. When using $scCO_2$ as the reaction medium, they report catalyst deactivation, presumably because of Pt poisoning by CO formed from the reverse water-gas shift reaction between CO_2 and H_2 ($CO_2 + H_2 \rightarrow CO + H_2O$). When using supercritical ethane ($P_c = 48$ bar; $T_c = 32.3$°C) as the reaction medium, the deactivation was not observed while high enantioselectivity was maintained. In contrast, none of the continuous catalytic hydrogenation studies in $scCO_2$ (9-13) report catalyst deactivation by CO.

Recently, Hutchenson and co-workers (33), showed that exposure of platinum catalysts to equimolar amounts of CO_2 and H_2 at 60 °C and 125 bar leads to the formation of a variety of surface species such as carbonates, formates, and CO. The CO was suggested to form from the reverse water-gas shift reaction. Clearly, a reliable and fundamental understanding of CO formation and its effects on catalyst activity at the operating conditions are essential to rationally evaluate the suitability of CO2 as a reaction medium for hydrogenation reactions.

We have recently started a detailed investigation of the formation of surface species on Pd/Al_2O_3 and other noble metal conditions upon exposure to CO_2+H_2 at reaction conditions (70 °C, 138 bar in the case of cyclohexene hydrogenation on Pd/Al_2O_3 catalysts). The first results on Pd/Al_2O_3 are shown in Figure 11. A peak corresponding to CO adsorbed on Pd (1955 cm^{-1}) was observed to evolve with time. This CO peak was not present for exposure time up to 20 minutes beyond which a weak, yet definite, peak evolved with time until the end of the experiment

(6 h). Hence, it is possible that the catalyst could deactivate on exposure to CO_2+H_2 in a batch reactor where sufficiently long exposure time leads to the formation of CO that strongly adsorbs to the active site at reaction conditions. The slow kinetics of the reverse water-gas shift reaction at the low operating temperatures (70 °C) might lead to a more stable catalyst in a short residence-time (~20 s) fixed-bed reactor compared to a batch reactor where the residence times are considerably longer. These results clearly bring out the value of *in situ* FTIR studies in understanding the formation of deactivating surface species when CO_2 is used in hydrogenation reactions. Such studies can provide valuable guidance to the type of reactor operation and the residence time to be employed while using $scCO_2$ as a solvent for hydrogenation reactions.

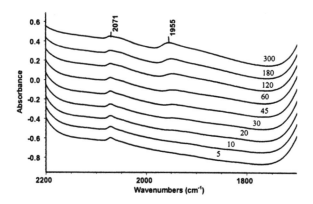

Figure 11 Temporal (5-300 min) evolution of CO (1955 cm^{-1}) upon exposure of Pd/Al$_2$O$_3$ to CO$_2$ (95%) and H$_2$ (5%) at 70 °C and 138 bar.

CONCLUSION

The hydrogenation of cyclohexene to cyclohexane over Pd/C was performed in a fixed-bed reactor employing $scCO_2$ to solubilize the reaction mixture consisting of the reactants (cyclohexene and hydrogen) and the product (cyclohexane) in a single supercritical phase surrounding the solid catalyst. The reaction was performed at a near critical temperature of 70 °C and at a pressure of 138 bar that was verified experimentally to permit operation in a single-phase. For an olefin space velocity of 20 h^{-1}, excellent temperature control around the set point (70 °C) and stable catalyst activity were demonstrated at cyclohexene conversion exceeding 80% throughout a 22 h run. The "hot spot" temperature in the bed was found to be about 16 °C when the fixed-bed reactor was operated at adiabatic conditions. Further, the reaction was found to be parametrically insensitive to temperature between 70 and 78 °C. Reducing the operating pressure led to an increase in "hot spot" temperature due to lowering of the heat capacity of the reaction medium.

This indicates that the near-critical reaction medium (possibly aided by the reactor material) has sufficient heat capacity to effectively remove the heat of reaction.

The selective hydrogenation of toluene to methylcyclohexene over Ru/Al$_2$O$_3$ was conducted in a fixed-bed reactor (60 °C, 138 bar) using scCO$_2$ as the solvent. It was shown that selectivities greater than 20% toward the cycloalkenes could be obtained without the addition of any modifiers. This increase in selectivity is attributed to the increased solubility and pore-diffusivity afforded by the supercritical reaction medium.

The important issue of possible deactivation of noble metal catalysts by CO formed from the reverse water gas shift reaction between CO$_2$ and H$_2$ was investigated using high-pressure transmission FTIR spectroscopy. It was shown that CO could be formed on Pd/Al$_2$O$_3$ when exposed to CO$_2$ (95%) and H$_2$ (5%) at the reaction condition (70 °C, 138 bar). This adsorbed CO evolved with time and was insignificant at short residence times, implying that short-residence time continuous reactors are preferred over batch reactors (with residence times > 20 min) to minimize the effects of possible catalyst deactivation by CO.

ACKNOWLEDGEMENTS

This study was supported in part by funds from EPA (R827034-01-0), NSF (CTS-9816969) and a DuPont Aid-to-Education (ATE) grant.

REFERENCES

1   RS Oakes; AA Clifford, CM Rayner. The use of supercritical fluids in synthetic organic chemistry. J Chem Soc Perkin Trans 1 9:917-941, 2001.

2   T Ikariya, Y Kayaki. Supercritical fluids as reaction media for molecular catalysis. Catal Surv Jpn 4:39-50, 2000.

3   RM Crooks, M Zhao, L Sun, V Chechik, LK Yeung. Dendrimer-Encapsulated Metal Nanoparticles: Synthesis, Characterization, and Applications to Catalysis. Acc Chem Res 34:181-190, 2001.

4   JR Hyde, P Licence, D Carter, M Poliakoff. Continuous catalytic reactions in supercritical fluids. Appl Catal A: 222:119-131, 2001.

5   B Subramaniam, CJ Lyon, V Arunajatesan. Environmentally benign multiphase catalysis with dense phase carbon dioxide. Appl Catal B: 37(4), 279-292 (2002).

6   PG Jessop, W Leitner (Eds.), Chemical synthesis using supercritical fluids. Weinheim: Wiley-VCH, 1999.

7   KW Hutchenson. Organic chemical reactions and catalysis in supercritical media. In: Y-P. Sun, Ed. Supercritical fluid technology in materials science and engineering. New York: Marcel Dekker, 2002, pp 87-187.

8   DJ Bochniak, B Subramaniam. Fischer-Tropsch synthesis in near-critical n-hexane: Pressure-tuning effects. AIChE J 44:1889-1896, 1998.

9   V Arunajatesan, B Subramaniam, KW Hutchenson, FE Herkes. Fixed-bed hydrogena-
      tion of organic compounds in supercritical carbon dioxide. Chem Eng Sci 56:1363-
      1369, 2001.
10   MG Hitzler, FR Smail, SK Ross, M Poliakoff. Selective catalytic hydrogenation of
      organic compounds in supercritical fluids as a continuous process. Org Process Res
      Dev 2:137-146, 1998.
11   L Devetta, A Giovanzana, P Canu, A Bertucco BJ Minder. Kinetic experiments and
      modeling of a three-phase catalytic hydrogenation reaction in supercritical CO_2. Catal
      Today 48:337-345, 1999.
12   R Wandeler, N Kunzle, MS Schneider, T Mallat, A Baiker. Continuous enantioselec-
      tive hydrogenation of ethyl pyruvate in "supercritical" ethane: Relation between phase
      behavior and catalytic performance. J Catal 200:377-388, 2001.
13   S van den Hark, M Härröd. Fixed-Bed Hydrogenation at Supercritical Conditions to
      Form Fatty Alcohols: The Dramatic Effects Caused by Phase Transitions in the Reac-
      tor. Ind Eng Chem Res 40:5052-5057, 2001.
14   WK Snavely. MS Thesis, University of Kansas, Lawrence, KS, 1998.
15   J Hanika, K Sporka, V Ruzicka, J Hrstka. Measurement of axial temperature profiles
      in an adiabatic trickle bed reactor. Chem Eng J 12:193-197, 1976.
16   J Hanika, V Stanek. Application of fixed-bed reactors to liquid-phase hydrogenation.
      In: L Cerveny, ed. Caltaltyic Hydrogenation. New York: Elsevier, 1986, pp 547-577.
17   EE Gonzo, M Boudart. Catalytic hydrogenation of cyclohexene 3. Gas-phase and liq-
      uid-phase reaction on supported palladium. J Catal 52:462-471, 1978.
18   FA Hessari, SK Bhatia. Reaction rate hysterisis in a single partially internally wetted
      catalyst pellet: Experiment and modelling. Chem Eng Sci 51:1241-1256, 1996.
19   NB Vargaftik, YK Vinogradov, VS Yargin Handbook of physical properties of liquids
      and gases – pure substances and mixtures. 3rd ed. New York: Begell House 1996.
20   MC Clark, B Subramaniam. 1-Hexene isomerization on a Pt/γ-Al_2O_3 catalyst: The
      dramatic effects of feed peroxides on catalyst activity. Chem Eng Sci 51:2369-2377,
      1996.
21   RL Augustine. Heterogeneous catalysis for the synthetic chemist. New York: Marcel-
      Dekker, 1996, pp 93-95.
22   JB Butt. Reaction kinetics and reactor design. 2nd ed. New York: Marcel Dekker, 2000,
      pp 477-480.
23   MC Shoenmaker-Stolk, JW Verwijs, JA Don, JJF Scholten. The catalytic hydrogena-
      tion of benzene over supported metal catalysts. I. Gas-phase hydrogenation of benzene
      over ruthenium-on-silica. Appl Catal 29: 73-90, 1987.
24   J Struijk, JJF Scholten. Selectivity to cyclohexenes in the liquid phase hydrogenation
      of benzene and toluene over ruthenium catalysts, as influenced by reaction modifiers.
      Appl Catal A 82:277-287, 1992.
25   T-K Ranatakyla, S Toppinen, T Salmi, J Aittamaa. Investigation of the hydrogenation
      of some substituted alkylbenzenes in a laboratory scale trickle-bed reactor. J Chem
      Tech Biotechnol 67:265-275, 1997.
26   Z Belholav, P Kluson, L Cerveny. Partial hydrogenation of toluene over a ruthenium
      catalyst – a model treatment of a deactivation process. Res Chem Intermed 23:161-168
      1997.
27   J Wang, L Huang, Q Li. Influence of different diluents in Pt/Al_2O_3 catalyst on the
      hydrogenation of benzene, toluene, and o-xylene. Appl Catal A 175:191-199, 1998.

28   AGA Ali, LI Ali, SM Aboul-Fotouh, AK Aboul-Gheit. Hydrogenation of aromatics on modified platinum-alumina catalysts. Appl Catal A 170:285-296, 1998.

29   H Takagi, T Isoda, K Kusakabe, S Morooka. Effects of solvents on the hydrogenation of mono-aromatic compounds using noble-metal catalysts. Energy Fuels 13:1191-1196, 1999.

30   MA Keane, PM Patterson. The role of hydrogen partial pressure in the gas-phase hydrogenation of aromatics over supported nickel. Ind Eng Chem Res 38:1295-1305, 1999.

31   P Kluson, J Had Z Belholav L Cerveny. Selective hydrogenation over ruthenium catalysts prepared by sol-gel method. Appl Catal A 149:331-339, 1997.

32   B Minder, T Mallat, KH Pickel, K Steiner, A Baiker. Enantioselective hydrogenation of ethyl pyruvate in supercritical fluids. Catal Lett 34:1-9, 1995.

33   KW Hutchenson, FE Herkes, DJ Walls, TK Das, JF Brennecke. Potential catalyst poisoning during hydrogenation reactions in $scCO_2$. Am Chem Soc National Meeting, San Diego, CA, 2001.

36

Direct Reduction of C=O and C=N Bonds by Polymethylhydrosiloxane Promoted by Zn-Diamine Catalysts in Alcohol Solvents

Virginie Bette, André Mortreux, and Jean-François Carpentier[*]
University of Lille, Villeneuve d'Ascq, France

Polymethylhydrosiloxane (PMHS), a safe and inexpensive polymer co-product of the silicon industry, is an efficient alternative reducing agent for C=O and C=N bonds when associated with catalysts (1). Mimoun *et al.* recently reported a new system based on zinc hydride catalysts which enables the chemoselective reduction of unfunctionalized and α,β-unsaturated- aldehydes, ketones and esters (2). Because gummy silicon residues, which are usually associated with silane reductions, do not form, this PMHS system is attractive for synthetic / industrial purposes. Nevertheless, in contrast to tin-catalyzed reductions of ketones with PMHS (1i), this system cannot be operated in protic solvents due to a dehydrogenative silylation of the solvent by PMHS; therefore, the recovered product is a silyl ether which must be subjected to a separate hydrolysis step (2). We report a diamine-modified zinc catalyst system that overcomes these limitations and broadens the scope of this process. The present system proceeds in alcoholic solvent and allows fast, chemoselective reduction of a large range of functionalized ketones and imines to the corresponding alcohols and amines in a one-step procedure.

---

[*] *Current affiliation*: University of Rennes, Rennes, France. Jean-francois.carpentier@univ-rennes1.fr.

Scheme 1 Carbonyl Reduction vs. Oxidative Coupling with ROH

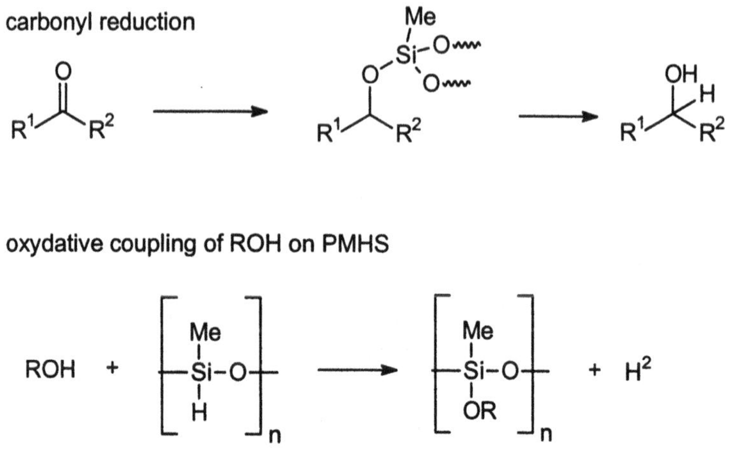

carbonyl reduction

oxydative coupling of ROH on PMHS

Mimoun *et al.* already described the use of diamines with zinc species in a chiral version of this system to reduce arylalkylketones (3). These experiments were carried out in toluene because of the anticipated side reaction between PMHS and protic solvent (Scheme 1).

However, we observed that the reduction of C=O and C=N bonds proceeds efficiently in an alcoholic solvent using a catalysts system of diethyl zinc (1.1M in toluene) and one equiv. of N,N'-dibenzylethylenediamine (dbea). The choice of dbea was motivated by its commercial availability and good performance of its chiral derivative, N,N'-ethylenebis(1-phenylethylamine) (ebpe), in the reduction of acetophenone (3). In the ZnEt$_2$/dbea system, ketones are chemoselectively reduced at room temperature in methanol solvent to give the corresponding alcohols in moderate (ca. 50%) to quantitative (>99%) yields (Scheme 2, Table 1). The reaction tolerates a variety of functional groups. In particular, this system selectively reduces aldehydes and ketones in the presence of esters, enabling the valuable conversion of α- and β-ketoesters (**1d-g**) to α- and β-hydroxyesters. Similarly, α-ketoamides (**1h-j**) are reduced into α-hydroxyamides which are readily recovered in >96% yield and high purity from the final reaction mixture, due to their very low solubility in hydrocarbon solvent in contrast to the silicon residues. For the ketones investigated, the reduction in methanol proceeds significantly faster and more selectively than using the two-step procedure in toluene. For instance, the reduction of acetophenone (**1a**) goes to completion within 1 h in methanol while only ca. 8% conversion is observed in toluene over the same reaction time. Also, the conversion of α-chloroacetophenone (**1b**) in toluene reaches a maximum of

Scheme 2 One-Step Reduction of Ketones with Zn-Diamine-PMHS-MeOH system.

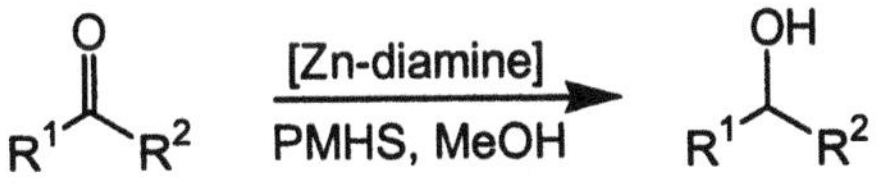

$$
\underset{\textbf{1a-i}}{R^1 \overset{\displaystyle O}{\underset{}{\bigparallel}} R^2}
\xrightarrow[\text{PMHS, MeOH}]{\text{[Zn-diamine]}}
\underset{\textbf{2a-i}}{R^1 \overset{\displaystyle OH}{\underset{}{\diagup}} R^2}
$$

a	$R^1 = Ph$	$R^2 = Me$
b		$R^2 = CH_2Cl$
c		$R^2 = CF_3$
d		$R^2 = CO_2Me$
e	$R^1 = Me$	$R^2 = CO_2Me$
f	$R^1 = Ph$	$R^2 = CH_2CO_2Me$
g	$R^1 = Me$	$R^2 = CH_2CO_2Me$
h	$R^1 = Ph$	$R^2 = CO_2NHBn$
i	$R^1 = Me$	$R^2 = CO_2NHPh$

Table 1 Zn-Catalyzed Reduction of Ketones and Imines with PMHS in Methanol[a]

entry	substrate	PMHS (equiv.)	time[b] (h)	Conversion[c] (mol %)
1	1a	2	1	>99
2	1b	2	1	>99
3	1c	2	1	>99
4	1d	2	1	>99 (67)[d]
5	1e	2	1	>99
6	1f	2	24	50
7	"	5	24	76
8	1g	2	6	68
9	"	5	1	>99
10	1h	2	1	>99 (98)[d]
11	1i	10	1	>99 (96)[d]
12	3a	2[e]	18	60[e]
13	3b	2	18	>99

[a] General conditions: **1** or **3** / ZnEt$_2$ / dbea = 2.75 : 0.055 : 0.055 mmol in MeOH / toluene (80:20 v/v, 2.5 mL), $T = 20$ °C.
[b] Reaction time was not optimized.
[c] Conversion of **1** and **3** into **2** and **4**, respectively. No side product was formed as determined by quantitative GLC and ^{1}H NMR.
[d] Isolated yields of spectroscopically pure products
[d] Similar results were obtained on using 4 equiv. of PMHS

43% after 24 h, but is quantitative within 1 h in methanol. Finally, when carried out in toluene, reduction using the Zn-diamine catalyst is not selective for methyl pyruvate (**1e**) and does not proceed at all with methyl benzoylacetate (**1f**). However, these substrates are rapidly and chemoselectively reduced with the ZnEt$_2$/dbea/methanol catalyst system.

While only 1.0 equiv. of PMHS is needed to complete the reduction of some ketones; e. g. α,α,α-trifluoroacetophenone (**1c**) and methyl phenylglyoxylate (**1d**), excess PMHS is necessary in most cases. As shown in Table 1, incomplete conversions are mostly observed for the reduction of relatively acidic substrates; i. e. β-ketoesters **1f** and **1g** and β-ketoamides **1h** and **1i** (pKa = 10–13). Therefore, a likely hypothesis is that the [Zn-diamine]/PMHS system is active not only for the reductive reaction of the carbonyl function, but also for the oxidative silylation of any enolisable group. Thus, the enol-silyl ether produced would hydrolyze back in methanol to the free enol, accounting for the consumption of extra equiv. of PMHS. Nevertheless, this hypothesis does not account for the reduction of imine **3a**, since no improvement in the conversion is noted on doubling the amount of PMHS (Scheme 3, Table 1). Other imines; e. g. **3b**, are readily reducible with the present Zn-diamine-methanol system.

Methanol is the most suitable protic solvent for this system. In ethanol, the reduction of the ethyl analogues of **1d-g** are less selective (4) and rates are slower. Though experiments were routinely carried out using 2.0 mol% of catalyst, the reduction of ketoester **1d** proceeds quantitatively (in 75 min) with 0.5 mol% of [Zn-diamine] catalyst.

Current efforts are directed towards (i) mechanistic issues to rationalize the effectiveness of the diamine-modified system in protic solvent in comparison to Zn-hydride catalyst, and (ii) the development of an efficient enantioselective version of this system with chiral diamines.

Scheme 3 One-Step Reduction of Imines with Zn-Diamine-PMHS-MeOH system.

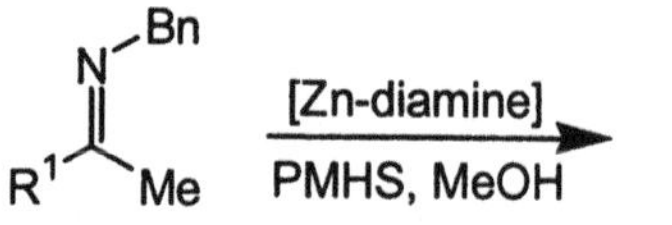

a   R^1 = Ph
b   R^1 = Cyclohexyle

EXPERIMENTAL SECTION

In a typical experiment (Table, entry 10), to a solution of dbea (13.2 mg, 0.055 mmol) in freshly distilled toluene (0.5 mL) under nitrogen, were successively added ZnEt$_2$ (50 μL of a 1.1 M solution in toluene, 0.055 mmol), a solution of **1h** (657 mg, 2.75 mmol) in MeOH (2.0 mL), and finally PMHS (0.32 mL, 5.0 mmol). The resulting solution was stirred with a magnetic stir bar and the reaction was monitored by GLC. After completion of the reaction, volatiles were removed under vacuum to give a white oil, which was triturated with pentane (2.0 mL). The resulting precipitate was separated off from the liquid phase, washed with a minimal amount of pentane, and dried under vacuum to give the expected hydroxyamide **2h** as a white powder (650 mg, 98 %).

ACKNOWLEDGMENT

This research was funded by the CNRS and PPG-SIPSY (grant to VB).

REFERENCES

1. (a) For a review on PMHS, see: NJ Lawrence, MD Drew, SM Bushell. J Chem Soc Perkin Trans 1 3381–3391, 1999. (b) S Nitzche, M Wick. Angew Chem 69:96, 1957. (c) M.T Reding, SL Buchwald. J Org Chem 60:7884–7890, 1995. (d) Y Kobayshi, E Takahisa, M Nakano, K Watatani. Tetrahedron 53:1627–1634, 1997. (e) MD Drew, NJ Lawrence, D Fontaine, L Sehkri. Synlett 989–991, 1997. (f) MD Drew, NJ Lawrence, W Watson, SA Bowles. Tetrahedron Lett 38:5857–5860, 1997. (g) X Verdaguer, MC Hansen, SC Berk, SL Buchwald, J Org Chem 62:8522–8528, 1997. (h) J Yun, SL Buchwald, J Am Chem Soc 121:5640–5644, 1999. (i) NJ Lawrence, SM Bushell, Tetrahedron Lett 41:4507–4512, 2000. (j) MC Hansen, SL Buchwald, Org Lett 2:713–715, 2000.
2. H Mimoun, J.Org Chem 64:2582–2589, 1999.
3. H Mimoun, JY de Saint Laumer, L Giannini, R Scopelliti, C Floriani, J Am Chem Soc 121:6158–6166, 1999.
4. Reduction of methyl ketoesters in ethanol and vice-versa afforded mixtures of methyl and ethyl hydroxyesters due to relatively slow transesterification.

37

Liquid-Phase Hydroxymethylation of 2-Methoxyphenol: Effect of Reaction Parameters on the Performance of Solid Acid Catalysts

F. Cavani[*] and R. Mezzogori
Dipartimento di Chimica Industriale e dei Materiali, Università di Bologna, Bologna, Italy.

The liquid-phase hydroxyalkylation of 2-methoxyphenol with aqueous solutions of formaldehyde (formalin), catalyzed by solid acid materials, was investigated. Zeolitic catalysts, especially H-mordenites, yield the best results in terms of reactant conversion and selectivity to 3-methoxy-4-hydroxybenzyl alcohol (p-vanillol). Important reaction parameters are the formaldehyde-to-guaiacol ratio, the stirring rate and the reaction temperature. The reaction scheme consists of parallel reactions for the formation of vanillic alcohol isomers and of monoarylic by-products, obtained by reaction between vanillols and methanol or hemiformal contained in formalin. The consecutive reactions mainly involve the transformation of p-vanillol to monoaryl and to diaryl compounds, the latter obtained by condensation of two vanillol molecules.

INTRODUCTION

The hydroxyalkylation of activated arenes (containing functional groups such as the hydroxy or methoxy groups) [1-4] with aldehydes and ketones is a reaction of

---

[*] *E-mail*: cavani@ms.fci.unibo.it

interest for the production of drugs, polymers, and food additives. For instance, the hydroxymethylation of 2-methoxyphenol (guaiacol) represents one-step in the multistep synthesis of the 3-methoxy-4-hydroxybenzaldehyde (vanillin), an environmentally friendly process for the production of this important food additive [5,6].

Hydroxyalkylations are catalyzed by Lewis-type acids, like $AlCl_3$, and by mineral Brönsted acids. Some papers and patents have appeared in recent years, where zeolitic materials are described as catalysts for this reaction [1-4,7]. Solid acid materials are highly desirable catalysts, since the environmental impact of the process benefits from easier separation of the catalyst, the absence of liquid wastes containing inorganic salts, and less severe corrosion problems.

Usually, the condensation between arenes and aldehydes is carried out in the liquid phase, and large pore zeolites are necessary in order to make the reaction occur at an acceptable rate in the condensed phase. When formaldehyde is the reactant, one main problem is the presence of water, since the aqueous solution of formaldehyde (formalin) is the simplest, cheapest and most widely available reactant for use on an industrial scale. This implies the need for hydrophobic zeolites, in order to avoid preferential filling of the pores by more polar water molecules rather than by the aromatic substrate [8,9]. The objective of the work reported here was to compare the performance of different solid acids, and to study the effect of operating parameters on catalytic performance in the hydroxymethylation of guaiacol with formaldehyde catalyzed by a commercial dealuminated, and thus more hydrophobic, H-mordenite zeolite.

EXPERIMENTAL

Catalytic tests were carried out in a glass, batch reactor, loading 48 ml of a commercial aqueous solution of formaldehyde, and 1 g of solid catalyst. The mixture was then heated at 80°C, and 4 ml of guaiacol was added. The reaction mixture was left at 80°C for 2 h under stirring (650 rpm). The commercial aqueous solution of formaldehyde (formalin) contains 37 wt.% formaldehyde and 10-15% methanol. Most of the tests were carried out using a commercial H-mordenite catalyst (HM-45), supplied by Engelhard. Other zeolites used were supplied by Tosoh, Zeolyst and Engelhard.

The products were analyzed by HPLC (TSP Spectra Series), equipped with a Alltech Hypersil ODS column, and with a UV-Vis TSP UV150 detector (λ 280 nm). Elution was done with a mixture of acetonitrile and water. The products were identified using GC-MS and comparison with the retention times of standard components (when available).

RESULTS AND DISCUSSION

Table 1 compares the catalytic performance of some solid acids in the hydroxymethylation of guaiacol. The following products were identified: (**a**) vanillic alcohols VAs: o-VA (2-hydroxy-3-methoxybenzyl alcohol), m-VA (3-hydroxy-4-methoxybenzyl alcohol, while the amount of 3-hydroxy-2-methoxybenzyl alcohol was negligble), and p-VA (3-methoxy-4-hydroxybenzyl alcohol); (**b**) monoaryl products other than vanillols, which are constituted of the products of etherification of vanillols by methanol or hemiformal (both present in formalin); (**c**) diaryl compounds of MW 260, which corresponds to the bisarylmethane obtained by condensation of one molecule of vanillol with one molecule of guaiacol, or by condensation of two molecules of vanillols and elimination of formaldehyde, and of MW 290, obtained by condensation of two molecules of vanillols. Traces of triaryl compounds (MW 396) were also found. The formation of high-molecular-weight phenolic resins was not observed under our reaction conditions.

Table 1 Comparison of the catalytic performance achieved with different solid acid catalysts. Temperature 80°C, reaction time 2 h, initial formaldehyde/guaiacol ratio 18/1.

Sample	Si/Al at. ratio	Conv %	Sel. to VA isomers %				Sel. %	
			o-VA	m-VA	p-VA	VAs	By-prod.	
Nafion	-	28	9	5	14	28	72	
H-ZSM5	30	10	7	4	35	46	64	
CP811E-150 Zeolyst (H-β)	75	30	12	8	27	47	53	
CP811C-300 Zeolyst (H-β)	150	21	11	7	26	44	56	
HM-45 Engelhard (H-mordenite)	23	41	15	5	50	70	30	
HSZ-690HOA Tosoh (H-mordenite)	100	54	18	5	34	57	43	
H-USY Engelhard	6	14	1	1	5	7	93	

The very low conversion achieved with H-Y is likely due to the high hydrophilicity of the zeolite [10], which in the presence of a large excess of water makes the adsorption and activation of organic molecules more difficult. The H-mordenites and H-β zeolites possess comparable activity, higher than that of the H-ZSM5. This is due to the fact that the small/medium-pore zeolites are not suitable for acid-catalyzed reactions occurring in the liquid phase and involving substituted aromatic compounds.

In regard to the distribution of the products, zeolitic catalysts yield a considerable fraction of p-VA, while this does not occur in the case of Nafion. This indicates an effect of shape-selectivity which favours the formation of the para isomer, which is less bulky than the ortho isomer (o-VA). The para (p-VA) and the meta (m-VA) isomers have similar bulkiness, but the formation of the meta isomer is less favoured due to the electronic effects of the hydroxy and methoxy substituents. Nafion yields the highest amount of by-products, as does the H-Y zeolite (even though in the latter case the low conversion makes comparisons with other zeolites less reliable).

Tests were then carried out using the commercial mordenite HM-45, supplied by Engelhard and having a Si-to-Al ratio equal to 23. This catalyst gives the best yield of p-VA. Preliminary tests were carried out to check the effect of stirring speed on catalytic performance, in order to exclude the presence of mass transfer effects which might limit the rate of reaction. The results are reported in Figure 1. The guaiacol conversion is only slightly affected by the stirring rate, showing a small decrease with increasing stirring rate, especially for rates higher than 650 rpm. The transfer of guaiacol from the bulk to the catalyst surface therefore is never the rate-limiting step. Correspondingly, the distribution of the products also changes slightly, as a consequence of the variation in conversion. However, when the stirring rate is lower than 240 rpm, the selectivity to p-VA is very low. Under these conditions, the main by-product is obtained by consecutive reaction of p-VA with methanol, with formation of the corresponding ether. This means that under conditions of low stirring rate the back-diffusion of p-VA (from the catalyst surface into the liquid phase) is slow, and the compound may further react at the external catalyst surface (i.e., in the boundary layer).

The reaction scheme was studied by carrying out tests at increasing reaction times, as shown in Figure 2. The guaiacol conversion increases as reaction time increases. The reaction scheme consists of parallel reactions for the formation of the three vanillol isomers and of by-products. The latter are mainly constituted of monoaryl products obtained by reaction between vanillols and methanol or hemiformal. These are consecutive reactions from a chemical point of view, but are parallel ones from a kinetic point of view, since they occur extensively even at low guaiacol conversion; therefore they likely occur in the zeolite pores before vanillols diffuse back into the liquid phase. Consecutive reactions mainly involve p-VA, with formation of by-products which consist of diaryl and of monoaryl compounds.

The higher reactivity of p-VA with respect to the other isomers derives from the higher stability of the corresponding $Ar\text{-}CH_2^+$ species formed by protonation and dehydration of the hydroxymethyl group [11] (see Scheme 1). The cation is effectively stabilized by delocalization of the charge in those positions which are more influenced by the mesomeric effect of the hydroxy group. The same occurs for the cation which originates from o-VA, but in this case the electron-attracting inductive effects of the hydroxy and methoxy groups make this cation less stable. Moreover, the energy barrier for the rate-determining step in p-VA transformation (the formation of the Ar-CH$_2^+$ species in a S$_N$1-type mechanism) is also decreased by the further stabilization of the cation by polar molecules (the reaction is carried out with water as the solvent). Less stable carbocations may prefer to react via a S$_N$2-type mechanism, and in this case the effect of water on the cation stability is less important [12].

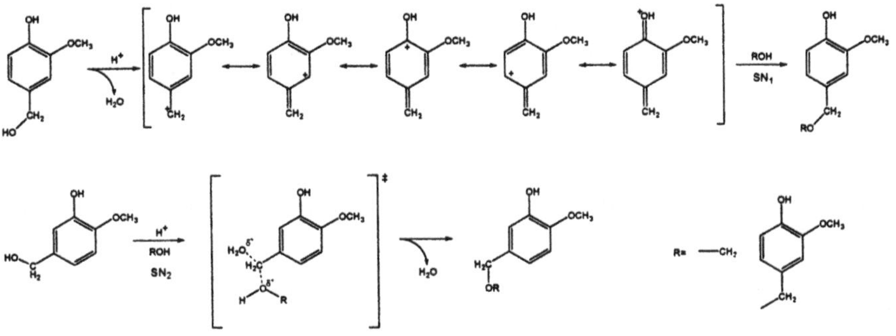

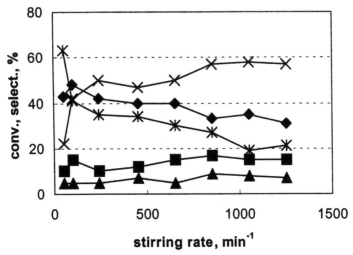

Scheme 1 Mechanism of transformation of vanillols to by-products

Figure 1 The effect of stirring rate on guaiacol conversion (♦), and on selectivity to p-VA (✕), o-VA (■), m-VA (▲) and by-products (✻). Temperature 80°C, reaction time 2 h, initial reactants ratio: formaldehyde/guaiacol 18/1.

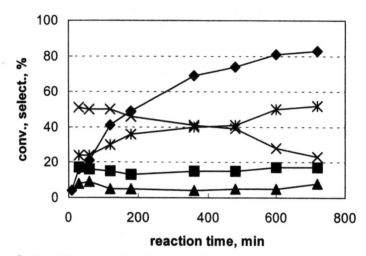

Figure 2 The effect of reaction time on catalytic performance. Symbols and reaction conditions as in Figure 1; stirring rate 650 rpm.

The effect of the reaction temperature is shown in Figure 3. The conversion of guaiacol increases with increasing temperature, reaching 41% at 80°C. The activation energy was calculated to be 72±2 kJ/mole, which confirms the absence of diffusional limitations for the rate-limiting step. The selectivity to vanillols decreases with increasing temperature. Higher temperatures favour the consecutive reactions of vanillol transformation to diaryl compounds [2].

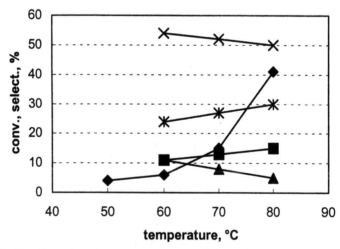

Figure 3 The effect of reaction temperature on catalytic performance. Symbols and reaction conditions as in Figure 1; stirring rate 650 rpm.

One of the most important parameters is the initial molar ratio between formaldehyde and guaiacol. Figure 4 shows that in the range of formaldehyde-to-guaiacol ratios between 1 (the stoichiometric requirement) and 35, the conversion increases with increasing values of this parameter, thus for a decrease in guaiacol concentration. It is thus confirmed that the reaction rate is approximately proportional to guaiacol concentration, and that this compound participates in the rate-determining step. The need for such a large excess of formaldehyde with respect to the stoichiometric requirement can be explained by considering that in the formalin solution the amount of monomeric formaldehyde (which is indeed present mainly as methylene glycol) is relatively low, due to the formation of oligomers, hemiformal and formal by reaction between methylene glycol and methanol itself. Another reason for the need for a large formaldehyde excess is the presence of water and methanol, which are nucleophilic reactants, and are protonated by the active sites. This may also cause a sort of levelling of the catalyst acidity, due to the fact that the true active sites are hydroxonium species [13].

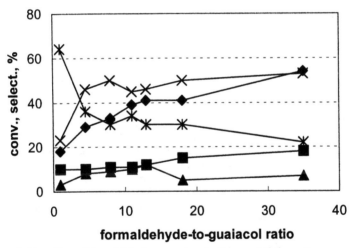

Figure 4 The effect of formaldehyde-to-guaiacol ratio on catalytic performance. Symbols and reaction conditions as in Figure 1; stirring rate 650 rpm. The initial composition was varied by decreasing the amount of guaiacol added to the reactant solution, while keeping the formaldehyde concentration and total volume constant.

The distribution of the products is also affected considerably by the formaldehyde-to-guaiacol ratio when the latter is lower than 5-8. Under these conditions the amount of by-products is very high, despite the low guaiacol conversion, mainly because of the formation of diaryl compounds. In our experimental range, an increase in the formaldehyde-to-guaiacol ratio leads to an increase in the selectivity to vanillols, even though an increase in conversion should result in the opposite effect, due to the greater contribution of consecutive reactions. Therefore, a large excess of formaldehyde is necessary to kinetically favour the electrophilic attack of formaldehyde on guaiacol, rather than the reactions between aromatic compounds to yield diaryl by-products. For formaldehyde-to-guaiacol ratios higher than 8-10, the effect on the distribution of products is less pronounced.

CONCLUSIONS

The reactivity of different solid acids in the hydroxymethylation of guaiacol with formaldehyde was studied. Mordenites and β zeolites gave the best results, as also suggested in the patent literature [7]. The best yields to vanillols and specifically to p-VA were obtained using a large excess of formaldehyde with respect to the stoichiometric requirement, and by maintaining the reaction temperature around 80°C. Reactions of transformation of p-VA to monoaryl and diaryl compounds were responsible for the decrease in selectivity to p-VA at both low and high guaiacol conversion. Specifically, parallel reactions involving the formation of monoaryl by-products limited the initial selectivity to vanillols.

REFERENCES

1. A. Corma, H. Garcia, J. Chem. Soc., Dalton Trans., (2000) 1381.
2. C. Moreau, F. Fajula, A. Finiels, S. Razigade, L. Gilbert, R. Jacquot, M. Spagnol, in "Catalysis of Organic Reactions", F.A. Herkes (Ed.), M. Dekker, 1998, p. 51.
3. M.J. Climent, A. Corma, H. Garcia, J. Primo, J. Catal., 130 (1991) 138.
4. M.H.W. Burgers, H. van Bekkum, Stud. Surf. Sci. Catal., 78 (1993) 647.
5. R.A. Sheldon, H. van Bekkum, in "Fine Chemicals through Heterogeneous Catalysis", R.A. Sheldon and H. van Bekkum (Eds.), Wiley-VCH, 2001, p. 1.
6. P. Metivier, in "Fine Chemicals through Heterogeneous Catalysis", R.A. Sheldon and H. van Bekkum (Eds.), Wiley-VCH, 2001, p. 173.
7. C. Moreau, S. Razigade-Trousselier, A. Finiels, F. Fazula, L. Gilbert, WO patent 96/37452 (1996), assigned to Rhone-Poulenc Chimie.
8. A. Finiels, P. Geneste, J. Lecomte, F. Marichez, C. Moreau, P. Moreau, J. Molec. Catal., Chemical, 148 (1999) 165.
9. J. Lecomte, A. Finiels, P. Geneste, C. Moreau, J. Molec. Catal., Chemical, 140 (1999) 157.
10. C.H. Berke, A. Kiss, P. Kleinschmit, J. Weitkamp, Chem. Ing. Tech., 63 (1991) 623.
11. N. Barthel, A. Finiels, C. Moreau, R. Jacquot, M. Spagnol, Topics Catal., 13 (2000) 269.
12. T.H. Lowry, K.S. Richardson, "Mechanism and Theory in Organic Chemistry", Harper Collins Publ.
13. C. Moreau, R. Durand, P. Geneste, S. Mseddi, J. Molec. Catal., Chemical, 112 (1996) 133.

38

Molecular Imprinting and Sol-Gel Encapsulated Rh Catalysts for Styrene Hydroformylation

Bei Chen and Steven S. C. Chuang[*]
The University of Akron, Akron, Ohio, U.S.A.

INTRODUCTION

Effective control of chemo and steric selectivity of a chemical reaction lies in our ability to manipulate nano-environment of active sites. Due to difficulty in manipulating the active site on the nanoscale and lack of fundamental nderstanding of the reaction mechanism, development of chemo- and stereo-selective catalysts has relied heavily upon empirical studies. One successful xample of fine-tuning steric environment of the active site is the use of chiral diphosphite ligands to control the selectivity of styrene hydroformylation on Rh omplex catalysts (1-3).

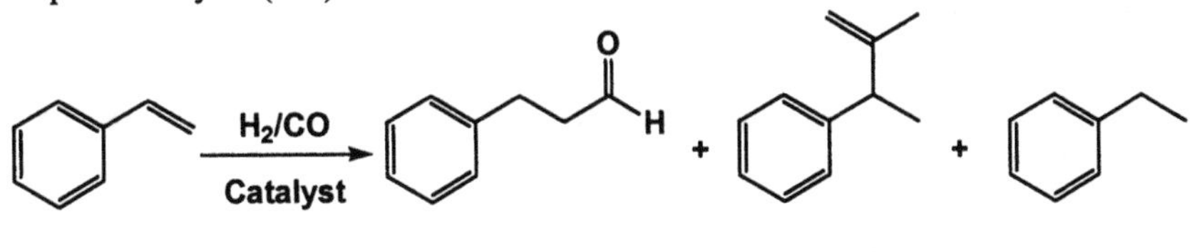

| Styrene | | 2-phenylpropanal | 3-phenylpropanal | Ethylbenzene |

These homogeneous metal complex-catalyzed reaction processes have several disadvantages: a low reaction rate, required use of high pressure, and involvement of an energy intensive step for catalyst recovery. Immobilization of

---

[*] *Corresponding author.*

the homogeneous metal complex into a carrier matrix may combine the benefits of both homogeneous and heterogeneous processes. Attempts to immobilize the metal complex into a zeolite cage and onto an oxide/polymer support have not been very successful due to low stability of these immobilized catalysts (4-7).

Hydroformylation reactions that convert alkenes to aldehydes involve a CO insertion step. The findings of the CO insertion activity of Rh^0 and Rh^+ sites on SiO_2 have led to exploitation of oxide supported metals and metal sulfides for hydroformylation (8-12). While the heterogeneously supported metals and metal sulfides exhibit activity for hydroformylation, their selectivities for aldehyde formation are generally lower than their homogeneous counterparts. Thus, immobilization of homogenous catalysts combined with controlling the nanoenvironment of the active site may promise a catalyst with high selectivity, high activity, and high durability for hydroformylation reactions.

The present study aims at investigating the effect of the nano-environment of (i) sol-gel encapsulated Rh catalyst, and (ii) imprinted polymer-supported Rh catalysts on styrene hydroformylation. The sol-gel encapsulation has been demonstrated to be an effective approach for immobilization of Rh complex for allylbenzene isomerization and dihydroarenes disproportionation (13-16). This approach has the advantage of keeping the Rh complex from leaching out of the sol-gel matrix. The results of this study show that sol-gel encapsulation is also effective in retaining the Rh complex for styrene hydroformylation. Molecular imprinting is a rational approach to constructing a receptor site for which its structure is complementary to that of the template (17-24). This approach was used to manipulate the nano-environment of Rh sites to control chemo selectivity of styrene hydroformylation. In situ infrared spectroscopy was used to determine the nature of adsorbed species on both catalysts. Results of in situ infrared observation are discussed in conjunction with the rate and selectivity data to shed light into the nano-environment of sol-gel encapsulated and polymer imprinted Rh catalysts.

EXPERIMENTAL

Chemicals. All chemicals were purchased from Aldrich and used as received without further treatment.

Preparation of sol-gel encapsulated Rh catalyst (i.e. 1.9 wt% Rh/ sol-gel catalyst). Preparation of sol-gel encapsulated Rh catalyst involves addition of 2.5 ml (0.0169 mol) $Si(OCH_3)_4$ into a completely miscible aqueous solution of 20 mg (7.59×10^{-2} mmol) $RhCl_3 \cdot 3H_2O$ with equal molar $[(C_8H_{17})_3N^+Me]Cl^-$ (Aliquat 336) in 2.4 ml (0.133 mol) H_2O, and 3.5 ml (0.0864 mol) methanol. Gelation was achieved by adding 4 ml 1 wt% aqueous NH_4OH as a catalyst. The gel was obtained by raising the temperature up to 318 K until constant weight was achieved. The sol-gel sample was ground into the granular form, which was then washed by CH_2Cl_2 and boiling water, and finally dried in a vacuum to obtain the sol-gel encapsulated Rh catalyst. Infrared (IR) analysis of the Rh/sol-gel catalyst exhibits C-H stretching vibration, C-H bending, δ_{as} scissoring and H_2O at 2847-2919, 1455, 1438 and 1647-1982 cm^{-1}, respectively, indicating that ion pairs (i.e. $[(C_8H_{17})_3NMe]^+[RhCl_3 \cdot nH_2O]$.) have been trapped inside the silica matrix. The Rh loading, based on the weight of the resulting SiO_2 matrix, is 1.9 wt%.

Preparation of Rh-molecular imprinted catalyst(i.e. 4.5 wt% Rh/imprinted polymer catalyst).   Preparation of 2-phenylpropanal-imprinted polymer started with a solution containing 1 ml of methylacrylic acid (MAA) as the functional monomer, 1 ml of 2-phenylpropanal as the template, and 5 ml of pentane. The solution allows the formation of the template-function monomer assembly in the pentane solution. Polymerization of the template-monomer assembly with 2 ml of ethylene glycol dimethacrylate EGDMA was carried out at 318 K for 12 h with 0.2 g 2,2'-Azobis(2-methylpropionitrile (AIBN) as an initiator. The resulting bulk imprinted polymer was ground into fine particles ranging from 50 to 85 µm. Following the removal of the template from the polymer particles by washing with methanol/acetic acid (v/v=4/1) and methanol, the imprinted polymer was impregnated with $RhCl_3$ to give a loading of 4.5 wt%. The Rh/blank polymer catalyst was prepared by the same procedure without use of a template.

Competitive adsorption.   Competitive adsorption was determined by measuring the difference in the concentration of 2-phenylpropanal (2-PPA) and 3-phenylpropanal (3-PPA) before and after 24 h of their adsorption on the imprinted and blank polymers at 298 K. Each measurement employed 0.1 g of polymer particles with 7 ml of the solution containing 1ml 2-phenylpropanal and 1 ml 3-phenylpropanal-pentane solution. The concentration of 2-/3-phenylpropanol was determined by a GC-17A gas chromatograph (Shimadzu Corp.) equipped with a capillary column Rtx-5 (15 m x 0.05 m). The selectivity of the polymers for 2-phenylpropanal was determined by:

Selectivity % = (adsorbed 2-PPA)/(adsorbed 2-PPA+ absorbed 3-PPA) x 100 %

Hydroformylation reaction.   Styrene hydroformylation was carried out in an *in situ* high pressure infrared (IR) cell at a total pressure of 0.83 MPa. 0.1 g catalyst was pressed into a self-support disk and placed inside the IR cell. The reactor was heated up to the desired temperature and 0.2 ml styrene was injected slowly into the cell in the CO and H_2 flow [CO/H_2 (v/v) =1/1, total flow rate = 30 ml/min] to give a reactants' molar ratio of CO:H2:styrene = 1:1:1148. The reactor was pressurized to 0.83 MPa and the reaction was carried out in a batch mode. The IR spectra were collected during the entire course of reaction studies by a Nicolet MAGNA 550 Series II Fourier transform infrared spectrometer equipped with a DTGS (Deuterated Tri-Glycine Sulfate) detector. The resolution of the IR spectrometer is set at 4 cm^{-1} with a range of 400 to 4800 cm^{-1}. The products of the reaction were flushed out and analyzed by a GC-17A gas chromatograph (Shimadzu Corp.) equipped with a capillary column Rtx-5 (15 m x 0.05 m). The selectivity for 2-phenylpropanal was obtained according to the follows equation:

$$2\text{-PPA Selectivity } \% = 2\text{-PPA/Conversion of styrene x } 100 \%$$

RESULTS AND DISCUSSION

The surface of 1.9 wt% Rh/sol-gel and 4.5 wt% Rh/polymer catalysts were characterized by CO adsorption. Figure 1 shows the IR spectrum of CO adsorption at 0.1 MPa and 298 K as well as the styrene hydroformylation temperature program reaction (TPR) over the Rh/sol-gel catalyst. Adsorption of CO on the catalyst produced a gem-carbonyl band at 2086 and 2015 cm^{-1}, indicating that highly dispersed oxide rhodium sites were formed on the catalyst surface. Styrene was brought to contact with adsorbed CO on the catalyst surface by CO and H$_2$ flow.

Appearance of styrene's bands at 1627 cm-1, 1599 cm-1, and the C=C aromatic ring stretching 1945-1686 cm-1 is an indication that the catalyst has been exposed to all of reactants. The reaction was carried out in a batch mode with a temperature ramping of 2 K/min from 298 K to 453 K. The formation of 2- and 3-phenylpropanal, as indicated by the C=O stretching vibration at 1717 cm-1, were observed at 353 K. The intensity of this band increased with temperature.

Figure 2 shows the in situ IR spectra of styrene hydroformylation over 1.9 wt% Rh/sol-gel catalyst at 373 K and 0.83 MPa. Phenylpropanals, indicated by the C=O stretching vibration at 1717 cm-1, formed after 2 min of the reaction in the batch mode. The IR intensity of the C=O band increased with the reaction time. The sol-gel encapsulated Rh catalyst exhibited high stability at all reaction conditions. No rhodium leakage has been observed for the entire reaction process. However, a higher loading Rh/sol-gel catalyst is difficult to obtain by the sol-gel encapsulation approach due to the limited solubility of the Aliquat 336, [(C$_8$H$_{17}$)$_3$N$^+$Me]Cl$^-$, which is needed to pair with RhCl$_3 \cdot$ 3H$_2$O.

Table 1 shows the results of competitive adsorption of 2-phenylpropanal and 3- phenylpropanal on both 2-phenylpropanal-imprinted and blank polymers. The blank polymer showed nearly equal adsorption capacity for both 2-phenylproanal via. 52%) while the imprinted polymer favored the template (i.e. 2-phenylproanal) adsorption (61% for the template). The difference in the adsorption selectivity between the blank and imprinted polymers indicated that a certain number of the molecular recognition cavities were obtained for 2-phenylpropanal (template) on the surface of the imprinted polymer. However, the selectivity of this imprinted polymer is lower than most of reported imprinted polymers for organic compounds with large molecular size, such as L-PheNHPh and steroids, and proteins (20-24). The lack of the selectivity for 2-phenlpropanal appears to be due to the low number of interactions and the lack of accuracy in the locations of functional group for interacting with templates.

Figure 3 shows the IR spectrum of CO adsorption at 0.1 MPa and 298 K as well as the styrene hydroformylation over 4.5wt% Rh 2-phenylpropanal-imprinted catalysts at 353 K and 0.83 MPa. Observation of gem-dicarbonyl at 2086 and 2015 cm-1 upon CO adsorption indicated that Rh sites are in the highly dispersed state. 2-phenylpropanal and 3-phenylpropanal were formed after 1 min of the reaction in the batch model as shown by the appearance of the phenylpropanals' carbonyl band at 1717 cm-1. The IR intensity of phenylpropanals increased with the reaction time. Table 2 summarizes the conversion, selectivity, and turnover frequency (TOF) of hydroformylation over 1.9 wt% Rh/sol-gel catalyst, 4.5 wt% Rh/blank polymer, and Rh/imprinted polymer. All catalysts showed excellent hydroformylation activity and selectivity toward aldehyde products (98%).

Hydrogenation reaction, which converted styrene to ethylbenzene under H_2 atmosphere is the only side reaction observed. The selectivity for ethylbenzene over 1.9 wt% Rh/sol-gel catalyst, 4.5 wt% Rh/blank polymer and Rh/imprinted polymer catalyst is 2.0%, 1.6% and 1.1%, respectively. Both 4.5 wt% Rh/blank polymer and Rh/imprinted polymer catalyst exhibited higher activity than 1.9 wt% Rh/sol-gel catalyst. The initial turnover frequency (TOF) of Rh/polymer catalysts is four times higher than that of the Rh/sol-gel catalyst (0.12 via 0.04). This difference may be attributed to the difference in chemical environment of Rh sites.

It is interesting to observe that the reaction selectivity of 1.9 wt% Rh/imprinted polymer catalyst in Table 2 is closely matched with its adsorption selectivity toward 2-phenylpropanals in Table 1. The consistency in selectivity indicated that the presence of the structurally complementary sites inside the polymer matrix allows Rh sites to catalyze the formation of 2-phenylpropanal. The selectivity of this catalyst may be further improved by the further improvement of the selectivity of the imprinted polymer and anchoring the Rh sites at the specific location.

Although 4.5 wt% Rh/imprinted polymer catalyst showed higher chemo selectivity than the 1.9 wt% Rh/sol-gel catalyst, the imprinted polymer catalyst tends to lose the activity due to leaching of Rh metal cations into the reactant mixture while the Rh/sol-gel catalyst traps the large ion pairs, $[(C_8H_{17})_3NMe]^+[RhCl_3 \cdot nH_2O]-$), inside the silica matrix. Lack of interaction between Rh+ cationic sites with polymer presents a challenge to immobilize Rh sites on the polymer surface.

CONCLUSION

1.9 wt% Rh/sol-gel and 4.5 wt% Rh/polymer catalysts are active in styrene hydroformylation reaction. The Rh/sol-gel catalyst effectively immobilizes Rh cationic sites into the silicate matrix while the Rh/imprinted polymer catalysts exhibit high chemoselectivity to the template. Properly choosing the functional monomer and the template is the key to preparing the selective imprinted polymer. Combining molecular imprinted approach with sol-gel encapsulation may lead to a selective heterogeneous catalyst which combines benefits of both homogeneous and heterogeneous catalysis.

ACKNOWLEDGEMENT

This work has been supported in part by the NSF Grant CTS 9816954 and the Ohio Board of Regents Grant R5538.

REFERENCES

1.   Sakai, N., Nozaki, K., Mashima, K., and Takaya, H. (1992) *Tetrahedron: Asymmetry* **3**, 583.
2.   Babin, J. E. and Whiteker, G. T., Patent WO 93/03830, 1992.
3.   Buisman, G. J. H., Vos, E. J, Kamer, P. C. and van Leeuwen, P. W. N. M. (1995) *J. Chem. Soc. Dalton Trans* 409.
4.   Shyu, S., Cheng, S. and Tzou, D. (1999) *Chem. Commun.* 2337–2338.
5.   Mulukutla, R., Asakura, K., Kogure, T., Namba, S. and Iwasawa, Y. (1999) *Phys. Chem. Chem. Phys.* **1**, 2027–2032.
6.   Schubert, U., Husing, N. and Lorenz, A. (1995) *Chem. Mater.* **7**, 2010.
7.   Wilkes, G., Orler, B. and Huang, H. (1985) *Polym. Prepr.* **26**, 300.
8.   Chuang, S. S. C. and Pien, S. I. (1992) *J. Catal.* **135**, 618–634.
9.   Srinivas, G. and Chuang, S. S. C. (1993) *J. Catal.* **144**, 131–147.
10. Chuang, S. S. C. (1990) *Appl. Catal.* **66**, L1–L6.
11. Pien, S. I and Chuang, S. S. C. (1991) *J. Mol. Catal.* **68**, 313–330.
12. Chuang, S. S. C. U.S Patent 5082977, 1992.
13. Sertchook, H., Avnir, D., Blum, J., Joó, F., Kathó, A., Schumann, H., Weimann, R. and Wernik, S. (1996) *J. Mol. Catal.* **108**, 153–160.
14. Rosenfeld, A., Blum, J. and Avnir, D. (1996) *J. Catal.* **164**, 363–368.
15. Rosenfeld, A., Anvir, D. and Blum, J. (1993) *J. Chem. Soc., Chem. Commun.* 583–584.
16. Blum, J., Rosenfeld, A., Polak, N., Israelson, O., Schumann, H., Avnir, D. (1996) *J. Mol. Catal. A: Chem.* **107**, 217.
17. Haupt, K. & Mosbach, K. (2000) *Chem. Rev.* **100**, 2495–2504.
18. Bartsch, R. A. & Maeda, M. (1998) *Molecular and Ionic Recognition with Imprinted Polymers* (Am. Chem. Soc., Washington, DC).
19. Wulff, G. (1998) *CHEMTECH* **28(11)**, 19–26.
20. Katz, A. & Davis, M. E. (1999) *Macromolecules* **32**, 4113–4121.
21. Mosbach, K., Cormack, P., Ramstrom, O. & Haupt, K. U.S. Patent 5,994,110. 1999.

22. Katz, A. & Davis, M. E. (2000) *Nature* **403**, 286–289.
23. Shea, K. J. & Sasaki, D. Y. (1991) *J. Am. Chem. Soc.* **113(11)**, 4109–4120.
24. Lanza, F. & Sellergren, B. (1999) *Anal Chem.* **71**, 2092–2096.

39

The Palladium Catalyzed Synthesis of Carboxylate-Substituted Imidazolines: A New Route Using Imines, Acid Chloride, and Carbon Monoxide

Rajiv Dhawan, Rania D. Dghaym, and Bruce A. Arndtsen
McGill University, Montreal, Quebec, Canada

ABSTRACT

The palladium catalyzed coupling of imine, carbon monoxide and acid chloride is reported as a new route to prepare peptide-based imidazoline-carboxylates. Mechanistic studies suggest this process proceeds via the palladium catalyzed generation of 1,3-oxazolium-5-oxide intermediates, which react with imine to generate the observed products.

INTRODUCTION

The design of new methods to prepare peptide analogues, i.e. peptidomimetics, is an area of growing importance in pharmaceutical design (1-2). The incorporation of peptidomimetics into peptides can lead to a number of beneficial features, including increased proteolytic stability, improved bioavailability, decreased side-effects and higher selectivity and potency compared to the parent peptide sequence (1). One such class of peptidomimetics are carboxylate-substituted imidazoline derivatives **1** (Figure 1). Compounds such as **1** have been shown to be suitable as peptidic amide bond replacements, as illustrated in **1a** (2). This not only modifies the structural properties of the amino acid residue, but also stabilizes it towards degradation, thereby potentially enhancing its therapeutic utility (2). Imidazoline-carboxylates can also be considered as non-naturally occurring proline derivatives (**1b**), and have been used as synthetic intermediates towards the synthesis of non-proteinogenic 1,2-diamino acids (3). The latter are useful building blocks for a variety of antibiotics, enzyme inhibitors and other biologically active peptide-based molecules.

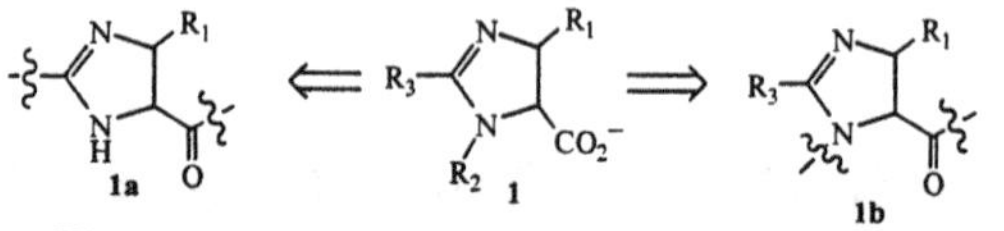

Figure 1 Utility of Imidazoline-Carboxylates.

The synthesis of carboxylate-substituted imidazoline derivatives has previously been accomplished by the condensation of presynthesized 1,2-diamines with amides, or through the transition metal catalyzed aldol-type reaction between isocyanates and imines (4,5). We have recently communicated an alternative palladium catalyzed route to synthesize a new class of imidazoline carboxylates, utilizing acid chloride, imines and carbon monoxide as starting materials (see Table 1) (6). This work was the first demonstration of a goal in our laboratory of developing metal-mediated routes to peptide-based molecules, using CO and imines as α-amino acid residue synthons. Notably, this process has the advantage of allowing for the synthesis of these heterocycles from readily available imine, acid chloride and carbon monoxide precursors, via a four component coupling methodology. As imines are derived from a large pool of commercially available aldehydes and amines, this methodology could lend itself to the synthesis of a number of new and structurally diverse imidazoline derivatives. In this report, we explore the scope and limitations of this catalytic process as a route to prepare imidazoline carboxylates. In addition, the mechanism by which the four separate components (imines, carbon monoxide and acid chloride) are coupled into the imidazoline product is also examined (7).

RESULTS AND DISCUSSION

Synthesis and Structure of Imidazole-Carboxylate 4

The reaction of a 1:1 mixture of Tol(H)C=NBn (**2**) (Tol= p-C$_6$H$_4$CH$_3$, Bn= CH$_2$C$_6$H$_5$) and PhCOCl (**3**)   with 1 atm CO in the presence of 5 mol% Pd$_2$(dba)$_3$·CHCl$_3$ (dba= *trans, trans*-dibenzylideneacetone) and 2,2'-bipyridine (10 mol%) in CH$_3$CN leads to the slow disappearance of starting materials over the course of 4 days at 55°C. Filtration of the reaction mixture, followed by acid and base washing and recrystallization, yields the imidazoline derivative **4** as a white solid (82% yield; Table 1, entry 1). To our knowledge, N,N-disubstituted, imidazoline-carboxylates such as **4** have not been reported prior to this work. In addition, the mechanism by which they might be formed from imine, carbon monoxide and acid chloride is not readily obvious (*vide infra*). Both of these features made the structural assignment of **4** challenging. Notably, both ^{1}H and ^{13}C NMR show two inequivalent imine units and one acid chloride phenyl ring incorporated into **4**. This is despite the presence of an equimolar amount of imine and acid chloride in the initial reaction mixture (8). The ^{13}C NMR shows the resonances for the imidazoline ring carbons (166.4, 81.4, and 72.2 ppm), along with an additional downfield peak for the carboxylate group (165.2 ppm). HMBC, HMQC and NOESY experiments are all consistent with the connectivity about the five-membered ring, with the substituents, as shown.

The structure of **4** has been definitively assigned by x-ray crystallography on the related N,N-dimethyl substituted imidazoline derivative (Table 1, entry 2). As illustrated in Figure 2, this clearly shows the incorporation of two imines, an acid chloride and carbon monoxide into the product, with an overall transoid orientation of the aromatic groups. The two imines are coupled through their carbons in the heterocyclic core, with one imine unit having its methine hydrogen replaced with the carboxylate group. In addition, the "CPh" unit of the benzoyl chloride bridges the two imine nitrogens, while the former acid chloride oxygen is coupled with CO into the carboxylate unit in the 4-position.

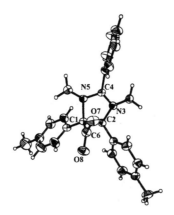

Figure 2 Crystal structure of N,N-dimethyl-2-phenyl-4,5-(p-tolyl)-imidazole-4-carboxylate (only one structure from unit cell shown). Selected bond lengths: C4-N5, 1.338 (8); C4-N3, 1.302 (8); C2-N3, 1.475 (7); C1-N5, 1.527 (7).

Scope of Catalytic Imidazoline Carboxylate Synthesis

The overall scope of this catalytic transformation is summarized in Table 1. In general, this palladium catalyzed process proceeds in high yield with a range of imines of aromatic aldehydes. In particular, functionality within the imine fragment is tolerated, including ethers (entry 3) and thioethers (entry 4) and aromatic halides (entry 5). More complex rings systems can also be generated via this process, including products such as the furfuryl- (entry 9) and piperonal-based imidazolines (entry 10). In addition, both aryl and alkyl acid chlorides can be utilized.

Table 1 Palladium Catalyzed Synthesis of Imidazoline-Carboxylates.

#	Ligand	R_1	R_2	R_3	% Yield[b]
1	bipy	PhCH$_2$	p-tolyl	Ph	82
2	bipy	CH$_3$	p-tolyl	Ph	92
3	bipy	CH$_3$OCH$_2$CH$_2$	p-tolyl	Ph	78
4	bipy	PhCH$_2$	p-CH$_3$SC$_6$H$_4$	Ph	73
5	bipy	PhCH$_2$	p-ClC$_6$H$_4$	Ph	62
6	bipy	PhCH$_2$	p-tolyl	CH$_3$	70
7	bipy	PhCH$_2$	p-NO$_2$C$_6$H$_4$	CH$_3$	-
8	bipy	Ph	p-tolyl	Ph	-
9[d]	-	2-furfuryl	p-tolyl	Ph	67
10[d]	-	PhCH$_2$	2-piperonyl	Ph	50
11	-	PhCH$_2$	i-butyl	Ph	-
12	-	PhCH$_2$	-CH=CH C$_6$H$_5$	Ph	-
13[c]	-	PhCH$_2$	p-tolyl	Ph	83
14[c]	-	PhCH$_2$	Ph	Ph	72
15	pyridine	PhCH$_2$	p-tolyl	Ph	87
16[c]	diphos	PhCH$_2$	p-tolyl	Ph	-

[a]0.57 mmol imine, 0.57 mmol acid chloride, 1 atm CO with 5 mol % Pd$_2$(dba)$_3$.CHCl$_3$ and 10 mol% ligand for 4 days at 55 °C. [b]Isolated yield. [c]24 h. [d]48 h.

The efficiency of this coupling does show some dependence upon the electronic nature of the imine substrate employed. While good to excellent yields are obtained with the relatively electron rich p-CH$_3$C$_6$H$_4$, and p-CH$_3$SC$_6$H$_4$ aldimines, limitations arise from imines that incorporate electron withdrawing groups. For example, the less electron rich p-ClC$_6$H$_4$(H)C=NBn gives slightly lower yields than its corresponding p-tolyl derivative, while the nitro substituted p-NO$_2$C$_6$H$_4$(H)C=NBn (entry 7) does not react to form imidazoline products. Similarly, N-aryl substituted imines fail to yield any imidazoline carboxylate product (entry 8). This electronic effect may be attributed to the lower nucleophilicity of these imines, which could inhibit their interaction with acid chloride (*vide infra*). In addition, neither alkyl (entry 11) nor alkenyl (entry 12) substituents at the R$_2$ position are tolerated under these reaction conditions. The former may be related to the lower stability of alkyl-imines in the presence of acid chlorides (9). In spite of these limitations, this four component coupling strategy provides easy access to a range of diversely substituted imidazoline derivatives from inexpensive, flexible and readily available starting materials.

Mechanistic Studies: Palladium Catalysis

This imidazoline-carboxylate synthesis involves the coupling of four separate components (two imines, an acid chloride and carbon monoxide), and the generation of at least five separate bonds, all via a one-pot, palladium catalyzed process. From an analysis of the structure of the imidazoline carboxylate, the individual constituents can be seen (Figure 3). This structure might be considered to arise from the dipolar cycloaddition of an imine with a mesoionic 1,3-oxazolium-5-oxide (5) intermediate, which itself could be generated from imine, acid chloride and carbon monoxide. Consistent with this potential formulation, performing the catalytic reaction with ^{13}CO leads to the incorporation of the carbon-13 label into the carboxylate position of 4.

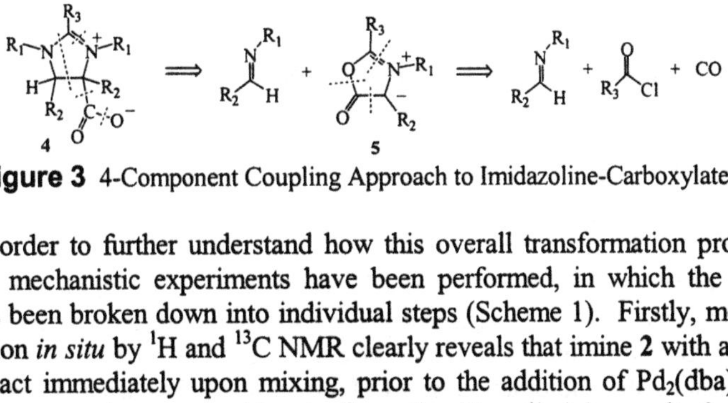

Figure 3 4-Component Coupling Approach to Imidazoline-Carboxylates.

In order to further understand how this overall transformation proceeds, a series of mechanistic experiments have been performed, in which the catalytic cycle has been broken down into individual steps (Scheme 1). Firstly, monitoring the reaction *in situ* by ^{1}H and ^{13}C NMR clearly reveals that imine 2 with acid chloride 3 react immediately upon mixing, prior to the addition of Pd$_2$(dba)$_3$·CHCl$_3$, bipyridine, or carbon monoxide, to form the N-acyliminium salt 6 (Step A, Scheme 1) (10). The structure of 6 can be confirmed by its independent synthesis from imine and acid chloride. The conversion of 2 and 3 to this iminium salt is essentially quantitative by ^{1}H NMR.

We have previously reported that N-acyliminium salts can undergo rapid oxidative addition to low valent metals to generate metal-chelated amides (11). Similarly, the addition of the catalyst (5 mol% Pd$_2$(dba)$_3$·CHCl$_3$ and 10% bipyridine) to this solution of 6 (Step B, Scheme 1) results in the rapid conversion of the palladium source into a new complex (7). This same complex 7 can be prepared and isolated upon the stoichiometric reaction of 6, 0.5 equiv. Pd$_2$(dba)$_3$·CHCl$_3$ and 0.5 equiv. bipyridine in CH$_3$CN, and has been characterized to be the amide-chelated palladium complex (bipy)Pd[η^2-CH(Tol)NBn(CO)Ph]$^+$Cl$^-$ (7). Of note, the ^{1}H NMR of 7 shows the presence of the methine iminium salt hydrogen singlet shifted upfield to 5.11 ppm, consistent with the reduction of the C=N upon addition to palladium, and the presence of diastereotopic benzylic hydrogens (δ 4.32, dd). In addition, the amide carbonyl resonance is shifted downfield of a free amide to 180.5 ppm in the ^{13}C NMR, consistent with its chelation to the palladium center. All other spectroscopic and mass spectrometry data are consistent with this structure, and directly analogous to the previously reported (bipy)Pd[η^2-C(H)TolNBnCOCH$_3$]$^+$ OTf (12).

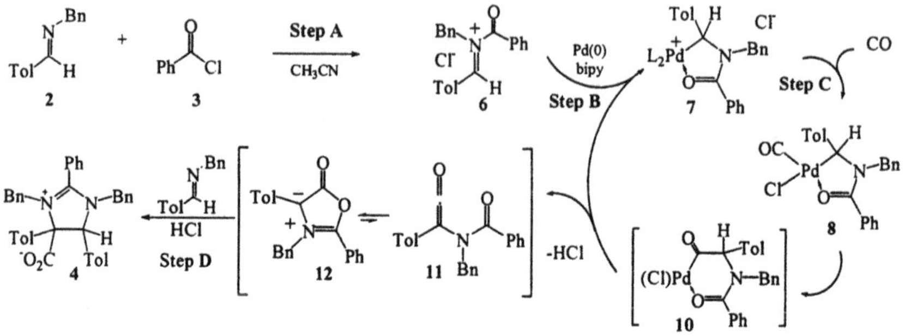

Scheme 1 Mechanism of Imidazoline-Carboxylate Formation.

The addition of 1 atm ^{13}CO to the CD$_3$CN solution of 6 and 7 generates the final catalysis mixture, and also allows the observation of a third potential intermediate in the catalytic cycle (Step C, Scheme 1). Examination of this solution by ^{1}H and ^{13}C NMR reveals the presence of a stoichiometric amount of 6, palladium catalyst 7, and the partial conversion (ca. 10%) of complex 7 into a new structure. This compound, 8 has ^{1}H and ^{13}C NMR resonances for an amide ligand analogous to that in complex 7 (δ 6.13 (s, 1H, C(H)Tol), 4.58 (dd, 2H, CH$_2$Ph), though shifted downfield from the original complex. In addition, the ^{13}C NMR of the catalytic mixture reveals the presence of a labeled carbonyl resonance at 174.9 ppm, suggesting the coordination of a CO ligand in 8. Our original postulate for the structure of 8 was that CO coordination to Pd occurred via dechelation of the amide oxygen, in analogy to the reactivity observed in palladium catalyzed sequential olefin/CO insertion (13). Once again, the identity of this reaction intermediate has been determined by its independent synthesis.

The reaction of equimolar amounts of imine 2 and acid chloride 3 with 0.5 equivalents of Pd$_2$(dba)$_3$·CHCl$_3$ in CH$_3$CN leads to the generation of {Pd(Cl)[η^2-CH(Tol)NBnCOPh]}$_2$ (9), which can be isolated as a yellow powder (Eq. 1). The addition of ^{13}CO to a CD$_3$CN solution of this complex results in its rapid reaction, and generation of the same complex observed in the catalytic reaction mixture. The formation of 8 in the absence of bipyridine demonstrates that this ligand has dissociated from the metal center in the catalytic reaction. IR (ν_{CO} = 2114 cm^{-1} in CH$_3$CN) and ^{13}C NMR (174.9 ppm) confirms the presence of a single coordinated CO ligand in 8, as well as the chelated amide ligand (179.1 ppm). In addition, no IR stretch was observed between 1650 and 1800 cm^{-1}, suggesting that CO has not undergone insertion into the palladium-carbon bond. The structure of 8 has been tentatively assigned as the CO coordinated amide-chelated complex shown. Thus, it appears that during catalysis the chloride and carbon monoxide together act to create an empty coordination site on the palladium center for CO via the displacement of the bipyridine ligand, rather than dechelation of the amide oxygen. Consistent with this hypothesis, imidazoline formation is inhibited by the use of a strongly coordinating diphos ligand (Table 1, entry 16), while the absence of any ligand (which should facilitate the generation of **8**) results in a significantly increased rate of catalysis (24h vs. 4 days, entry 13).

While warming the catalysis mixture to 55°C (Step **D**, Scheme 1) leads to no other observable reaction intermediates, the generation of intermediate **8** would allow the series of steps shown in Scheme 1. Insertion of the coordinated CO into the palladium-carbon bond would lead to the overall coupling of acid chloride, imine and carbon monoxide in complex **10**. The subsequent loss of HCl from **10**, either via direct deprotonation or β-H elimination, would form the α-amide substituted ketene **11**. The latter is known to be in rapid equilibrium with its cyclic mesoionic 1,3-oxazolium-5-oxide tautomeric **12** (14). These steps would lead to the liberation of the Pd(0) catalyst, which can return to the catalytic cycle.

Mechanistic Aspects: Münchnone Reactivity

1,3-oxazolium-5-oxides, commonly referred to as Münchnones, are well-known substrates in 1,3-dipolar addition reactions (14, 15). This reactivity has been extensively exploited in the cyclization of alkynes and alkenes to access pyrrole and pyrroline derivatives, respectively (15). However, to our knowledge, N-alkyl substituted imines have not been previously reported to undergo dipolar cyclization with **14**, and instead typically react the ketene valence tautomer **15** in a formal [2+2] cycloaddition to generate β-lactams, **16**. In order to explore the potential intermediacy of Münchnones in this catalytic imidazoline synthesis, **14** has been generated independently upon the dehydration of PhCON(Bn)CH(Tol)CO₂H (**13**) with dicyclohexyl carbodiimide (Scheme 2) (14, 15). Consistent with previously reports, the addition of imine to this Münchnone and heating to 55 °C for 24 hours leads to the formation of the amide-substituted β-lactam **16** (14). However, the presence of acid, which is generated in the palladium catalyzed synthesis of imidazolines, has been found to have a dramatic affect upon the cyclization chemistry of imines with **14**. Thus, the reaction of the independently formed **14** with imine in the presence of 1 equiv. HCl results in the extremely rapid (< 5 min) formation of imidazoline-carboxylate **17** in high yield (81%). To our knowledge, the ability of HCl to divert the reaction of imines with Münchnones from β-lactam to imidazoline formation has not been previously reported.

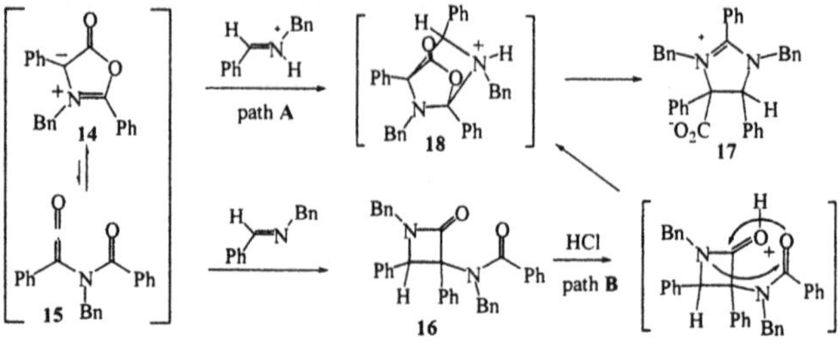

Scheme 2 Reactivity of Münchnone 14 with Ph(H)C=NBn.

The role of acid in influencing the cyclization of **14** with imines towards imidazolines products is at present unclear. One possibility is suggested by the work of Ferraccioli and Croce (16), who have shown that the electronic nature of the imine can have a significant influence upon its reactivity with Münchnone. In particular, while N-alkyl substituted imines react with Münchnones to form β-lactams, more electron poor imines, such as the N-tosyl substituted substrates, have been found to undergo a 1,3-dipolar cyclization with **14** to form imidazoles. (16) In our case, the role of acid may be in protonation of the imine substrate, thereby creating a more electrophilic C=N which can undergo a dipolar cycloaddition with **14** (path **A**, Scheme 2). Subsequent heterolysis of the C-O bond in **18**, would yield the observed imidazoline-carboxylate **17**.

Alternatively, control experiments show that the addition of HCl to the β-lactam product formed from reaction of imines with **14** leads to its quantitative rearrangement into the imidazoline-carboxylate product **17** (path **B**, Scheme 2). This acid induced reaction likely proceeds in a similar fashion to previously reported amide-substituted β-lactam rearrangements, (17) whereby lactam protonation induces C-N bond-cleavage to reduce ring-strain, in this case leading to the generation of intermediate **18**. While at present we cannot rule out either of these mechanistic possibilities, it is notable that imine cyclization with **14** to generate β-lactams is a slow transformation (>24h at 55°C), relative to the almost instantaneous addition of imine to **14** in the presence of HCl to form imidazolines, arguing against the intermediacy of β-lactams in this transformation. Regardless of the precise role of HCl, these experiments confirm the plausible intermediacy of Münchnones in this palladium catalyzed imidazoline-carboxylate synthesis.

CONCLUSION

These studies have shown that the palladium catalyzed coupling imines, CO and acid chlorides is a viable and general route for the synthesis of a new class of peptide-based imidazolines. Mechanistic studies suggest this four-component coupling reaction proceeds via the *in situ* formation of Münchnone intermediates, which undergo cyclization with imines in the presence of acid to yield the observed product. The use of this chemistry to access other amino acid and/or heterocyclic target molecules is currently the subject of research in our laboratories.

ACKNOWLEDGEMENTS

We thank Dr. Francine Blanger-Garipy for the determination of the crystal structure, and NSERC of Canada and Fonds FCAR du Quebec for their financial support. R.D and R.D.D thank McGill for McGill Major Awards. R.D. thanks NSERC for a Postgraduate Fellowship.

EXPERIMENTAL

General Procedure for Catalytic Formation of Imidazolines

Imine (0.57 mmol) and acid chloride (0.57 mmol) were stirred in CH_3CN (5 mL) for 15 min. This solution was added to $Pd_2(dba)_3 \cdot CHCl_3$ (5 mol%) in CH_3CN (10 mL) and stirred for 30 min in a 100 mL reaction bomb. CO (g) (760 torr) was added, and the reaction mixture warmed to 55°C. The resulting solution was filtered through celite, and washed with dilute HCl, saturated $NaHCO_3$, saturated NaCl, and dried over Na_2SO_4. Filtration, followed by evaporation of solvent, addition of diethyl ether, and cooling at –40°C afforded the imidazoline product as a white solid (18).

REFERENCES

1. (a) GL Olson, DR Bolin, MP Bonner, M Bos, CM Cook, DC Fry, BJ Graves, M Hatada, DE Hill, M Kahn, VS Madison, VK Rusiecki, R Sarabu, J Sepinwall, GP Vincent, ME Voss. Concepts and progress in the development of peptide mimetics. J Med Chem 36:3039-3049, 1993. (b) M Goodman, J Zhang. Peptidomimetic building blocks for drug design. Chemtracts-Org Chem 10:629-645, 1997.

2. (a) RCF Jones, GJ Ward. Amide bond isosteres: imidazolines in pseudopeptide chemistry. Tet Lett 29:3853-3856, 1988. (b) IH Gilbert, DC Rees, RS Richardson. Amide bond replacements: incorporation of a 2,5,5-trisubstituted imidazoline into dipeptides and into a CCK-4 derivative. Tet Lett 32:2277-2280, 1991. (c) IH Gilbert and DC Rees. Imidazolines as amide bond replacements. Tetrahedron 51: 6315-6336, 1995.

3. XT Zhou, YR Lin, LX Dai, J Sun, LJ Xia, MH Tang. A Catalytic Enantioselective Access to Optically Active 2-Imidazoline from *N*-Sulfonylimines and Isocyanoacetates. J Org Chem 64:1331-1334, 1999.

4. T Hayashi, E Kishi, VA Soloshonok, Y Uozumi. Erythro-selective aldol-type reaction of N-sulfonylaldimines with methyl isocyanoacetate catalyzed by gold(I). Tet Lett 37:4969-4972, 1996.

5. YR Lin, XT Zhou, LX Dai, J Sun. Ruthenium Complex-Catalyzed Reaction of Isocyanoacetate and *N*-Sulfonylimines: Stereoselective Synthesis of *N*-Sulfonyl-2-Imidazolines. J Org Chem 62: 1799-1803, 1997.

6.   RD Dghaym, R Dhawan, BA Arndtsen. The use of carbon monoxide and imines as peptide derivative synthons: a facile palladium-catalyzed synthesis of α-amino acid derived imidazolines. Angew Chem Int Ed 40:3228-3230, 2001.

7.   Some of the data in this manuscript has been communicated previously (6).

8.   0.50 equiv. of the acid chloride starting material is regenerated at the completion of the reaction.

9.   (a) H Hiemstra, N Speckamp. Additions to N-Acyliminium Ions. In: BM Trost, I Fleming, ed. Comprehensive Organic Synthesis. New York: Pergrammon Press, pp 1047-1081. (b) BR Brown. Chemistry of Aliphatic Carbon-Nitrogen Compounds. New York: Oxford Science Publications, 1994.

10. K Ratts, J Chupp. Trimethyl phosphite displacement on mucochloryl chloride. J Org Chem 39:3300-3301, 1974.

11. D. Lafrance, JL Davis, R Dhawan, BA Arndtsen. Insertion of Imines and Carbon Monoxide into Manganese-Alkyl Bonds: Synthesis and Structure of a Manganese-α-Amino Acid Derivative. Organometallics 20:1128-1136, 2001.

12. RD Dghaym, KJ Yaccato, BA Arndtsen. The Novel Insertion of Imines into a Late-Metal-Carbon σ-Bond: Developing a Palladium-Mediated Route to Polypeptides. Organometallics 17:4-6, 1998.

13. E Drent, PHM Budzelaar. Palladium-Catalyzed Alternating Copolymerization of Alkenes and Carbon Monoxide. Chem Rev 96:663-681, 1996.

14. (a) R Huisgen, H Gotthardt, HO Bayer, FC Schaefer. A new class of mesoionic arenes and their 1,3-dipolar cycloadditions with acetylene derivatives. Angew Chem 76:185-186, 1964. (b) R Huisgen, H Gotthardt, HO Bayer, FC Schaefer. 1,3-Dipolar cycloadditions. Synthesis of N-substituted pyrroles from mesoionic oxazolones and alkynes. Chem Ber 103: 2611-2624, 1970.

15. HL Gingrich, JS Baum. In: IJ Turchi. ed. The Chemistry of Heterocyclic Compounds: Oxazoles. New York: Wiley, 1986, pp 731-961.

16. R Consonni, PD Croce, R Ferraccioli, C La Rosa. A new approach to imidazole derivatives. J Chem Research 7:188-189, 1991.

17. PD Croce, R Ferraccioli, C La Rosa. Cycloaddition reactions of 5H,7H-thiazolo[3,4-c]oxazolium-1-oxides with imines. Tetrahedron 51: 9385-9392. 1995.

18. For spectral data, see Reference 6.

40

Assymmetric Hydroboration–Homologation: Towards the Synthesis of Gliflumide

Meredith Fairgrieve and Cathleen M. Crudden
University of New Brunswick, Fredericton, New Brunswick, Canada

INTRODUCTION

In 1999, we reported a new catalytic asymmetric one-carbon homologation strategy that employed a rhodium catalyzed hydroboration reaction to introduce stereo chemistry.[1,2] When followed by homologation with LiCHCl$_2$ and oxidation, 2-arylpropionic acids of high enantiomeric purity are generated (Scheme 1). Previously reported catalytic carbon–carbon bond forming reactions for the synthesis of these acids include hydrocarbonylation techniques employing carbon monoxide[3,4] or cyanation with hydrogen cyanide.[5] These are one-pot procedures; both the hydrometalation and the carbon–carbon bond forming steps are incorporated into a single catalytic cycle. Our approach is unique in that the generation of the required stereo center is separated from the second, stereo specific carbon–carbon bond forming reaction. This has the ad vantage that each step can be optimized individually, and dichloromethane can be used as the carbon source instead of carbon monoxide or hydrogen cyanide. We are currently working towards diversifying this two-step procedure with the aim to extend its scope to the use of different carbon sources.

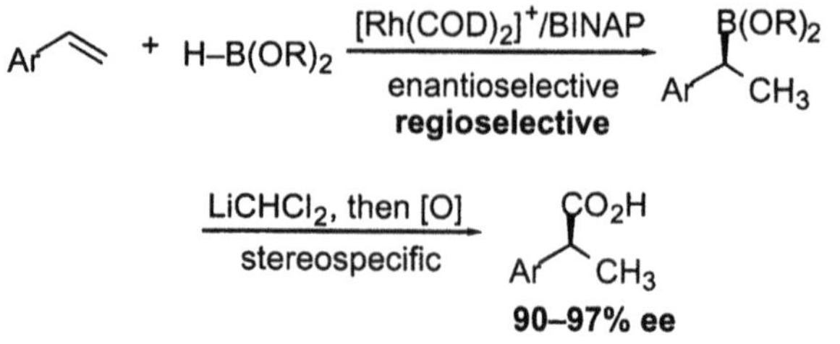

Scheme 1 Hydroboration-Homologation Strategy

APPROACH TO TOTAL SYNTHESIS: RESULTS AND DISCUSSION

In order to examine the scope and limitations of our recently developed hydroboration-homologation strategy, we have begun the synthesis of a more complex synthetic target, the antidiabetic compound Gliflumide (1) (Figure 1). Gliflumide is a

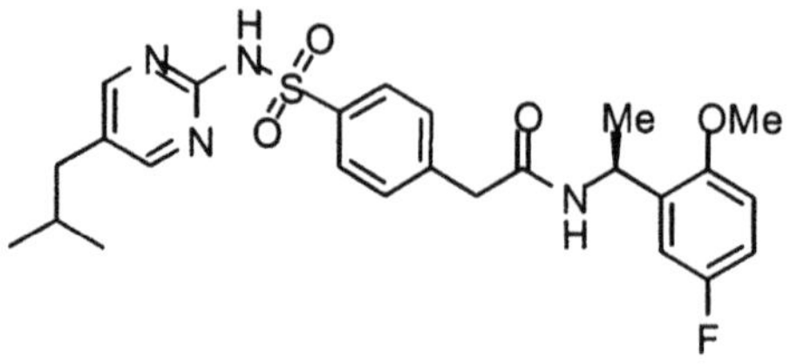

Figure 1 Gliflumide (1)

hypoglycemic compound with a blood glucose lowering activity in the dose range
of 0.1mg/kg.[6] It was first prepared in 1974 by Rufer, who used a classical resolution procedure to obtain the desired enantiomer.[6,7] The S enantiomer is from 30–
300 times more potent in rabbits than the R enantiomer, depending on the exact
substituents. We envisioned that an enantioselective synthesis of 1 could be affected using the protocol developed in our labs. From a retrosynthetic standpoint,1
can be divided into two fragments (Scheme 2), allowing for a convergent approach.
The left-hand portion of the molecule (compound 4) is accessible by various possible routes. Reaction of this compound with 3 gives ketone 2. Curtius,Schmidt or
Beckman rear rangments[8] will permit the regioselective introduction of the nitrogen
atom of the amide functionality. Compound 3 is prepared by hydroboration-
homologation of vinyl arene 5.

Our first challenge arose during the preparation of the hydroboration substrate, 5. Classical synthetic methods beginning either with p-fluoroanisole (9) or 2-
bromo-4-fluoroanisole (10) proved low yielding (Scheme 3). As 5 could be pre-

pared directly from **10** via cross-coupling chemistry, we began a careful survey of various methods and conditions.

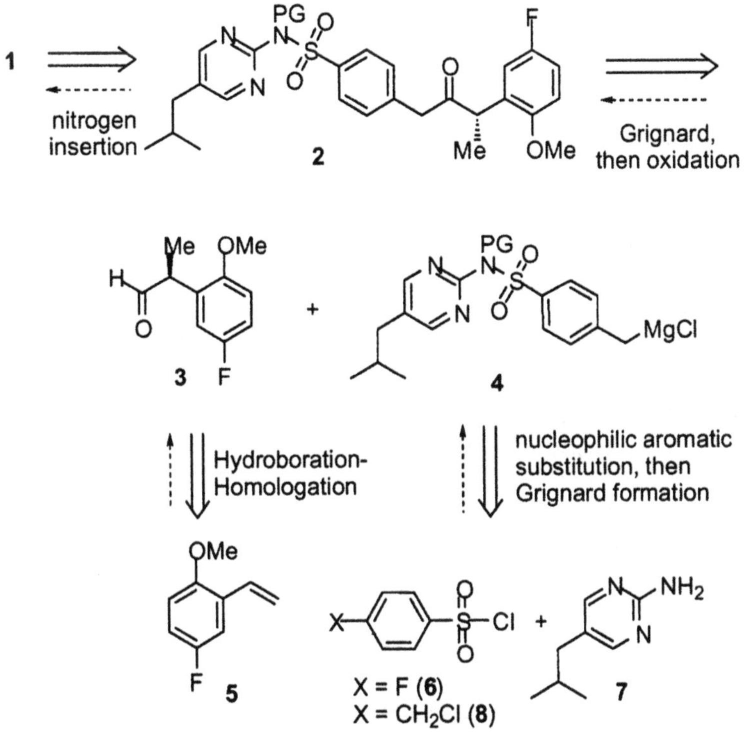

Scheme 2 Analysis of Synthetic Plan

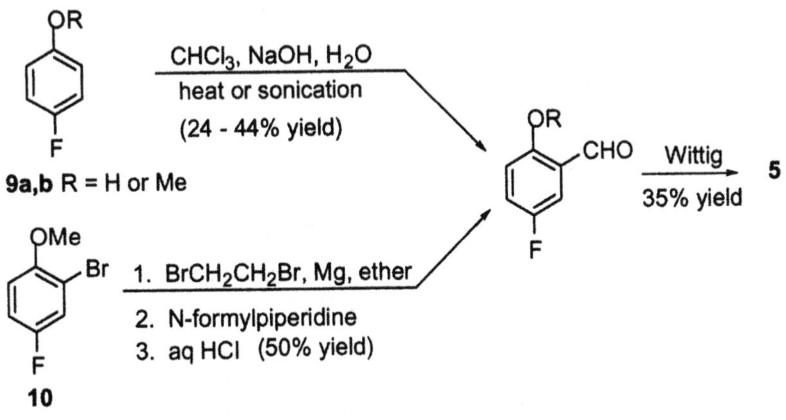

Scheme 3 Initial attempts at Synthesis of Vinyl Arene 5.

Hiyama coupling[9] proved unsatisfactory (eq. 1,Table 1). Under various conditions, **5** was formed in only low to moderate yields, and heating of the reaction
was necessary. This initially presented a problem because of the low boiling point
of the silane, but when the reactions were performed in an autoclave under 50psi of
argon, the desired product could be obtained, albeit in low yield. Difficulties sur-
faced during attempts to separate **5** from unreacted **10**; both distillation and column
chromatography were unsuccessful.

Table 1 Hiyama coupling for the synthesis of 5.

$$10 \ + \ SiR_3 \diagdown \quad \xrightarrow[\text{TBAF, THF}]{\text{Pd cat}} \quad 5 \quad \text{(eq. 1)}$$

R	TBAF (eq.)	Catalyst	Temp(°C)	Press (Ar, psig)	Yield (%)
CH_3	1	$Pd(PPh_3)_4$	25	1	0
CH_3	2	$Pd(dba)_2$	50	50	trace
CH_3	3	$[H_2CC(Me)_2PdCl]_2$	80	50	26
OCH_3	4	$[H_2CC(Me)_2PdCl]_2$	100	50	40

Stille coupling[10] gave better yields (eq. 2), but again the conversion was less
than one hundred percent, necessitating the separation of starting material from **5**.
Neither extended reaction times, higher catalyst loadings, nor the employment of
tri-2-furylphosphine, a ligand known to enhance the rate of the Stille reaction,[11] led
to complete conversion. Although toxic tin by-products could be removed from the
reaction mixture by conversion to the more polar tributyltin hydroxide/bistributyltin
oxide derivatives followed by flash chromatography,[12] we preferred to avoid the
use of tin altogether. A brief attempt at a Kumada coupling,[13] employing vinylmag-
nesiumbromide and 10 mol% $NiCl_2$ (dppe), was unsuccessful.

$$10 \ + \ Bu_3Sn\diagdown \quad \xrightarrow{\substack{\text{1. 2-5 mol \% Pd(PPh}_3)_4, \\ \text{toluene, 120}^0\text{C, 16-24h} \\ \text{2. aq NaOH, 2h, then column} \\ \text{(up to 89\% conversion)}}} \quad 5 \quad \text{(eq. 2)}$$

Using a recently published modification of the Heck reaction[14], we were de-
lighted to find that starting material was completely consumed under these conditions, giving **5** cleanly in 65% isolated yield (eq. 3).

$$10 \quad \xrightarrow{\substack{\text{200 psi ethylene, 5 mol\% Pd(OAc)}_2, \\ \text{16.5 mol\% (o-tol)}_3\text{P} \\ \text{1.25 mol\% BINAP, 1.8 eq NEt}_3 \\ \text{CH}_3\text{CN, 48h} \\ \text{(100\% conversion, 65\% yield)}}} \quad 5 \quad \text{(eq. 3)}$$

With **5** in hand, we were ready to test our hydroboration-homologation protocol (eq. 4). We were pleased to find that, using unoptimized conditions, hydroboration proceeded as expected to give the corresponding boronate ester **11** in 89%
ee and greater than 99% regioselectivity. We are currently working towards improving this already high yield and enantioselectivity with longer reaction times
and more selective chiral ligands.[15]

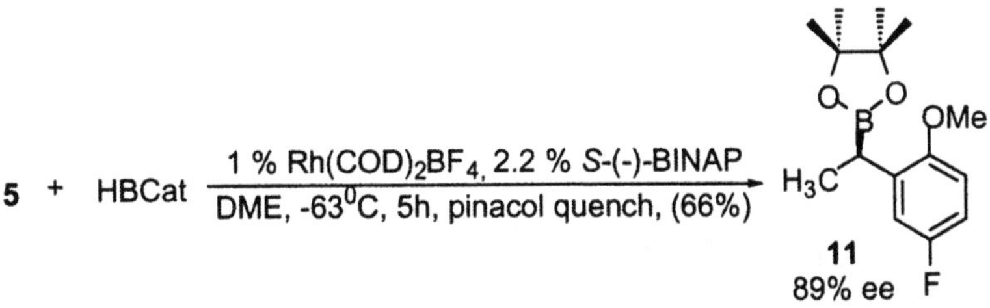

5 + HBCat $\xrightarrow[\text{DME, -63}^0\text{C, 5h, pinacol quench, (66\%)}]{1 \% \text{ Rh(COD)}_2\text{BF}_4, 2.2 \% \text{ S-(-)-BINAP}}$

11
89% ee   F   (eq. 4)

The synthesis of the second part of the molecule, is relatively straightforward
(Scheme 4). Diethylmalonate (**12**) was alkylated with 1-bromo-2-methylpropane on
a 109 mmol scale to give the corresponding alkyldiester (**13**) in 77% yield. Saponification followed by continuous extraction gave **14** in 93% yield. This product was
then coupled with guanidine-hydrochloride using oxalylchloride and dimethylformamide in CH_2Cl_2 to give the pyrimidine **7**.[7] In order to conserve the relatively
precious compound **7**, we employed a niline as a model of **7** in subsequent reactions (Scheme 6,below).

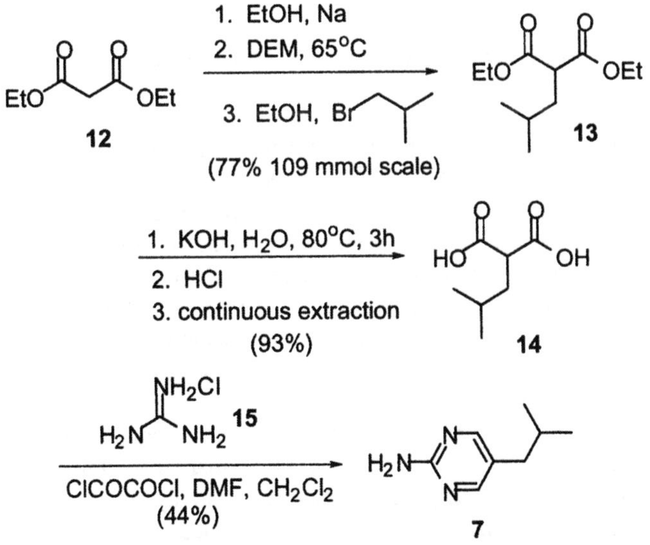

EtO   OEt
12

1. EtOH, Na
2. DEM, 65°C
3. EtOH, Br
(77% 109 mmol scale)

EtO   OEt
13

1. KOH, H$_2$O, 80°C, 3h
2. HCl
3. continuous extraction
(93%)

HO   OH
14

NH$_2$Cl
H$_2$N   NH$_2$ **15**

ClCOCOCl, DMF, CH$_2$Cl$_2$
(44%)

H$_2$N
7

Scheme 4 Synthesis of Heterocycle 7.

Two routes are currently being investigated for the synthesis of **4** (Scheme 5). The key step of the first method (Route A) involves nucleophilic aromatic substitution on compound **16**. We know of no literature precedents for this specific transformation, but fluorine is known to be a very effective leaving group for this reaction, and the sulfonyl group is a strong electron withdrawing group. The other route (B) involves chlorination of *p*-toluenesulfonylchloride with *N*-chlorosuccinimide (NCS) or bromination with NBS, followed by reaction with **7** to give **17**.

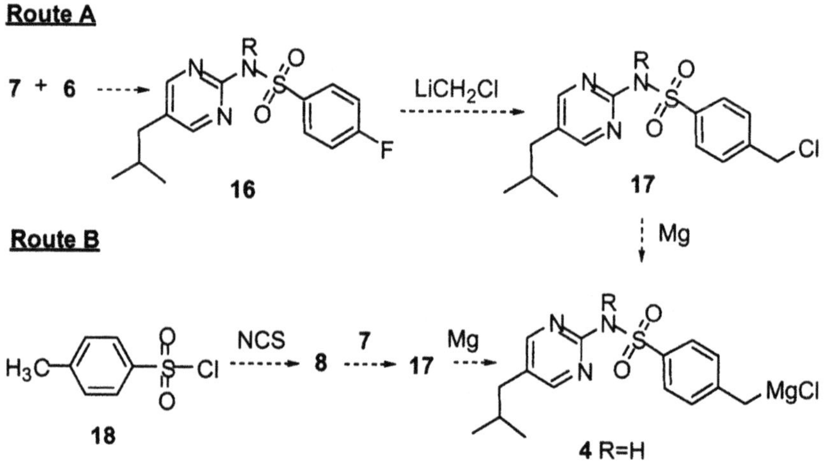

Scheme 5 Synthetic Routes to Grignard Reagent **4**.

We have begun examining Route A using a niline (**19**) as a model for **7** (Scheme 6). Reaction of **19** with commercially available 4-fluorobenzene-sulfonylchloride (**6**) gave **20** in 90% yield. Compound **20** results from selective reaction at the sulfonyl chloride in the presence of the aromatic fluoride. The temperature of the reaction was crucial for success; at 40°C, a mixture of **20** and **21** was obtained, but at 0°C, only **20** was formed. Compound **21** arises from a double addition of an iline. This latter reaction gives us reason to believe that the nucleophilic aromatic substitution needed to prepare our key model compound **22**, representing **17** in Route A (Scheme 5), will be possible.

We are currently investigating the use of $LiCH_2Cl$ as a homologating agent in nucleophilic aromatic substitution reactions. Preliminary experiments suggest the need for a nitrogen protecting group. Compound **4** is only one step away from **17**, and once **4** has been prepared we will be ready to couple **4** with **3** to give **2**. This leaves us one step away from Gliflumide. Model studies for the desired nitrogen insertion reaction are in progress.

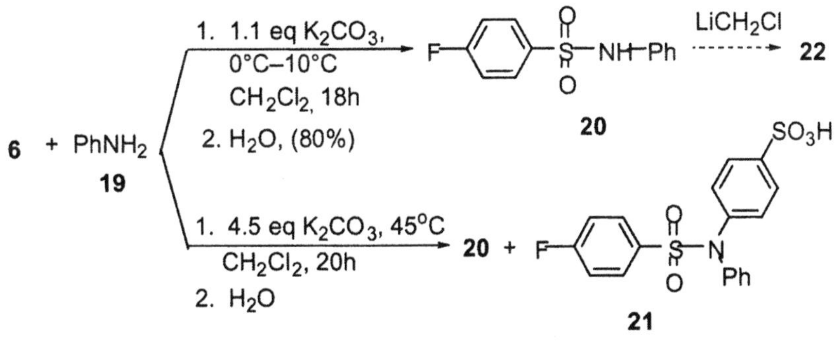

Scheme 6 Synthesis of Model Compound **20**.

CONCLUSIONS

The asymmetric synthesis of Gliflumide is still underway. We are near the completion of the synthesis, and thus far we have definitively shown that the hydroboration protocol can be used with more complicated aryl systems than simple monosubstituted styrenes.

REFERENCES

(1) Chen, A.; Ren, L.; Crudden, C.M. *Chem. Commun.* **1999**, 611.
(2) Chen, A.; Ren, L.; Crudden, C.M. *J. Org. Chem.* **1999**, 64, 9704.
(3) Alper, H.; Hamel, N. *J. Am. Chem. Soc.* **1990**, 112, 2803.
(4) Sakai, N.; Mano, S.; Nozaki, K.; Takaya, H. *J. Am .Chem. Soc.* **1993**, 115, 7033.
(5) RajanBabu, T.V.; Casalnuovo, A.L. *J. Am. Chem. Soc.* **1992**, 114, 6265.
(6) Rufer, C.; Biere, H.; Ahrens, H.; Loge, O.; Schroder, E. *J. Med. Chem.* **1974**, 17, 708.
(7) Gutsche, V.K.; Schroder, E.; Rufer, C.; Loge, O. *Arzneim.-Forsch.* **1974**, 7, 1028.
(8) Zabicky, J. The Chemistry of Amides. New York: John Wiley & Sons Ltd., 1970.
(9) Hatanaka, Y.; Hiyama, T. *J. Org. Chem.* **1988**, 53, 918.
(10) McKean, D.R.; Parrinelo, G.; Renaldo, A.F.; Stille, J.K. *J. Org. Chem.* **1987**, 52, 422.
(11) Farina, V.; Krishnan, B. *J. Am. Chem. Soc.* **1991**, 113, 9585.
(12) Renaud, P.; Lacote, E. *Tetrahedron Lett.* **1998**, 39, 2123.
(13) Tamao, K.; Sumitani, K.; Kumada, M. *J. Am. Chem. Soc.* **1972**, 94, 4374.
(14) Raggon, J.; Snyder, W. *Org. Proc. Res. Devel.* **2002**, 6, 67.
(15) Demay, S.; Volant, F.; Knochel, P. *Angew. Chem. Int. Ed.* **2001**, 40, 1235.

41

Reductive Amination of Poly(vinyl Alcohol)

G. A. Vedage, M. E. Ford, and J. N. Armor
Air Products and Chemicals, Allentown, Pennsylvania, U.S.A.

ABSTRACT

Reductive amination of poly(vinyl alcohol) in solutions of anhydrous amines with Pd/alumina as catalyst provides partially aminated poly(vinyl alcohol) in a single step. With preferred conditions, moderate degrees of substitution are obtained, and chain scission is minimized.

INTRODUCTION

Amine functional polymers (AFPs) have a wide range of potential applications, which include wet strength additives for nonwovens, water treatment chemicals, and curing agents for epoxies and polyurethanes. Recent routes to AFPs have centered on (co)polymerization of a derivative of vinylamine, such as N-vinylformamide, and subsequent removal of the derivatizing group [1-3]. While effective, such processes often generate stoichiometric quantities of formate salts which require catalytic decomposition [4-5] or physical removal and disposal. In contrast, reductive amination of poly(vinyl alcohol) (PVOH) with polyfunctional amines might offer direct access to a family of novel AFPs, with water as the only coproduct.

While many techniques for derivatization of PVOH to higher value products have been reported [6], few methods are available for introduction of amine functionality. Although reductive amination of low molecular weight alcohols to provide the corresponding amines is well established, application of this technique to PVOH is relatively unknown. An extensive literature survey found only a single, expired US patent [7] on catalytic amination of polymeric polyols (specifically PVOH, poly(allyl alcohol), and their copolymers) with ammonia or mono- or dialkylamines. Heterogeneous nickel catalysts (Raney® nickel or nickel on kieselguhr) were used, often under forcing conditions (200°C). Catalytic reductive amination of PVOH with polyfunctional amines, such as ethylenediamine (EDA), ethanolamine, or diethylenetriamine (DETA) has not been reported in either the journal or patent literature. Realization of the latter transformation could provide a uniquely flexible route to AFPs, in which not only the molecular weight of the polymer, but also the structure and degree of functionality of the pendant amine group could be varied by appropriate choice of starting materials.

RESULTS AND DISCUSSION

We now report preparation of new AFPs by reductive amination of PVOH with polyfunctional amines. The process consists of heating a solution of PVOH in the desired amine under hydrogen in the presence of Pd/Al$_2$O$_3$. With high molecular weight PVOH and hexamethylenediamine (HMDA) as a representative amine, reaction temperatures of at least 160°C, and preferably 180°C, were needed to obtain meaningful conversions (Table 1, runs 1 – 4). Somewhat higher conversions were obtained at longer reaction time (Table 1, cf runs 2 – 4 and 5 – 7, respectively). Isolation of products from the higher temperature reactions showed that significant amounts of low molecular weight, isopropanol-soluble coproducts had been formed. Specifically, although identical conversions were obtained in runs 5 – 7, significantly higher amounts of low molecular weight byproducts were formed in runs 5 and 6 vs run 7. A similar observation applies to runs 2 – 4. Since these reactions were done with increasing pressures under otherwise identical conditions, they demonstrate that chain scission is inhibited by high partial pressures of hydrogen. Closer comparison of runs 2 – 4 and 5 – 7 also shows that while operation at 1000 psig significantly retards chain scission, slightly higher degrees of chain scission are observed upon increasing the reaction pressure from 200 to 400 psig. The reason for this observation is unclear. However, this effect is relatively small, and its presence does not contravene the conclusion that high hydrogen pressures are needed to minimize chain scission. Chain scission may result from (base-catalyzed) retro-aldol reaction of ketone or imine intermediates in the reductive amination pathway (eq. 1). Increasing hydrogen pressure would

(1)

X = O, NR

slow the rate of formation of the ketone/imine intermediate, and increase the rate of conversion of the imine to amine product. These results clearly show the advantage of a high partial pressure of hydrogen. Consequently, subsequent reactions were carried out at 1000 psig. Comparable degrees of substitution of high molecular weight PVOH were also obtained with EDA and DETA (Table 1, runs 8 and 9, respectively).

An attempt to scale up run 9 by a factor of five with a somewhat more concentrated solution of PVOH (20 wt%, vs 15 wt% for run 9), gave a noticeably lower conversion (2%). The higher viscosity of the reaction mixture may have been sufficient to slow hydrogen mass transfer and thus limit conversion. To test this, high molecular weight PVOH was reacted with either EDA or DETA for either 4 or 18 hrs (Table 1, runs 10 – 13); 10 wt% solutions were used to facilitate viscosity measurements. While extended reaction of DETA with high molecular weight PVOH provided little additional conversion (Table 1, runs 12 – 13), the analogous reaction of EDA provided double the amount of amine incorporation (Table 1, runs 10 – 11). These results parallel the trend in viscosities of the initial reaction mixtures (Table 2), and indicate that the viscosity of the medium presents an additional barrier to hydrogen mass transfer. Further reactions were therefore done with 10 wt% solutions of PVOH.

Under the latter conditions, similar conversions have also been obtained from reactions of low molecular weight grades of PVOH with several ethyleneamines (Table 1, runs 14 – 19), and medium molecular weight PVOH with either EDA or DETA (eg, Table 1, runs 20 and 21, respectively).

In conclusion, reductive amination of PVOH with anhydrous amines provides partially aminated PVOH directly. With appropriate choice of reaction conditions (and specifically, concentration, pressure, and time) moderate conversions can be obtained with minimal chain scission.

EXPERIMENTAL

Screening reactions were carried out with 10 – 15 wt% solutions of PVOH in a 100 mL stirred stainless steel Parr autoclave. Catalyst (5% Pd/Al$_2$O$_3$; 5 wt%, based on PVOH) was added, the reactor was purged and pressure checked with nitrogen, purged and pressurized with hydrogen, and heated with stirring to the desired temperature for the desired time (see Table 1) at the desired pressure (typi-

Table 1 Reductive Amination of Poly(vinyl alcohol)[a]

Run	PVOH MW (x10³)	Amine (wt% PVOH)[b]	P (psig)	T (°C)	t (hr)	Conv (%)[c]	Yield (%)[d]
1	124-186	Hexamethylene-diamine (HMDA; 15)	400	160	18	5	e
2	124-186	HMDA (15)	200	180	4	1	18
3	124-186	HMDA (15)	400	180	4	3	10
4	124-186	HMDA (15)	1000	180	4	6	48
5	124-186	HMDA (15)	200	180	18	8	37
6	124-186	HMDA (15)	400	180	18	8	23
7	124-186	HMDA (15)	1000	180	18	8	62
8	124-186	EDA (15)	1000	180	4	7	60
9	124-186	DETA (15)	1000	180	4	5	39
10	124-186	EDA (10)[f]	1000	180	4	7	82
11	124-186	EDA (10)[f]	1000	180	18	15	81
12	124-186	DETA (10)[f]	1000	180	4	5	70
13	124-186	DETA (10)[f]	1000	180	18	7	70
14	13-23	EDA (10)	1000	180	4	5	87
15	13-23	DETA (10)	1000	180	4	10	16
16	13-23	Tris(aminoethyl)amine (TAEA; 10)	1000	180	4	13	7
17	31-50	EDA (10)	1000	180	4	6	66
18	31-50	N-(2-aminoethyl)piperazine (AEP; 10)	1000	180	4	6	56

Table 1 (cont'd)

Reductive Amination of Poly(vinyl alcohol)[a]

Run	PVOH MW (x10³)	Amine (wt% PVOH)[b]	P (psig)	T(°C)	t(hr)	Conv (%)[c]	Yield (%)[d]
19	31-50	TAEA	1000	180	4	6	16
20	85-146	EDA (10)	1000	180	4	14	91
21	85-146	DETA (10)	1000	180	4	13	96

[a] All reactions carried out with 5% Pd/Al$_2$O$_3$; 5 wt% based on PVOH.
[b] Wt% PVOH dissolved in the indicated amine.
[c] Mole percent of alcohol aminated; balance remains as pendant alcohol.
[d] Wt% of high molecular weight amine functionalized polymer. Excludes very low molecular weight, isopropanol-soluble byproducts.
[e] Not determined.
[f] Reaction carried out on 5X scale in 1 L reactor.

Table 2 Brookfield Viscosities of Pvoh/Amine Solutions

PVOH MW (x10³)	Amine	Viscosity (cps)[a]
124-186	EDA	1150
124-186	DETA	12100

[a] Measured on a Brookfield viscometer with 9.1 wt% solutions at 23°C.

cally 200 – 1000 psig). Products were isolated by addition of the minimum amount of deionized water needed for solubilization, followed by precipitation with isopropanol, and were subsequently analyzed by ^{1}H and ^{13}C NMR. The same procedure was followed with runs 10 – 14 with a 1 L stainless steel reactor

ACKNOWLEDGMENTS

We thank J. Cunningham and J. Sychterz for technical assistance, and Air Products and Chemicals, Inc. for permission to publish this work.

REFERENCES

1.   R. K. Pinschmidt and D. J. Sagl, The Polymeric Materials Encyclopedia, Synthesis, Properties and Applications, J. C. Salamone, ed., CRC Press, Boca Raton, in press.
2.   T. Itagaki, M. Siraga, S. Sawayama, and K. Satoh, US Pat. 4,808,683, to Mitsubishi Chemical Industries Limited (1989).
3.   D. J. Dawson and P. J. Brock, US Pat. 4,393,mDynapol (1983).
4.   F. Brunmueller, M. Kröner, and F. Linhart, US Pat. 4,421,602, to BASF (1983).
5.   M. E. Ford and J. N. Armor, Chem. Ind. (Dekker), 68, 367 (1996).
6.   See, eg, Polyvinyl Alcohol, C. A. Finch, ed., Wiley, New York, pp. 183-202, 391-411 (1973); Polyvinyl alcohol – Developments, C. A. Finch, ed., Wiley, New York, pp. 157-194 (1992).
7.   C. F. Hobbs, US Pat. 4,070, 530 to Monsanto Co. (1978).

42

Anchored Homogeneous Catalysts: Some Studies into the Nature of the Catalytically Active Species

Pradeep Goel, Stephen Anderson, Jayesh J. Nair, Clementine Reyes, Yujing Gao, Setrak K. Tanielyan, and Robert L. Augustine
Center for Applied Catalysis, Seton Hall University, South Orange, New Jersey, U.S.A.

Robert Appell, Christian Goralski, and David Hawn
The Dow Chemical Company, Midland, Michigan, U.S.A.

ABSTRACT

While the model used to describe our anchored homogeneous catalyst system is a simple one, it was sufficient for the general development of these catalysts. However, a more complete understanding of the types of interactions occurring in these catalysts is important for their full optimization. Data have been obtained which indicate that the catalytically active species in these anchored homogeneous catalysts is not present as a Rh^+ species in an ion pair but, rather, is attached to the heteropoly acid through the metal atom of the complex. Results have been obtained which indicate that the nature of the observed catalyst activation on hydrogenation may be the result of the removal of the diene group on the catalyst during the preparation of the anchored species. Preliminary XPS data show that there appears to be no change in the nature of the tungsten of the PTA but there were some oxidation state changes in the Rh atom of the complex. Further work in this area is on-going at present.

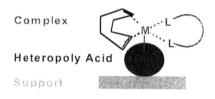

Complex

Heteropoly Acid

Support

Figure 1   A cartoon depicting the mode of complex
anchoring currently thought to be correct.

INTRODUCTION

We have recently described a novel catalyst system comprised of a
homogeneous complex anchored to a solid support using a heteropoly acid as
the anchoring agent (1-5). This type of arrangement is particularly effective for
anchoring enantioselective homogeneous catalysts since the anchored species
generally have at least the same activity and selectivity as the homogeneous
analogs. Further, loss of the complex is minimal even on multiple re-use or
continuous use of these catalysts. We have applied a simplistic approach in
describing the nature of these catalysts as depicted in Fig. 1 showing the
complex, the heteropoly acid and the support (5). While this has served as a
reasonable working hypothesis, a deeper understanding of this system is needed
to enhance the utility of such catalysts.

DISCUSSION

Even though one can get a precipitate by treating a solution of a complex with a
solution of a heteropoly acid such as phosphotungstic acid (PTA), all three
components, the active complex, the heteropoly acid and the support, are needed
for a successful catalyst. In the hydrogenation of methyl 2-acetamidoacrylate (**1**)
over a Rh(DiPamp)/PTA precipitate with the catalyst being used three times, the
rate of hydrogenation appears to be fairly constant. However, the liquid from
the third run after being separated from the solid material promoted hydrogen
uptake when further substrate was added to this liquid and hydrogenation
initiated. These data show that an active species from the complex/HPA has
dissolved in the solvent, methanol. When this same procedure is repeated using
an alumina anchored catalyst, the reaction mother liquors show no catalytic
activity (5).

A recent paper describing the use of these anchored complexes for
aldehyde hydrogenation stated that cationic complexes were required for the
formation of the anchored species and that this anchored catalyst was,
essentially, an ion pair comprised of a rhodium cationic species and an oxygen

anion on the surface of the heteropoly acid (6). However, we have anchored neutral complexes such as the Wilkinson catalyst onto PTA/Al$_2$O$_3$ and used the resulting material for the hydrogenation of 1-hexene with a TOF of 4.7 mmoles H$_2$/mmole Rh/min. This anchored catalyst was also used to hydrogenate carvone in absolute ethanol at 50 psig and room temperature with the catalyst re-used several times. Analysis of the reaction mixture from the first reaction found only 0.3 ppm of Rh and less than 0.2 ppm of W present. This very low level of leaching was typical of most of the reactions run using these anchored complexes. A more complete description of the solvent effect on the rate and leaching of these catalysts can be found elsewhere in this volume (7).

Some evidence supporting the covalent attachment of the complex to the HPA is the uv absorption data presented previously (5). Here it was shown that Rh(DiPamp) complexes having various counterions gave significantly different uv spectra depending on whether the complex contained a Rh$^+$ or a Rh-X covalent bond. The Rh$^+$ species with BF$_4^-$, SbF$_6^-$ and CF$_3$SO$_2^-$ counterions had spectra distinguished by absorption doublets at 340 and 460 nm. The Rh-Cl complex had a single absorption band at 280 nm while the Rh-OAc species had a single band at 285nm. When each of these complexes were treated with PTA, washed thoroughly and then dissolved they all showed only a single absorption band at 285nm. This similarity with the Rh-OAc spectrum suggests the presence of a Rh-O bond between the complex and the PTA but does not support the presence of a Rh$^+$ species. For comparison the PTA absorption curve has a maximum at 267nm. Similar data were obtained from treatment of these complexes with silicotungstic acid (STA).

It was considered that if these anchored species were ion pairs, then one should be able to remove them by treatment with an excess of another anion. To test this, a sample of Rh(dppb) anchored onto PTA/Al2O3 was stirred for 30 hours with a large excess of LiBF4 in ethanol under an inert atmosphere. Analysis of the reaction liquid showed that, at most, 2-3% of the Rh present on the catalyst was lost. However, tungsten was also detected with a Rh:W ratio of 1:12. Since there are twelve tungsten atoms in a PTA molecule, these data indicate that this procedure apparently did not remove only the Rh complex from the PTA but, instead, somehow removed the Rh/PTA from the alumina.

Another factor which needs some explanation is the fact that after the first use of these anchored species, the catalyst frequently becomes more active and selective (5). It was considered that this increase in activity was caused by the exposure of the catalyst to hydrogen during the first use and that this 'hydrogenated' species was, somehow, more effective. This conclusion was verified by using pre-hydrogenated catalysts which were more active and selective from the start. The data discussed above were obtained using pre-hydrogenated catalysts. The question remains, however, as to what was causing this change in activity and selectivity.

While it has been considered that the diene present on these complexes was easily removed during the early stages of a hydrogenation, a recent paper presented data which showed that the nature of the diene on the homogeneous catalyst had a significant effect on the rate of hydrogenation (8). It was suggested that the decreased rate and low selectivity observed during the first use of the anchored catalyst might have been caused by the slow removal of the COD diene on the complex and that once this was removed the catalyst became more active and selective.

Table 1 Rate data for the hydrogenation of 2 over homogeneous and anchored, Rh(DiPamp)(diene) catalysts.

Catalyst	Diene	Run #	Rate[a]
Homogeneous	COD		0.90
	NBD		3.31
Anchored	COD	1	1.26
		2	1.15
	NBD	1	1.52
		2	1.31

[a] mmole H_2 / mmole Rh / min.

We repeated the published work which was concerned with the hydrogenation of methyl 2-acetamidocinnamate (3) using a Rh(DiPamp) catalyst with either COD or NBD as the diene. The results are very similar to those published (8) with the NBD containing complex more active than that with the COD. However, when these complexes were anchored onto PTA/Al$_2$O$_3$, the reaction rates were much closer (Table 1). In light of these findings, extension of this concept to other ligands and substrates seemed reasonable. In the hydrogenation of 1-hexene over Rh(dppb) complexes having either the COD or NBD diene the rates were similar. When the corresponding anchored complexes were used, the COD containing material was more active than that containing NBD (Table 2). These data indicate that, while diene hydrogenation may play a role in the activation of these catalysts, the extent to which it is involved appears to be dependant on the substrate and/or the ligand.

Table 2 Rate data for the hydrogenation of 1-hexene over homogeneous and anchored Rh(dppb)(diene) catalysts.

Catalyst	Diene	Run #	Rate[a]
Homogeneous	COD		0.89
	NBD		1.29
Anchored	COD	1	1.98
		2	1.98
		3	1.98
	NBD	1	1.34
		2	1.34
		3	1.34

[a] mmole H2 / mmole Rh / min.

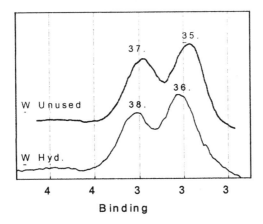

Figure 2 XPS determined binding energies for W (7f) in new and hydrogenated Rh(DuPhos)/PTA/Al$_2$O$_3$.

It was noted that the anchored catalysts would frequently change color during the pre-hydrogenation step and that the original color would come back on exposure of the catalyst to air. The unused catalysts were usually a yellow color with the color change ranging from a red-orange to green to a light purple depending on the ligand present on the complex. It was thought that this color change could be the result of a change in the oxidation state of the tungsten of the PTA used as the anchoring agent. To see if this could be the case XPS spectra were obtained from an unused Rh(DuPhos)/PTA/Al$_2$O$_3$ and the catalyst isolated in an inert atmosphere after use in a hydrogenation. Fig. 2 shows the binding energies for W (4f) for these two samples. There is no shift in the binding energies as a result of the hydrogenation. A shift is, however, seen in the binding energies of the Rh (3d) (Fig. 3). Also included in Fig. 3 are the Rh binding energy data for the Rh(DuPhos)(COD)$^+$ BF$_4^-$ homogeneous complex used to prepare the anchored species. These data show that not only have the binding energies shifted on hydrogenation but, in the case of the Rh data, the peak shapes have also changed. This latter factor indicates the presence of species having different oxidation states, particularly in the hydrogenated material. This XPS investigation is on-going and a full discussion of these results will be presented at a later time.

CONCLUSIONS

Data have been obtained which indicate that the catalytically active species in these anchored homogeneous catalysts is not present as a Rh$^+$ species in an ion pair but, rather, is attached to the heteropoly acid through the metal atom of the

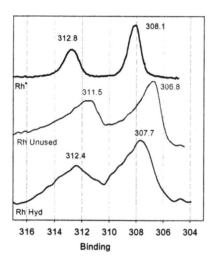

Figure 3 XPS determined binding energies for Rh(3d) in new and hydrogenated Rh(DuPhos)/PTA/Al₂O₃.

complex. Results have been obtained which indicate that the nature of the observed catalyst activation on hydrogenation may be the result of the removal of the diene group on the catalyst during the preparation of the anchored species. Preliminary XPS data show that there appears to be no change in the nature of the tungsten of the PTA but there were some oxidation state changes in the Rh atom of the complex. Further work in this area is on-going at present.

REFERENCES

1) R. Augustine and S. Tanielyan, *Chem. Commun.*, 1257 (1999).
2) S.K. Tanielyan and R.L. Augustine, US Patents 6,005,148 (1999), 6,025,295 (2000); Australian Patent 737536 (2001).
3) S.K. Tanielyan and R.L> Augustine, PCT Int. Appl., WO-9828074: *Chem. Abstr*, **129**, 109217 (1998).
4) S.K. Tanielyan and R.L. Augustine, *Chem. Ind. (Dekker)*, **75** (Catal. Org. React.), 101 (1998)
5) R.L. Augustine, S.K. Tanielyan, S. Ancerson, H. Yang and Y. Gao, *Chem. Ind. (Dekker)*, **82** (Catal. Org. React.), 497 (2000).
6) M.J. Burk, A. Gerlach and D. Semmerl, *J. Org. Chem.*, **65**, 8933 (2000).
7) C.R. Reyes, Y. Gao, A. Zsigmond, P. Goel, N. Mahata, S. Tanielyan and R. Augustine, This Volume.
8) A. Borner and D. Heller, *Tetrahedron Letters*, 223 (2001).

43

Selectivity Effects in the Hydrogenation of 4-t-butyl Phenol over Rhodium and Platinum Heterogeneous Catalysts

K. G. Griffin, S. Hawker, P. Johnston, and M-L. Palacios-Alcolado
Johnson Matthey, Royston, Hertfordshire, United Kingdom

C. L. Clayton
University of Cambridge, Cambridge, United Kingdom

ABSTRACT

This work illustrates the effect of catalytic metal, support, solvent, temperature and hydrogen pressure on the activity and selectivity of the ring hydrogenation of 4-t-butyl phenol.

The results obtained show that the most active metal is Rhodium, particularly when supported on alumina, with carbon and graphite giving lower activity catalysts, which consequently give different selectivities. Rh/C, particularly in hexane solvent gave the highest selectivities to cis-cyclohexanol. Of the other metals evaluated (Pt, Pd, Ru, Ir) only Pt showed significant activity when used in hexane solvent, but this gave the reverse selectivity to the Rh/C catalyst.

Interesting changes in selectivity were observed between the use of isopropanol (polar) and hexane (non-polar) as solvents with hexane giving higher levels of the cis-cyclohexanol product (6:1 compared to 3:1 cis:trans ratio).

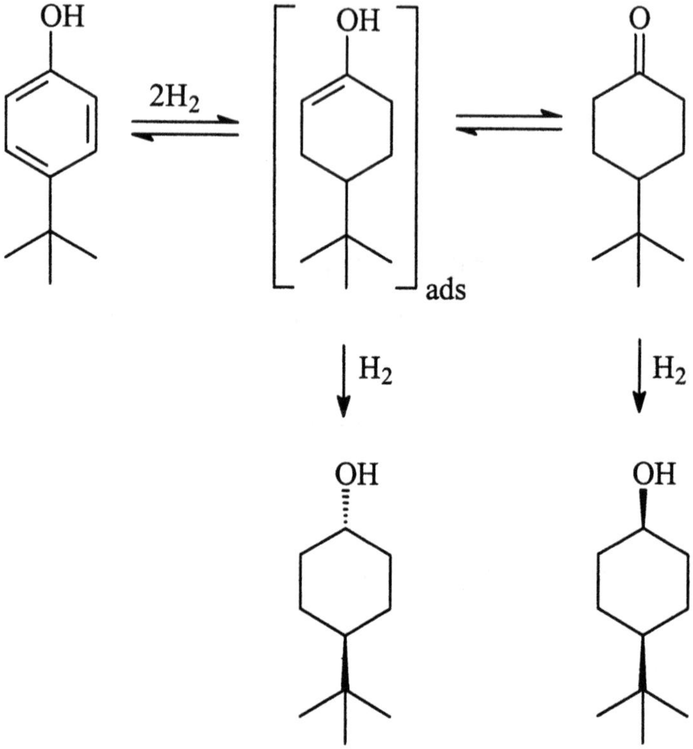

Figure 1 Reaction scheme for the hydrogenation of 4-t-butyl phenol to cis and trans-4-t-butyl cyclohexanol.

Temperature (20 – 70EC) and pressure (1 – 10 bar) had little effect on the selectivity of the reaction, but gave faster reaction rates when both were increased.

INTRODUCTION

The selective hydrogenation of 4-t-butyl phenol is of commercial interest for the manufacture of cis-4-t-butyl cyclohexanol (1), which as is an important product used in the Flavours and Fragrance industry. The reaction is also of interest in that several compounds are potential products of the reaction, namely the cis and trans-t-butyl cyclohexanol as well as the 4-t-butyl cyclohexanone. Previous studies (2, 3) have shown rhodium to be the most active catalyst for this reaction and depending upon the conditions used can give high selectivities to the cis-isomer. These studies also proposed a reaction scheme as shown in Figure 1, whereby the reaction proceeds via an intermediate enol and ketone, where the enol gives rise to the trans-isomer and the ketone the cis-isomer.

The aim of this work was to study the effect of solvents on the activity and selectivity with a range of Johnson Matthey supported heterogeneous Platinum Group Metal catalysts.

EXPERIMENTAL

Catalysts

The following range of commercial heterogeneous Platinum Group Metal catalysts were used throughout these studies:
5% Rh/Carbon, 5% Rh/Graphite, 5% Rh/Alumina
5% Pt/Carbon, 5% Pt/Graphite
5% Pd/Carbon
5% Ru/Carbon
5% Ir/Carbon

Reactor and Analytical Procedures

All the catalyst test work was performed in a 50ml batch stirred autoclave using 30ml of solvent containing 1g of the 4-t-butyl phenol with 50mg of dry powder catalyst. The reactions were performed at 50°C and 10 bar hydrogen pressure unless otherwise stated (when the effect of varying temperature and pressure were investigated). The stirrer speed used throughout the experiments was 1,000 rpm. The solvents used in the studies were hexane, decane, isopropanol, octanol and water.

The reactions were monitored by recording the hydrogen uptake to allow calculation of the reaction rate as mol/hr/gm cat. The hydrogen uptake was then compared to the theoretical amount of hydrogen (0.448 litres) required to convert the phenol to the cyclohexanol. The selectivity was determined by GC analysis of the reaction products after filtering off the catalyst.

RESULTS AND DISCUSSION

Testing of the full range of catalysts in a range of solvents showed that only the Rhodium and Platinum catalysts were active for this reaction at 50°C and 10 bar, as all the other catalysts either showed no reaction at all or were extremely slow, such that no conversion was detected by either hydrogen uptake or GC analysis.

Table 1 Results with Rh and Pt catalysts in various solvents.

Catalyst	Solvent	Rate (mol/hr/gm cat)	Cis:trans ratio
5% Rh/C	Hexane	175	5.9
5% Rh/Graphite	Hexane	80	1.5
5% Rh/Al$_2$O$_3$	Hexane	445	1.0
5% Pt/C	Hexane	64	0.3
5% Pt/Graphite	Hexane	Very Slow	-
5% Rh/C	Decane	216	4.9
5% Rh/Al$_2$O$_3$	Decane	599	0.9
5% Rh/C	Water	163	1.7
5% Rh/Al$_2$O$_3$	Water	274	1.0
5% Rh/C	Isopropanol	313	4.3
5% Rh/Graphite	Isopropanol	127	1.0
5% Rh/Al$_2$O$_3$	Isopropanol	366	1.1
5% Pt/C	Isopropanol	Very Slow	-
5% Pt/Graphite	Isopropanol	Very Slow	-

Results for the Rh and Pt catalysts reported in Table 1 show that the 5% Rh/Al$_2$O$_3$ catalyst is the most active in all solvents, but 5% Rh/C is the most selective to the cis-isomer particularly when using hexane (or other non polar) solvents. Interestingly Pt/C has a reasonable activity in hexane, but almost no activity in isopropanol (this was checked by also running in octanol solvent and again it was inactive). Also the selectivity was the reverse of the Rh/C catalyst in that more trans than cis was formed with the Pt/C catalyst.

Analysis of the reaction mixture during the reaction is plotted in Figure 2. This shows that the reaction goes through the intermediate 4-t-butyl cyclohexanone (as illustrated in Figure 1), which shows a maximum concentration at approximately 70% conversion of the 4-t-butyl phenol in both hexane and isopropanol solvents. This intermediate is hydrogenated to the cis-isomer as can be seen from Figure 2; as the level of the intermediate decreases the cis level increases significantly, but the trans level barely changes.

The reaction profile with the Rh/C catalyst in isopropanol solvent also follows a similar trend, but the level of cis-isomer product is less than in the case with the hexane solvent.

The reaction profile with Pt/C in hexane shown in Figure 2 is totally different to that for Rh/C in that no build up of the 4-t-butyl cyclohexanone is observed, which leads to the enhanced levels of trans-isomer product as the reaction proceeds through the enol intermediate as shown in Figure 1.

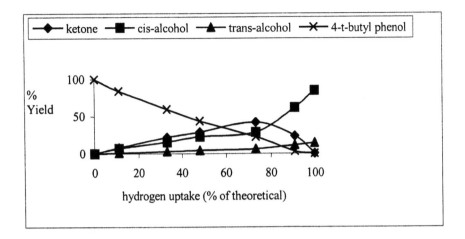

Figure 2 Selectivity profile during the hydrogenation of 4-t-butyl phenol over Rh/C in hexane solvent at 50°C and 10 bar.

The effect of temperature on activity and selectivity was studied in decane solvent in the temperature range of 20 - 70°C with the 5% Rh/C catalyst at 10 bar. The results presented in Table 2 show that temperature has little effect on the selectivity as the cis:trans ratio remains constant at 4-5:1, but the reaction rate as expected increased (from 108 to 259 mol/hr/gm cat.).

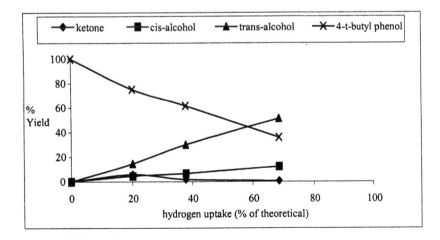

Figure 3 Selectivity profile during the hydrogenation of 4-t-butyl phenol over 5% Pt/C in hexane solvent at 50°C and 10 bar

534

Table 2 Effect of temperature on activity and selectivity with 5% Rh/C in decane solvent at 10 bar.

Temperature (°C)	Rate (mol/hr/gm cat)	Cis:trans ratio
20	108	4.2
30	132	4.6
40	156	4.6
50	175	4.9
60	197	4.4
70	259	4.3

These results show an apparent activation energy for the hydrogenation of 4-t-butylphenol of 15kJ/mol, which indicates that the reaction is controlled by liquid/solid mass transfer.

Similar results were obtained when studying the effect of pressure between 1 and 10 bar again with the 5% Rh/C catalyst in decane solvent as the cis:trans ratio remained unchanged at 4:1 whilst the reaction rate increased.

MECHANISM

The results reported in this paper agree with previous data (1, 3) and are consistent with the mechanism of the hydrogenation in that the reaction proceeds via an enol which is in equilibrium with the cyclohexanone. The enol can react with hydrogen to give the trans-isomer and the cyclohexanone gives the cis-isomer as shown by conformational analysis of the enol and ketone intermediates shown in Figure 4, where the preferred direction of addition of the hydrogen to the C=C and C=O bonds is illustrated.

The non-polar hexane solvent in conjunction with the 5% Rh/C catalyst appears to promote the conversion of the enol to the ketone, which gives rise to the higher cis:trans ratios observed with hexane and decane (compared to the polar isopropanol). However, with the Pt/C catalyst in hexane this conversion is not promoted and consequently higher levels of the trans-isomer are observed.

CONCLUSIONS

Of the catalysts evaluated in this programme by far the best for selectivity to cis-4-t-butylcyclohexanol product is the 5% Rh/C catalyst.

The best solvents to use in conjunction with the Rh/C catalyst are non polar such as hexane and decane, which give significantly higher cis:trans ratios than polar solvents such as isopropanol or octanol.

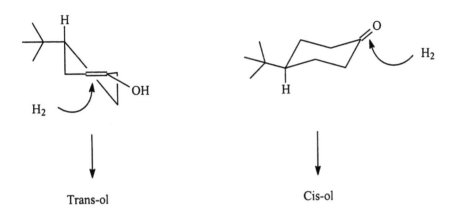

Trans-ol   Cis-ol

Figure 4 Conformational analysis showing the enol and ketone intermediates leading to formation of the trans and cis isomers.

The Pt/C catalyst is only active in hexane solvent, but gives the opposite selectivity to the Rh/C catalyst, as the trans-isomer is the major product.

Increasing both temperature (20 - 70°C) and pressure (1 – 10 bar) has little effect on the selectivity, but increased rates are observed. Activated carbon is the best support for the rhodium catalyst as switching to either graphite or alumina gives catalysts with lower selectivities to the cis-isomer.

The mechanism for the hydrogenation proceeds via the enol, which is in equilibrium with the cyclohexanone. To obtain high selectivities to the cis-isomer the reaction needs to be 'pushed' through the ketone intermediate i.e. the conversion of the enol to the ketone needs to be enhanced, whilst conversion of the enol to the cyclohexanol needs to be depressed.

REFERENCES

1. D. Y. Murzin, A. I. Allachverdiev, N. V. Kul'kova. Kinetics of liquid phase stereoselective hydrogenation of 4-tert-butylphenol over rhodium catalysts. In: Heterogeneous Catalysis and Fine Chemicals III. Elsevier 1993, pp 243-250
2. S. R. Konuspaev, K. N. Zhanbekov, N. V. Kul'kova, D. Y. Murzin. Kinetics of 4-tert-butylphenol hydrogenation over rhodium. *Chemical Engineering Technology* 20:144-148, 1997
3. D. Y. Murzin. Liquid phase hydrogenation of 4-tert-butylphenol I. The kinetic model. *Kinetics and Catalysis* 34 (3):437-441, 1993

44

Stoichiometry and Kinetics of Gas Phase Cyclohexene Epoxidation by a Silica-Supported *tert*-Butylperoxidititanium Complex

Abdillah O. Bouh, Azfar Hassan, and Susannah L. Scott
University of Ottawa, Ottawa, Ontario, Canada

INTRODUCTION

Peroxo and alkylperoxo metal complexes are presumed intermediates in the homogeneous and heterogeneous catalytic epoxidations of olefins. Both solid and soluble versions containing high valent, d^0 transition metal ions activate peroxides heterolytically, thereby facilitating oxygen atom transfer to electron-rich substrates (1). However, mechanistic studies have been undertaken largely with the more readily characterized soluble catalysts (2–4). For heterogeneous catalysts, the fraction of metal sites which participate in selective oxidation is generally unknown. Although measurements and comparisons of overall activity and selectivity are common (5), it is not generally possible to distinguish between mechanisms or measure elementary rate constants for individual sites.

Titanium alkoxides are effective homogeneous catalysts for the epoxidation of substituted olefins. Their propensity for association to multinuclear species is also well-established (6). The active form of a homogeneous, enantioselective titanium-tartrate catalyst was demonstrated to be dinuclear in titanium (7). In contrast, heterogeneous catalysts consisting of titanium embedded in an aluminosilicate framework contain mostly isolated titanium sites (8), although the assertion that such sites are uniquely responsible for catalyst activity is based on indirect (and disputed) spectroscopic evidence (9). Since the stoichiometry of an epoxidation reaction can be established only if the compo-sition o f the *active* site is known, and since the active site is not necessarily a major (spectroscopically observable) component of a heterogeneous system, there is considerable advantage in mechanistic studies of solid catalysts in which all metal sites have the same or very similar properties.

Grafting of metal complexes onto oxide surfaces can lead to supported catalysts with a high degree of uniformity in the surface organometallic fragments, facilitating the interpretation of kinetic studies. The preparation of solid catalysts by reaction of $Ti(O^iPr)_4$ with silica has been reported by several groups (10, 11). This material appears to have several desirable properties compared to other heterogeneous Ti/silicate materials, *viz.*, lower moisture-sensitivity and an absence of steric constraint on substrate size (12). Armed with knowledge of the composition of its uniform active sites (13), w e now report detailed information about the stoichiometry and kinetics of olefin epoxidation over the solid catalyst.

EXPERIMENTAL

Preparation of Silica-Supported Titanium Catalyst

The catalyst was prepared by the reaction of $Ti(O^iPr)_4$ (99.999%, Aldrich) with a non-porous, pyrogenic silica, as described previously (13). Degussa Aerosil™-200 (surface area 183 m^2/g) was rehydrated then partially dehydroxylated *in vacuo* at 200°C, after which treatment the hydroxyl content is reproducible at 2.6 OH/nm^2 (14). All subsequent manipulations were performed *in situ* in the absence of inert gases or solvents, using standard breakseal and high vacuum techniques.

Anhydrous *tert*-butylhydroperoxide (Aldrich, 10 M in decane) was dried over $MgSO_4$, degassed by three freeze-pump-thaw cycles and stored in a glass bulb under vacuum. Titanium-modified silica pellets or powder were exposed to an excess of ^tBuOOH vapor in order to generate the supported *tert*-butylperoxide complex, followed by evacuation to remove ^iPrOH and unreacted ^tBuOOH .

Stoichiometric and Kinetics Measurements

Cyclohexene was dried and vacuum-distilled before use. It was degassed by three freeze-pump-thaw cycles and stored over activated molecular sieves in a glass bulb. It was introduced into the reactor via vapor phase transfer through a high vacuum manifold (base pressure < 10^{-4} Torr). After 30 mins, the epoxide yield was quantified on an HP 6890 GC/MS equipped with a J&W Scientific DB1 capilary column. At the end of each experiment, Ti analysis was performed (15) and epoxide/Ti ratios were calculated. For kinetics experiments, silica powder containing the *tert*-butylperoxotitanium complex was prepared in an *in situ* reactor and the reaction initiated by addition of olefin. The IR spectrum of the gas phase above the silica was recorded at timed intervals. Pseudo-first-order rate constants were extracted from non-linear least-squares fits of the integrated rate equation to the change in absorbance in the $v(C=C)$ region vs. time.

RESULTS AND DISCUSSION

Composition of the Titanium Sites in the Grafted Catalyst

We previously reported the preparation and quantitative characterization of well-defined, uniform titanium alkoxide complexes on a silica surface, via the grafting of $Ti(O^iPr)_4$. We demonstrated, via the mass balance, that the reaction on Aerosil-200 silica pretreated at 200°C results exclusively in the formation of dinuclear oxo-bridged complexes regardless of Ti loading (13), eq 1.

$$2 \equiv SiOH + 2\ Ti(O^iPr)_4 \quad (\equiv SiO)_2TiOTi(O^iPr)_4 + 3\ ^iPrOH + C_3H_6 \qquad (1)$$
$$\mathbf{1}$$

The dinuclear Ti complex assembles spontaneously on the silica surface via a non-hydrolytic alkoxide condensation reaction which generates the Ti-O-Ti bridge as well as C_3H_6 (16). Confirmation of this condensation process was obtained via the exposure to $Ti(O^iPr)_4$ of a silica fully loaded (*i.e.*, with no unreacted hydroxyl groups) with a mononuclear titanium alkoxide complex. The latter was prepared by grafting of $Ti(NEt_2)_4$ followed by ligand metathesis with ^iPrOH, eq 2. The alkoxide condensation reaction proceeds quantitatively (13), eq 3:

$$(\equiv SiO)_2Ti(NEt_2)_2 + 2\ ^iPrOH \rightarrow (\equiv SiO)_2Ti(O^iPr)_2 + 2\ HNEt_2 \qquad (2)$$
$$(\equiv SiO)_2Ti(O^iPr)_2 + Ti(O^iPr)_4 \rightarrow (\equiv SiO)_2TiOTi(O^iPr)_4 + ^iPrOH + C_3H_6 \qquad (3)$$

The 2-propoxide ligands of **1** are displaced quantitatively in the presence of excess tert-butylhydroperoxide, eq 4:

$$(\equiv SiO)_2TiOTi(O^iPr)_4 + 4\ ^tBuOOH \rightarrow (\equiv SiO)_2TiOTi(OO^tBu)_4 + 4\ ^iPrOH \quad (4)$$
$$\mathbf{1} \qquad\qquad\qquad\qquad\qquad\qquad \mathbf{2}$$

That ^iPrOH is completely displaced from the coordination sphere of Ti, and can be removed from the reaction vessel by simple evacuation, was previously shown by IR experiments involving D-labelling and ^{13}C CP/MAS NMR (13). Th e silica-supported Ti complexes remain firmly anchored to the support in the presence of excess *tert*-butylhydroperoxide vapor: no regeneration of surface hydroxyls (corresponding to protonolysis of the Si-O-Ti linkages (17)) was observed.

Reaction of the Alkylperoxodititanium Complex with Cyclohexene

The dinuclear *tert*-butylhydroperoxo complex **2** exhibits a vibration at 855 cm^{-1} assigned to $v(O-O)$, Figure 1a. Upon exposure of **2** to excess cyclohexene vapor at room temperature, this peak disappeared completely, Figure 1b. After reaction of **2** with cyclohexene for 30 mins, cyclohexene oxide was the exclusive volatile product. Upon evacuation to remove physisorbed cyclohexene oxide, the low frequency IR spectrum is qualitatively and quantitatively identical to that of a sample of $(\equiv SiO)_2TiOTi(O^tBu)_4$, **3**, prepared independently by reaction of **1** with *tert*-butanol.

Epoxidation Stoichiometry

Using techniques previously described to quantify gaseous products of surface reactions (13,14), we measured the quantity of epoxide formed by reaction of the *tert*-butylperoxotitanium complex **2** with an excess of cyclohexene in the absence of *tert*-butylhydroperoxide (*i.e.*, stoichiometric gas phase reaction conditions). Complex **2**, with four *tert*-butylperoxo ligands per TiOTi unit, generated (4.11 ± 0.08) equiv. of cyclohexene oxide (average of four experiments), demonstrating that all of its Ti sites are active and all the peroxidic oxygens are utilized. The stoichiometry of the epoxidation can therefore be described precisely, eq5.

$$(\equiv SiO)_2TiOTi(OO^tBu)_4 + xs\ C_6H_{10} \rightarrow (\equiv SiO)_2TiOTi(O^tBu)_4 + 4\ C_6H_{10}O \quad (5)$$
$$\mathbf{2} \qquad\qquad\qquad\qquad\qquad\qquad \mathbf{3}$$

This finding of complete transfer of peroxidic oxygens to substrate is consistent with IR spectroscopic evidence (Figure 1). Furthermore, regeneration of the *tert*-butylperoxo complex **2** by treatment of **3** with *tert*-butylhydroperoxide, followed by its subsequent reaction with cyclohexene, resulted in the formation of another four equiv. of cyclohexene oxide, eq 6. Its active sites are therefore not susceptible to deactivation either during or after epoxidation.

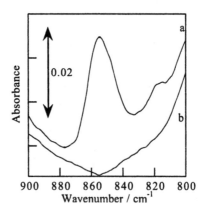

Figure 1 Transmission IR spectra of $(\equiv SiO)_2TiOTi(OO^tBu)_4$, **2**, (a) before, (b) after reaction with cyclohexene.

$$(\equiv SiO)_2TiOTi(O^tBu)_4 + tBuOOH + xs\ C_6H_{10} \rightarrow 4\ C_6H_{10}O \qquad (6)$$

Kinetics of Epoxidation

The rate of expoxidation was measure *in situ* via the uptake of cyclohexene vapor by the catalyst. At low pressures (*ca.* 10 Torr), cyclohexene does not adsorb on the unmodified silica surface nor on either of the silica-supported 2-alkoxide complexes **1** and **3**. However, the addition of cyclohexene vapor to **2** resulted in a rapid, exponential loss of $v(C=C)$ intensity in the IR spectrum of the gas phase above the silica-supported complex, Figure 2. We infer that epoxidation results in adsorption of cyclohexene oxide on the catalyst surface.

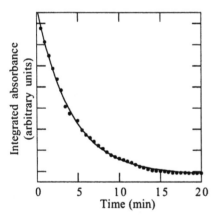
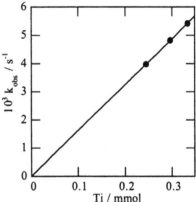

Figure 2 Rate of dissappearence of cyclohexene above **2**. The solid line is the fit to the first-order rate law.

Figure 3 Dependence of pseudo-first-order rate constants for epoxidation of cyclohexene at 25°C on the amount of Ti in **2**.

The uptake of cyclohexene by the Ti-modified silica is pseudo-first-order, as shown by the fit to the exponential curve in Figure 2. Furthermore, the measured pseudo-first-order rate constants are linearly dependent on the amount of Ti present in the reactor, Figure 3, consistent with the rate law shown in eq 7:

$$-d[C_6H_{12}]/dt = 2k[C_6H_{12}]n_{Ti} \qquad (7)$$

where the factor of 2 arises from the conversion of two equiv. of cyclohexene on each Ti site (see above). The second-order rate constant k for cyclohexene epoxidation is 8.1 s^{-1} (mol Ti)$^{-1}$ at 25°C.

CONCLUSION

Measurement of the stoichiometry of cyclohexene epoxidation over a grafted Ti-silica catalyst demonstrates that (1) all Ti sites are active; and (2) all peroxidic oxygens bound to Ti react. The rate law for the gas-solid reaction is mixed-second-order. Because of the compositional uniformity of the Ti sites and their quantitative participation in the epoxidation reaction, the measured second-order rate constant accurately reflects the activity at each Ti site.

ACKNOWEDGEMENT

This work was supported by a Strategic Research Grant from NSERC (Canada) and a Cottrell Scholar Award from Research Corporation.

REFERENCES

1.   KA Jørgensen. Transition metal catalized epoxidations. Chem Rev 89:431–458, 1989.
2.   H Mimoun, M Mignard, P Brechot, L Saussine. Selective epoxidation of olefins by oxo[N-(2-oxidophenyl)salicylidenaminato]vanadium(V)alkylperoxides. On the mechanism of the Halcon epoxidation process. J Am Chem Soc 108:3711–3718, 1986.
3.   MK Trost, RG Bergman. Cp*MoO₂Cl-catalyzed epoxidation of olefins by alkylhydroperoxides. Organometallics 10:1172–1178, 1991.
4.   SS Woodard, MG Finn, KB Sharpless. Mechanism of asymmetric epoxidation. 1. Kinetics. J Am Chem Soc 113:106–113, 1991.
5.   IWCE Arends, RA Sheldon, M Wallau, U Schuchardt. Oxidative transformations of organic compounds mediated by redox molecular sieves. Angew Chem Int Ed Eng l36:1144–1163, 1997.
6.   DC Bradley, RC Mehrotra, DP Gaur. Metal Alkoxides. New York: Academic Press, 1978.
7.   MG Finn, KB Sharpless. Mechanism of asymmetric epoxidation. 2. Catalyst structure. J Am Chem Soc 113:113–126, 1991.
8.   JM Thomas, G Sankar, MC Klunduk, MP Attfield, T Maschmeer, BFG Johnson, RG Bell. The identity in atomic structure and performance of active sites in heterogeneous and homogeneous titanium-silica epoxidation catalysts. J Phys Chem B 103:8809–8813, 1999.
9.   AS Soult, DD Pooré, EI Mao, AE Stiegman. Electronic spectroscopy of the titanium centers in titanium silicalite. J Phys Chem B 105:2687–2693, 2001.
10. C Bland, J-L Pellegatta, R Choukroun, B Gilot, R Guiraud. Esterification de l'acid estearique catalysée par les titanates greffés. Can J Chem 71:34–37, 1993.
11. C Cativiela, JM Fraile, JI García, JA Maoral. A new titanium-silica catalyst for the epoxidation of alkenes. J Mol Catal A: Chem 112:259–267, 1996.
12. JM Fraile, J García, JA Maoral, MG Proietti, MC Sánchez. Titanium catalysts supported on silica. X-ray absorption investigation on their structures and

comparison of their catalytic activities in Diels-Alder and epoxidation reactions. J Phys Chem 100:19484–19488, 1996.

13. AO Bouh, GL Rice, SL Scott. Mono- and dinuclear silica-supported titanium(IV) complexes and the effect of Ti-O-Ti connectivity on reactivity. J Am Chem Soc 121:7201–7210, 1999.

14. GL Rice, SL Scott. Characterization of silica-supported vanadium(V) complexes derived from molecular precursors and their ligand exchange reactions. Langmuir 13:1545–1551, 1997.

15. S Haukka, A Saastamoinen. Determination of chromium and titanium in silica-based catalysts by ultraviolet/visible spectrophotometry. Analyst 117:1381–1384, 1992.

16. NY Turova, EP Turevskaa, MI Yanovskaa, AI Yanovsky, VG Kessler, DE Tcheboukov. Physicochemical approach to the studies of metal alkoxides. Polyhedron 17:899–915, 1998.

17. HCL Abbenhuis, S Krijnen, RA van Santen. Modelling the active sites of heterogeneous titanium epoxidation catalysts using titanium silasequioxanes: Insight into specific factors that determine leaching in liquid-phase processes. Chem Commun 331–332, 1997.

45

Dye-Labeled PNIPAM [Poly(N-isopropylacrylamide)]-Supported Acylation Catalysts

Chunmei Li and David E. Bergbreiter
Texas A&M University, College Station, Texas, U.S.A.

ABSTRACT

An azo benzene derivative was incorporated into a soluble PNIPAM-supported catalyst to study the phase preference, concentration, and recoverability of the catalyst.

RESULTS AND DISCUSSION

The separation step in organic synthesis and catalysis is of increasing importance in academic and industrial chemistry. Polymer-supported reagents and catalysts received special attention in this aspect because using a 'phase tag' for a substrate or a catalyst simplifies product isolation and allows for catalyst recovery. This is especially important in high throughput chemistry and Green chemistry. Fluorous supports, [1,2,3] soluble polymer supports and dendrimer supports have all been used for this purpose. [4,5,6] Small molecules such as the recently developed 'precipiton' or metal complexing groups can also serve a similar purpose. [7,8,9,10]

Our group has extensively used PNIPAM as a soluble polymer support for catalysts.[11] Such supports can be easily prepared as copolymers with an active ester. Such copolymerization provides an easy way to attach a catalyst onto this polymer support by allowing an amine-terminated catalyst to react with poly (N-isopropyl acrylamide)-co-poly(N-acryloxy succinimide) (PNIPAM-NASI). The active ester of this copolymer reacts with the amine to form an amide bond between the catalyst and the polymer, as shown in Figure 1. We have also shown that azo dyes can also be attached to a polymer in this way and have used such dye-labeled polymers as probes of separability of a polymer support under liquid/liquid conditions.[11,12,13]

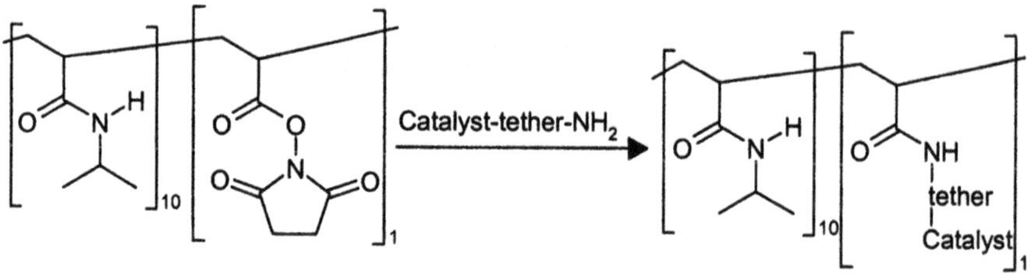

Figure 1 Attaching Catalysts to PNIPAM Support.

An important consideration in catalyst, reagent or substrate recovery is measuring and verifying how effective such recovery actually is. While we have modeled such recovery using dye-modified polymers, analyses of catalysts typically requires additional analytical work. For example, ICP analysis for residual metal can be used as a quantitative and sensitive assay. Such assays are however more problematic with non-metallic catalysts. In this paper, we show that bifunctional polymers where both a catalyst and a colorimetric label are included in the same polyacrylamide polymer provide a simple way to monitor separability *and* catalyst recovery for non-metallic polymer-bound catalysts.

There are several different methods to separate PNIPAM-supported catalysts from the reaction mixtures. Both liquid-solid separations and liquid-liquid separations can be used. The most frequently used liquid-solid separation method takes advantage of the varying solubility of polymers in different solvents. For example, PNIPAM can be precipitated from THF into hexanes. PNIPAM copolymers also exhibit lower critical solution temperature (LCST) behavior. Specifically, PNIPAM and its copolymers can be prepared such that these polymers are soluble in water at low temperature but precipitate when heated up. This property may be used as either a purification method or a separation technique.[11] A thermomorphic system is a liquid-liquid biphasic system developed in our group. It uses various solvent mixtures with temperature-dependent miscibility to effect separation of catalysts from substrates and products, as shown in Figure 2.

The solvent mixtures are not miscible at room temperature, and the catalyst has different phase preference from the substrate/product. At elevated temperatures, the solvent mixture can become a single phase. Reactions thus occur under homogeneous conditions but the catalyst phase can be easily separated from the product phase on cooling. The catalyst phase can then be mixed with fresh substrate solution and recycled.[12,14] To illustrate the potential of dye-labeled polymers in monitoring and verifying catalyst reuse/recovery, we opted to prepare a soluble polymer-bound version of 4-*N,N*-dimethylpyridine. This bifunctional polymer (PNIPAM-DAAP-MR) was accessible using the synthetic scheme shown below in Scheme 1. The catalyst loading of **13** was determined by ^{1}H NMR spectroscopy and the methyl-red loading was determined by UV analysis ($\lambda_{max} = 430$ nm) assuming that the polymer-bound catalyst and the low molecular weight amine-terminated methyl red **12** have the same extinction coefficient. Their UV spectra are shown in Figure 3.

Figure 2 Thermomorphic Catalysis.

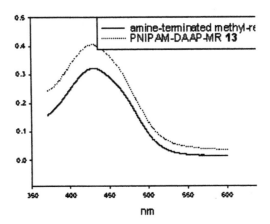

Figure 3 UV Spectra of **12** and **13**.

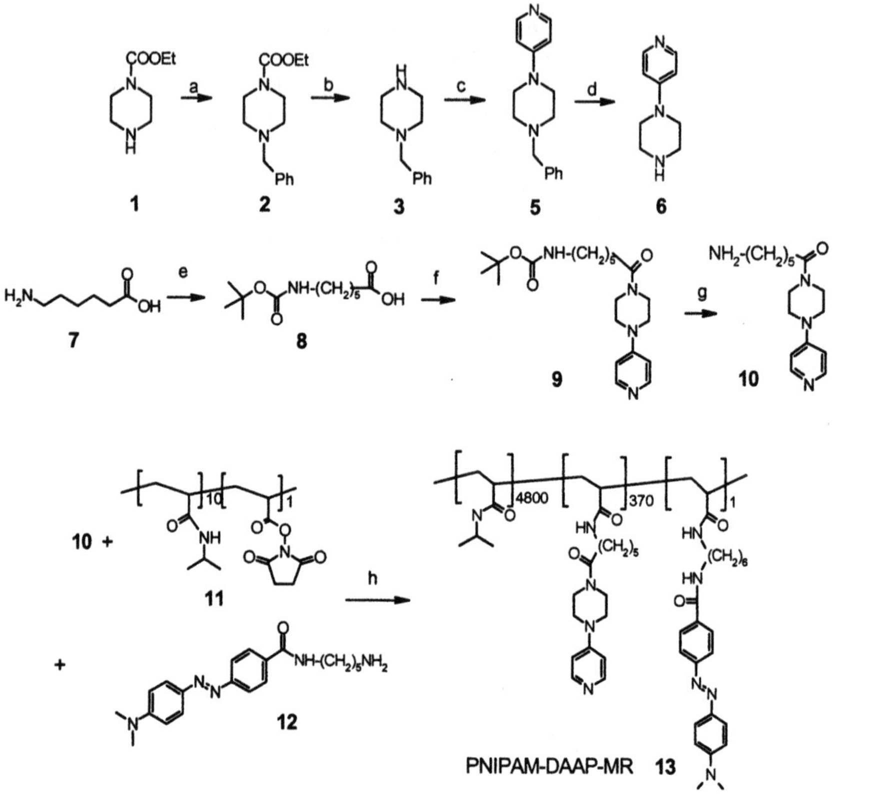

a. NaHCO$_3$, PhCH$_2$Cl, 95% EtOH, reflux, 18h, 99%; b. KOH, MeOH, reflux, 69%; c. 4-chloropyridine **4**, xylene, reflux, 15h, 93%; d. 2N HCl, H$_2$, Pd/C, MeOH, 35 psi, 99%; e. (Boc)$_2$O, NaOH, THF-H$_2$O, 88%; f. **6**, DCC, CH$_2$Cl$_2$, 84%; g. TFA, CH$_2$Cl$_2$, 68%; h. CH$_2$Cl$_2$, i-PrNH$_2$

Scheme 1 Synthesis of PNIPAM-DAAP-MR.

Dialkylaminopyridine (DAAP) catalysts are effective catalysts for acylation reactions. The acetylation of 1- methylcyclohexanol by acetic anhydride in the presence of triethylamine (TEA) was used as a test reaction for the polymer-bond DAAP. The conversion of the starting alcohol was measured by GC analysis using ethyl benzene as an internal standard. The initial rates were studied and a first order analyses showed that the rate constant for reaction (1) was 0.178 h^{-1} for 4-dimethylaminopyridine (DMAP), and 0.120 h^{-1} for PNIPAM-DAAP-MR.

$$DAAP = DMAP \text{ or } PNIPAM\text{-}DAAP\text{-}MR$$

Although our polymer-bound catalyst was not as active as DMAP in a single cycle, it can be recovered and reused. Pouring the dichloromethane solution into hexanes induces precipitation of the polymer which can then be used again. The hexanes solution had no UV-visible absorption, suggesting that the catalyst precipitates quantitatively. The recovered catalyst was used for 3 more cycles. The rate constant obtained for cycle 2, 3 and 4 are 0.117 h^{-1}, 0.112 h^{-1} and 0.110 h^{-1}, respectively. To test whether the small decrease in the rate constants is caused by decrease of catalytic activity or by mechanical loss of polymer during handling, the correlation between rate constants in each cycle and the catalyst concentration were studied by UV analysis, which showed that the catalyst retains the same activity during the 4 cycles This catalyst also worked for other acylation reactions. Moreover, these polymers can be used, separated and recovered in thermomorphic system. In all these cases, incorporating a colorimetric compound into the polymer support allowed us to show that PNIPAM-DAAP-MR can be recovered quantitatively and reused with no diminution in activity.

ACKNOWLEDGMENTS

We gratefully acknowledge the Petroleum Research Fund and the National Science Foundation for funding this research.

REFERENCES

1. IT Horvath, J Rabai. *Science* **266**: 72-75, 1994.
2. JJJ Juliette, IT Horvath, JA Gladysz. *Angewandte Chemie International Edition* **36**: 1610-1612, 1997.
3. C Rocaboy, W Bauer, JA Gladysz. *European Journal of Organic Chemistry* **14**: 2621-2628, 2000.
4. B Clapham, TS Peger, KD Janda. *Tetrahedron* **57**: 4637-4662, 2001.
5. A Kirschning, H Monenschein, R Wittenberg. *Angewandte Chemie International Edition* **40**: 650-679, 2001.
6. R Haag. *Chemistry: A European Journal* **7**: 327-335, 2001.

7.   T Bosanac, CS Wilcox. *Chemical Communications* **17**: 1618-1619, 2001.

8.   T Bosanac, CS Wilcox. *Tetrahedron Letters* **42**: 4309-4312, 2001.

9.   T Bosanac, J Yang, CS Wilcox. *Angewandte Chemie International Edition* **40**: 1875-1879, 2001.

10. SV Ley, A Massi, F Rodriguez, DC Horwell, RA Lewthwaite, MC Pritchard, AM Reid. *Angewandte Chemie International Edition* **40**: 1053-1055, 2001.

11. DE Bergbreiter, BL Case, YS Liu, JW Caraway. *Macromolecules* **31**: 6053-6062, 1998.

12. DE Bergbreiter, PL Osburn, A Wilson, EM, Sink. *Journal of the American Chemical Society* **122**: 9058-9064, 2000.

13. DE Bergbreiter, JG Franchina, BL Case, LK Williams, JD Frels, N. Koshti. *Combinatorial Chemistry & High Throughput Screening* **3**: 153-164, 2000.

14. DE Bergbreiter; YS Liu, PL Osburn. *Journal of the American Chemical Society* **120**: 4250-4251, 1998.

46

Oxidation of Primary Alcohols with *tert*-Butylhydroperoxide in the Presence of Ruthenium Complexes

Nagendranath Mahata, Setrak K. Tanielyan, and Robert L. Augustine
Center for Applied Catalysis, Seton Hall University, South Orange, New Jersey, U.S.A.

Indra Prakash
The NutraSweet Company, Mount Prospect, Illinois, U.S.A.

ABSTRACT

Primary alcohols are selectively oxidized to aldehydes by the $RuCl_2(TPP)_3$/ tert-butylhydroperoxide (TBHP) system. The reaction takes place at room temperature using catalytic amounts of the ruthenium complex. Specific conditions are found leading to improved efficiency in the TBHP decomposition and in reduction of the level of secondary products originating from the aldehyde post-reactions. The effect of some free-radical scavengers and the structure of the ruthenium complex will also be presented.

INTRODUCTION

The catalytic oxidation of alcohols selectively to carbonyl compounds is one of the more important transformations in the synthetic organic chemistry. A large number of oxidants have been reported in the literature and most of them are based on transition metal oxides such as chromium and manganese [1-3]. A serious drawback to these reagents is the need to use them in large amounts, very often in excess. Since most of them and the products of their transformations are toxic species, their use also creates serious problems concerning their handling and disposal. The search for efficient, easily accessible catalysts and "clean" industrially viable oxidants such as hydrogen peroxide, hydroperoxides or molecular oxygen is still a challenge [4,5]. A large number of transition metal complexes and oxidants have been reported to catalyze the selective oxidation of primary alcohols to aldehydes with varying levels of effectiveness. Materials such as $RuCl_3$-$NaBrO_3$ [6], Bu_4NRuO_4-4-methylmorpholine N-oxide (MMO) [7], $RuCl_2(PPh_3)_3$ in combination with $(Me_3SiO)_2$ [8], H_2O_2 and tert-butylhydroperoxide (t-BuOOH) [9] and MMO [10] have been used.

An extensive study of the activity of the $RuCl_2(TPP_3)_3$ based systems has shown that the use of MMO as an external oxidant is more effective than when t-BuOOH is used at ambient temperatures [10]. Since the t-BuOOH is a much cleaner oxidizing agent, it is readily available in large quantities and its handling at high concentration does not create any safety concerns, we decided to re-visit the ruthenium catalyzed oxidation of a model primary alcohol using t-BuOOH. The reaction sequence involves the formation of t-butyl peroxoruthenium species which selectively decomposes to an oxo-ruthenium intermediate, which, after abstraction of α-hydrogen from the alcohol, forms an alkyl ruthenium complex. A two electron transfer within the complex (hydride transfer) produces the carbonyl compound, water and Ru(II) which returns into the catalytic cycle.[10]

We present here some results obtained for the selective oxidation of a primary alcohol, with t-BuOOH catalyzed by some commercially available ruthenium complexes. In addition, the effect of some free radical scavengers on the reaction selectivity is also described as well as the use of other transition metal complexes as oxidizing agents.

EXPERIMENTAL

Apparatus Experiments were carried out in the low pressure glass volumetric system described elsewhere in this volume [11].

Procedures In a representative run the reaction flask was charged with the alcohol (8.5mmol), ruthenium complex (0.1 mmol) and solvent (20 ml) in argon atmosphere. The mixture was stirred at 30°C over 15 min and t-BuOOH (0.85 mmol) was added over 1 hour at a controlled rate using a syringe pump. The progress of the reaction was monitored by taking aliquots at predetermined time intervals. The liquid samples were filtered through a Celite pad and analyzed by GLC.

Table 1 Transition metal complex catalyst screening[a]

Run	Catalyst	Reaction characteristics					
		Based on converted t-BuOOH			Based on converted alcohol		
		Conversion[c] %	Efficiency[d] %	Yield[e] %	Conversion[f] %	Selectivity[g] %	Yield[h] %
1	$MoO_2(AcAc)_2$	99	11	11	30	38	12
2	$VO(AcAc)_2$	47	11	5	22	25	6
3	$TiO_2(AcAc)_2$	41	0	0	3	0	0
4	WO_2Cl_2	21	4	1	4	7	1
5	$ZrO(OAc)_2$[b]	45	79	35	54	68	37
6	$ZrO(ClO4)_2$	72	3	2	20	10	2
7	$ZrO(AcAc)_4$	54	1	1	2	5	1
8	$ZrO(OCOCF_3)_2$	44	7	3	25	13	3
9	ZrO_2	27	6	1	10	16	2

[a] Reaction conditions: 8.5 mmol substrate, 0.4 mmol catalyst, 20 ml chloroform, 0.85mmol t-BuOOH added over 1h and the solution stirred for an additional 2 h. Reaction temperature 80°C; [b]solvent acetonitrile 20 ml; [c]determined by GC based on the amount of t-BuOOH used; [d] the ratio between the aldehyde formed and the converted t-BuOOH; [e] yield based on the t-BuOOH used; [f] determined by GC based on the alcohol used; [g] the ratio between the aldehyde formed vs converted alcohol; [h] based on the amount converted alcohol.

RESULTS AND DISCUSSION

1. Effect of the catalyst system

We first screened several transition metal complexes which have been shown to have activity in transferring oxygen from hydroperoxides onto organic substrates via peroxometal intermediates. If such an intermediate is capable of coordinating an alcohol moiety, this may produce a potentially active catalyst for alcohol oxidation. The results of the oxidation of the alcohol with t-BuOOH are listed in Table 1. The molybdenum complex effected a complete conversion of the t-BuOOH to t-BuOH (Run 1) at very low efficiency towards the formation of the aldehyde with the alcohol converted primarily to the acid which further reacted with the t-BuOH or the starting alcohol to give the esters. The use of the other complexes, with the exception of the ZrO(OAc)₂ (Run 5), led to low conversions and low selectivity to the aldehyde.

The ZrO(OAc)$_2$ has been known to promote the oxidation of some primary alcohols with t-BuOOH in CCl$_4$ with reported yields as high as 80-85% with a two hour reaction time [12]. The material we used was a commercially available 22% ZrO(OAc)$_2$ in aqueous acetic acid which was used as received without further purification (Run 5). The oxidation was run in acetonitrile and produced the aldehyde in 37% yield and 68% selectivity with regard to the alcohol. The selectivity with respect to the t-BuOOH converted was also high.

Table 2 Ruthenium catalyzed oxidation with t-BuOOH. Activity of some ruthenium complexes[a]

Run	Catalyst	Reaction characteristics					
		Based on converted t-BuOOH			Based on converted alcohol		
		Conversion[c] %	Efficiency[d] %	Yield[e] %	Conversion[f] %	Selectivity[g] %	Yield[h] %
10	RuCl$_2$(PPh$_3$)$_3$	98	40	39	59	69	41
11	RuCl$_2$(PPh$_3$)$_3$[b]	85	46	39	51	79	40
12	Ru(AcAc)$_3$	75	22	17	25	70	17
13	RuHCl(CO)$_2$(PPh$_3$)$_3$	77	18	14	27	58	15
14	RuO$_4$.xH2O	13	48	6	8	84	6
15	RuCl$_2$(CO)$_2$(PPh$_3$)$_2$	24	16	4	10	40	4
16	RuH$_2$(CO)$_2$(PPh$_3$)$_3$	8	40	3	10	35	4
17	RhCl(PPh$_3$)$_3$	10	30	3	3	95	3

[a] Reaction conditions: 8.5 mmol substrate, 0.1 mmol catalyst, 20 ml chloroform, 0.85mmol t-BuOOH added over 1h and the solution stirred for additional 2 h at 30°C; [b] temperature 10°C ; [c] determined by GC based on the amount of t-BuOOH used; [d] the ratio between the aldehyde formed and the converted t-BuOOH; [e] yield based on the t-BuOOH used; [f] determined by GC based on the alcohol used; [g] the ratio between the aldehyde formed vs converted alcohol; [h] based on the amount converted alcohol.

2. Activity of ruthenium complexes

We also studied the activity of several ruthenium complexes listed in Table 2 (Run 10-16). The oxidation was run under standard conditions in THF at a temperature of 30°C. The RuCl$_2$(PPh$_3$)$_3$ complex showed higher activity at relatively low catalyst concentrations (Run 10) as compared to the rest of the complexes studied. The yield based on the consumed t-BuOOH and the alcohol was 39 and 41% respectively. Some over-oxidation to the acid also took place leading to the formation of esters. In THF, a third by-product, 2,4,6-trialkyl-

Table 3 Ruthenium catalyzed oxidation with t-Bu OOH. Solvent effect[a]

Run	Solvent	Reaction characteristics					
		Based on converted t-BuOOH			Based on converted alcohol		
		Conversion[c] %	Efficiency[d] %	Yield[e] %	Conversion[f] %	Selectivity[g] %	Yield[h] %
18	THF	100	38	38	55	72	40
19	1,4-Dioxane	94	30	28	49	61	30
20	1,1,1-$C_2H_4Cl_3$	95	23	22	64	36	23
21	Acetone	59	29	17	24	75	18
22	Chloroform	91	4	4	5	9	4
23	Acetonitrile	65	11	7	36	21	8

[a] Reaction conditions: 8.5 mmol substrate, 0.1 mmol catalyst, 20 ml solvent, 0.85mmol t-BuOOH added over 0.5h and the solution stirred for additional 2.5 h at 30°C; [c] determined by GC based on the amount of t-BuOOH used; [d] the ratio between the aldehyde formed and the converted t-BuOOH; [e] yield based on the t-BuOOH used; [f] determined by GC based on the alcohol used; [g] the ratio between the aldehyde formed vs converted alcohol; [h] based on the amount converted alcohol.

1,3,5-trioxane was also produced. The extent of these side reactions affects the overall selectivity to the desired aldehyde. It is worthy to note that the first side products, which appear almost simultaneously, are the ester, and the trioxane. In a 1.5 ratio which did not change with the temperature. When the temperature was reduced to 10°C (Run 11), the degree of over-oxidation to acid and the cyclization to the trioxane was diminished. While the degree of over-oxidation could possibly be reduced by modifying the ligand environment around the Ru complex, the cyclization is most likely a function of the acidity of the complex and/or the nature of the solvent used. Both variables were considered important and attempts were made to optimize the $RuCl_2(PPh_3)_3$ based system. The next two complexes listed in Table 2, $Ru(AcAc)_3$ and $RuHCl(CO)_2(PPh_3)_3$, showed increased affinity toward the inefficient decomposition of t-BuOOH and very low activity in oxidizing the alcohol. The ruthenium carbonyl complexes were inactive in both the decomposition of the hydroperoxide and in the oxidation step. It was also surprising that the Wilkinson complex, $RhCl(PPh_3)_3$, when performing at low activity under the test conditions showed the highest selectivity towards the alcohol conversion. Additional studies are under way to determine the potency of this catalyst for the current application.

Table 4 Ruthenium catalyzed oxidation with t-BuOOH. Additive effect[a]

Run	Additive	Reaction characteristics					
		Based on Converted t-BuOOH			Based on converted alcohol		
		Conver- sion[e] %	Efficiency[f] %	Yield[g] %	Conver- sion[h] %	Selectivity[j] %	Yield[j] %
23	None	85	46	39	51	79	40
24	DTBMP[b]	100	42	42	59	74	44
25	TEMPO[c]	96	41	39	53	77	41
26	NPBA[d]	97	38	36	53	71	38

[a] Reaction conditions: 8.5 mmol substrate, 0.1 mmol catalyst, 0.01mmol additive, 20 ml THF, 0.85mmol t-BuOOH added over 0.5 h and the solution stirred for additional 3 h at 30°C; [b] 2,6-Di-*tert*-Butyl-4-Methylphenol; [c] 2,2,6,6-Tetramethylpiperidineoxyl; [d] N-Phenyl –2-Naphthylamine; [e] determined by GC based on the amount of t-BuOOH used; [f] the ratio between the aldehyde formed and the converted t-BuOOH; [g] yield based on the t-BuOOH used; [h] determined by GC based on the alcohol used; [i] the ratio between the aldehyde formed vs converted alcohol; [j] based on the amount converted alcohol.

3. Solvent effect in the ruthenium catalyzed oxidation with t-BuOOH

A number of solvents were screened in order to determine the optimum conditions for suppressing the side product formation. The results are collected in Table 3. The oxidations were run under standard conditions with the only difference being the reduced time for the t-BuOOH addition. In Run 18 with THF as a reaction solvent the major products again being the ester and the trioxane. In 1,4-dioxane (Run 19) complete conversion of the t-BuOOH to t-butanol was observed while the alcohol was converted to the aldehyde and the corresponding acid without formation of the ester. In 1,1,1-$C_2H_4Cl_3$ or acetone, the hydroperoxide conversion went to completion to give t-BuOH and the main side products were again the acid and the ester. With chloroform as the solvent decomposition of t-BuOOH without formation of the aldehyde was observed. In acetonitrile both reactions were strongly inhibited and no aldehyde was formed in detectable quantities. Based on the limited number of solvents studied it appeared that THF was the solvent of choice for this particular set of oxidation conditions.

4. Additive effect

It was speculated that the second step of over-oxidation to acid might take place via a free radical pathway, arising from the catalytic decomposition of the t-BuOOH. It was further thought that the use of a free radical inhibitor might reduce the extent of the acid formation and improve the overall aldehyde selectivity. The use of free radical scavengers such as 2,6-di-*tert*-butyl-4-methylphenol (Table 4, Run 24), the stable free-radical, TEMPO, (Run 25) or the amine type inhibitor, N-Phenyl –2-Naphthylamine (Run 26), did not show any improvement in the reaction selectivity towards the formation of the aldehyde. The lack of any significant reduction in the amounts of ester formed when using these modifiers showed that both steps of aldehyde and acid formation most likely do not include the involvement of free radical intermediates.

CONCLUSIONS

The $RuCl_2(PPh_3)_3$ complex is shown to be an efficient catalyst for the selective oxidation of primary alcohols with t-BuOOH at ambient temperatures. The major by-products are the corresponding acids and their esters with the alcohol and with the t-BuOH. The formation of the by-products takes place through a molecular or ionic mechanism since the rate of their formation is not affected by the presence of free radical inhibitors.

REFERENCES

1.   R.A Sheldon, J.K. Kochi, *Metal Catalyzed Oxidation of Organic Compounds*, Academic Press, New York, 1981.
2.   S.V. Ley and A. Madin, In *Comprehensive Organic Synthesis*, (B.M. Trost and I. Fleming, eds.) Pergamon Press, Oxford, 1991, Vol 7, p.251.
3.   W.J. Mijs and C.R.H.I. DeJonge, *Organic Synthesis by Oxidation with Metal Compounds*, Plenum Press, New York, 1968.
4.   A. Dijksman and. I.W.C.E Arends. and R. Sheldon, *Chem. Commun.*, 1591 (1999).
5.   I.E. Marko, P.R. Giles, M. Tsukazaki, S.M. Brown and C.J. Urch, *Science*, **274**, 2044 (1969).
6.   S. Konemoto, S. Tomoioka and K. Oshima, *Bull. Chem. Soc. Japan*, **59** 105 (1986).
7.   W.P. Griffith, S.V. Ley, G.P. Whitcombe and A.D. White, *J. Chem. Soc., Chem. Commun.*, 1625 (1987)
8.   S. Konemoto, S. Mutsubara, K. Takai and K. Oshima, *Bull. Chem., Soc. Japan.*, **61**, 3607 (1988).
9.   Y. Tsuji, T. Ohta, and T. Ido, *J.Organometalic Chemistry*, **280**, 280 (1984).
10. K.B. Sharpless, K. Akashi and K. Oshima, *Tetrahedron Lett.*, **29**, 2503 (1976).

11.   R.L. Augustine and S.K. Tanielyan, This volume.
12.   K. Kanada, Y. Kawanishi and S. Teranishi, *Chem. Lett..,* 1481 (1984).

47

A Selective Process to Prepare Triacetone Amine

Young-Chan Son and Steven L. Suib
University of Connecticut, Storrs, Connecticut, U.S.A.

Russell E. Malz, Jr.
Uniroyal Chemical Division, Crompton Corporation, Naugatuck, Connecticut, U.S.A.

ABSTRACT

2,2,6,6-tetramethy-4-oxopiperidone (TAA) is an intermediate for polymer light stabilizer, for drugs, and for nitroxyl radicals of piperidine and pyrolidone derivatives [1,2]. Numerous methods have been reported for the synthesis of TAA [3,4]. Perhaps the widest scope of any patent pertaining to the synthesis of TAA was issued to Ciba Geigy [4] which claims the broad use of any acid in the condensation of $\geq$ 4 moles of acetone with 1 mole of ammonia. The process leads to a mixture of many products as shown in Scheme 1.

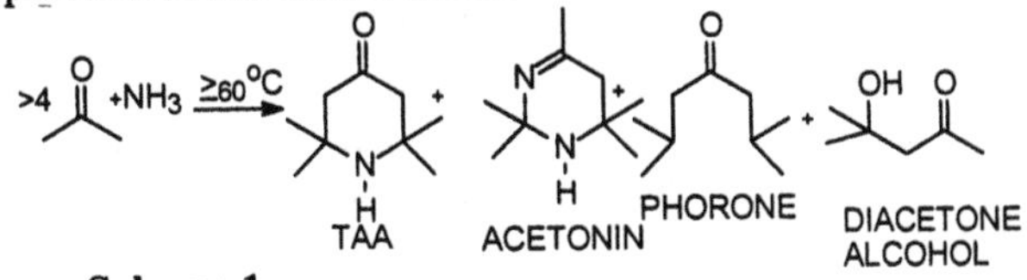

Scheme 1

Another route to triacetone amine that involves the conversion of acetone and ammonia to acetonine [5] (2,2,4,4,6-pentamethyl-1,2,5,6-tetrahydro-pyrimidine) is made from acetone, ammonia and a catalyst at room temperature [6]. After purification, acetonine is mixed with more than six equivalents of acetone and refluxed, yielding TAA in the presence of Lewis acids such as calcium, zinc, ammonium chloride, aluminum, or boron trifluride. This two step process does not yield a much improved selectivity as shown in Scheme 2.

Scheme 2

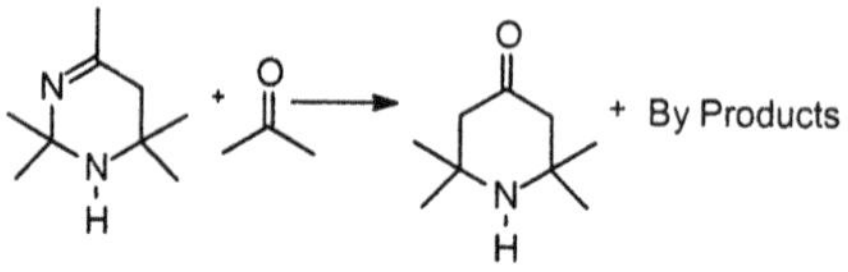

TAA can also be obtained from phorone, diacetone alcohol, diacetone amine instead of the condensation of acetone and ammonia or the condensation of ace-tonine and ammonia, but not economically.

Included among the catalysts for the condensation of acetone with ammonia are several using calcium chloride as a catalyst leading to acetonine [5] and TAA [7].

We wish to report a efficient, environmentally friendly, simple and eco-nomical route to TAA.

RESULTS AND DISCUSSION

We discovered that we could maintain the reaction at below 60°C, often carrying the reaction out at room temperature, by using ammonia nitrate as a transient source of ammonia. This is shown is Scheme 3.

Scheme 3

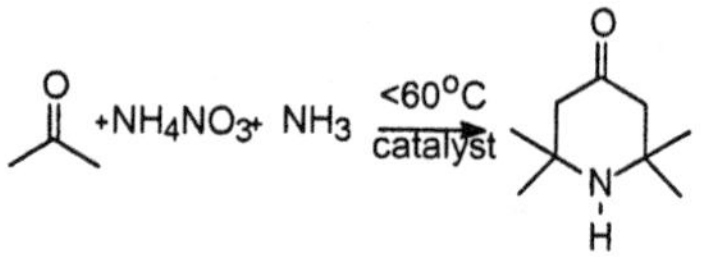

In evaluating catalysts, as shown in Table 1, we started with a 6:1 molar ra-tio of acetone to ammonium nitrate. We would then add ammonium hydroxide over 20 minutes until we obtained a ratio of 3:1 of acetone to total ammonia from ammonium nitrate and ammonium hydroxide. The reaction mixture was stirred for four hours at a temperature of 20°C to 25°C. The highest yields obtained were with $Sr(NO_3)_2$ and $Ca(OAc)_2$, but both of these showed lower selectivity. The best combination of yield and selectivity was obtained from a CaY prepared from Engelhard USY by three exchanges at either room temperature or 80°C.

Table 1   The effect of different catalysts[a]

Catalyst	% Yield[b]	% Selectivity[c]	Side products[D]
Na₂SiO₃	9.0	85	A,U
Zn(OAc)₂	2.6	99	-
(AcO)₂Mn·4H₂O	No Reaction	-	-
BaO	7.1	98	-
Sr(NO₃)₂	16	84	A,D,U
Ca(OAc)₂	14	91	A
Mg(NO₃)₂	No Reaction	-	-
Ca(NO₃)₂	12	91	A,U
CaY[e]	12	99	-
CaCl₂	13	94	A

[a] Each reaction was run in a 100-mL 3-necked glass round bottom flask equipped with a magnetic stirrer, thermometer and 25-mL dropping funnel. The flask contained 17.4 g (0.30 mole) of acetone and 4.0 g (0.05 mole) of ammonium nitrate and 0.0045 mole of the indicated catalyst. To this was added 3.1 mL (0.05 mole) ammonium hydroxide (28-30 vol. %) over a period of 20 minutes. The reaction mixture was stirred for 4 hours at a temperature of 20°C to 25°C. The acetone was evaporated, and the residue was extracted with ethyl ether. The ether layer was separated, dried with MgSO₄, and evaporated. The ether layer was separated, dried with MgSO₄, and evaporated. The product was weighed and analyzed by gc.
[b] The weight percent yield based on ammonia from ammonia hydroxide and ammonium nitrate was determined from an area percent gc of the isolated product.
[c] Determined from an area percent gc of the isolated product.
[d] A=Acetonine, U=unknown, D=diacetone alcohol. These were determined by GC-MS.
[e] Used 0.5 g of CaY in this experiment.

We repeated this experiment four times using CaY as the catalyst with reaction times of 2, 4, 6, and 17 hours. In all cases we were able to detect only TAA in the extracted layer. The results are shown in Figure 1.   A break in the rate appears to occur after five hours. We have many interrelated variables possibly contributing to the diminished rate. The ratio of acetone to ammonia is diminishing, the concentration of ammonia is diminishing, and the water content is increasing. All of these contribute to the diminished rate. We decided to look at decreasing the water in the reaction.   We did this by studying the effect of adding ammonia as a gas in the presence and absence of a desiccant. The data is found in Table 2.

Decreasing the water greatly increases the yield the reaction. Increasing the ratio of acetone at the end of the addition of the ammonia also increases the yield significantly, but this has less of an effect than removing water.

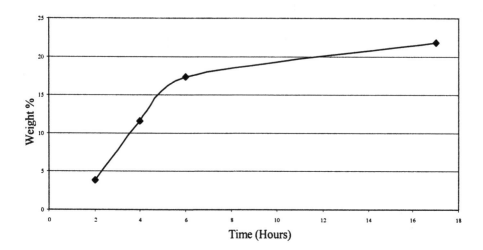

Figure 1 The rate of reaction to TAA with CaY catalyst and a 3:1 ratio of acetone total ammonia.

Table 2 The effect of lower water content on yield of TAA[a]

Desiccant	Mole % Yield[b]	Selectivity[c]
None	38	98
Na_2SO_4	41	98
$MgSO_4$	43	98
None	47	98
$MgSO_4$	55	98

[a] Each reaction was run in a 100 mL 3-necked glass round bottom flask equipped with a magnetic stirrer, thermometer and Y-tube with a 25 mL addition funnel and an addition port. The flask contained 17.4 g (0.30 mole) of acetone and 4.0 g (0.05 mole) of ammonium nitrate and 0.5 g of CaY catalyst. We added 1.0 g of dessicant in the indicated experiments. The vessel was sealed, agitation was begun and a total of 0.05 moles of ammonia added by flowing 0.026 moles of gas for a two hour period at a temperature of 20°C to 25°C. The reaction mixture was stirred for an additional 16 hours. The acetone was evaporated, and the residue was extracted with ethyl ether. The ether layer was separated, dried with $MgSO_4$, and evaporated. The product was weighed and analyzed by gc.
[b] The weight percent yield based on total of ammonia from ammonia nitrate and ammonia gas was determined from an area percent gc of the isolated product.
[c] Determined from an area percent gc of the isolated product.

We compared the effect of other ammonia sources to ammonium nitrate. The data are shown in Table 3. Acetonine was the major product on ammonium sulfate and ammonium chloride. When we use ammonia nitrate as the only ammonia source, we obtained a 10 to 15% conversion to TAA after 20 hours at room temperature with at least 98% selectivity.

We compared ratios of acetone to final total ammonia from 2:1 to 10:1. The data are found in table 4. When the ratio of acetone to total ammonia is less than 3:1, acetonine, pherone, diacetone alcohol and unknown product formed with the major product being acetonine.

Table 3 Effect of Different Ammonia Sources.[a]

Ammonia Source	Weight % Yield[b]	Selectivity[c]	Side Product[e]
NH_4NO_3	12	99	N.D.
$(NH_4)_2SO_4$	3.2	25	D,P,A,U
NH_4Cl	3.9	40	D,P,A,U

a. Each reaction was run in a 100 mL 3-necked glass round bottom flask equipped with a magnetic stirrer, thermometer and 25 mL dropping funnel. The flask contained 17.4 g (0.30 mole) of acetone and 0.05 mole) of ammonium nitrate and 0.50 g of CaY catalyst. To this was added 3.1 mL (0.05 mole) ammonium hydroxide (28-30 vol. %) over a period of 20 minutes. The reaction mixture was stirred for 4 hours at a temperature of 20°C to 25°C. The acetone was evaporated, and the residue was extracted with ethyl ether. The ether layer was separated, dried with $MgSO_4$, and evaporated. The ether layer was separated, dried with $MgSO_4$, and evaporated. The product was weighed and analyzed by gc.
b. The weight percent yield based on ammonia from ammonia hydroxide and ammonium nitrate was determined from an area percent gc of the isolated product.
c. Determined from an area percent gc of the isolated product.
d. N.D.=None Detected, A=Acetonine, U=unknown, D=diacetone alcohol. These were determined by GC-MS.

Table 4 Effect of the Ratio of Acetone to Ammonia[a]

Acetone:Ammonia[b]	Mole % Yield[c]	Selectivity[d]	Side Products[e]
10:1	15	99	N.D.
5:1	12	99	N.D.
3:1	12	99	N.D.
2:1	6	55	D,P,A,U

[a] Each reaction was run in a 100 mL 3-necked glass round bottom flask equipped with a magnetic stirrer, thermometer and 25 mL dropping funnel. The flask contained 17.4 g (0.30 mole) of acetone and 0.05 mole) of ammonium nitrate and 0.50 g of CaY catalyst. To this was added a molar quantity of ammonium hydroxide (28-30 vol. %) to reach the indicated ratio over a period of 20 minutes. The reaction mixture was stirred for 4 hours at a temperature of 20°C to 25°C. The acetone was evaporated, and the residue was extracted with ethyl ether. The ether layer was separated, dried with $MgSO_4$, and evaporated. The ether layer was separated, dried with $MgSO_4$, and evaporated. The product was weighed and analyzed by gc.
[b] This represents the ratio of total ammonia from both ammonium hydroxide and ammonium nitrate.
[c] The weight percent yield based on ammonia from ammonia hydroxide and ammonium nitrate was determined from an area percent gc of the isolated product.
[d] Determined from an area percent gc of the isolated product.
[e] N.D.=None Detected, A=Acetonine, U=unknown, D=diacetone alcohol. These were determined by GC-MS.

We carried out one experiment to discover the mass balance of acetone. In this experiment we achieved a yield of 0.030 moles of TAA and recovered, by distillation, 0.207 mole of acetone, from a reaction in which we started with 0.30 moles of acetone, for a mass balance of 99%.

CONCLUSION

We have developed an economical and environmentally friendly method of producing TAA with extremely high selectivity using ammonium nitrate, ammonium hydroxide, and acetone in the presence of a calcium containing acid catalyst.

ACKNOWLEDGEMENNTS

We acknowledge support of Engelhard Corp. and the Geosciences and Biosciences Division, Office of Energy Sciences, Office of Science, U. S. Department of Energy.

REFERENCES

1. (a) Dagonneau, M.; Kagan, E. S.; Mikhailove, V. I.: Rosantev, E. G.: Sholle, V. D.; *Synthesis*, **1984**, 895-916 and references cited therein. (b) Keana, J. F. W. *Chem Rev*, **1977**, *78*, 37-64.
2. (a) Volodarsky, L; Kosover, V. *Tetahedron Lett.* **2000**, *41*, 179-181. Huang, W-1; Chiarelli, R; Rassaat, A.; *Tetrahedron Lett.* **2000**, *41*, 8787-8789.
3. (a) Hall, H. K. *J. Am. Chem. Soc.* **1957**, *79*, 5444-5447. (b) Lutz, W. B.; Lazarus, S.; Meltzer, R. I. *J. Am. Chem. Soc.* **1962**, *27*, 1695-1703. (c) Sosnovsky, G.; Konieczny, M. *Synthesis*, **1976**, *11*, 735-736. (d) Wu, A.; Wang, W.; Pan, X. *Synth. Commun.* **1996**, *26*, 3565-3569.
4. Cantatore, G; Di Reno, C.; Cassandri, P., US Patent 4,536,581; Aug. 20, 1985.
5. (a) Matter, E. *Helv. Chim. Acta.* **1947**, *XXV*, 1114-1122. (b) Bradbury, R. B.; Hancock, N. C.; Hatt, H. H. *J. Chem. Soc.* **1947**, 1939.
6. a) Murayama et al., US Patent 3, 513,170; July 19, 1967. (b) Orban et al., US Patent 3,959,298, May 25, 1976, IBID, US Patent 3,960,875, June 1, 1976.
7. Sosnowsky, G.; Koniecny, M. *Z. Naturforsch* **1977**, *32b*, 328.

48

Hydrogenolysis of Ethyl Laurate to Dodecanol on Ru-Sn/Al$_2$O$_3$ Catalysts

S. Göbölös, N. Mahata, M. Hegedüs, I. Borbáth, and J. L. Margitfalvi
Institute of Chemistry, Hungarian Academy of Sciences, Budapest, Hungary

ABSTRACT

Three series of catalysts with Ru- and Sn-loading ranging from 1.0 to 6.9 and 0 to 17.5 wt. %, respectively, were prepared by consecutive or co-impregnation varying the impregnation sequence with different precursor salts (RuCl$_3$, Ru(acac)$_3$, SnCl$_2$.xH$_2$O and C$_{12}$H$_{24}$O$_4$Sn), metal loading and Sn:Ru atomic ratio. The calcined catalysts were characterized by Temperature Programmed Reduction (TPR). Co-impregnation of alumina with SnCl$_2$xH$_2$O and RuCl$_3$ at 5 wt.% Ru loading and Sn:Ru atomic ratio of 2 resulted in the most active catalyst. Ethyl laurate was hydrogenolyzed to dodecanol and ethanol on the 5%Ru-11.6%Sn/Al$_2$O$_3$ catalyst at 250 °C and 10 MPa H$_2$ pressure with 95 % conversion and 83 % selectivity. Correlation was found between the yield of dodecanol and the amount of hydrogen consumed for the reduction of tin species in the TPR experiment above 260 °C. Co-impregnation provides close interaction of ruthenium with tin, resulting in the formation of highly active "Sn^{n+}-Ru ensemble sites" to the maximum extent. The formation of a "metal ion-metal nanocluster active site ensemble" is required to activate the carbonyl group of the substrate.

INTRODUCTION

Fatty alcohols and their derivatives are widely used as surfactants. In recent years, besides other bimetallic systems (PdRe, ReSn, RhSn, CoSn) Ru-Sn/Al$_2$O$_3$ catalysts have gained much attention in studying the hydrogenolysis of fatty esters to fatty alcohols [1]. It has been proposed that the active centers are metallic Ru particles in interaction with tin oxide acting as Lewis acid centers involved in the activation of the carbonyl group [2]. However, systematic studies on the effect of catalyst preparation and pretreatment parameters on the catalytic performance are rather scarce. Also, the relationship between catalyst structure and performance, and the mode of action of the tin modifier are not clear yet.

In this work the hydrogenolysis of ethyl laurate (EL) to dodecanol (ROH) and ethanol has been studied on different Ru-Sn/Al$_2$O$_3$ catalysts. Systematic studies have been made to investigate the influence of precursor compounds, sequence of impregnation, metal loading, Sn:Ru atomic ratio, catalyst pretreatment (calcination, reduction) and reaction conditions (temperature, H$_2$ pressure). The calcined catalysts were characterized by Temperature Programmed Reduction (TPR). Correlation between the activity and TPR characteristics of Ru-Sn/Al$_2$O$_3$ catalysts was also demonstrated.

EXPERIMENTAL

Catalyst Preparation

Alumina supported Ru-Sn catalysts were prepared by co-impregnation or successive impregnation. Calculated amount of metal precursor salt(s) were dissolved and the support (γ-alumina, Ketjen CK 300, particle size = 0.05-0.1 mm, S = 180 m^{2}g^{-1}) was added to the solution and the slurry was left for 12 h. In order to avoid the reduction of catalyst particle size the slurry was stirred in each half an hour for 5 minutes at a rate of 60 rpm. After evaporating the solvent the catalyst was dried overnight at 120 °C and subsequently calcined in air at 400 °C, unless otherwise specified, for 4 h. Various metal precursor salts were used: RuCl$_3$xH$_2$O, Ru(acac)$_3$, SnCl$_2$x2H$_2$O, C$_{12}$H$_{24}$O$_4$Sn (dibutyltin diacetate). Water was used as solvent for inorganic precursors, toluene and ethanol was used to dissolve Ru(acac)$_3$ and C$_{12}$H$_{24}$O$_4$Sn, respectively. Catalysts with Ru- and Sn-loading ranging from 1.0 to 6.9 and 0 to 17.5 wt. %, respectively, and Sn:Ru atomic ratio ranging from 0 to 3 were prepared by co-impregnation. The catalyst precursors were calcined (T$_c$) in air and reduced (T$_r$) in flowing hydrogen (60 ml/min) at 400 °C unless otherwise mentioned. The duration of both calcination and reduction was 4 hours. The catalysts prepared by co-impregnation were designated as xM$_1$yM$_2$, where x and y were weight content (%) of the corresponding metal M$_1$ and M$_2$. Catalysts prepared by successive impregnation were designated as e.g. xM$_1$/yM$_2$, in this case M$_2$ was first impregnated onto Al$_2$O$_3$ followed by the impregnation of M$_1$ to the previously calcined monometallic sample.

TPR Measurement

TPR experiments were carried out in ASDI RXM-100 Multifunctional Catalyst Testing and Characterization Machine. In the TPR experiments 0.08 g catalyst was heated up to 700 °C at a rate of 10 °C/min in flowing 5%H$_2$-Ar (40 ml/min). The

TPR curves were deconvoluted, assuming that the individual TPR peaks had a Gaussian form. The height and the position of the peaks were determined. The position of individual peaks was allowed to fluctuate in a range of ± 5 °C, whereas the half width of each peak was fixed at a value of 10 °C. Stepwise appearance of the new individual TPR peaks was used in the consecutive deconvolution procedure.

Catalytic Test

The hydrogenolysis of EL was carried out in SS batch reactor (volume = 100ml). Prior to reaction catalysts were reduced at 400 °C, unless otherwise specified, for 4 h under H_2 flow (60ml/min). Catalysts were transferred into the batch reactor with exclusion of air. The amount of catalyst and ester was 0.5 g (2.8 wt. %) and 20 ml (75 mmol), respectively. The reaction was carried out under optimal reaction condition at T = 250 °C H_2 and pressure of 9.5 MPa. The reaction time (t) was 8 hours. Liquid reaction products were analyzed by gas chromatography.

RESULTS AND DISCUSSION

Catalytic Activity

Influence of Impregnation Sequence and Precursor Compounds
Co-impregnation and successive impregnation were used to prepare tin modified Ru/Al$_2$O$_3$ catalysts. Results shown in Table 1 indicated that co-impregnated Ru-Sn/Al$_2$O$_3$ catalyst (row 3) was the most active and selective in producing ROH. Comparable conversion, but a lower selectivity of ROH was observed over the catalyst prepared by successive impregnation of Sn followed by Ru (row 1). In contrast, the catalyst prepared by reverse sequence of impregnation was comparatively less active with an in-between selectivity of ROH (row 2).

Table 1 Effect of impregnation sequence and precursor compounds of Ru and Sn

Catalyst	Conversion %	ROH	Selectivity % HC	HE	Yield % ROH
5Ru/11.6Sn[a,d]	93	76	4	1	71
11.6Sn/5Ru[a,e]	86	79	2	5	68
5Ru11.6Sn[a,f]	94	81	3	1	76
5Ru11.6Sn[b,f]	94	74	3	3	70
5Ru11.6Sn[c,f]	84	76	2	1	64

[a] Ru precursor: RuCl$_3$xH$_2$O; Sn precursor: SnCl$_2$x2H$_2$O; [b] Ru precursor: Ru(acac)$_3$; Sn precursor: SnCl$_2$x 2H$_2$O; [c] Ru precursor: Ru(acac)$_3$; Sn precursor: C$_{12}$H$_{24}$O$_4$Sn (Di-butyl tin di-acetate); Sn:Ru atomic ratio = 2; [d] consecutive impregnation, Sn introduced first; [e] consecutive impregnation, Ru introduced first; [f] co-impregnation.

These results suggest that an intimate mixing of Ru with Sn is preferable for the creation of catalytic sites. Covering of tin by ruthenium resulted in less selectivity of ROH, while covering of Ru by Sn resulted in lower conversion. These results suggest that co-impregnation provides mixed Ru-SnO$_x$ sites to the maximum extent.

Results shown in Table 1 also indicated that the combination of RuCl$_3$xH$_2$O and SnCl$_2$x2H$_2$O gave the highest EL conversion and ROH selectivity (row 3). The combination of Ru(acac)$_3$ and SnCl$_2$x2H$_2$O resulted in a comparable conversion but considerably lower selectivity of ROH (row 4). Much lower conversion and selectivity of ROH in-between the former two catalysts was obtained over the catalyst prepared by the combination of Ru(acac)$_3$ and C$_{12}$H$_{24}$O$_4$Sn (row 5). Recent studies have indicated that in the hydrogenolysis of dicarboxylic acid ester chlorine poisons the activity of only reduced Ru-Sn/Al$_2$O$_3$ catalysts prepared from Ru(NO)(NO$_3$)$_3$ and SnCl$_2$ precursors. However, the calcination at 400 °C or above removed much of the chlorine and resulted in high yield of alcohol [3]. Residual amounts of chlorine promoted the hydrogenation of carbonyl group in the hydrogenation of α,β-unsaturated aldehyde over supported Pt and Ru catalysts [4]. Therefore, it can be speculated from our study that small amount of residual chlorine could be beneficial for a better selectivity of ROH in the hydrogenation of EL.

Influence of Ru Loading

Results obtained in studying the effect of Ru loading on the hydrogenolysis activity of Ru-Sn/Al$_2$O$_3$ catalysts prepared by co-impregnation are given in Table 2. Upon increasing the Ru content from 1 to 2 wt. % both the EL conversion and the selectivity of ROH significantly increased. Further increase of Ru loading up to 5wt. % only slightly increased the conversion of EL, but the selectivity of ROH remained the same, resulting in the highest yield of ROH at 5wt. % Ru. Conversion as well as selectivity of ROH considerably decreased with Ru loading above 5wt. %. Selectivity of HE decreased sharply with the increase of Ru loading above 2wt. %.

Table 2 Effect of Ru loading

Catalyst	Conversion %	Selectivity %			Yield %
		ROH	HC	HE	ROH
1.0Ru2.3Sn	9	43	11	31	4
1.3Ru3.0Sn	44	55	10	21	24
2.0Ru4.7Sn	85	78	6	3	66
3.0Ru7.0Sn	88	76	2	4	67
5.0Ru11.6Sn	94	81	3	1	76
6.9Ru16.3Sn	87	70	2	5	61

Ru precursor: RuCl$_3$xH$_2$O; Sn precursor: SnCl$_2$x2H$_2$O; Sn:Ru atomic ratio = 2 in all cases. Catalysts were prepared by co-impregnation.

Influence of Sn/Ru Atomic Ratio

Ru-Sn/Al$_2$O$_3$ catalysts with various Sn:Ru atomic ratios were prepared by co-impregnation keeping the Ru loading constant at 5wt. %. Activity data summarized in Table 3 indicated that the conversion as well as the selectivity of ROH was very low over the Ru/Al$_2$O$_3$ catalyst (Sn:Ru=0). The main products over this catalyst were various hydrocarbons. Upon incorporating tin the conversion rapidly increased as the Sn:Ru atomic ratio increased from 0 to 1, at higher Sn:Ru atomic ratio the conversion decreased slowly. The selectivity of ROH increased significantly with Sn:Ru atomic ratio up to a value of 1.5, above this ratio the increase in the selectivity of ROH was only small. Thus, the yield of ROH attained a maximum value of 76% at a Sn:Ru atomic ratio of 2. The selectivity of HCs and the HE is small over the catalysts having a Sn:Ru atomic ratio above 1.5.

Table 3 Effect of Sn:Ru atomic ratio

Catalyst	Sn:Ru at. ratio	Conversion %	Selectivity %			Yield %
			ROH	HC	HE	ROH
5Ru0.0Sn	0.0	18	4	67	14	<1
5Ru2.9Sn	0.5	70	15	56	11	11
5Ru5.8Sn	1.0	95	60	10	6	57
5Ru8.7Sn	1.5	92	79	3	4	73
5Ru11.6Sn	2.0	94	81	3	1	76
5Ru14.6Sn	2.5	91	80	3	4	73
5Ru17.5Sn	3.0	88	83	3	4	73

Ru precursor: RuCl$_3$xH$_2$O; Sn precursor: SnCl$_2$x2H$_2$O. Catalysts were prepared by co-impregnation.

Influence of Catalyst Pretreatment

In the present study Ru-Sn/Al$_2$O$_3$ catalysts were calcined in air and reduced under H$_2$ flow at various temperatures. The effect of calcination temperature on the performance of catalysts is shown in Table 4. Calcination was performed for 4 h at the specified temperatures. After calcination the catalysts were subsequently reduced under the flow of H$_2$ at 350 °C for 4 h. Results listed in Table 4 indicated that the calcination temperature hardly affected the conversion of EL, while it showed clear influence on the product selectivity. Upon increasing the calcination temperature the selectivity of ROH passed through a maximum at 350 °C. Appreciable selectivity of hydrocarbons along with comparatively lower selectivity of ROH over the catalysts, calcined at 250 °C or at a lower temperature, can be explained by a large amount of residual chlorine. On the other hand, calcination of the catalyst at 400 °C or at a higher temperature may result in the segregation of surface Ru and Sn species, and hence lower selectivity of ROL [4].

The effect of catalyst reduction temperature on the conversion of EL and product selectivity is demonstrated Table 5. The catalysts were reduced for 4 h at the specified temperatures. Prior to reduction, the catalysts were calcined at 400 °C for 4 h in air. Reduction of the catalyst at low temperature (250 °C) might have resulted in the formation of some isolated tin-oxygen species, which were responsible for the formation of the HE to an appreciable extent [1,5]. However, the selectivity pattern only slightly varied with the reduction temperature at 300 °C or higher. 350 °C was found to be the optimum catalyst reduction temperature, which gave 91% EL conversion with 77% ROH selectivity.

Table 4 Influence of calcination temperature

Calcination Temperature °C	Conversion %	Selectivity %			Yield %
		ROH	HC	HE	
120	91	78	9	<1	71
250	91	81	6	1	74
300	90	83	3	1	75
350	90	86	2	<1	77
400	91	80	3	2	73
450	92	80	3	2	74

Catalyst: 5Ru11.6Sn, prepared by co-impregnation; Ru precursor: $RuCl_3xH_2O$; Sn precursor: $SnCl_2x2H_2O$; Sn:Ru atomic ratio = 2; T_r = 350 °C.

Table 5 Influence of reduction temperature

Reduction Temperature °C	Conversion %	Selectivity %			Yield %
		ROH	HC	HE	
250	86	77	2	6	66
300	86	80	3	1	69
350	91	80	3	2	73
400	86	78	3	2	67

Catalyst: 5Ru11.6Sn, prepared by co-impregnation; Ru precursor: $RuCl_3xH_2O$; Sn precursor: $SnCl_2x2H_2O$; Sn:Ru atomic ratio = 2.

TPR Measurements

Results of TPR measurements are shown in Figures 1–4. The effect of Sn:Ru atomic ratio on TPR spectra of calcined at 400 °C Ru-Sn/Al$_2$O$_3$ catalysts prepared by co-impregnation is shown in Figure 1. As seen in Figure 1A Ru/Al$_2$O$_3$ catalyst was completely reduced between 100 and 220 °C. The small reduction peak at 120 °C can be attributed to non-decomposed RuCl$_3$, whereas the shoulder at ca. 150 °C can be assigned to different highly dispersed ruthenium-oxygen species [6]. The large narrow peak observed at 190 °C can be attributed to the reduction of bulk RuO$_2$ crystallites [6]. Figure 1A shows also the TPR spectrum of bimetallic catalyst with a Sn:Ru atomic ratio of 2. The low temperature part of the spectrum (below ca. 250 °C) can be attributed to the reduction of Ru-containing surface species, whereas the high temperature part of the spectrum can be related to the reduction of different tin-containing species. TPR spectra of Ru-Sn/Al$_2$O$_3$ catalysts with Sn:Ru atomic ratios 1, 2 and 3 shown in Figure 1B indicate that the higher the Sn:Ru ratio the higher the temperature of the reduction of Ru-containing species. The broadening peaks and the complexity of the low temperature part of the TPR curve of bimetallic catalysts indicates that the reduction of RuO$_2$ is affected by tin. This suggests a close interaction between highly dispersed tin oxide species and RuO$_2$ crystallites. The presence of highly dispersed tin oxide species and replacement of about 6 % tin by ruthenium in the lattice of RuO$_2$ was evidenced by X-ray diffraction measurements [7]. The broad overlapping peaks in the high temperature part of the spectra suggest that ionic tin species have different forms and environment. Furthermore one can assume that metallic ruthenium formed at low temperature during TPR by activating hydrogen facilitates the reduction of tin species at higher temperatures [1,2].

The effect of calcination temperature on TPR of Ru-Sn/Al$_2$O$_3$ catalysts prepared by co-impregnation (Ru-loading: 5wt. %, Sn:Ru atomic ratio: 2) is shown in Figure 2. The complicated shape of TPR curves shown in Figure 2 indicates that different types of oxidized Ru and Sn species exist in these catalysts. The higher the temperature of calcination the higher the temperature at which the most intense sharp peak appears in the TPR spectra. These peaks can be attributed to the reduction of well crystallized RuO$_2$ supported on tin oxide covered alumina. Well crystallized RuO$_2$ and the presence of amorphous tin-oxide in the catalysts was evidenced by XRD [7]. In agreement with XRD results, the larger the average crystallite size of RuO$_2$ the higher the temperature of reduction. In the temperature range of 150-350 °C the broad TPR curve of the catalyst calcined at 250 °C indicates the presence of non-decomposed RuCl$_3$ species in the sample [6]. In TPR spectra of both Ru-Sn/Al$_2$O$_3$ catalysts calcined at 400 and 450 °C the small peaks and shoulders in the temperature range of 100-160 °C can be assigned to tiny ruthenium oxychloride and ruthenium-oxygen species.

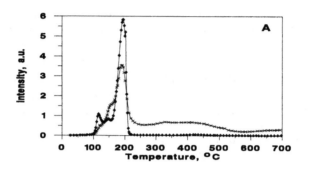

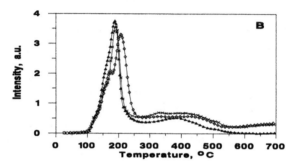

Figure 1 Effect of Sn:Ru atomic ratio on TPR of Ru-Sn/Al$_2$O$_3$ catalysts. (A) - Sn:Ru atomic ratio: (♦) - 0; (✖) - 2. (B) - Sn:Ru atomic ratio: (▲) -1; (✖) - 2; (◊) - 3. Preparation: coimpregnation, Ru loading: 5 wt.%, calcination temperature: 400 °C

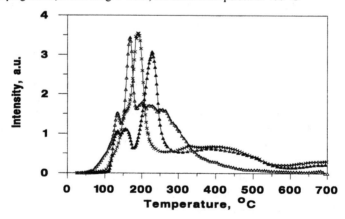

Figure 2 Effect of calcination temperature on TPR of Ru-Sn/Al$_2$O$_3$ catalysts. Calcination temperature: (Δ) - 250 °C; (✖) - 400 °C; (▲) - 450 °C. Preparation: coimpregnation, Ru loading: 5 wt.%, Sn:Ru atomic ratio: 2.

The effect of preparation method on the TPR spectra of Ru-Sn/Al$_2$O$_3$ catalysts is shown in Figure 3. The temperature of the most intense peak attributed to the reduction of RuO$_2$ crystallites increased in the order: RuSn(co-impregnation) < Ru/Sn(Sn impregnated first) < Sn/Ru(Ru impregnated first). The highest reduction temperature of the Sn/Ru type catalyst can be explained by coverage of Ru-containing species with tin oxide. Shoulders or small intensity developed peak for Sn/Ru type catalyst can be attributed to the reduction of pure RuO$_2$ particles. The high temperature part of the TPR spectra, above 300 °C, assigned to the reduction of various tin species is very similar for the three catalysts prepared by different methods.

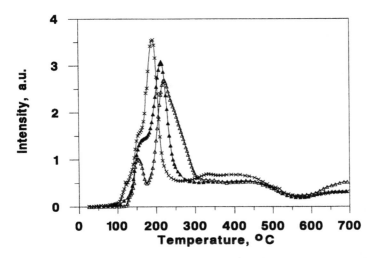

Figure 3 Effect of preparation method on TPR of Ru-Sn/Al$_2$O$_3$ catalysts. (Δ) - consecutive impregnation, Ru first (Sn/Ru); (▲) - consecutive impregnation, Sn first (Ru/Sn); (✖) - coimpregnation (Ru + Sn). Ru loading: 5 wt.%, Sn:Ru atomic ratio: 2, calcination temperature: 400 °C.

The complexity of the deconvoluted TPR spectra of alumina supported Ru and Ru-Sn catalysts is demonstrated in Figure 4. In order to distinguished between Ru- and Sn-containing species TPR spectra of the catalysts were deconvoluted. As seen in Figure 4A TPR spectrum of the Ru/Al$_2$O$_3$ catalyst can be decomposed into four peaks indicating the presence of ruthenium(IV)-oxide species in different form and environment. One can assume that the higher the temperature of the reduction the larger the size of RuO$_2$ crystallites and/or the stronger the interaction between Ru species and the support. Figure 4B shows that the low temperature part (below ca. 250 °C) of the TPR spectrum of bimetallic catalyst is significantly different from that of Ru/Al$_2$O$_3$ catalyst. This may indicate that the reduction of

Ru-containing phase is affected by tin replacing ruthenium in the lattice of RuO_2 [7], or tin oxide species decorating RuO_2 nanoclusters. The comparison of TPR spectra in Figure 4A and Figure 4B indicates that the high temperature part of the spectra can be attributed to the partial reduction of X-ray amorphous [7] tin-oxide species.

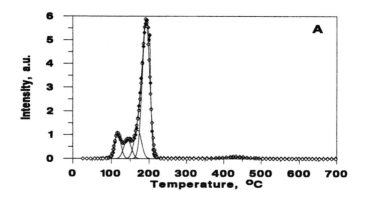

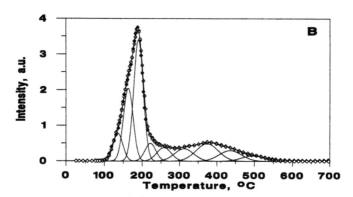

Figure 4 Deconvolution of the TPR peaks of Ru/Al_2O_3 (A) and $Ru-Sn/Al_2O_3$ (B) catalysts; (♦) measured and (◊), fitted. Preparation of $Ru-Sn/Al_2O_3$: coimpregnation, Ru loading: 5 wt.%, Sn:Ru atomic ratio: 1, calcination temperature: 400 °C.

The activity, i.e. dodecanol yield of the catalysts (5wt% Ru; Sn:Ru atomic ratio: 2) prepared and pre-treated in different ways was correlated with their TPR characteristics. As seen in Figure 5 correlation was found between the yield of dodecanol on $Ru-Sn/\gamma-Al_2O_3$ catalysts and the amount of hydrogen consumed above 260 °C for the reduction of tin oxide species in the TPR experiment. It is noteworthy, that the higher the amount of hydrogen consumed in the TPR experiment above 260 °C, the higher the degree of reduction of oxidized tin species in the catalysts. We assume that the higher degree of tin-oxide reduction in the bimetallic catalysts can be explained by a stronger interaction between ruthenium- and tin-containing species. Therefore, the higher the reducibility of tin oxide species in the catalysts, the more intimate contact between ruthenium- and tin-containing species, and as a consequence, the higher the catalytic activity. On the basis of our finding we suggest that calcination and reduction of the bimetallic catalysts prepared by impregnation results in the formation of Ru-SnO$_x$ mixed sites. It is suggested co-impregnation provides intimate contact of ruthenium with tin resulting in the formation of active "Sn^{n+}-Ru, metal ion-metal metal nanocluster" ensemble sites to the maximum extent.

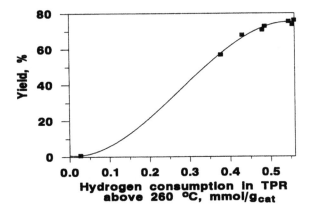

Figure 5 Correlation between the yield of dodecanol over Ru-Sn/Al$_2$O$_3$ catalysts and the amount of hydrogen consumed in TPR above 260 °C.

CONCLUSIONS

In summary, the strength of parameters determining the performance of Ru-Sn/γ-Al$_2$O$_3$ catalysts in the hydrogenolysis of ethyl laurate changed in the order: Ru-loading ~ Sn:Ru atomic ratio > method of preparation > calcination temperature > reduction temperature. The highest ethyl laurate conversion and dodecanol yield, 95 % and 79 %, respectively, was obtained on the 5Ru11.6Sn/Al$_2$O$_3$ catalyst (Sn:Ru atomic ratio = 2) at 250 °C and 9.5 MPa H$_2$ pressure. Correlation was found between the yield of dodecanol on Ru-Sn/γ-Al$_2$O$_3$ catalysts and amount of hydrogen consumed above 260 °C for the reduction of tin-containing species in the TPR experiment. It is suggested that co-impregnation provides intimate contact of ruthenium with tin resulting in the formation of highly active "Sn^{n+}-Ru, metal ion-metal nanocluster" ensemble sites to the maximum extent.

ACKNOWLEDGEMENT

Financial support of the Hungarian Scientific Research Fund (OTKA Grant N° 32065) is gratefully acknowledged.

REFERENCES

1. Y. Pouilloux, F. Autin, C. Guimon and J. Barrault, "Hydrogenation of Fatty Esters over Ruthenium-Tin Catalysts; Characterization and Identification of Active Centers" *J. Catal.*, **176**, 215 (1998).
2. V.M. Deshpande, K. Ramnarayan and C. S. Narasimhan, "Studies on Ruthenium-Tin Boride Catalysts II. Hydrogenation of Fatty Acid Esters to Fatty Alcohols" *J.Catal.*, **121**, 174 (1990).
3. K. Tahara, H. Tsuji, H. Kimura, T. Okazaki, Y. Itoi, S. Nishiyama, S. Tsuruya and M. Masai, "Liquid-phase hydrogenation of dicarboxylates catalyzed by supported Ru-Sn catalysts" *Catal. Today,* **28**, 267 (1996).
4. B. Bachiller-Baeza, A. Guerrero-Ruiz and I. Rodrigez-Ramos, "Role of the residual chlorides in platinum and ruthenium catalysts for the hydrogenation of α,β-unsaturated aldehydes" *Appl. Catal. A: General,* **192**, 289 (2000).
5. Y. Pouilloux, A. Piccirilli and J. Barrault, "Selective hydrogenation into oleyl alcohol of methyl oleate in the presence of Ru-Sn/Al$_2$O$_3$ catalysts" *J. Mol. Catal. A: Chemical,* **108,** 161 (1996).
6. P. Betancourt, A. Rives, R. Hubaut, C. E. Scott and J. Goldwasser, "A study of the ruthenium-alumina system" *Appl. Catal. A: General,* **170,** 307 (1998).
7. S. Göbölös, N. Mahata, I. Sajó, M. Hegedüs, I. Borbáth and J. L. Margitfalvi, "Hydrogenolysis of Ethyl Laurate on Ru-Sn/γ-Al$_2$O$_3$ Catalysts Prepared by Impregnation II. X-ray Diffraction and Temperature Programmed Reduction Study" submitted to *Catal. Lett.*

49

Crotonaldehyde Hydrogenation on Ir/TiO$_2$

Mohamed Abid and Raymonde Touroude
Université Louis Pasteur, Strasbourg, France

Dmitry Yu. Murzin
Åbo Akademi University, Åbo/Turku, Finland

INTRODUCTION

One of the challenging tasks in catalytic hydrogenation is selective reduction of C=O double bond in α,β –unsaturated aldehydes. In this field of research, monometallic catalysts supported on Al$_2$O$_3$ or SiO$_2$ lead mostly to the formation of saturated aldehydes [1], e.g. only C=C bond is hydrogenated. To improve the selectivity towards unsaturated alcohols, the use of additives (promoters), bi-metallic catalysts or easily reducible supports like TiO$_2$, Nb$_2$O$_5$, Y$_2$O$_3$, ZrO$_2$, CeO$_2$, ZnO, [2-5] was reported. In the latter case it is believed that strong metal support interactions could play a decisive role in creating new and favourable catalytic sites, responsible for selective hydrogenation of the carbonyl bond.

Although most attention was devoted to Pt, other metals, like Ir [6,7] also showed rather high selectivity towards unsaturated alcohol. Thus liquid phase hydrogenation of acrolein and crotonaldehyde over Ir/C afforded selectivities to unsaturaed alcohol in the range of 50-60% [6, 7]. Gas-phase hydrogenation of crotonaldehyde on Ir/TiO$_2$ resulted in selectivity to crotyl alcohol of ca. 35% [8].

Taking into account the difficulty to control strong metal support interactions (SMSI), it is easy to imagine that the performances of these systems are very changeable and depend on a vast number of factors intervening during the preparation of the catalysts and even during the reaction. That is the reason why apparent discrepancy exists in the literature with respect to selectivity values on the most studied Pt/TiO$_2$, namely the selectivity values spread in a large range on catalysts reduced at the same temperature depending, among other factors, on the nature of the metallic salt precursors and on the catalyst pretreatments.

This study is devoted to investigation of title reaction on Ir/TiO$_2$ catalyst, which was previously scarcely studied [8]. Catalytic results, as well as catalyst characterization data will be presented and compared with Pt/TiO$_2$ catalyst.

CATALYST PREPARATION AND CHARACTERIZATION

5% Ir/TiO$_2$ catalyst was prepared by impregnation of TiO$_2$-Eurotitania, made of anatase-(Tioxide, UK limited) at 398 K with an aqueous solution of H$_2$IrCl$_6$ with further drying at 393K and calcination at 673K.

X-ray Photoelectron Spectroscopy (XPS) analyses were performed using a VG ESCA III spectrometer with Mg K$_\alpha$ radiation (1253.6 eV) as incident beam without a monochromator. Before conducting the analysis, the samples were treated at atmospheric pressure in hydrogen, at 773K in a preparation space connected to the analysis chamber. In order to obtain the Ir 4f signals, the deconvolution procedure was applied in 45-75eV binding energy (BE) range, taking into account the Ti3s signal and its satellites (due to Mg K$_{\alpha2,\alpha3}$ X-ray excitation source). The Ir 4f BE was referred to Ti 2p$^{3/2}$ BE at 458.8 eV. Theoretical curves were adjusted to fit the peaks by means of an in-house computer program. This procedure has been applied to the raw data points. A Doniach Sunjic Lorentzian asymmetric function [9] was convoluted with an experimental Gaussian curve (G=0.8) and a Shirley background [10] was subtracted. Then values of BE, γ (half-width at half maximum of the Lorentzian curve), α (asymmetric parameter) were deduced, they were equal respectively to 60eV for 4f 7/2, 0.113 and 0.171. Relative element intensities (Ir, Ti, Cl) were calculated from peak surface ratio measurements, corrected by differences in escape depths (a root square approximation was used) and in cross section (using Scofield's data [11]). For the Ir 4f lines, the curve fitting is shown in Fig. 1 and the quantitative results of the sample is reported in the Table 1.

Table 1 Surface atomic ratios evaluated from XPS.

Ir/Ti	0.20
Cl/Ir	0.20
Cl/Ti	0.04

((Ir/Ti)$_{nominal}$ = 0.022)

The position in binding energy and the asymetric shape of the Ir 4f $_{7/2}$ peak indicate the presence of Ir^0 species. Moreover one notices that some amount of chlorine remained into the catalyst after the reduction treatment.

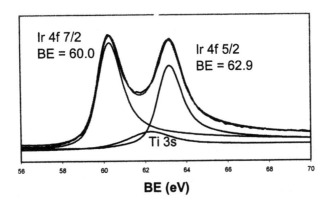

Figure 1 XPS analysis of 5% Ir/TiO$_2$ reduced at 773 K.

The metal particle size is an important parameter which usually controls the activity of metal catalysts and sometimes the selectivity as in Au/TiO$_2$[12]. Therefore, Transmission Electron Microscopy, coupled with an High Resolution observation was applied to measure the particle size. The HRTEM images were recorded on a TOPCOM 002B electron microscope, operating at 200kV, with a structural resolution of 0.18 nm. Catalyst grains were ground and diluted in an ethanolic solution. One drop of this solution, previously dispersed in an ultra-sonic tank was deposited onto a Cu grid coated by an holey carbon film and dried in air. Various regions of the grid were observed and the particle sizes were measured from the observation of 250 to 500 particles.

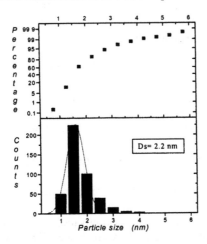

Figure 2 HRTEM particle size histogram of 5% Ir/TiO$_2$.

The following formula was used to calculate the mean surface diameter: $d_s = \Sigma n_i d_i^3 / \Sigma n_i d_i^2$, where n_i is the number of particles of diameter d_i. The cumulative curve, as a function of particle size (d_k) was also presented, from which the percentage of particles with size below d_k could be estimated. Fig.2. shows particle size histogram of the studied catalyst. The particle size distribution showed, that majority of particles are below 3 nm with a mean diameter equal to 2.2 nm.

CATALYTIC TESTS

Catalytic tests were carried out in a glass reactor, operating at atmospheric pressure. The total gas flow, controlled by a flowmeter, was varied by changing the pumping rate at the end of the flow line. The crotonaldehyde (CROTAL) supplied by FLUKA puriss and stored in argon was used as received. A known quantity of CROTAL was drawn up from the bottle using a tight syringe and introduced through a vaccin cap into a reservoir, installed on-line and maintained at 273 K ; therefore CROTAL at constant partial pressure (8 Torr) was carried over the catalyst by the hydrogen flow. The H_2 gas was first purified by passing through a trap, maintained at room temperature, containing Pt/Al_2O_3 catalyst mixed with zeolite to remove oxygen and water. A further purification was made through a MnO trap at 293K, installed just before the CROTAL reservoir. Beyond the CROTAL reservoir, the gas line was thermostated at about 333 K to avoid any condensation. The stability of the CROTAL pressure and the duration of the experiment were controlled by two catharometers inserted upstream and downstream with respect to the reactor, enabling the CROTAL flow rate ($1.25 \ 10^{-7}$ mol s^{-1}) to be measured. The reaction products were drawn off the flow line at different time during the catalytic run and analysed by gas-liquid chromatography (GLC, with a 30m long, 0.5461×10^{-3}m diameter DB-Wax column (J&W Scientific), at 353 K and using a flame ionisation detector).

Before catalytic experiments, the catalyst (50 mg) was reduced at 773K for 1 h and cooled down to the reaction temperature (353 K) under H_2 flow.

The selectivity to the different products, crotyl alcohol (CROTOL), butanal, butanol, hydrocarbons was calculated as the molar ratio of the selected product to the total products formed. The sensitivity factors are taken as 1 for CROTAL, CROTOL, butanal, butanol, side products and 1.4 for hydrocarbons.

It was checked that the support alone had no activity for crotonaldehyde hydrogenation.

The overall activity was found to decrease during the time on stream with profound deactivation (Fig.3).

Catalyst deactivation in gas-phase hydrogenation reaction of crotonaldehyde is frequently observed [2, 13]. Supposing that coke (or carbon deposit) is produced from adsorbed crotonaldehyde. Deactivation and self-regeneration proceed simultaneously, moreover, these steps are essentially slower than the reaction

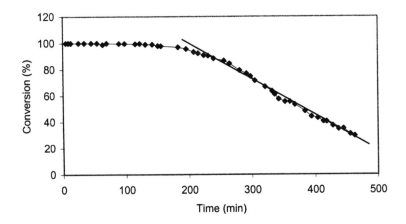

Figure 3 Hydrogenation of crotonaldehyde on Ir/TiO₂. Time–on-stream behaviour.

steps. This consideration allows to apply steady-state hypothesis only to hydrogenation, but not to deactivation steps. Detailed kinetic analysis was developed for Pt catalysts and presented in [13], leading to a following expression for conversion of crotonaldehyde (α)

$$\alpha = p_3 + p_1 \exp(-p_2 t) \tag{1}$$

where t – is time on stream. Compared to supported Pt catalysts deactivation over iridium was not profound to the same extent, therefore eq. (1) could be approximated with

$$\alpha = 100 \text{ at } t < t_1, \; \alpha = a_1 - a_2 t \text{ at } t > t_1 \tag{2}$$

with $t_1 = 200$ min in case crotonaldehyde hydrogenation of 5% Ir/TiO₂.

Eq. (1) was used for the description of experimental data generated in the present study. Sum of squares of relative deviations served as an object function. Results of simulations (with a_1 and a_2 equal to 152 and 0.27 correspondingly) are presented in Fig.3.

SELECTIVITY

For further discussion let us briefly analyze the reaction network and reaction kinetics.

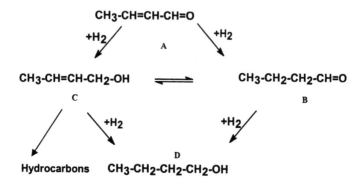

Figure 4 Hydrogenation of crotonaldehyde on Ir/TiO_2. Reaction network.

The reaction network is given in Fig.4, where A, B, C and D denote crotonaldehyde, butyraldehyde, crotyl alcohol (desired product) and butanol respectively.

In contrast to platinum catalysts, large amount of hydrocarbons (mainly butane and small amount of propane) were observed as products. Butane is produced essentially from dehydration of butanol and further hydrogenation; butanol being obtained in large amount on Ir/TiO_2 catalyst. Experiments with butanol as a starting reactant on Ir/TiO_2 catalyst showed high selectivity towards butane formation, moreover small amount of propane were also produced. Contrary to Ir catalysts, hydrogenation of crotonaldehyde over Pt/TiO_2 did not result in such extensive formation of hydrocarbons, although butanol on Pt/TiO_2 undergoes dehydration similarly to Ir/TiO_2. It indicates that Pt is much less active than Ir in hydrogenation of butanal and/or crotyl alcohol to butanol.

Selectivity towards crotyl alcohol was ca. 70% (Fig.5) at 20-80% conversion, which was substantially higher than ca. 30-40% reported in the literature for Ir/TiO_2 [8]. It is remarkable that this selectivity remained so high until 80% conversion. A more detailed comparison between the two catalysts is presented in Table 2.

At the present moment we can only speculate about the differences in performance of these two catalysts. Presumably, the differences in metal loading and the nature of the interactions between the metal and the support are responsible for such a profound difference in selectivity.

Table 2 Comparison between iridium catalysts.

	This work	[8]
Support, TiO₂	Eurotitania, 100 % anatase	Degussa P 25 80% anatase
Surface area, m²/g	70	50
Metal loading, wt%	5	0.5
Precursor	H_2IrCl_6	$IrCl_4$
Calcination T, K	673	673
Reduction T, K	773	773
Particles size	<3 nm, mean 2.2 nm	3 nm
Reaction T, K	353	323-373
Reaction set-up	Fixed bed with constant crotonaldehyde pressure	Pulses of 1 μl

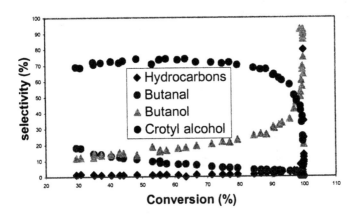

Figure 5 Hydrogenation of crotonaldehyde on 5%Ir/TiO₂. Selectivity *vs* conversion.

Detailed derivation of kinetic equations is given in [2], where it was shown, that the mole fraction of crotylacohol is defined by the following equation

$$N_C = L_C/(1-M_C)*(N_A^{M_C} - N_A) \qquad (3)$$

where

$$L_C = \cfrac{1}{1 + \cfrac{k_B(K_{AB}+K_{ABC})}{k_C(K_{AC}+K_{ABC})}} \qquad (4)$$

and

$$M_C = L_C \frac{k_{CD}}{k_C(1 + \frac{K_{ABC}}{K_{AC}})} \tag{5}$$

Here L_C and M_C value are combination of constants. Adsorption constants represent adsorption of crotonaldehyde via the olefinic bond (K_{AB}), via the carbonyl bond (K_{AC}) and via both bonds (K_{ABC}), while k_C, k_B and k_{CD} – rate constants of crotonaldehyde hydrogenation to crotyl alcohol and butanal and of crotyl alcohol to butanol. Analysis of the equation (3) reveals that the initial selectivity (at low conversions) depends on the L_C value. Selectivity profile as a function of conversion depends more on the M_C value. For parallel-consecutive reactions the lower value of M_C, the less pronounced is the crotyl alcohol selectivity dependence on the conversion.

Application of a regression program to experimental data shows (Fig.6), that eq. (3) can adequately describe dependence of crotyl alcohol mole fraction as a function of crotonaldehyde mole fraction. The object function (sum of squares) was 0.00011 with the following values of parameters: L_C =0.225±0.013, M_C = 0.230±0.03.

Crotyl alcohol mole fraction

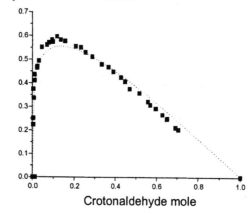

Figure 6 Hydrogenation of crotonaldehyde on 5%Ir/TiO$_2$. Crotyl alcohol *vs* crotonaldehyde mole fractions.

CONCLUSIONS

Ir/TiO$_2$ catalyst was studied in gas-phase crotonaldehyde hydrogenation at atmospheric pressure and 353 K. Strong deactivation was observed. However up to 80% conversion, the crotyl alcohol selectivity was high (70%) and remained constant. Butanol formed at the expense of butanal when the conversion increased from 0 to 80%. At conversion higher than 80% crotyl alcohol transformed into butanol and butane. This behavior differ from the ones observed on Pt/TiO₂ and Ir/Al₂O₃ [14]. Kinetic model, which describes activity and selectivity behaviour was advanced.

REFERENCES

1. V Ponec, On the role of promoters in hydrogenations on metals; α,β-unsaturated aldehydes and ketones. Applied Catalysis A. General: 149, 27-48, 1997.
2. M Consonni, D Jokic, DYu Murzin, R Touroude. High performances of Pt/ZnO catalysts in selective hydrogenation of crotonaldehyde, Journal of Catalysis: 188, 165-175, 1999.
3. RM Makouangou, DYu Murzin, AE Dauscher, RA Touroude. Kinetics of crotonaldehyde hydrogenation over titania supported platinum catalyst, Industrial & Engineering Chemistry Research: 33, 1881- 1888, 1994.
4. H Yoshitake, Y Iwasawa. Active sites and reaction mechanisms for deuteration of acrolein on titania-, yttria-, zirconia-, ceria- and sodium/silica-supported platinum catalysts. Journal of Chemical Society. Faraday Transactions: 88, 503-510, 1992.
5. M Abid, R Touroude. Pt/CeO₂ catalysts in selective hydrogenation of crotonaldehyde: high performance of chlorine-free caralysts, Catalysis Letters: 69, 139-144, 2000.
6. ML Khidekel, EN Bakhanova, AS Astakhova, KhA Brikenshtein, VI Savchenko, IS Monakhova, VG Dorokhov. Preparation of unsaturated alcohols by the hydrogenation of α,β-unsaturated aldehydes in the presence of an iridium catalyst. Izvestiya Akademii Nauk SSSR, Seria Khimicheskaya: 499, 1970.
7. EN Bakhanova, AS Astakhova, KhA Brikenshtein, VG Dorokhov, VI Savchenko, ML Khidekel, Selective hydrogenation of the carbonyl group of α,β-unsaturated aldehydes in the presence of an iridium catalyst. Izvestiya Akademii Nauk SSSR, Seria Khimicheskaya: 1993-1998, 1972.
8. P Reyes, M C Aguirre, G Pecchi, JLG Fierro, Crotonaldehyde hydrogenation on Ir supported catalysts: Journal of Molecular Catalysis, 164, 245- 251, 2000.
9. S Doniach and M Sunjic, Many-electron singularity in X-ray photoemission and X-ray line spectra from metals, Journal of Physics C: 3, 285-291, 1970.
10. DA Shirley, High resolution X-ray ptotoemission spectrum of the valence band of gold. Physics Review B: 5, 4709-4714 , 1972.
11. JH Scofield, Hartree-Slater photoionization cross-sections at 1254 and 1487 eV. Journal of Electron Spectroscopy and Related Phenomena: 8, 129-137, 1976.
12. P Claus, A Bruckner, C Mohr and H Hofmeister, Supported Gold Nanoparticles from Quantum Dot to Mesoscopic Size Scale: Effect of Electronic and Structural Properties on Catalytic Hydrogenation of Conkugated Functional Groups, J. Am. Chem. Soc., 122, 11430-11439, 2000.
13. M Consonni, R Touroude, DYu Murzin, Deactivation and selectivity pattern in crotonaldehyde hydrogenation. Chemical Engineering & Technology, 21, 605-609, 1998.
14. M Abid, Thesis, Selective Hydrogenation of Crotonaldehyde, Strasbourg, 2001.

50

Catalyst Deactivation in Selective Hydrogenation of β-Sitosterol to β-Sitostanol over Palladium

Minna Lindroos, Päivi Mäki-Arvela, Narendra Kumar, Tapio Salmi, and Dmitry Yu. Murzin
Åbo Akademi University, Åbo/Turku, Finland

Tapio Ollonqvist and Juhani Väyrynen
University of Turku, Turku, Finland

INTRODUCTION

In the modern days the awareness is increasing about the impact of cholesterol on human health. Although a certain concentration of it is necessary for a living organism, high levels of cholesterol in the bloodstream cause atherosclerosis. Deposits of cholesterol and other fatty substances can build up, eventually converting the blood vessel walls and narrowing the channels, thus leading to heart attacks and strokes [1].

Phytosterols, especially sitostanol, are known inhibitors of cholesterol absorption [2]. The major sterols belonging to the group of 4-desmethylsterols are campesterol (24α-methyl-5-cholesten-3β-ol), sitosterol (24α-ethyl-5-cholesten-3β-ol), campestanol (24α-methyl-5α-cholestan-3β-ol) and sitostanol (24α-ethyl-5α-cholestan-3β-ol). The saturated counterparts of campesterol and sitosterol are campestanol and sitostanol, commonly known as stanols.

Sitostanol can be obtained by hydrogenation of the double bond in sitosterol over Pd catalysts [3, 4] according to the following scheme:

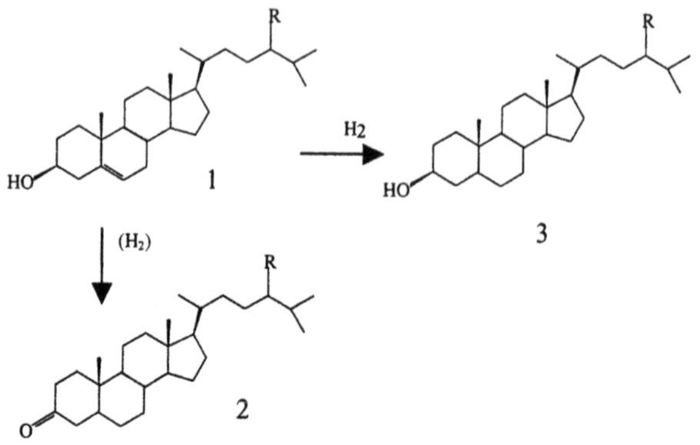

Figure 1 Reaction scheme of β-sitosterol (1) transformations into β-sitostanol (3): R=C₂H₅.

However, palladium catalysts did not exhibit very high selectivity, moreover they are prone to extensive deactivation.

The aim of this work was to investigate catalyst deactivation in selective hydrogenation of β-sitosterol to β-sitostanol over a Pd supported catalyst, as well as to find ways to regenerate the deactivated catalyst.

EXPERIMENTAL

Hydrogenation reactions of β-sitosterol were performed in a 300 ml autoclave (Parr) using 1-propanol as a solvent. The stirring rate was 1000 rpm. The hydrogen (AGA, 99.999% and AGA, 99.99%) pressure was 4 bar and temperature was 80°C. The fresh, deactivated and regenerated catalysts were characterized by BET and XPS techniques.

The raw material sitosterol is a product of the pulp and paper industry and contains besides β-sitosterol (94%) also ca. 6 % campesterol (structure 1 in Fig. 1 with R=CH₃) and several impurities such as sulfur (< 50 mg/kg), chloride and phosphorus (<20 mg/kg).

Pd supported catalysts were used. The metal content was 2 wt %, BET specific surface area 184.7 m²/g and mean metal particle size 1.78 nm (XRD).

Samples were withdrawn from the reactor at different time intervals. Prior to analysis samples were silylated with l-bis-(trimetylsilyl)trifluoracetamide. GC analysis was applied (Hewlett Packard 6890 with HP 5 column: 30m/0.32 mmID/0.25μm df, FID detector at 300 °C, T $_{column}$ = 300 °C).

CATALYTIC ACTIVITY

Some of the experimental results are presented in Fig. 2. The catalysts used in the
hydrogenation reaction were from the same batch and the raw material (sitosterol)
used was also the same. The main product was sitostanol, while sitostanone (2 in
Fig. 1) was obtained in inferior quantities (below 1%).

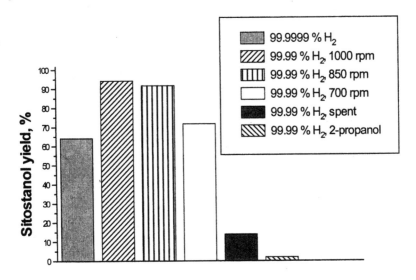

Figure 2 Yields of sitostanol at different experimental conditions after 1 h reaction time.

Preliminary experiments demonstrated that the stirring speed of 1000 rpm
was necessary to overcome possible impact of external mass transfer, therefore
throughout the text this value is implicitly assumed, if not stated otherwise. Application of 2-propanol as a solvent resulted in extremely low catalyst activity, thus
only 1-propanol was further utilized. It was also noticed, that preheating and stirring of the mixture of 1-propanol and sitosterol was essential to improve conversion of sitosterol.

It became apparently clear, that hydrogen purity is important for the catalyst
to exhibit sufficient activity. Surprisingly lower hydrogen purity, which means the
presence of oxygen, was found to provide a beneficial effect on catalyst activity.
The reason for this effect is not fully understood, however, it is known that exposure to oxygen is essential for obtaining high reaction rates and product enantioselectivity in the hydrogenation of phenyl-1,2-propanedione [5]. In case of enantioselective hydrogenation several explanations are possible, one being, that oxygen

helps to restore the original adsorption properties of the metal, which is necessary for the metal to be active.

This conclusion is perfectly in line with another observation, that follows from Fig. 2, namely that a spent catalyst shows much lower catalytic activity; than the fresh one.

In order to address the deactivation issue in more detail kinetic experiments with different amount of catalysts (from 0.5 to 1.5 g) were performed (Fig. 3).

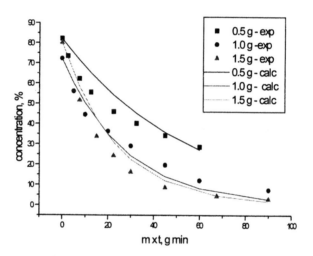

Figure 3 Kinetics of sitosterol hydrogenation.

The catalyst after working for few hours deactivated. Interestingly, the higher amounts of catalyst exhibited higher hydrogenation rates when plotted against the normalized abscissa (mass of the catalyst x time).

The possible reason for the deactivation of Pd surface could be poisoning by S, Cl and P which are present in the feed, blockage and coke formation. XPS results which will be discussed in more detail later, unequivocally demonstrated presence of Cl (1%) and carbon deposit (12.1%) in the spend catalyst.

In case of first order reaction kinetics (or proportionality of the reaction rate and sitosterol concentration) the following equation holds

$$-\frac{1}{m}\frac{dC}{dt} = kC \tag{1}$$

However, when a poison is present in the reactant feed, a certain part of catalytically active sites could be blocked. The number of these sites is proportional to the amount of poisons. Therefore, instead of m (catalyst mass) an effective catalyst mass ($m^* = m - f$), should be used, where f is the amount of catalyst irreversibly blocked by the poisons.

Assuming that f is independent on sitosterol concentration, one arrives at the following equation

$$C = C_o e^{-k(m-f)t} \tag{2}$$

Where k and f are parameters obtained via nonlinear regression analysis and

$C = C_o$ at t=0.

Results of parameter estimation presented in Fig. 3 indicate good agreement between experimental data and calculated values, providing additional support for the concept of poisoning as the reason for catalyst deactivation.

REGENERATION OF DEACTIVATED CATALYSTS

Several regeneration procedures are known to restore activity of deactivated catalysts. Among other methods catalyst regeneration either with oxygen or air or hydrogen were reported [6-9], depending on the type of deactivation. In the present study several procedures were applied and the results are presented in Fig. 4.

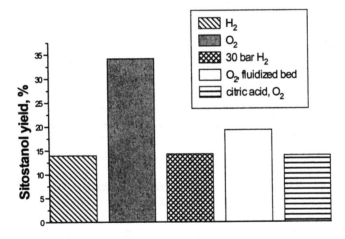

Figure 4 Yields of sitostanol over regenerated at after 1 h reaction time.

The deactivated catalyst was regenerated in oxygen/air and/or in hydrogen in a fixed and a fluidized bed reactor at different temperatures and tested in successive hydrogenation runs. The activity tests clearly demonstrated that the activity of the used catalysts in the second hydrogenation is quite low.

In order to get a deeper insight in the deactivation process, BET specific surface area measurements and XPS studies were undertaken. X-ray diffraction was utilized for the purpose of monitoring the possible changes in Pd particle size due to possible sintering.

Table 1 Catalyst characterization data

Catalyst		XPS Data		Particle size /nm	BET/$m^2 g^{-1}$
Fresh	8.10%	Pd 3d5/2	335.2 eV	1.78	185
	77.60%	PdO	336.5 eV		
	14.30%	PdO2	338.1 eV		
Used		Pd 3d5/2	334.6 eV	3.14	176
O$_2$ regenerated	48%	Pd 3d5/2	334.53 eV	3.18	180
	52%	PdO	335.83 eV		

Metal particle size in the spent catalyst did differ from the fresh one, indicating the importance of sintering. Interestingly specific surface area of used catalysts did not change very much providing additional support to the supposition that catalyst poisoning is mainly responsible for catalyst deactivation. At the same time importance of coke formation cannot be underestimated. In order to assess catalyst deactivation by this route further, extraction of hydrocarbons from the spent catalyst sample was undertaken. The deactivated catalyst was dissolved at room temperature in hydrofluoric acid solution 40 % and soluble coke from the catalyst was extracted using methylene chloride (CH_2Cl_2). The soluble coke was analyzed using GC-MS. Preliminary it can be reported that the hydrocarbons with mass numbers 141, 156, 170 and 185 are present, exact chemical nature of these compounds still remain unknown.

Moreover, presence of carbon containing species in the spent catalyst sample was proved by TPD studies when CO_2 content in effluent gas was monitored by MS. It was demonstrated that the temperature when the amount of released CO_2 reached maximum is around 400°C. Such thermal treatment during the regeneration procedure restores to some extent catalytic activity at the expense of metal sintering. The used catalyst was heated in O$_2$ flow with the temperature program 5 °C/min –400 °C (6h). The catalyst regenerated in presence of oxygen exhibited 11.1 % of carbon deposit. According to XPS P, Cl and S were not observed. Pd content was lower for regenerated catalyst than for the fresh one.

PROCESS ANALYTICS

Efficient analysis plays a crucial role in chemical process industry. It should provide timely, precise and accurate measurements of the process parameters. For
fine chemicals synthesis it usually means that extent of reaction in a batch reactor
should be effectively determined. Although gas chromatography is a very reliable
tool for accurate determination in sitosterol hydrogenation, rather cumbersome
analytical procedure calls for another more fast way to monitor progress of the
reaction.
    Detailed understanding of this reaction cannot be achieved without profound
kinetic studies, which require a necessity to process a large number of reaction
samples. Complicated analytics could be seen as a clear obstacle.
To address these issues in the present study UV spectroscopy as an alternative to
GC analysis was applied. UV-Vis spectrophotometry uses ultraviolet and visible
electromagnetic radiation to energetically promote valence electrons in a molecule
to an excited energy state. The UV-Vis spectrophotometer then measures the absorption of the energy to promote the electron by the molecule at a specific wave-
length or over a range of wavelengths. UV and Visible spectra can be used to determine the concentration of a chromophore compound in a mixture using the
Beer-Lambert law. Sitosterol contrary to sitostanol is a chromophore as it contains
π-electrons. Therefore UV spectrophotometry is a technique especially suited to
monitor the disappearance of the double bond in sitosterol. Varian DMS 90 UV-
VIS Spectrophotometer with an UV lamp (190 - 350 nm) was utilized in the present study. UV spectra of several samples taken during the course of reaction are
given in Fig. 5, while the corresponding kinetic curve is displayed in the box in the
same figure.

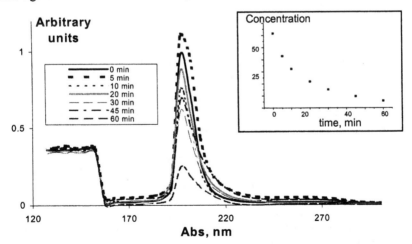

Figure 5 UV spectra of reaction samples taken at different reaction time.

It follows from Fig. 5, that UV spectroscopy proved to be an efficient analytical tool in β-sitosterol hydrogenation.

CONCLUSIONS

The selective hydrogenation of β-sitosterol was performed at 80 ^{0}C and hydrogen pressure of 4 bar using 1-propanol as a solvent. The fresh, deactivated and regenerated catalysts were characterized by XRD, BET and XPS techniques. Fresh catalyst exhibited high activity and selectivity (over 99 %) to sitostanol, while profound deactivation was observed several hours of operation. The reasons for the deactivation of the catalysts are poisoning of the Pd surface by S, Cl and P, palladium sintering, pore blockage and coke formation. A model is proposed which describes deactivation behaviour. UV spectroscopy was proved to be an efficient analytical tool in studying β-sitosterol hydrogenation kinetics. Several regeneration procedures, including regenerated in oxygen / air and or in hydrogen in a fluidized bed reactor, were applied to restore activity of deactivated catalysts. The most probable reasons for the catalyst deactivation is the formation of coke and poisoning of the Pd catalyst surface.

REFERENCES

1. P H Jones. Cholesterol: precursor to many lipid disorders. The American Journal of Managed Care 7: S289-S298, 2001.
2. T Miettinen, P Puska, H Gylling, H Vanhanen, E Vartiainen. Reduction of serum cholesterol with sitostanol-ester margarine in a mildly hypercholesterolemic population. The New England Journal of Medicine 333: 1308-1312, 1995.
3. J Helminen, U Hotanen, E Paatero, M Hautala, A Kärki. Process for hydrogenating unsaturated plant-based compounds and regeneration of used catalyst. WO 97/34917, 1997.
4. A Wong, W Boguski. Preparation of saturated phytosterols. WO 00/59921, 2000.
5. E Toukoniitty, P Mäki-Arvela, A. Nunes Villela, A. Kalantar Neyestanaki, T Salmi, R Leino, R Sjöholm, E Laine, J Väyrynen, T Ollonqvist, P J Kooyman. The effect of oxygen and the reduction temperature of the Pt/Al$_2$O$_3$ catalyst in enantioselective hydrogenation of 1-phenyl-1,2-propanedione. Cat.Today 60: 175-184, 2000.
6. J B Butt, EE Petersen. Activation, deactivation and poisoning of catalysts. Academic Press, 1988.
7. T Hattori, R L Burwell, Jr., Role of carbonaceous deposits in the hydrogenation of hydrocarbons on platinum catalysts, Journal of Physical Chemistry 83: 241-249, 1979.
8. C A. Querini, S C Fung, Temperature programmed oxidation technique: kinetics of coke-O$_2$ reaction on supported metal catalysts, Applied Catalysts A 117:53-74, 1994.
9. J Edvardsson, P Rautanen, A. Littorin, M Larsson, Deactivation and coke formation on palladium and platinum catalysts in vegetable oil hydrogenation, Journal of American Oil Society 78: 319-327, 2001.

51

Kinetics of the Hydrodechlorination of Dichlorobenzenes

Mark A. Keane
University of Kentucky, Lexington, Kentucky, U.S.A.

Dmitry Yu. Murzin
Åbo Akademi University, Åbo/Turku, Finland

INTRODUCTION

The presence of non-biodegradable chlorinated organic compounds in industrial effluent discharges is of increasing concern due to the mounting evidence of adverse ecologic/public health impacts (1). Catalytic hydro-dechlorination (HDCl), the focus of our studies, represents an innovative "end-of-process" strategy that offers a means of recovering valuable raw material from chlorinated waste. Incineration of chloroarenes is problematic as complete combustion occurs at prohibitively high temperatures (> 1700 K) while the formation of hazardous dioxins/furans can result from incomplete incineration (2). Catalytic oxidation represents a more progressive approach, requiring lower operating temperatures (523-823 K) and fuel/air ratios (3). Nevertheless, by-products include CO, Cl_2 and $COCl_2$ that are difficult to trap while complete oxidation generates unwanted CO_2. Biological oxidation can be effective when dealing with biodegradable organics but as chloroaromatics are used in the production of herbicides and pesticides, they are very resistant to biodegradation (4). The possibility of achieving dechlorination by electrochemical means has been addressed in the literature (5) but high dechlorination efficiency typically necessitates the use of non-aqueous reaction media and environmentally destructive cathode materials that has mitigated against practical application.

Catalytic hydrodehalogenation is established for homogeneous systems (6) and while high turnovers have been achieved, this approach is not suitable for environmental remediation purposes due to the involvement of additional chemicals (solvents/hydrogen donors) and the often difficult product/solvent/ catalyst separation steps. HDCl in heterogeneous systems has been viewed in terms of both nucleophilic (7) and electrophilic (8,9) attack and, in common with most hydrogenolysis reactions, dechlorination rate is strongly influenced by the electronic structure of the active sites (10). Chlorobenzene (ClB) has been the most widely adopted model reactant to assess catalytic HDCl activity in both the gas (7-9,11-18) and liquid (19,20) phases using Pd (7,18-20), Pt (19), Rh (17-19) and Ni (8,9,11-17) catalysts. The removal of multiple Cl atoms from an aromatic host has been studied to a lesser extent (9,14-17,21). In the treatment of polychlorinated aromatics, a range of partially dechlorinated isomers has been isolated where product composition depends on the nature of the catalyst and process conditions, i.e. temperature, concentration, residence time etc. HDCl kinetics has largely been based on pseudo-first order approximations (13,16,17) and there have been few attempts (7,9,18,22) to construct kinetic models from mechanistic considerations. The studies cited above represent a compilation of rate data that needs to be extended in a strategic fashion in order to facilitate a determination of the most active HDCl catalysts under clearly defined operating conditions. The mechanism of C-Cl bond hydrogenolysis is still open to question and this must be established and combined with a robust "predictive" kinetic model in order to inform reactor design and facilitate process optimization. In a previous study (23), we considered the applicability of a number of mechanistically sound kinetic models to describe the gas phase HDCl of ClB promoted by Ni/SiO_2 over a range of reaction conditions. The kinetic model that best reproduced the experimental data involved reaction between non-competitively and dissociatively adsorbed ClB and spillover hydrogen on a non-uniform surface. We have extended that kinetic study to consider the HDCl of dichlorobenzene (DCB) isomers and report herein the results of our kinetic modeling.

CATALYST PREPARATION AND ACTIVATION

A 1.5 % w/w Ni supported on silica (Cab-O-Sil 5M, 194 m^2 g^{-1}) catalyst precursor was prepared by homogeneous precipitation/deposition as described in detail elsewhere (24). The hydrated sample, sieved in the 150-200 μm range, was reduced, by heating directly in a 100 cm^3 min^{-1} stream of dry H_2 (99.9%) at 5 K min^{-1} to 673 ± 1 K which was maintained for 18 h. Under these activation conditions the catalyst supports 1.1×10^{20} exposed Ni atoms/g catalyst with a surface weighted average Ni diameter = 1.4 nm; these values are based on chemisorption measurements (12).

CATALYTIC REACTOR SYSTEM

All the catalytic reactions were carried out under atmospheric pressure, in situ immediately after activation, in a fixed bed glass reactor where $523\ K \leq T \leq 573$ K. The catalytic reactor has been described previously in some detail (25) but the pertinent features are given below. A microprocessor controlled infusion pump (kd Scientific) was used to deliver the DCB feed via a glass/teflon air-tight syringe and teflon line and the vapor was carried through the catalyst bed in a stream of dry H_2. The catalytic measurements were made at W/F (catalyst weight/DCB molar flow rate) values in the range 33-167 g mol^{-1} h. DCB was fed to the reactor either undiluted or as n-hexane solutions (1:3 v/v); n-hexane as solvent had no effect on HDCl rate. The reactor was operated in the differential mode with fractional conversion of the inlet DCB feed < 0.07. The catalytic measurements were made at an overall gas space velocity of 2250 h^{-1} where the H_2 partial pressure was varied from 0.35 to 0.90 atm by dilution in He while the DCB partial pressure spanned the range 0.01-0.1 atm. One standard set of reaction conditions (T = 548 K, P_{DCB} = 0.03 atm, P_{H2} = 0.9 atm) was routinely repeated to ensure that the catalyst had not sustained any long-term loss of activity. The catalytic system has been shown (12) to operate with negligible diffusion constraints (effectiveness factor (η) > 0.99 at 573 K). Heat transport effects can also be disregarded; the temperature differential between the catalyst particles and bulk fluid phase was < 1 K. The ratio of reactor to catalyst particle diameter = 90, exceeding the lower limit of 10 set by Froment and Bischoff (26), satisfying the application of plug-flow conditions. The reactor effluent was frozen in a liquid nitrogen trap for subsequent analysis by capillary GC as described elsewhere (12). A chlorine (in the form of HCl product) mass balance was performed by passing the effluent gas through an aqueous NaOH (3-8 × 10^{-3} mol dm^{-3} agitated at $\geq$ 300 rpm) trap with pH and potentiometric analyses of the scrubbing solution (12); Cl mass balance was complete to better than ± 10%. At the end of each catalytic run, the catalyst was heated (5 K min^{-1}) in flowing dry H_2 to 673 K (12 h). The catalyst exhibited the same level of activity after this regeneration step and all the data presented here have been generated in the absence of any significant catalyst deactivation. Each catalytic run was repeated (up to six times) using different samples from the same batch of catalyst and the measured rates did not deviate by more than ± 7%.

KINETIC MODELING

The HDCl of the three DCB isomers generated ClB (partial dechlorination) and benzene (complete dechlorination) as the only detectable products; there was no evidence of any aromatic ring reduction. The kinetic patterns were found to be broadly the same for each DCB isomer. Let us consider first the dependence of HDCl rate on DCB partial pressure (P_{DCB}), which exhibits a maximum (see

Figures 1–3) that was particularly pronounced in the case of 1,2-DCB. Taking the rate dependence at the lower P_{DCB} range (below the rate maximum), the reaction orders with respect to DCB are greater than unity and close to 2. Any kinetic model based on mechanistic assumptions must explain the second order

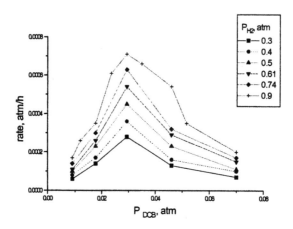

Figure 1 Dependence of the experimentally determined HDCl rates on 1,2-DCB partial pressure: T=523 K, legends correspond to different H_2 partial pressures.

behavior in the domain of low P_{DCB} and the occurrence of a P_{DCB} related rate maximum. Any expression used to capture the observed P_{DCB}/HDCl rate relationships must exhibit a reaction order in DCB in the nominator equal to 2. The rather steep descent at higher P_{DCB} necessitates the incorporation of a P_{DCB} term in the denominator that is raised to the power of four. Such a rate expression is provided below

$$r = \frac{kK_{DCB}P_{DCB}^2}{1 + K_{DCB}P_{DCB}^2 + K_{oli}K_{DCB}P_{DCB}^4} \tag{1}$$

where K_{DCB} is the equilibrium adsorption coefficient for DCB. The mechanism that corresponds to this kinetic expression assumes that an adsorption of two DCB molecules on adjacent sites is required to create catalytically active species. This intermediate complex can oligomerize on the surface, *i.e.* two complexes give one oligomer. This step is taken to be equilibrated and K_{oli} in eq. (1) denotes the associated equilibrium constant. There was no evidence of any site blocking by surface oligomer leading to irreversible catalyst deactivation under the differential reaction conditions that were employed. The applicability of this Model can be assessed in Figure 2 for the HDCl of 1,2-DCB at 523 K and in Figure 3 wherein an increase in reaction temperature can be seen to elevate the HDCl of 1,3-DCB.

Kinetic analysis revealed that variations in the gas phase H_2 partial pressure had little effect on the dependence of reaction rate on P_{DCB}, an effect that is supported by previous work (12,23) wherein we proposed a non-competitive adsorption of H_2 and chloroaromatic. Moreover, the decreasing values of H_2 reaction order with increasing temperature is taken to be indicative that H_2 adsorption (leading to HDCl) is endothermic. The reactive form of H_2 was assigned (12), based on H_2 TPD data for freshly activated and used catalysts

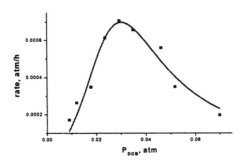

Figure 2 Variation of the experimental (■) and calculated (eq. 1, solid line) HDCl rates with varying 1,2 -DCB partial pressure: P_{H_2} = 0.9 atm, T = 523 K.

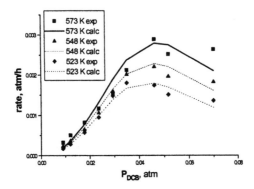

Figure 3 Variation of the experimental and calculated (eq. (1), solid/dashed/dotted lines) HDCl rates with varying 1,3-DCB partial pressure at fixed temperatures: P_{H_2} = 0.9 atm.

(22), to spillover species on the support and metal/support interface. The involvement of spillover hydrogen in HDCl has been proposed elsewhere (27) for supported Pd catalysts. As the reactive spillover hydrogen does not compete with the incoming DCB for surface adsorption sites, surface coverage by spillover hydrogen can be accounted for by the addition of a second term in Model (1). Our previous calculations for ClB HDCl (23) have revealed that the incorporation of coverage dependent kinetic and equilibrium constants provide a better reproduction of the experimental data. Although both sites associated with reactive DCB and H_2 can be inhomogeneous, the multi-centered nature of "bulky" organic molecule interactions with metal surfaces can mask the influence of non-uniformity (28). Therefore, for the sake of simplicity we assume here that only H_2 adsorption on the support or support/metal interface is coverage dependent. Dependence of adsorption enthalpy on coverage for an "evenly non-uniform" surface (discussed in Ref. 28) yields a logarithmic adsorption isotherm and the kinetic expression takes the form

$$r = \frac{kK_{DCB}P_{DCB}^2}{1+K_{DCB}P_{DCB}^2+K_{ol}K_{DCB}^2P_{DCB}^4}\ln(K'_H P_{H_2}) \qquad (2)$$

where K_H is the equilibrium constant for spillover hydrogen adsorption. The fit that this expression provides to the experimental rate dependence can be seen in Figure 4 where P_{DCB} is varied at a constant H_2 partial pressure and in Figure 5 where rate dependence on H_2 partial pressure is plotted at fixed P_{DCB} values. The generation of surface active hydrogen should involve two steps, i.e. the exothermic dissociative chemisorption of H_2 on Ni and the endothermic spillover onto SiO_2. Spillover at the metal/support interface requires a breaking of bonds with the adsorbing surface and the formation of new bonds with the accepting surface

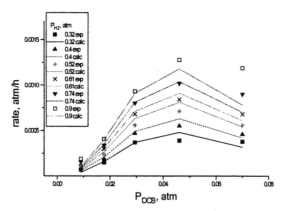

Figure 4 Variation of the experimental (symbols) and calculated (eq. 2, solid/dashed/dotted lines) *HDCl* rates with varying *1,4-DCB* partial pressure at fixed H_2 partial pressures: *T*=548 K.

in an endothermic step, which is facilitated by the increase in entropy of the spillover species. Parameter estimation for equation (2) was performed *via* minimization of the sum of residual squares. An equal weight was given to every experimental point in the estimation. The minimization was carried out numerically using the Marquardt-Levenberg method adapted to the kinetic non-linear regression program package Reproche (29). The algebraic model option was adopted where the partial pressures of H_2 and DCB were the independent variables and calculations were performed for the entire data set (generated by varying temperature and partial pressures) assuming an Arrhenius and van't Hoff temperature dependence for kinetic and equilibrium constants, respectively; the extracted adsorption/kinetic parameters are listed in Table 1.

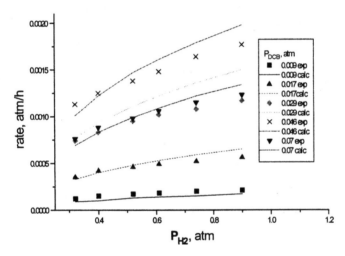

Figure 5 Variation of the experimental and calculated (eq. 2, solid/dashed/dotted lines) HDCl rates with varying H_2 pressure at fixed 1,2-DCB partial pressures (atm): T=523 K.

Table 1 Values of the calculated parameters.

Reactant	k_1K_{DCB}, h⁻¹	$(E_a$-$\Delta H_{DCB})/R$, K	$K_{oli}K_{DCB}$, atm	$(\Delta H_{oli}$-$\Delta H_{DCB})/R$, K
1,2-DCB	34	1.6×10^3	117	1.4×10^3
1,3 DCB	168	1.6×10^3	4	470
1,4 DCB	114	1.2×10^3	1.4×10^3	2.8×10^3

	K'_H, atm⁻¹	$\Delta H_H/R$, K	Mean residual, %	
1,2 DCB	0.3×10^{-2}	1.1×10^5	15.7	
1,3 DCB	7.5×10^3	2.9×10^3	15.5	
1,4 DCB	1.0×10^2	4.5×10^3	18.4	

Taking an overview of the Model fit to the experimental data plotted in Figures 4 and 5, it is fair to state that the overall agreement is satisfactory and even in the case with the greatest mean residual, systematic deviations are only really pronounced at the highest rates. The catalytically significant heats of hydrogen adsorption are positive, indicative of an overall endothermic process where the dissociative energy of hydrogen exceeds the heat released in the formation of spillover species, as noted previously (23).

SELECTIVITY

Selectivity in terms of the HDCl of DCBs refers to the degree of partial (to ClB) as opposed to total dechlorination (to benzene). The relationship between DCB conversion and product distribution is shown in Figure 6 where the data were generated by varying temperature and H_2/DCB partial pressures. The ClB mole fraction (at a fixed P_{DCB} and T) was essentially invariant over the range of H_2 partial pressures that were considered which supports the concept of separable kinetics. The anti-sympathetic relationship between conversion and ClB selectivity is to be expected in a sequential HDCl scheme where complete dechlorination is enhanced at high conversions. Both the 1,3- and 1,4-DCB isomers exhibited essentially the same selectivity behavior but deviated somewhat from that associated with 1,2-DCB. In the case of the latter isomer, a concerted removal of both Cl substituents is preferred to individual C-Cl bond scission with a resultant higher proportion of benzene in the product at any given DCB conversion. Where the Cl substituents are separated further apart on the aromatic ring, partial dechlorination is more significant, as observed elsewhere (13,15). The mole fraction of the partially

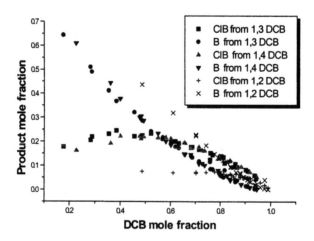

Figure 6 Dependence of product mole fraction on *DCB* conversion.

dechlorinated product (N_{CIB}) formed in the possible parallel/consecutive reaction pathways,

$$DCB \xrightarrow{k1} CIB \xrightarrow{k2} B$$

(with $k3$ pathway from DCB to B)

can be described by the following equations (30)

$$-\frac{n}{m}\frac{dN_{DCB}}{dt} = (k_1 + k_3)N_{DCB} \qquad (3)$$

$$\frac{n}{m}\frac{dN_{CIB}}{dt} = k_1 N_{DCB} - k_2 N_{CIB} \qquad (4)$$

where n is the number of moles of DCB contacting the catalyst bed of mass m, k_i the effective rate constants, which are composite terms that also include adsorption contributions. A combination of equations (3) and (4) with integration yields

$$N_{CIB} = \frac{L}{1-M}(N^{M}_{DCB} - N_{DCB}) \qquad (5)$$

where the parameters L and M are given by:

$$L = \frac{k_1}{k_1 + k_3}; M = \frac{k_2}{k_1 + k_3} \qquad (6)$$

and can be determined by numerical parameter estimation. Results of the calculations are presented in Figure 7, which reveals a more than adequate description of selectivity trends at three temperatures; the estimated values of $L=0.8$ and $M=1.4$.

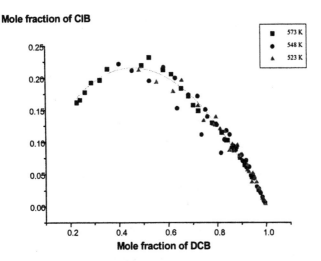

Figure 7 Dependence of *CIB* mole fraction on *1,4-DCB* conversion at three reaction temperatures: experimental data (symbols); fit to Model (3) (dotted line).

A strictly consecutive reaction network should yield a value of L close to unity (k_3=0). However, numerical analysis demonstrated that with this value of L, systematic deviations are observed, suggesting the involvement of two quite distinct *DCB* hydrodechlorination routes to benzene, one involving ClB as a reactive intermediate. Taking equation (5), ClB selectivity at low DCB conversions depends mostly on the L value, confirming a significant contribution of sequential dechlorination *via* ClB. The M parameter reflects the selectivity response over the range of conversions, where partial HDCl is favored at lower M values. The ratio of L/M equals the ratio of N_{ClB}/N_{DCB} at the maximum N_{ClB}.

CONCLUSIONS

The dependence of DCB HDCl rates over Ni/SiO_2 on H_2 and DCB pressure can be described in terms of the involvement of two types of active sites with a logarithmic dependence on H_2 partial pressure. A rate maximum is observed with respect to DCB partial pressure where, under conditions of ascending rate, the reaction order with respect to DCB exceeds unity. The mechanism, which explains such kinetic regularities, assumes adsorption of two DCB molecules on adjacent sites and reaction with spillover hydrogen. Selectivity towards ClB was seen to be independent of H_2 pressure but dependent on the nature of the isomer; 1,2-DCB undergoes complete dechlorination to a greater extent and HDCl selectivity trends for 1,3-DCB and 1,4-DCB coincide.

ACKNOWLEDGEMENTS

The authors wish to thank G. Tavoularis and C. Menini for help with HDCl measurements.

REFERENCES

1. Toxics Release Inventory, Public Data Release. Washington DC: USEPA, Office of Pollution Prevention and Toxics, 1991.
2. A Converti, M Zilli, DM De Faveri, G Ferraiolo. Hydrogenolysis of organochlorinated pollutants: kinetics and thermodynamics. J. Hazard. Mater. 27: 127-135, 1991.
3. JR González-Velasco, A Aranzabal, R López-Fonseca, R Ferret, JA González Marcos. Enhancement of the catalytic oxidation of hydrogen-lean chlorinated VOCs in the presence of hydrogen-supplying compounds. Appl. Catal. B: Environmental 24: 33-43, 2000.
4. RB Clark. Halogenated Hydrocarbons in Marine Pollution. Oxford: Oxford Science Publ., 1989.
5. MS Mubarak, DG Peters. Electrochemical reduction of di-, tri-, and tetrahalobenzenes at carbon cathodes in dimethylformamide. J. Electroanal. Chem. 435: 47-53, 1997.

6. CM King, RB King, NK Bhattacharyya, MG Newton. Organonickel chemistry in the catalytic hydrodechlorination of polychlorobiphenyls (PCBs). J. Organomet. Chem.600: 2000, 600, 63-70.

7. M Kraus, V Bazant. In: J.W. Hightower, ed. Proceedings of the 5th International Congress on Catalysis. New York: North-Holland, 1973, pp. 1073-1083.

8. E-J Shin, MA Keane. Detoxifying chlorine rich gas streams using solid supported nickel catalysts. J. Hazard. Mater. B 66: 265-278, 1999.

9. BF Hagh, DT Allen. Catalytic hydroprocessing of chlorinated benzenes. Chem. Eng. Sci. 45: 2695-2701, 1990.

10. A Yu Stakheev, LM Kustov. Effects of the support on the morphology and electronic properties of supported metal clusters: modern concepts and progress in 1990s Appl. Catal. A: General 188: 3-35, 1999.

11. E-J Shin, MA Keane. Gas phase catalytic hydrodechlorination of chlorophenols using a supported nickel catalyst. Appl. Catal. B: Environmental 18: 241-250, 1998.

12. G Tavoularis, MA Keane. Gas phase catalytic hydrodechlorination of chlorobenzene over nickel/silica. J. Chem. Technol. Biotechnol. 74: 60-70, 1999.

13. E-J Shin, MA Keane. Detoxification of dichlorophenols by catalytic hydrodechlorination of chlorophenols using nickel/silica. Chem. Eng. Sci. 54: 1109-1120, 1999.

14. E-J Shin, MA Keane. Gas phase catalytic hydrodechlorination of pentachlorophenol over supported nickel Catal. Lett. 58: 141-145, 1999.

15. E-J Shin, MA Keane. Gas phase catalytic hydroprocessing of chlorophenols. J. Chem. Technol. Biotechnol. 75: 159-167, 2000.

16. BF Hagh, DT Allen. Catalytic hydroprocessing of chlorobenzene and 1,2-dichlorobenzene. AIChE J. 36: 773-778, 1990.

17. J Frimmel, M Zdražil. Comparative study of activity and selectivity of transition metals in parallel hydrodechlorination of dichlorobenzene and hydrodesulphurization of methylthiophene. J. Catal. 167: 286-295, 1997.

18. B Coq, G Ferrat, F Figueras. Conversion of chlorobenzene over palladium and rhodium catalysts of widely varying dispersion. J. Catal. 101: 434-445, 1986.

19. Y Ukisu, TJ Miyadera. Hydrogen transfer hydrodechlorination of aromatic halides with alcohols in the presence of noble metal catalysts. Mol. Catal. A: Chemical 125: 135-142, 1997.

20. JL Benitez, G Del Angel. ^{27}Al, ^{1}HNMR studies of supported LiAlH$_4$ and Pd-LiAlH$_4$ as reagents for dechlorination of chlorobenzene. React. Kinet. Catal. Lett. 66: 13-18, 1999.

21. Y Cesteros, P Salagre, F Medina, J-E Sueiras. Synthesis and characterization of several Ni/NiAl$_2$O$_4$ catalysts active for the 1,2,4-trichlorobenzene hydrodechlorination. Appl. Catal. B: Environmental 25: 213-227, 2000.

22. E-J Shin, A Spiller, G Tavoularis, MA Keane. Chlorine-nickel interactions in gas phase catalytic hydrodechlorination: catalyst deactivation and the nature of reactive hydrogen. Phys. Chem., Chem. Phys. 1: 3173-3181, 1999.

23. MA Keane, DYu Murzin. A kinetic treatment of the gas phase hydrodechlorination of chlorobenzene over nickel/silica: beyond conventional kinetics. Chem. Eng. Sci. 56: 3185-3195, 2001.

24. MA Keane. The role of catalyst activation in the enantioselective hydrogenation of methyl acetoacetate over silica-supported nickel catalysts. Can. J. Chem. 72: 372-381, 1994.

25. MA Keane, PM Patterson. The compensation effect in the hydrogenation of benzene and its derivatives over supported nickel catalysts. J. Chem. Soc., Faraday Trans. 92: 1413-1421, 1996.

26. G Froment, KB Bischoff. Chemical Reactor Analysis and Design. New York: John Wiley, 1990.

27. S Kovenklioglu, Z Cao, D Shah, RJ Farrauto, EN Balko. Direct catalytic hydrodechlorination of toxic organics in wastewater. AIChE J. 38: 1003-1012, 1992.

28. D Yu Murzin, T Salmi. Isothermal multiplicity in catalytic surface reactions with coverage dependent parameters: case of polyatomic species. Chem. Eng. Sci. 51: 55-62, 1996.

29. S Vajda, P Valko, Reproche, Regression program for chemical engineers. Budapest: Eurecha, 1985.

30. DYu Murzin, NV Kul'kova, MI Temkin, Kinetics of hydrogenation of benzene into cyclohexene and cyclohexane. Kinet. Katal. 31: 983-987, 1990.

52

Effect of Reaction Parameters on the Hydrogenation of Nitrobenzene to p-Aminophenol

Jayesh J. Nair, Setrak K. Tanielyan, and Robert L. Augustine
Center for Applied Catalysis, Seton Hall University, South Orange, New Jersey, U.S.A.

Robert J. McNair and Dingjun Wang
Johnson Matthey, West Deptford, New Jersey, U.S.A.

ABSTRACT

The hydrogenation of nitrobenzene to para-aminophenol (PAP) takes place by way of an initial partial hydrogenation to phenylhydroxyl amine which then undergoes an *in situ* acid catalyzed rearrangement to PAP. This reaction is most commonly run over Pt/Carbon catalysts in the presence of aqueous sulfuric acid and surfactant to assist in dispersing the nitrobenzene throughout the reaction medium. The yield of PAP is closely related to those reaction parameters which facilitate, first, the partial hydrogenation step and, second, the acid promoted rearrangement before further hydrogenation to aniline can take place. We examined a number of reaction parameters such as hydrogen pressure, temperature, nitrobenzene concentration and the amount of surfactant present in the reaction mixture. High selectivity to PAP formation (>88%) and catalyst productivity (>220 g PAP / g catalyst / hr) at 100% conversion were achieved at a hydrogen pressure of 50 psig and 90°C.

INTRODUCTION

The catalytic hydrogenation of nitrobenzene to p-amionophenol (PAP) under phase-transfer conditions is an industrially important process that was first reported in 1940 (1). PAP is an important intermediate which has been used in the production of analgesic drugs, (2) photographic developers and dyes (3). The

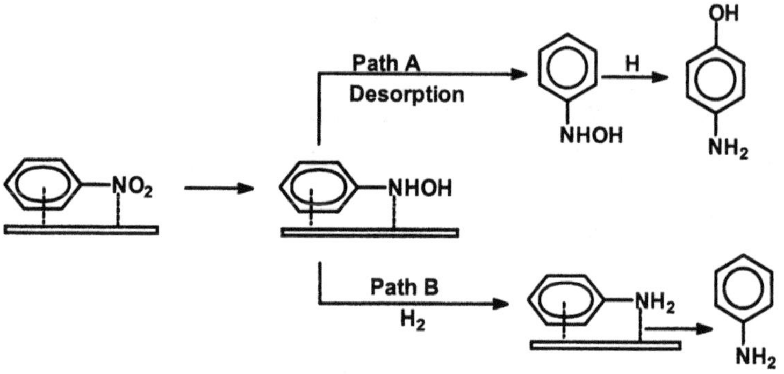

Path A
Desorption

Path B
H_2

hydrogenation is conducted in an acidic medium in the presence of a supported platinum catalyst. A significant rate enhancement and improvement in selectivity to PAP formation was further achieved by using quaternary salts or polyether surfactants as phase transfer catalysts (PTC) (4,5).

While most of the information regarding this reaction is well documented in patents, very little has been published in the open literature (6-8). The commonly accepted reaction mechanism involves the partial hydrogenation of nitrobenzene to give phenylhydroxylamine (PHA) which desorbs from the catalyst into the aqueous phase where it is converted to p-aminophenol (PAP) through an acid catalyzed rearrangement (8-12) (Scheme 1). Some of the PHA is not desorbed and is hydrogenated to aniline.

We present here details regarding the influence of the reaction variables on the outcome of nitrobenzene hydrogenation in a biphasic acid system in presence of a platinum catalyst and N,N-dimethyldodecylamine as a PTC.

EXPERIMENTAL

Standard Reaction Conditions

The reaction flask was charged with 33.7 mg of a wet 1.5% Pt/C (JM C5105) (13.5mg dry basis, 1.04 μmole Pt)), 30 ml of a 10% w/w aqueous H_2SO_4, 0.75 ml of the phase transfer catalyst solution (40 μl/ml, 0.145M, N,N-dimethyldodecyl amine dissolved in 10% aqueous sulfuric acid) and 4.05 ml (39.3 mmole) of freshly distilled nitrobenzene. The flask was connected to the reaction system, and alternately filled and purged with nitrogen five times to 15 psi pressure. The reactor temperature was then slowly raised to the set value under constant stirring. The stirring rate was accurately maintained at 1700 rpm using a microprocessor controlled magnetic stirrer (VWbrand Model 400S). Once the temperature reached 90°C, the nitrogen was replaced by hydrogen using multiple flush cycles, and the flask was then pressurized to 10 psi with hydrogen. The stirring was re-initiated and the computer monitoring of the reaction parameters was begun. The progress of the reaction was followed by the rate of hydrogen uptake; and after the reaction was completed, the reaction flask was cooled to ambien, and a sample was extracted for analysis.

Data Processing and Analysis

The following parameters were used to evaluate the reaction:

Selectivity to PAP	% ([PAP])*100 /([PAP]+[Aniline])
Rate of H_2 uptake, mmol/min	From H_2 uptake, to 30% conversion
Productivity to PAP, g /g catalyst/h	g PAP/g catalyst/reaction time (hr)

RESULTS AND DISCUSSION

Characteristic Regimes in the Hydrogen Uptake Curve

Generally, the platinum catalyzed hydrogenation of nitrobenzene in aqueous sulfuric acid has been carried out using a relatively low concentration of nitrobenzene (1-2 M), a small amount of Pt/C catalyst, low hydrogen pressures (10-30 psi) and a temperature of 80-90 °C. Four distinct regimes are seen the hydrogen uptake curve for a standard hydrogenation run (Fig. 1). The first region, A, is characteristic of the linear rate of hydrogen uptake, which continues to a 70-75% conversion level. In the second region, B, the rate of hydrogen uptake slows but the acid/nitrobenzene/water emulsion is still intact. In these two regions the catalyst is well dispersed in the organic phase, and the emulsion is stable. It takes an average of 10 – 15 minutes after the stirring has stopped to see a phase separation with a bottom organic and top aqueous phases. At these two stages the catalyst is wetted in the organic phase. During the third stage, C, the emulsion collapses, and the catalyst "precipitates" with the remaining nitrobenzene and reaction intermediates in a separate liquid phase. The fourth stage, D, is characterized by the disappearance of the organic phase, re-dispersion of the catalyst back into the aqueous phase and the slow hydrogenation of the remaining dissolved nitrobenzene, primarily to aniline.

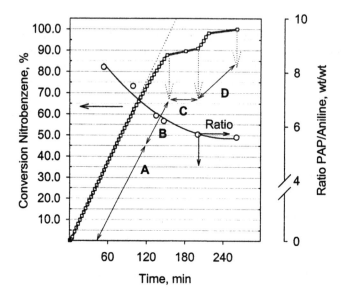

Figure 1 Conversion and PAP selectivity curves recorded during the hydrogenation of nitrobenzene under standard conditions.

The PAP/aniline ratios for samples withdrawn at different stages of these four characteristic regions are also shown in this figure. The selectivity to PAP declines throughout the hydrogenation but the rate of the decrease slows toward the end of the reaction. In all subsequent runs the reaction was stopped at about 80% completion.

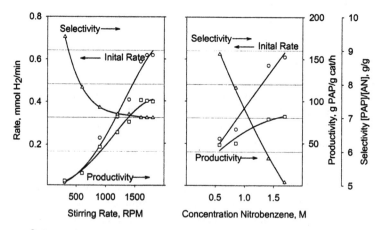

Figure 2 Dependence of the hydrogen uptake rate, the selectivity to PAP and the productivity on the stirring rate (**A**) and on the nitrobenzene concentration (**B**) under standard reaction conditions.

Stirring Rate and Nitrobenzene Concentration Effect

The stirring rate was varied between 300 to 2000 RPM using the standard reaction conditions. The results are shown in Figure 2A. These data show that the most dramatic changes occurred when the stirring rate was between 300 and 1400 RPM. Here PAP selectivity gradually declined to a PAP/aniline ratio of 7 and the hydrogen uptake rate increased in a second order relationship. When the stirring rate was further increased from 1400 to 1800 RPM, the selectivity remained essentially constant. All further studies were conducted at fixed stirring rate of 1700 RPM at which point the reaction rate and the PAP selectivity apparently became free of any diffusion limitation.

The effect of the initial nitrobenzene concentration was studied over the range 0.5 – 1.6 M. The results are shown in Figure 2B. When the substrate concentration was increased over this range, the hydrogen uptake rate increased linearly, the productivity showed a tendency of leveling, while the PAP selectivity declined continuously to a final PAP/aniline ratio of 5 at high substrate concentrations.

Catalyst Quantity and Hydrogen Pressure Effect

The effect of the amount of catalyst used on the hydrogen uptake rate and on the selectivity was studied using 10 to 60 mg portions of a 1.5% Pt/C catalyst. The initial rate increased in a linear fashion with the increase in the quantity of catalyst. The selectivity to PAP deteriorates at high catalyst loading.

The effect of the hydrogen pressure on the initial rate and the selectivity at 90 °C and stirring rate of 1700 RPM is shown in Figure 3B. With an increase

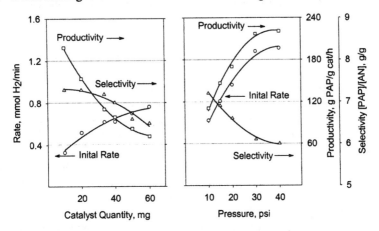

Figure 3 Dependence of the hydrogen uptake rate, the selectivity to PAP and the productivity on the catalyst quantity (A) and on hydrogen pressure (B) under standard reaction conditions.

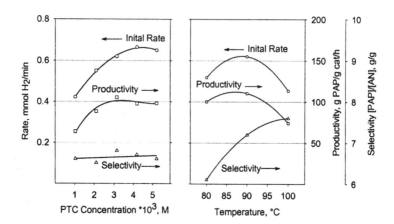

Figure 4 Dependence of the hydrogen uptake rate, the selectivity to PAP and the productivity on the PTC concentration (A) and on temperature (B) under standard reaction conditions.

in the hydrogen pressure both the initial rate and the productivity increased almost linearly up to 30 psi beyond which point they reached constant levels of 1.4mmol/min and 225 g PAP/g cat/h, respectively. This increase in the reaction rate was accompanied by only a minor loss in selectivity. The catalyst productivity, which is a measure of the catalyst performance to selectively form PAP, increased two-fold over the range of hydrogen pressures shown in Figure 3B.

Phase Transfer Catalyst Concentration and Temperature Effect

The PTC concentration was varied over the range of 1×10^{-3} to 5×10^{-3} M with the results shown in Figure 4A. The hydrogen uptake rate increases linearly with increasing PTC concentrations up to a point. But after exceeding a certain critical concentration the reaction enters into an inhibited regime. Surprisingly, the selectivity to PAP recorded in the same region was not significantly affected.

The last parameter studied was the reaction temperature, which was varied in a relatively narrow range between 80-100°C. Outside this temperature range the rate of hydrogenation was either low or the selectivity deteriorated.

The data in Figure 4B show that 90 C is the temperature at which the reaction rate and the PAP productivity pass through a maximum under this set of conditions.

CONCLUSIONS

The hydrogenation of nitrobenzene to PAP in a biphasic system is taking place in four characteristic regions. In the first two the colloidal system appears to be stable, the hydrogenation rate remains relatively constant and the PAP selectivity shows a continuous decline. In the next two stages in which the colloidal system has collapsed, the nitrobenzene which is dissolved in the aqueous phase is hydrogenated and this leads to the formation of aniline and PAP at constant rates. In these regions the selectivity remains relatively constant. The variables which affect most the reaction efficiency are the stirring rate, the nitrobenzene concentration and the hydrogen pressure. It is suggested that under the conditions studied that the colloidal system is stable and the hydrogen transport rate is such that it creates a balanced hydrogen "poor" environment within the micelle sufficient to sustain a high reaction rate for the first stage of PHA formation and, at the same time, "deficient" enough to suppress the formation of aniline.

REFERENCES

1.   C.O. Henke, J.V. Vaughen, U.S.Patent, 2,198,249, (1940).
2.   S. Kirk-Othmer Mitchell, *Encyclopedia of Chemical Technology, 4th ed.*, Wiley – Interscience, New York, Vol.2, p 481, (1992).
3.   C. Raavichandran, S. Chellammal and P.N. Analtharman, *J. Applied Electrochemistry*, **19**, 45, (1989).
4.   S.S. Sathe, US Patent, 4,176,138 (1979)
5.   D.C. Miller, US Patent, 5,312,991 (1994)
6   T.M. Jiang, J.C. Hwang, H.O. Ho, C.Y. Chen, *J. Chin. Chem. Soc.*, **35** , 135 (1988).
7.   J.C. Hwang, K.A. Chang, C.Y. Chen, T.M. Juang, *Chin. Pharm. Journal*, **44** (6), 475 (1992)
8.   C.V. Rode, M.J. Vaidya, R.V. Chaudhari, *Organic Process Research & Development*, **3**, 465 (1999)
9.   L.T. Ternery, *Contemporary Organic Chemistry*, W.B. Saunders Co. Ltd., 1976, p.661.
10.   J. March, *Advanced Organic Chemistry, 3rd ed.*, Wiley, New York, 1984, p.606
11.   A.M. Stratz, *Chem. Ind (Dekker)*, **5** (Catal. Org. React.) 335 (1984).
12.   R. Augustine, *Heterogeneous Catalysis for the Synthetic Chemist*, Marcel Dekker, Inc., New York, p.490 (1995)

53

Organic Synthesis via Catalytic Distillation

Flora T. T. Ng, Yuxiang Zheng, and Garry L. Rempel
Department of Chemical Engineering, University of Waterloo, Waterloo, Ontario, Canada

ABSTRACT

Catalytic distillation (CD) is a green reactor technology which combines reaction and separation in a single reactor/distillation column. CD has a number of advantages including increased conversion for equilibrium limited reactions. It is also more energy efficient and reduces the need for waste treatment due to the high selectivity to desired products. CD has been applied successfully in the petroleum and petrochemical industry, notably the production of methyl-tertiary butyl ether. However, there are limited examples using CD for the synthesis of organic chemicals and solvents. This paper will discuss the aldol condensation of acetone and the synthesis of ethylene glycol monoethyl ether to illustrate the applications of CD for organic synthesis.

INTRODUCTION

Catalytic distillation (CD) is an unit operation combining reaction and separation in a single reactor/distillation column. CD belongs to the general class of two-phase flow fixed-bed catalytic reactor. An upward flow of vapor and downward flow of liquid comprise the two flowing phases in the CD reactor. Solid catalyst packed in a distillation column not only accelerates a chemical reaction but also supplies a packing surface for vapor-liquid mass transfer to separate the reactants

615

and products. There are many advantages offered by CD, such as high selectivity, high yield, energy savings, reductions of the capital and operating costs [1]. CD is an emerging green process technology since the high product yield and selectivity reduces the waste stream and the energy savings results in the reduction of green house gases.

The aldol condensation is a class of reactions widely used in organic synthesis for the production of various oxygenated compounds (Figure 1). The reactions may be conducted in the liquid or vapor phase with a variety of catalysts. We are interested in the application of CD to the aldol condensation of acetone (Ac) to examine the factors that determine the yield and selectivity in reactions involving a number of consecutive steps. Under ambient reaction conditions, diacetone alcohol (DAA) is readily formed. At higher temperatures DAA readily undergoes dehydration to mesityl oxide (MO) [2].

DAA is an environmentally friendly cleaning solvent, which is being marketed as an alternative to Ac [3]. Stricter environmental regulations concerning solvent use provide the incentive to seek alternative solvents with low volatility. In addition, DAA is a useful intermediate product in the aldol condensation of acetone. Perhaps the most common chemical manufactured from DAA is methyl isobutyl ketone (MIBK). Once MO is produced from the dehydration of DAA; hydrogenation of MO produces MIBK production. MO is also used in the production of methyl isobutyl carbinol and isophorone. The addition of alcohol to MO leads to diacetone alcohol ethers. In addition, hydrogenation of the ketone group of DAA leads to hexylene glycol, which is widely used in hydraulic brake fluids.

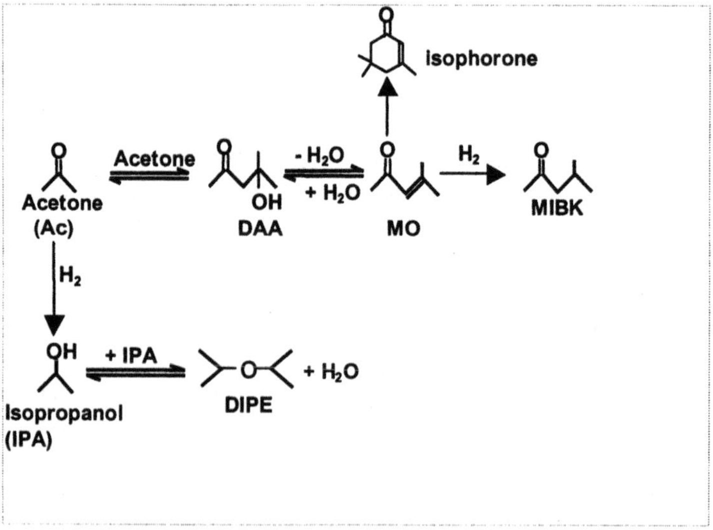

Figure 1 Reactions for the aldol condensation of acetone.

The self-condensation of Ac was the focus of our initial CD experiments since the formation of DAA is strongly limited by chemical equilibrium [4]. The equilibrium conversion of Ac to DAA is only about 4.3% at 54 °C which is the normal boiling point of Ac. In order to produce DAA in a reasonable yield in a fixed-bed reactor, refrigeration is required to cool the Ac feed to 5-20 °C [5]. It also requires recycling a large quantity of Ac. Using the CD technology, there is no need to cool the Ac feed and the overall conversion of Ac could also be increased greatly. This is because the reaction heat can be used to provide the energy to vaporize the liquid for distillation and in situ separation can drive the chemical equilibrium to produce more DAA.

Another common organic solvent is ethylene glycol ether, which is used in large quantities in the ink, resist and coating manufacture and other branches of the chemical industry. However, the major use of glycol ether is in brake fluid whose demand is growing greatly in the world market. The worldwide consumption of ethylene glycol ether is about 750,000 tons per annum.

Currently, a batch or fixed bed reactor is used in the synthesis of ethylene glycol monoethyl ether (EGME) from ethanol and ethylene oxide (EO). Since the reaction is highly exothermic, the reactor has to be cooled constantly. In addition, the reaction is carried out with an excess of ethanol. Therefore a separation unit has to be used to separate the excess ethanol from the products so that it can be recycled into the reactor. The quantity of ethanol fed to the reactor also influences the distribution of the final product. However, the reaction of EO with ethanol cannot be controlled in such a manner that only the EGME is formed due to the consecutive reactions. The main reaction to produce higher oligomers of EGME can be represented by the following equation:

$$C_2H_5OH + CH_2\overset{O}{-}CH_2 \rightleftharpoons C_2H_5OCH_2CH_2OH \tag{1}$$

The EGME will continue to react with EO and produce higher oligomers:

$$C_2H_5OCH_2CH_2OH + CH_2\overset{O}{-}CH_2 \rightleftharpoons C_2H_5(OCH_2CH_2)_2OH \tag{2}$$

and

$$C_2H_5(OCH_2CH_2)_nOH + CH_2\overset{O}{-}CH_2 \rightleftharpoons C_2H_5(OCH_2CH_2)_{n+1}OH \tag{3}$$

Due to the high boiling point of the di-or tri-glycol ethers, production of EGME requires its separation from the glycol ether mixture in a vacuum distillation process. In a CD process, the cooling system and vacuum distillation process are not required because of the utilzation of the reaction heat for in situ separation. The reaction heat can also be used to vaporize the liquid for distillation and high selectivity to EGME could be achieved due to in situ separation of products from the reactants.

In this paper, the aldol condensation of acetone and the synthesis of ethylene glycol monoethyl ether are used to show the applications of CD technology for organic synthesis. The predictions of product yield and selectivity and the design of a CD process using a non-equilibrium three-phase model developed by our laboratory [6] will also be discussed.

DESCRIPTION OF A CD COLUMN

Figure 2 is a schematic diagram of the CD pilot column used for the production of DAA and EGME. The column consists of three parts, namely rectifying, stripping and reactive sections. The non-reactive sections of the column are packed with 6 mm Intalox saddles, a typical inert packings used in the traditional distillation col-

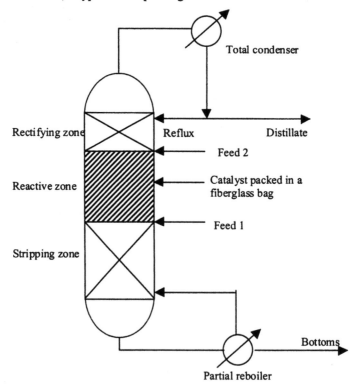

Figure 2 Typical configuration of a CD column.

umn. In the reactive section of the column, the solid catalyst is placed inside a fiberglass bag, which is wrapped with demister wire. The bag is approximately 1.5 cm in diameter and arranged in the reaction zone of the column. This method of packing catalyst for a CD column is similar to that developed by Smith [7]. It provides sufficient voids for vapor flow so that the pressure drop inside the column is minimized. A drawback of this packing method is that the influence of diffusion through the catalyst bag cannot be avoided [1,8].

The column has a total height of 5.4 m and a diameter of 2.54 cm. Initially, 450 ml of reactants was charged into the reboiler. A Milton-Roy LCD mini-pump was used to provide a continuous feed to the column. The bottom product flow rate was adjusted so that the liquid level in the reboiler remained constant. The operation was controlled by a computer system. The details of the pilot plant and the experimental procedure for the operation of the pilot plant have been reported by Podrebarac et al. [4].

RESULTS AND DISCUSSION

The Aldol Condensation of Acetone

The experiments were carried out at atmospheric pressure in a 1 inch packed column under total reflux. Feed 1 was used to introduce the acetone continually. The Amberlite IRA-900 anion exchange resin catalyst was used in these experiments [9]. Under the reaction conditions, the temperature at the catalyst zone was about 54 °C.

Several CD experimental runs were conducted with varying reflux flow rate and different amount of catalyst. The productions of DAA and MO were determined using gas chromatographic techniques. The experimental conditions and comparison of model predictions based on our recent three-phase non-equilibrium

Table 1 Comparison of simulated results with experimental data

catalyst ml	reflux g/min	DAA productivity mol/mol OH catalyst h		MO productivity mol/mol OH catalyst h		DAA selectivity DAA/(DAA+MO)	
		measured	predicted	measured	predicted	measured	Predicted
43	22.00	13.35	13.53	3.33	3.12	0.83	0.84
43*	25.40	14.92	14.54	4.76	4.53	0.79	0.79
43*	15.60	12.05	11.67	4.76	4.76	0.75	0.74
93	22.90	8.61	8.78	2.91	2.91	0.78	0.78
131	23.60	7.20	7.55	3.12	2.91	0.73	0.75
131	16.30	5.10	5.27	3.12	2.91	0.66	0.68

* Another batch of catalyst

model [6] are summarized in Table 1. Clearly, the model predictions are in good agreement with the experimental date. The DAA productivity is dependent on the number of moles of OH⁻ on the catalyst and the reflux flow rate while the MO productivity is dependent only on the number of moles of OH⁻ on the catalyst and independent of the reflux flow rate which influences the mass transfer. This result indicates that the production of DAA is dependent on the mass transfer while the production of MO is kinetically controlled. The selectivity to DAA, defined as the ratio of the productivity of DAA/(MO+DAA), is dependent on the reflux flow rate and the number of moles of OH⁻ on the catalyst. At a high reflux flow rate and a low amount of catalyst, the selectivity to DAA is the highest. This shows that the selectivity to a desired product could be fine-tuned based on the process parameters such as reflux flow rate and the amount of catalyst. High reflux flow rate results in high mass transfer rate and also more facile separation of the intermediate DAA resulting in a high selectivity to DAA. As the amount of catalyst was increased from 2 bag (43 ml) to 6 bags (131 ml), i.e. the length of the catalyst increases along the distillation column, the contact time or residence time of DAA with the catalyst also increases, therefore further reaction of DAA to MO occurs along the CD column resulting in a decreased selectivity to DAA.

Figures 3 and 4 are the predicted profiles of vapor and liquid composition along the column with 43 ml of catalyst and a reflux flow rate of 22 g/min. It is important to note that both the liquid and vapor concentration profiles for acetone in the column are relatively high and hence it is favorable for the formation of DAA. The equilibrium constants calculated from the equilibrium conversion data [9,10] are given in Figure 5, which indicates that at 54 °C, the Ac conversion at equilibrium conversion is only 4.3 wt %. In order to carry out the aldol condensation of acetone in the CD column, the temperature at the reaction zone of the CD column will be near the boiling point of Ac in order to maintain liquid vapor equilibrium. Our CD experimental results show that a maximum concentration of 55 wt% of DAA concentration was obtained which clearly exceeds the equilibrium conversion. The aldol condensation of Ac to produce DAA is an excellent example to demonstrate that in situ separation in a CD column results in an increased yield for equilibrium limited reactions.

Other Applications of CD for Aldol Condensation of Acetone

As shown in Fig 1, aldol condensation of acetone produces a variety of products. Besides the selective production of DAA, the one step direct synthesis of methyl isobutyl ketone could be achieved via CD. This process combines the aldol condensation and hydrogenation of MO in a CD column. A patent has been issued on this process [11], but the selectivity to MIBK is only 67%. No other report on the one step synthesis of MIBK via CD has been published. We are interested to improve the selectivity of this process to MIBK and have initiated a study of the hydrogenation of MO to MIBK. Our initial results indicate a combination of the IRA-900 anion exchange resin and a hydrogenation catalyst such as Ni or Pd on γ–

alumina are potential catalysts for such a process. High selectivity to MIBK will require not only suitable catalysts, but also the process design for carrying out the CD experiments. In this respect, detailed knowledge of the kinetics of the reaction together with the mass transfer characteristics of the catalyst and the catalyst packing will help to determine the most appropriate process configuration for the process. Another new application of CD related to using acetone as a feedstock is the one step synthesis to the produce diisopropyl ether (DIPE). This involves the hydrogenation of acetone to isopropyl alcohol followed by condensation and dehydration. DIPE has potential application as a high octane fuel, but the recent concern about the leakage of MTBE into ground waters could limit the application of DIPE as a high octane fuel.

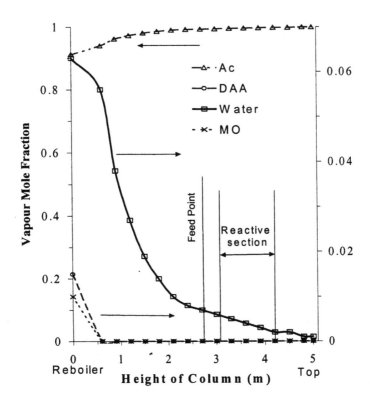

Figure 3 Profiles of vapor composition along the column (catalyst: 43 ml; reflux: 25.4 g/min; feed: 152 ml/h)

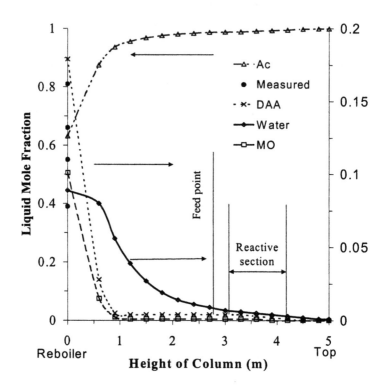

Figure 4 Profiles of liquid composition along the column (catalyst: 43 ml; reflux: 25.4 g/min; feed: 152 ml/h).

Applicability of CD for the EGME synthesis

The reaction between ethanol and ethylene oxide to form EGME is a highly exothermic reaction with a number of consecutive reactions. CD technology could be used to selectively produce EGME if in situ separation could be achieved prior to the consecutive reaction of EGME with another molecule of ethanol. Therefore selectivity to EGME is a competition between the mass transfer rate and the kinetics of the consecutive reaction to form the higher oligomers (see equations 2 and 3). Xu et al. [12] has studied the kinetics of the reaction using a solid acid catalyst, as well as the mass transfer characteristics of the catalyst packing [13] and also carried out the CD experiments. EO is introduced to the column at the bottom of reaction zone (Feed 1) and ethanol is fed at the top of the reaction zone (Feed 2).

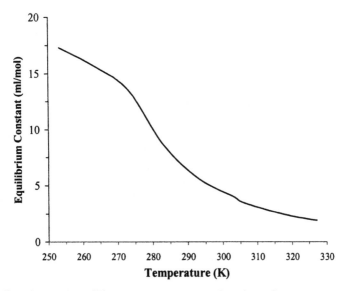

Figure 5 Chemical equilibrium constants as a function of temperature for the dimerization of Ac to DAA.

The three-phase nonequilibrium model developed in our laboratory which includes kinetics, mass transfer and heat transfer models [6] is used to predict the yield and selectivity of EGME from the reaction of ethanol and EO. The model predicts that the conversion of EO would reach 94 % and 99 % selectivity to EGME at an operating pressure of 235 kPa and a reflux ratio of 2. The model predictions are in excellent agreement with experimental data (Table 2). This result shows that our three-phase non-equilibrium model could be used for the prediction of yield and selectivity in a CD process.

Table 2 Comparison of the EC yield and selectivity between the experimental data and the model predictions

Catalyst ml	feed ratio Al/EO mol/mol	feed rate of EO ml/h	EC conversion		EC selectivity	
			measured	predicted	measured	Predicted
100	2	3500	0.895	0.902	0.98	0.99
100	4	3500	0.917	0.921	0.99	0.99
100	4	1600	0.945	0.940	0.99	0.99

CONCLUSION

CD is a novel green reactor technology that offers high selectivity and yield for equilibrium limited and complex reaction pathways involving consecutive reactions. CD has attracted attention from both academia and petrochemical industries. Publications on new process development and novel design of CD packings are increasing very rapidly. However, the application CD for organic synthesis is still relatively new and clearly there are many opportunities for the development of more selective processes for organic synthesis and the synthesis of pharmaceuticals. However, successful development of new CD processes will require more detailed knowledge of the key parameters that govern the conversion, yield and selectivity. Our detailed study on the aldol condensation of acetone to DAA and MO showed that the feed rate, catalyst mass, reflux flow rate and operating pressure are the most important parameters regarding the operation, conversion and yield. The three-phase nonequilibrium model developed in our laboratory could be used to predict and simulate the steady state operation of a CD process. Once the kinetics and mass transfer of the catalyst and the catalyst packing are known, the three-phase non-equilibrium model could be used for the design of a new CD process.

ACKNOWLEDGEMENT

Financial support from the Natural Science and Engineering Research Council of Canada, strategic projects program, is gratefully acknowledged.

REFERENCES

1.   GG Podrebarac, FTT Ng, GL Rempel. More uses for catalytic distillation. CHEM-TECH. 27: 37-45, 1997.
2.   FM Scheidt. Vapor-phase aldol condensation over heterogeneous catalysts. J Catal. 3: 372-378, 1964.
3.   EM Kirschner. Environment, health concerns force shift in use of organic solvents. C&EN. 20: 13-20, 1994.
4.   GG Podrebarac, FTT Ng, GL Rempel. The production of diacetone alcohol with catalytic distillation, Part I: Catalytic distillation experiments. Chem Eng Sci. 53: 1067-1075, 1998.
5.   J Braithwaite. Kirk-Othmer Encyclopedia of Chemical Technology. 4th ed. New York: Wiley, 1995.
6.   Y Zheng, FTT Ng, GL Rempel. Catalytic distillation: A three-phase nonequilibrium model for the simulation of the aldol condensation of acetone. Ind Eng Chem Res. 40: 5342-5349, 2001.
7.   LA Smith Jr. Catalyst system for separating isobutene from C4 streams. US Patent. 4,215,011. 1980.

8. X Xu, Y Zheng, G Zheng. Kinetics and effectiveness of catalyst for synthesis of methyl *tert*-butyl ether in catalytic distillation. Ind Eng Chem Res. 34: 2232-2236, 1995.

9. GG Podrebarac, FTT Ng, GL Rempel. A kinetic study of the aldol condensation of acetone using an anion exchange resin catalyst. Chem Eng Sci. 52: 2991-3002, 1997.

10. EC Craven. The alkaline condensation of acetone. J Appl Chem. 13: 71-76, 1963.

11. KH Lawson, B Nkosi. Production of MIBK using catalytic distillation technology. US Patent. 6,008,416. 1999.

12. X Xu, W Zhang, Y Liu, W Dong. Study on synthesis of cellosolve with catalytic distillation. J Chem Eng Chinese Universities. 4(4): 374-380, 1990.

13. Y Zheng, X Xu. Study on catalytic distillation processes, Part 1. Mass transfer characteristics in catalyst bed within the column. Trans Inst Chem Engrs. Part A, 70: 459-464, 1992.

54

Solvent Effects in the Use of Anchored Homogeneous Catalysts for Hydrogenation Reactions

Clementine Reyes, Yujing Gao, Agnes Zsigmond, Pradeep Goel, Nagendranath Mahata, Setrak K. Tanielyan, and Robert L. Augustine
Center for Applied Catalysis, Seton Hall University, South Orange, New Jersey, U.S.A.

ABSTRACT

Anchored homogeneous catalysts have been used to promote the hydrogenation of substrates in a number of different types of solvents. The results attained show that, while the solvent can have an effect on the reaction rate and selectivity, little, if any, metal loss was observed an almost every case. Comparison is made between catalysts prepared on different support materials, heteropoly acids, ligands and substrates. For added interest, the solvent effect observed with homogeneous Wilkinson catalyst and the commercially available polymer supported Wilkinson was also studied.

INTRODUCTION

Some questions have been asked concerning the stability of our anchored catalysts (1-5) in solvents other than absolute ethanol, the material we have commonly been using in all of our preliminary work. To examine this problem, we have run a number of hydrogenations with anchored complexes in various solvents to determine the effect of the solvent on both catalyst stability and activity.

DISCUSSION

Since the use of absolute ethanol may present a problem in large scale use, one of the first such survey involved the comparison of absolute EtOH with some denatured alcohols. The two chosen were EtOH-3C (denatured by 0.5% toluene) and EtOH-2B (denatured by 0.5% i-PrOH). These solvents were used in the hydrogenations of methyl 2-acetamidocinnamate (**1**) over Rh(DiPamp)PTA/Al$_2$O$_3$ with the results listed in Table 1.

Table 1 Solvent effect in the hydrogenation of 1 over h(DiPamp)/PTA/Al$_2$O$_3$[a].

Solvent	Run #	Rate[b]	% ee
Abs. EtOH	1	0.85	89
	2	0.81	89
	3	0.79	90
	4	0.85	94
EtOH-3C	1	0.38	89
	2	0.39	90
	3	0.38	90
EtOH-2B	1	0.72	89
	2	0.76	90
	3	0.70	90

[a] Run at 50°C and 1 atm H$_2$.
[b] mmole H$_2$ / mmole Rh / min.

While the product ee remained constant in all of these hydrogenations, there was a definite solvent effect on the reaction rate. The use of EtOH-3C resulted in an almost 3-fold decrease in rate while with EtOH-2B the activity was only slightly decreased.

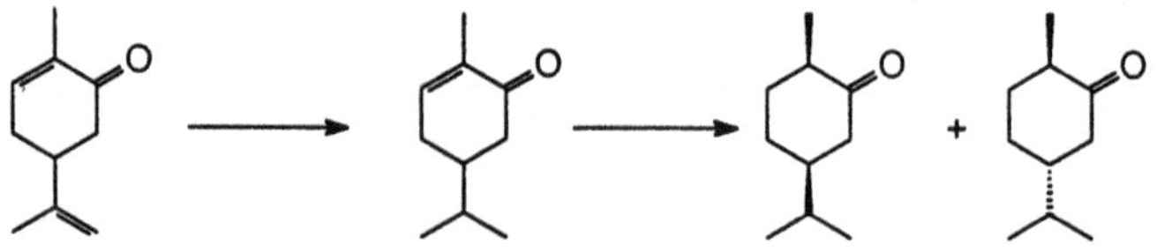

(Eqn. 1)

2   **3**

Attention was then turned to the use of some of the more common solvents and their effect on catalyst activity and stability and to extend the study to include heteropoly acids other than PTA. The partial hydrogenation of carvone (**2**) (Eqn. 1) was run over Wilkinson's catalyst anchored to PTA and STA on alumina. As a further comparison, the commercially available polymer supported Wilkinson's was also included in this study.

Figure 1 shows the effect of solvent on the rates of carvone hydrogenation over the homogeneous Wilkinson's catalyst and the different supported

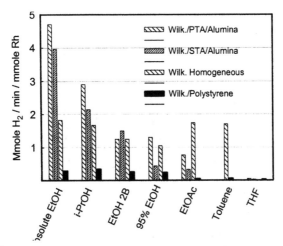

Figure 1 Solvent effect on the rate of hydrogenation of carvone over homogeneous, anchored, and supported Wilkinson's catalyst.

Wilkinson catalysts. It can be seen that the solvent has virtually no effect on those reactions run using the homogeneous catalyst but has a fairly significant effect on the heteropoly anchored species. The commercially available phosphinated polystyrene supported material is relatively inactive with the solvent having only a slight effect on the rate of reaction. These reactions were stopped after one equivalent of hydrogen was taken up so they represent the rate of formation of the dihydrocarvone (**3**). The solid catalysts were used several times and the rates are the average of these reactions. Also measured was the effect of solvent on the amount of Rh lost in each of those reactions run using a solid catalyst. Figure 2 shows the ppm of Rh found in the <u>first</u> reaction mixture for each solvent and catalyst. Thus, this number represents the <u>maximum</u> amount of Rh in these reaction mixtures. It can be seen that the maximum amount of loss is found with the solvents in which the reaction is the slowest. Further, the polystyrene material shows significant Rh loss.

   We also looked at the solvent effect in the use of anchored Rh(dppb) for the hydrogenation of carvone. The anchored catalysts were prepared using PTA on both neutral and basic alumina as well as STA and PMA on basic alumina. Figure 3 shows the hydrogen uptake data observed with these catalysts in several solvents. It would appear from these data that the STA anchored catalyst is the best in all of the solvents except i-PrOH.

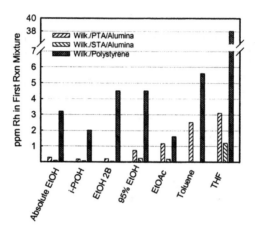

Figure 2 Solvent effect on the loss of Rh in the hydrogenation of carvone over homogeneous, anchored, and supported Wilkinson's catalysts.

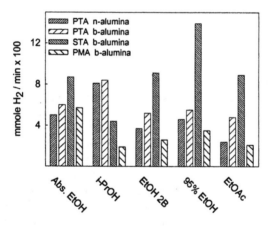

Figure 3 Solvent effect on the rate of hydrogenation of carvone over anchored Rh(dppb) catalysts.

The next step was to look at the composition of the product mixture at various stages of reaction to examine more closely the effect of solvent and catalyst on the selectivity of the reaction. To this end we ran the hydrogenation of carvone for 24 hours over both Rh(PPh$_3$)$_3$/PTA/Al$_2$O$_3$ and Rh(dppb)/PTA/Al$_2$O$_3$. The solvents used were abs. EtOH, IPA and EtOAc. Samples were taken every two hours for the first eight hours and then at 24 hours. The hydrogenations over the anchored Wilkinson went on to the tetrahydro species regardless of the solvent, even though in IPA some selectivity could have been obtained if the reaction were stopped after one equivalent of hydrogen was taken up. On the other hand, the anchored Rh(dppb) was more selective toward dihydrocarvone formation, particularly in IPA and EtOAc.

A recent paper has described the use of an anchored Rh(DiPFc) catalyst for the hydrogenation of a variety of functional groups with particular emphasis on the saturation of aldehydes to the primary alcohol (Eqn. 2) (6). One problem, however, was the acetal formation, **4**, apparently caused by the PTA catalyzed

$$\text{R-C}\overset{O}{\underset{H}{{<}}} \xrightarrow[\text{H}_2 \text{ R.T. } 50-100 \text{ psig}]{\text{Rh(DiPrFe)/PTA/Al}_2\text{O}_3} \text{R·CH}_2\text{·OH}$$

PTA, R'OH

$$\text{R-C}\overset{OR'}{\underset{H}{{-}}}\text{OR'}$$

(Eqn. 2)

reaction between the solvent alcohol and the aldehyde. This was negated by the use of a 50% water in i-propyl alcohol solvent. While no details of the hydrogen uptake were given, it was reported that the reactions were run for between 16 and 22 hours with hydrogen monitoring. In one case the catalyst was reused with some loss of activity between use and the report of about 2% of the metal being lost during the first use and less in subsequent uses.

We have prepared the supported Rh(DiPFc) on alumina using PTA as the anchoring agent and have used this catalyst for the hydrogenation of benzaldehyde and several substituted benzaldehydes. Table 2 lists the rate data for the benzaldehyde hydrogenations run in both ethanol and i-propanol. While

Table 2 Reaction data for the hydrogenation of benzaldehyde over Rh(DiPFc)/PTA/Al2O3 run at room temperature and 50 psig in different solvents.

Ethanol				isopropanol			
% H_2O	Use #	Rate [a]	% 4	% H_2O	Use #	Rate [a]	% 4
0	1	0.063	12	0	1	0.042	6
	2	0.030	21		2	0.035	3
	3	0.025	31		3	0.037	0
1	1	0.069	1				
	2	0.046	1				
	3	0.037	2				
2	1	0.082	<1	2	1	0.162	0
	2	0.067	<1		2	0.131	0
	3	0.060	<1		3	0.112	0

[a] mmoles H_2 uptake / min

the previous work used a 50% water-i-propanol solvent to prevent acetal formation, we decided to more carefully examine this factor and have used, in addition to the anhydrous solvents, those containing only 1% and 2% water. The hydrogenations were run at room temperature and 50 psig with the hydrogen uptake curves recorded for each run. The catalysts were re-used three times with each of the solvents. In every case there was a loss of activity with re-use but since no color was observed in any of the reaction mixtures, it is was thought that this loss of activity may be caused by something other than the leaching of the catalyst. Similar data were obtained on hydrogenation of the substituted benzaldehydes.

In all of these reactions 2 mmoles of aldehyde were used which corresponded to a substrate/catalyst ratio of about 100. When 6 mmoles of substrate were used in a series of hydrogenations, the reaction rates fell off more rapidly. It was suspected that during these reactions the aldehydes were being decarbonylated by the Rh catalyst producing the inactive Rh carbonyl species. To test this hypothesis, the reaction mixtures were carefully analyzed and a trace of benzene was found in the benzaldehyde reaction mixture and chlorobenzene in the p-chorobenzaldehyde reaction mixture. These materials were not detected in the starting solutions. Further examination of the used catalyst by diffuse reflectance FTIR showed the presence of a Rh-CO band near 2000cm^{-1}. The FTIR spectrum of the homogeneously catalyzed reaction mixture also showed a band at 2000cm^{-1}.

REFERENCES

1. R. Augustine and S. Tanielyan, *Chem. Commun.*, 1257 (1999).
2. S.K. Tanielyan and R.L. Augustine, US Patents 6,005,148 (1999), 6,025,295 (2000); Australian Patent 737536 (2001).
3. S.K. Tanielyan and R.L. Augustine, PCT Int. Appl., WO-9828074: *Chem. Abstr.*, **129**, 109217 (1998).
4. S.K. Tanielyan and R.L. Augustine, *Chem. Ind. (Dekker)*, **75** (Catal. Org. React.), 101 (1998)
5. R.L. Augustine, S.K. Tanielyan, S. Ancerson, H. Yang and Y. Gao, *Chem. Ind. (Dekker)*, **82** (Catal. Org. React.), 497 (2000)
6. M.J. Burk, A. Gerlach and D. Semmerl, *J. Org. Chem.*, **65**, 8933 (2000).

55

A Novel Large Scale Hydrogenation Strategy of N,N-Dimethyl 4-(Acetoxy-3-Nitrophenylmethyl)-1-H-Imidazole-1-Sulfonamide, a Key Intermediate in the Synthesis of ABT-866

Louis S. Seif, Steven M. Hannick, Daniel J. Plata, Howard E. Morton, and Padam N. Sharma
Abbott Laboratories, Abbott Park, Illinois, U.S.A.

ABSTRACT

The large scale cGMP synthesis of ABT-866 required reduction of the nitro group and hydrogenolysis of the acetate moiety of intermediate **I**. The palladium catalyzed hydrogenation of compound **I** was carefully studied in a systematic fashion for the effect of solvent, catalyst, pH, temperature, and concentration. Key to the success of our approach was the combination of rapid screening techniques and HPLC analysis.

| I | II | III |

Surprisingly, while nitro reduction occurred smoothly, deoxygenation of the benzylic position was sluggish under neutral or acidic conditions. However, both reductions occurred readily in basic media. These reactions were essentially complete within four hours in ethanol or isopropyl acetate using 0.5 equivalent of triethylamine and 10 weight percent of wet 5% palladium hydroxide on carbon at 50°C. The novelty of this reaction is the use of base to accelerate the deoxygenation. The one-pot procedure afforded product III in an isolated yield of 95% on lab scale. This strategy performed well in the large-scale campaign (on 2kg) and shortened the existing synthesis by one step.

INTRODUCTION

New equipment as well as new chemical and analytical methods are necessary to accommodate today's research needs.   Scheme 1 summarizes the practical synthesis of α-1 agonist ABT-866 (1-4), which includes a catalytic hydrogenation. In the preparation ABT-866, we were required to rapidly evaluate an array of catalysts and conditions for a hydrogenation step in the synthesis of this clinical candidate. The limited amount of starting material in this project restricted the use of standard hydrogenation apparatus such as pressure bottles, autoclaves, atmospheric hydrogenation, and shakers.

A simple and versatile 4-station multiple hydrogenator was constructed to effectively analyze a broad range of reaction conditions, Figure 1 (5-6). The four individual reactors are equipped with argon and hydrogen supplies, as well as sampling and venting valves. Each reaction port has an attached glass pressure flask containing a magnetic stir bar, as well as an external water bath with temperature control. The delivery of different pressures of reactive gas to the individual reactors was supplied from a main source via a regulator, a relief valve set for 70 psi, and a gas manifold. The apparatus is compact, safe, and suitable for hydrogenations on scales ranging from micromoles up to a mole of material.

BACKGROUND

The synthesis of ABT-866 requires the nitro group reduction and benzylic deoxygenation of intermediate I. The goal was to define conditions for dependable catalytic reduction of I for use in a cGMP delivery of ABT-866.

Catalytic reduction of I in tetrahydrofuran or ethanol yielded mixtures of II and III under standard conditions with Raney nickel or supported Pd. The reaction typically stalled after a few hours, and additional catalyst was necessary for further conversion.

I   II   III

Scheme I   A practical synthesis of α-1 agonist ABT-866

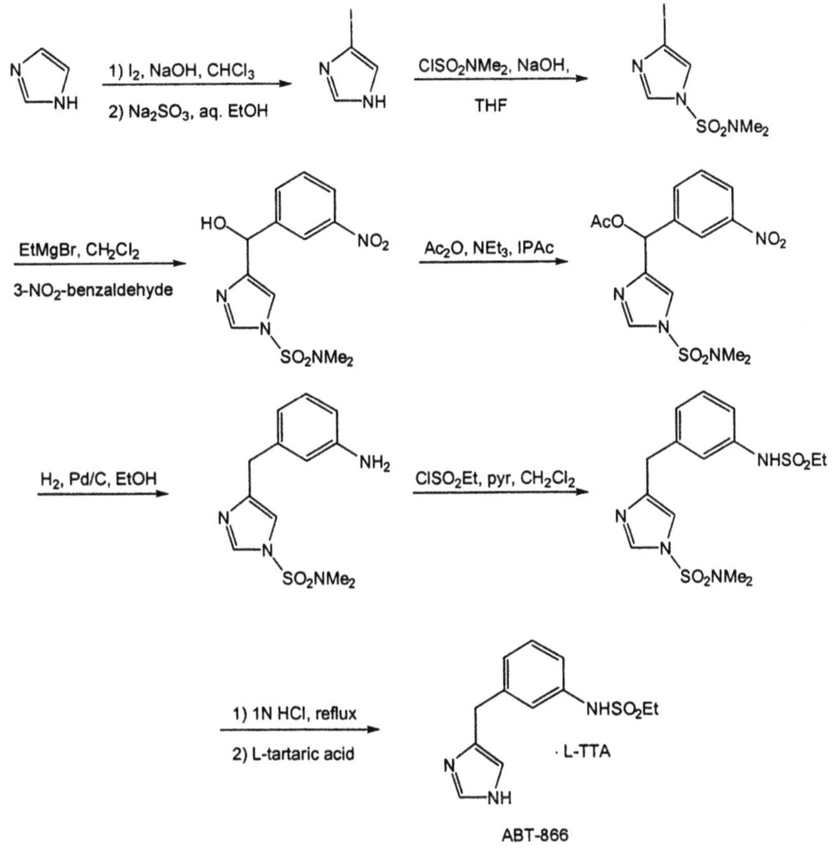

ABT-866

RESULTS AND DISCUSSION

The palladium catalyzed hydrogenation of compound **I** was carefully studied for the effect of pH, solvent, catalyst load, temperature, and concentration. Standard reaction conditions described in the experimental section were used for all screening reactions.

The reaction pH was adjusted with a range of additives. Neutral reaction conditions yielded incomplete conversion, Table 1. Acids such as acetic acid, sulfuric acid, phosphoric acid, and methanesulfonic acid gave poor conversion and many unknowns, Table 2. In contrast, runs with the addition of triethylamine (NEt₃) led consistently to complete conversion, although a stoichiometric amount of base was unnecessary (7). Reduction of **I** in the presence of 1.0, 0.5, or 0.25

equivalents of NEt$_3$ in isopropyl acetate or anhydrous ethanol with 10% load of 5% Pd(OH)$_2$/C or 5% load of 5% Pd/C gave excellent results, Tables 3-4.

Table 1 ABT 866 Nitro Reduction and Deoxygenation, Neutral Conditions, 5% Pd/C

Solvent	Time (hr)	HPLC Area % I : II : III	Unknowns
THF	3	0 : 9 : 88	3
	6	0 : 6 : 90	4
	23	0 : 2 : 95	3
MeOH	3	0 : 2 : 83	15
	6	0 : 2 : 80	18
	23	0 : 0 : 83	17
MeOH/THF	3	0 : 6 : 92	2
	6	0 : 6 : 92	2
	23	0 : 3 : 93	4
EtOH	1.4	0 : 14 : 85	1
	3.1	0 : 2 : 92	6
	21	0 : 5 : 85	10
^{i}PrOH	1.5	0 : 24 : 76	0
	3.3	0 : 5 : 92	3
	21	0 : 0 : 89	11
EtOAc	1.8	47 : 24 : 5	24
	4	0 : 0 : 88	12
^{i}PrOAc	1.3	0 : 10 : 82	8
	3.1	0 : 0 : 87	13
	21	0 : 0 : 84	16

The solvents evaluated were methanol, ethanol, isopropanol, ethyl acetate, isopropyl acetate, and tetrahydrofuran. Isopropyl acetate and ethanol were identified as the solvents of choice because they led to the lowest levels of impurities.

With isopropyl acetate, the method of agitation had an influence on conversion. The reaction stalled in shaken pressure bottles, whereas magnetic stirring in glass and mechanical stirring in stainless steel reactors led to efficient

Table 2 ABT 866 Nitro Reduction and Deoxygenation, Acidic Conditions, 5% Pd/C

Solvent	Additives	Time (hr)	HPLC Area % I : II : III	Comments
THF	0.1 ml H_3PO_4	1.3	2 : 70 : 20	7% unknowns
		4.4	0 : 34 : 37	29% unknowns
THF	0.05 ml H_2SO_4	1.3	93 : 2 : 0	5% unknowns
		5	84 : 1 : 2	13% unknowns
THF	0.1 ml H_2SO_4	1.3	51 : 13 : 0	36% unknowns
THF	0.1g $MeSO_3H$	1.3	82 : 5 : 0	13% unknowns
		4.5	77 : 3 : 0	20% unknowns
MeOH	0.2 ml AcOH	1.3	0 : 10 : 83	7% unknowns
		4	0 : 3 : 80	17% unknowns
MeOH	0.1 ml H_2SO_4	1	0 : 0 : 76	24% unknowns
		4	0 : 0 : 60	40% unknowns
EtOH	0.05ml H_2SO_4	1.2	0 : 75 : 14	11% unknowns
		4.2	0 : 34 : 56	10% unknowns
		21	0 : 3 : 83	14% unk, 23°C
EtOH	0.1 ml H_3PO_4	1.3	0 : 54 : 46	
		4.2	0 : 32 : 49	19% unknowns
		21	0 : 16 : 33	51% unknowns

reduction. The utilized apparatus effectively modeled the large-scale hydrogenation in the pilot plant.

The catalysts evaluated were 5% Pd/C (dry) and 5% Pd(OH)$_2$/C (63% water content). These materials were available internally for cGMP runs. Equal weights of the catalysts gave comparable results. Wet 5% Pd(OH)$_2$/C was chosen for ease of handling and lower cost.

The rate of reaction completion was related to catalyst loading. A 10 wt. % load of wet 5% Pd(OH)$_2$/C gave completion after one hour at 50°C. As little as 2.5 wt. % of catalyst gave complete conversion within 24 hours. Below that level, intermediate **II** remained after 24 hours.

All probe reactions were run at 50°C. Under the optimal conditions, the initial reaction exotherm heated the mixture to around 40°C, but extended agitation at 50°C was essential for completion. Reaction temperatures as high as 80°C gave comparable results. Temperatures below 40°C led to incomplete conversion.

Substrate concentrations higher than 10 wt. % were not feasible because the product crystallized out, which made the work-up difficult during hot filtration.

Table 3 ABT 866 Nitro Reduction and Deoxygenation, Basic Conditions, 5% Pd/C

Solvent	Additives	Time (hr)	HPLC Area % I : II : III	Comments
THF	1 eq NEt₃	1.5	0 : 0 : 98	2% unknowns
		5	0 : 0 : 97	3% unknowns
		21	0 : 0 : 82	18% unknowns
				10% catalyst load
EtOH	1 eq NEt₃	1.3	4 : 33 : 28	35% unknowns
		2.1	0 : 68 : 31	1% unknowns
		21	0 : 1 : 99	10% catalyst load
EtOH	1 eq NEt₃	1.3	0 : 0 : 100	5% catalyst load
		2	0 : 0 : 100	
		21	0 : 0 : 99	1% unknowns
EtOH	0.5 eq NEt₃	1	0 : 5 : 95	25% catalyst load
		2	0 : 2 : 97	1% unknowns
		4	0 : 0 : 97	2% unknowns,
				One gram scale-up,
				97% yield
EtOH	0.5 eq NEt₃	2.3	0 : 1 : 99	5% catalyst load
		5	0 : 0 : 100	
		21	0 : 0 : 96	4% unknowns
EtOH	2 eq NEt₃	2.1	0 : 3 : 97	5% catalyst load
		5	0 : 0 : 100	
		21	0 : 0 : 100	
EtOH	1 eq NEt₃	3.5	0 : 0 : 99	5% catalyst load,
				1% unknowns, 80°C
ⁱPrOAc	1 eq NEt₃	2.4	0 : 0 : 99	1% unknowns
		5.3	0 : 0 : 100	5% catalyst load
		21	0 : 0 : 100	
ⁱPrOAc	1 eq NEt₃	2	0 : 0 : 100	10% catalyst load
		4	0 : 0 : 100	
ⁱPrOAc	0.5 eq NEt₃	2	0 : 0 : 100	5% catalyst load
		4	0 : 0 : 100	

The 37 probe reactions that led to the optimized conditions required 7.4 g of starting material **I**. It took 3 weeks to identify the best conditions using parallel screening.

Table 4 ABT 866 Nitro Reduction and Deoxygenation, Basic Conditions, 5% Pd(OH)$_2$/C

Solvent	Additives	Time (hr)	HPLC Area % I : II : III	HPLC Std Yield	Comments
100%	0.5 eq NEt$_3$	2	0 : 2 : 98		10% catalyst load
EtOH		4	0 : 0 : 100		
95%	0.5 eq NEt$_3$	2	0 : 0 : 99		1% unknowns
EtOH		5	0 : 0 : 98		2% unknowns
		24	0 : 0 : 97		2% unknowns, 10% catalyst load
80%	0.5 eq NEt$_3$	2	75 : 4 : 6		15% unknowns
EtOH		5	0 : 0 : 92		8% unknowns
		24	0 : 0 : 87		13% unknowns, 10% catalyst load
^{i}PrOAc	1 eq NEt$_3$	2	0 : 0 : 100		10% catalyst load
		4	0 : 0 : 100		
^{i}PrOAc	0.5 eq NEt$_3$	2	0 : 0 : 100		10% catalyst load
		4	0 : 0 : 100		
^{i}PrOAc	0.5 eq NEt$_3$	1	0 : 47 : 53		5% catalyst load
		6	0 : 0 : 100		
		22	0 : 0 : 100	100	
^{i}PrOAc	0.5 eq NEt$_3$	1	0 : 53 : 22		1.25%
		6	0 : 31 : 69		catalyst load
		22	0 : 12 : 88	90	
^{i}PrOAc	0.5 eq NEt$_3$	1	92 : 6 : 2		0.63%
		6	21 : 64 : 15		catalyst load
		22	0 : 78 : 22	10	
^{i}PrOAc	0.25 eq NEt$_3$	1	0 : 1 : 98		5% catalyst load
		4	0 : 0 : 99	99	
^{i}PrOAc	0.5 eq NEt$_3$	1	0 : 0 : 99		4 eq water
		4	0 : 0 : 99	93	
^{i}PrOAc	1 eq NEt$_3$	2	0 : 0 : 92		1.0 g sm, Parr
		5	0 : 0 : 91	100	stirred reactor
^{i}PrOAc	1 eq NEt$_3$	2	0 : 38 : 61		1.0 gm sm,
		22	0 : 39 : 61		reaction stalled
^{i}PrOAc	0.5 eq NEt$_3$	1	0 : 56 : 44		16g scale
		2	0 : 41 : 59		5% catalyst load
		3	0 : 1 : 99		add 5% cat.
		4	0 : 0 : 100	99	@ 2 hr

EXPERIMENTAL

The hydrogenation of **I** was carried out in an Ace Glass 10 ml threaded hydrogenation vessel equipped with a Teflon bushing, ferrule, and magnetic stirrer. The vessel was charged with catalyst (5% by weight of 5% Pd/C or 10% by weight of 5% Pearlman type, 63% wet). The catalyst was wetted with solvent, and 200 mg of compound **I** was added to the vessel. The total volume was 2 ml. The vessels were then attached to the multiple hydrogenator and stirred under 4 atmospheres of hydrogen at 50 C. Samples were taken at various time periods and analyzed on a Hewlett Packard HPLC using a Zorbax SB-C8 column (4.6x250 mm, 5-micron).

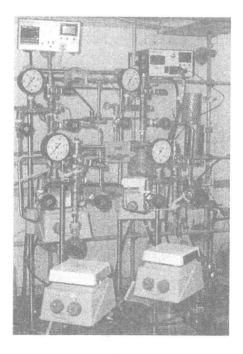

Figure 1

CONCLUSION

In ethanol, reduction of 1 gram of **I** with 0.5 equivalent of NEt_3 and 5% by weight of 5% Pd/C for two hours gave a 97% yield of **III**, one percent of aminoacetate **II**, and 2 percent of an unknown. In isopropyl acetate, the conversion and yield were 100%. When 16 g of **I** was reduced in isopropyl acetate using 0.5 equivalents of

NEt$_3$ and 5% by weight of wet 5% Pd(OH)$_2$/C, the reaction was 59% complete in two hours; an additional 5% catalyst was added to speed the rate, yielding 100% conversion and 99% yield in a total of four hours. These optimized conditions were transferred to the development group for scale-up.

REFERENCES

1. L Bhagavatula, RH Premchandran, DJ Plata, SA King, HE Morton. Efficient route to 1-dimethylsulfamoyl-4-iodo-imidazole, isomerisation of 1-dimethylsulfamoyl-5-iodoimidazole to 1-dimethylsulfamoyl-4-iodo-imidazole. Heterocycles 53:729-732, 2000.

2. RL Cournoyer, PF Keitz, C O'Yang, DM Yasuda. Phenyl-and aminophenyl-alkylsulfonamide and urea derivatives, their preparation and their use as alpha1A/1L adrenoceptor agonists. EP 0887346 A2, 1998.

3. RM Turner, SD Lindell, SV Ley. A facile route to imidazol-4-yl anions and their reactions with carbonyl compounds. J Org Chem 56: 5739-5740, 1991.

4. JHM Lange, HC Wals, AVD Hoogenband, AVD Kullen, J Hartog. Two novel syntheses of the histamine H$_3$ antagonist thioperamide. Tetrahedron 51:13447-13454, 1995.

5. DA Bradley, CR Schmid. A simple low-to medium-pressure hydrogenation manifold. Org Process Res & Develop 1:179-181, 1997.

6. LS Seif, DA Dickman, DB Konopacki, BS Macri. The use of a Fischer-Porter apparatus for chiral homogenous catalytic hydrogenation. In: JR Kosak, TA Johnson, ed. Catalysis of Organic Reactions. New York: Marcel Dekker, 1994, pp 69-79.

7. JS Tou, BD Vineyard. Novel synthesis of L-phenylalanine. J Org Chem 49:1135-1136, 1984.

56

Catalyst Recycling in Batch and Semi-Continuous, Fixed-Bed Processes for Suzuki Cross-Coupling Reactions

Frank P. Gortsema, Carl LeBlond, Luwam Semere, Arthur T. Andrews, and Yongkui Sun[*]
Merck Research Laboratories, Rahway, New Jersey, U.S.A.

John R. Sowa, Jr.[†]
Department of Chemistry and Biochemistry, Seton Hall University, South Orange, New Jersey, U.S.A.

ABSTRACT

Six consecutive Suzuki coupling reactions of aryl bromides were run with a recycled Pd/C catalyst producing yields of 70 – 92 % without apparent loss of activity. On the basis of total palladium in the initial run, the turnover number was 20/cycle or 120 over the entire sequence. Next, a semi-continuous, fixed-bed, flow system was built using palladium on 1/8" carbon extrudates. Suzuki coupling reactions of aryl bromides containing electron withdrawing groups were successfully accomplished in reasonable reaction times (2 – 12 h) and levels of conversion (> 98 %) with fresh catalyst. However, subsequent cycles lead to decreasing levels of activity. Nevertheless, turnover numbers for the fixed bed system based on total palladium reached up to 250 after four cycles.

[*] *E-mail*: yongkui_sun@merck.com

[†] *E-mail*: sowajohn@shu.edu

INTRODUCTION

Recently there has been considerable interest in the use of supported catalysts in fine chemical synthesis.[1] Compared to homogeneously catalyzed systems, it is easier to separate the product from a supported system. This leads to faster processing, higher purity products and the possibility of re-using the catalyst in batch or continuous flow systems. These factors are environmental, economic and labor-saving advantages on multi-kilo manufacturing scale syntheses.

The Suzuki cross-coupling reaction is frequently employed in laboratory and manufacturing scale syntheses of natural products, pharmaceuticals, liquid-crystal materials and polymers.[2] While the cross-coupling is usually catalyzed with homogeneous palladium catalysts with aryl phosphine ligands, there has been considerable interest in heterogeneously-supported palladium (Pd/C, Pd/Al$_2$O$_3$)[3] and nickel (Ni/C)[4] catalysts. However, in contrast to Pd/C, the Ni/C system requires additional phosphine ligands.[5] Recently, we reported Suzuki cross-coupling reactions of aryl chlorides using Pd/C without additional ligands.[6]

Although a recent report by Hirao and co-workers presents Pd/C as a reusable catalyst for Suzuki coupling of iodophenols in aqueous media,[7] studies that examine reuse of Pd/C for coupling of aryl bromides have not been reported. In this paper we describe both a batch Suzuki cross-coupling reaction with catalyst recycling and a semi-continuous fixed bed system for Suzuki coupling of aryl bromides.

RESULTS AND DISCUSSION

Batch Process with Catalyst Recycling

The reaction conditions for the batch reactions (eq 1) were similar to those previously reported by Marck and Buchecker[3a] except, in our hands, replacing sodium carbonate by potassium carbonate resulted in higher yields. The catalyst chosen for these studies was powdered Pd/C (5 wt %) supplied as a dry, egg-shell, unreduced catalyst with an initial catalyst charge of 5 mol %. Chemisorption studies indicate 20 % dispersion of the palladium on the carbon which corresponds to a surface active concentration of palladium of 1 mol %.

Using the same Pd/C sample, a total of six cross-coupling reactions were successfully performed (Table 1). The first four reactions between phenylboronic acid and 4-bromocyanobenzene gave consistent yields of 70 – 79 % in reaction times of 1 – 2.5 h at 50 °C. However, the first reaction was slowest (2. 5 h vs. 1 h) and gave the lowest yield (70 % vs. 75 – 79 %). This indicates that the initial catalyst is slightly deactivated but becomes activated after initial use. The final two cycles were conducted with 4-bromoanisole which shows that the recycled material can be used for a variety of substrates including aryl bromides with electron withdrawing (CN) and electron donating (OMe) groups. On the basis of total palladium in the initial cycle, the turnover number was 20/cycle or 120 for the entire sequence.

Table 1 Recycling of Pd/C Catalyst in the Suzuki Cross-coupling Reaction of Aryl Bromides with Phenylboronic Acid[a]

Cycle	4-RC$_6$H$_4$Br	Rxn time (h)[b]	Catalyst mixture (g)	Time between use (h)	Iso-lated Yield (%)	Purity (%), [Pd] (ppm) of crude product
1	CN	2.5	2.9	-	70	98.5, (< 3)
2	CN	1	3.3[c]	17	75	99.0, (< 3)
3	CN	1	3.8[c]	3	79	98.9, (< 3)
4	CN	1	2.47[c,d]	20	78	99.3, (< 3)
5	OMe	3.25	2.26[c]	116	80	92.0, (7)
6	OMe	-	2.10[c]	2	92	94.5, (7)

[a] Reaction conditions: 5 mol % Pd/C, ArBr (1 equiv), PhB(OH)$_2$ (1.3 equiv), K$_2$CO$_3$ (2.6 equiv), EtOH/H$_2$O solvent (5/1), 50 °C.
[b] Reaction completion was determined by TLC.
[c] These values are the amount of catalyst mixture recovered from the previous run. This amount varies due to Pd/C lost during recovery or salts, water and solvent sorbed by the catalyst.
[d] Significant amount of solid catalyst lost during transfer (ca. 0.5 g).

Product purity was monitored in each cycle by analyzing the crude product for organic impurities (HPLC) and palladium contamination (GFAA). The first four reactions showed outstanding purity profiles of > 98 % (area %) by HPLC and < 3 ppm palladium. Switching to 4-bromoanisole resulted in slightly less pure crude product (92.0 – 94.5 % purity, 7 ppm Pd). Thus, even after six cycles of the same Pd/C material, the crude product maintains excellent purity with minimal palladium contamination.

The catalyst isolation procedure simply involved filtration of the reaction solution, washing with acetone and water and drying in air. The drying time was allowed to vary from as short as 2 h and as long as 116 h with no apparent affect on catalyst performance. In addition, significant fluctuations in the mass of the recovered catalyst (relative to the initial mass) were observed from a 31 % increase to a 28 % decrease (Table 1). These fluctuations are attributed to adsorption of water, inorganic salts and organic materials (the catalyst isolated after the first cycle contained 14 % water) and physical loss of the Pd/C material. Nevertheless, in spite of these handling problems the catalyst remains sufficiently active for at least six cycles and likely many more cycles are possible.

Semi-Continuous Fixed-Bed Process

Another approach we explored was the use of Pd/C in a fixed-bed, tubular, flow reactor system. To prevent high back pressure buildup, we chose a Pd/C catalyst in the form of 1/8" extrudates (0.8 wt %, dry, pre-reduced, egg-shell type). Although, preliminary studies indicate that unreduced catalysts perform better in Suzuki coupling reactions, an unreduced form of this catalyst was unavailable. The initial catalyst charge was 0.1–0.2 mol %.

A schematic of the fixed-bed reactor system is shown in Figure 1. In the reactions described below, the solution in the holding reservoir remained colorless with suspension of small amount of fine, white solids. High flow rates (200 to 400 mL/min) prevented salt encrustation in the flow reactor system. Reactions employing this design were successfully run for several days.

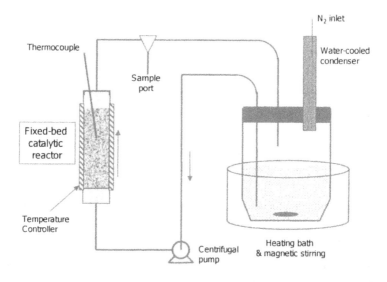

Figure 1 Schematic of the semi-continuous, fixed-bed catalytic reactor system for the Suzuki cross coupling reaction.

The Suzuki coupling reaction was examined between four different aryl halide substrates and phenylboronic acid on fresh catalyst. Initially, 4-bromobenzotrifluoride was successfully coupled at 25 °C with 98 % conversion in 12 h. However, at 50 °C the reaction was much faster with 99 % conversion in 1.8 h (curve 1 in Figure 2). Ethyl 4-bromobenzoate was also successfully coupled at 25

°C with 98 % conversion in 12 h. Thus, the fixed-bed system works well with fresh catalyst and electron deficient aryl bromides.

In contrast, coupling reactions with aryl chlorides and electron rich aryl bromides are not as well optimized in the fixed-bed system. For example, with 4-bromoanisole only 85 % conversion was observed at 80 °C after 52 h. Unfortunately, we were unable to achieve reasonable reactivity (25 % conv., 125 h, 80 °C) with an aryl chloride substrate (4-chlorobenzotrifluoride) using conditions that we previously developed for single-cycle batch reactions.[6] In both cases, the catalyst activity seemed to erode over time and reaction stalled before complete conversion. The difference in reactivity compared to the batch studies may be attributed to the low Pd loading of the catalyst (*i.e.*, 0.8 w/w% vs. a typical 5 w/w %) and the pre-reduced nature of the catalyst which may not be optimized for the Suzuki coupling reaction.

Catalyst Recycling and Regeneration Studies

Conditions were investigated to recycle and regenerate the catalyst. Since the reaction formally generates potassium bromide and boric acid, the initial approach was to wash the catalyst bed with recirculating water. Three successive cycles were run with 4-bromobenzotrifluoride by washing the catalyst with warm water (50 °C) after each cycle. Although the conversion in the first cycle was 99 % in 1.8 h, the second and third cycles showed decreasing levels of reactivity (curves 2 and 3, Figure 2).

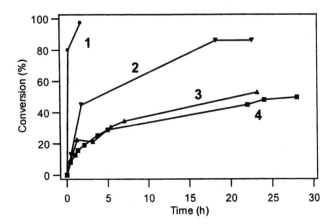

Figure 2 Suzuki coupling in a semi-continuous, fixed-bed reactor. Four cycles of reaction with 4-bromobenzotrifluoride at 50 °C. 1) fresh catalyst (initial rate, 0.22 mol/h); 2) water-washed catalyst, reagents replenished (initial rate, 0.017mol/h); 3) water-washed catalyst, fresh reaction solution (initial rate, 0.012 mol/h); 4) water-washed catalyst, hydrogen reduction, fresh reaction solution (initial rate, 0.01 mol/h).

The second cycle had a lower initial rate, *i.e.,* 0.017 mol/h as compared with 0.22 mol/h in the first cycle, and reaction stalled at ca. 85% conversion. Interestingly, water washing the catalyst prior to the third cycle led to partial recovery of the catalyst activity. The third cycle had a initial rate of 0.012 mol/h that is comparable to the second cycle, but the reaction again stalled, but at a lower conversion as compared with the second cycle. We also tested an additional regeneration strategy employing hydrogen gas in an attempt to remove possible carbonaceous deposits on the Pd catalyst. Prior to the fourth cycle, the catalyst was reduced in H_2 at 100 °C for 1 h after the standard catalyst wash by water. Retesting with fresh reagents at 50 °C following cooling catalyst in N_2 gave a conversion profile (curve 4, Figure 2) that was similar to that of the third cycle with catalyst wash by only water. The initial rate was slightly lower (0.01 mol/h), and the coupling reaction stalled at ca. 45% conversion. Reduction of the catalyst by hydrogen after the standard water wash did not seem to result in additional recovery in catalyst activity. Furthermore, regeneration with solution-based reducing agents such as sodium formate and hydrazine resulted in no improvement in catalyst performance. Including the fourth cycle, the turnover number for this sequence based on total palladium is 210.

We were interested if catalyst deactivation was substrate dependent. After an initial reaction with ethyl 4-bromobenzoate the catalyst was washed with water. However, retesting with ethyl 4-bromobenzoate lead to 33 % conversion over 22 h (25 °C); down from a conversion of 98 % after 12 h (25 °C) with the fresh catalyst. In the third cycle, the catalyst was washed with warm water, aqueous NaOH (5 %) and ethanol/water but the conversion of the ethyl 4-bromobenzoate was only 25 % after 66 h. With 4-bromoanisole, retesting at 80 °C gave only 32 % conversion down from 85 % for the fresh catalyst. Using the same catalyst but switching to the more reactive 4-bromobenzotrifluoride (50 °C) resulted in improved conversion of 60 % after 40 h. This indicates that the catalyst is still active toward electron deficient aryl bromides but after three cycles its activity is considerably reduced. These studies show that catalyst deactivation beyond the first cycle is independent of the aryl bromide substrate.

Effect of Solvent System on Catalytic Activity

To determine how the solvent system effected catalyst performance, the ethanol/water (5/1) solvent was circulated through the catalyst bed for 4 h at room temperature before adding reactants. This pretreatment resulted in 85 % conversion of 4-bromobenzotrifluoride after 12 h at 50 °C which is slower than the untreated catalyst (99 % conv, 1.8 h). Treating fresh catalyst with ethanol/water as above followed by a warm water wash for 2 h resulted in 99 % conversion of 4-bromobenzotrifluoride at 50 °C (curve 1, Figure 3). However, a reaction time of 23.5 h was required to reach this level of conversion compared to 1.8 h for an untreated catalyst. Thus, these studies show that high levels of conversion can be obtained on pretreated catalysts but there is a deleterious effect on rate.

The high levels of conversion with the pretreated catalyst were encouraging and suggested that the performance of the recycled catalyst might improve. Subsequent cycles on the above ethanol/water conditioned, water-washed catalyst gave conversions of 89 % for both the second and third cycles and 72 % for. the fourth cycle (Figure 3). Overall, the turnover number based upon total palladium reached 250 with this system. When compared to Figure 2 in which the catalyst was not pretreated, these results indicate that pretreatment enhances catalyst performance in later cycles relative to the untreated catalyst.

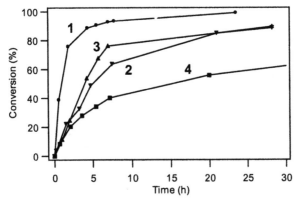

Figure 3 Suzuki coupling of 4-bromobenzotrifluoride at 50 °C in the semi-continuous, fixed-bed system with pretreated Pd/C 1/8" extrudates: 1) fresh catalyst, solvent conditioned and water washed; 2) water wash, second reaction; 3) water wash, third reaction, 4) water wash, fourth reaction.

Rationalization of Catalyst Activity Trends

The deterioration in performance of the Pd/C extrudates catalyst over time/batch may be attributable to catalyst deactivation through clogging of catalyst pores and Pd leaching. The reaction generated salts that are not fully soluble in the reaction solution. Conceivably the salts generated in the pores of the catalyst extrudates may crystalize out, clog the pores, and as a result limit reagent access to the palladium inside the pores. This hypothesis is supported by the partial success of water wash in recovering activity of deactivated catalyst as shown in Figures 2 and 3, and by the gain in catalyst weight post reaction measured after extensive catalyst wash followed by drying.

We have also determined that palladium leaching took place as the percent palladium decreases by ca. 20% after two cycles.[8] Since the extrudate catalyst initially contains a small amount of Pd (0.8 wt %), any inhibition through clogging of pores and loss through leaching will have dramatic affect on the amount of Pd available for catalysis. This also gives insight into the excellent recycling performance of the powdered catalyst used in the batch runs. Since the catalyst is in powdered form, it is easier to wash away adsorbed materials that occlude the Pd. In addition, the initial Pd content is much higher (5 wt %) than that of the extrudate which suggests that the catalyst can better tolerate loss of Pd than the extrudate. Furthermore, the high surface area, powdered catalyst is in constant contact with the reactants which promotes better mass diffusion and possibly enables re-deposition of leached Pd back onto the carbon support.

CONCLUSION

We have demonstrated that the Suzuki cross-coupling reaction can be carried out in a batch mode with catalyst recycling using Pd/C in powder form. Six cycles for a total of 120 turnovers were run using the same Pd/C catalyst without loss of activity. It is likely that many more cycles are possible. The excellent performance of the powdered catalyst in recycling during batch runs is attributed to its ability to be thoroughly washed, its high surface area which allows access to more Pd as well as the high initial Pd loading (5 wt %).

A fixed-bed, semi-continuous flow system was also developed using Pd on 1/8" carbon extrudates. The reaction with the fresh catalyst gives high conversions in reasonable rates with electron deficient aryl bromides. However, poorer performance is observed with aryl chlorides and electron rich aryl bromides. The catalyst can be recycled by circulating water through the catalyst bed to remove salts formed during the reaction. However, the maximum conversion in subsequent cycles is 80 % and the rate of conversion is substantially reduced. Nevertheless, turnover numbers of 210-250 are obtained with this system. The poorer performance of the extrudate catalyst in the fixed-bed reactor relative to the powdered catalyst in the batch reaction is predominantly attributed to lower loading of the Pd coupled with deactivation through clogging and leaching.

EXPERIMENTAL SECTION

All reagents were purchased from commercial sources and used without further purification. Potassium carbonate in granular form was obtained from EM Science or Aldrich. Ethanol (200 proof) was obtained from Aaper Alcohol and Chemical Co. The 5% Pd/C used in the batch runs was purchased from PMC (type 1610C, dry, shell-type, unreduced). The fixed-bed catalyst, 0.8% Pd/C, was obtained from Engelhard (Escat 132, dry, shell-type, reduced, 1/8" carbon extrudates). Metal dispersion measurements on the PMC 1610C catalyst and the Escat 132 carried out using an Altamira catalyst characterization apparatus gave Pd dispersions of 20 and 23 %, respectively. Palladium analyses were performed on a graphite furnace atomic absorption spectrometer (GFAA) or an inductively coupled plasma-mass spectrometer (ICP-MS). Organic samples were digested in conc. HNO_3 and Pd/C samples were digested in aqua regia. By this method, the PMC 1610C catalyst registered 4.7 % Pd and the Escat 132 catalyst registered 0.9 % Pd. HPLC analyses were performed on a HP 1100 HPLC with an ODS stationary phase and an acetonitrile/water (0.1 H$_3$PO$_4$) mobile phase gradient from 20 – 100 % acetonitrile.

Batch Procedure

To a 250-mL three-necked round-bottomed flask containing a 1" Teflon coated magnetic stirbar was added Pd/C (1.5 g, 5 mol %), phenylboronic acid (2.2 g, 0.018 mol), K$_2$CO$_3$ (4 g, 0.029 mol) and ethanol/water (5:1, 200 mL). The reaction mixture was degassed with three pump-N$_2$ fill cycles and aryl bromide (0.027 mol) was added and the reaction mixture heated to 50 °C. Reaction progress was monitored by thin-layer chromatography (fluorescent Silica gel plates, hexanes/EtOAc (4/1) eluent). When complete, the reaction solution was gravity filtered in air (Whatman #1 qualitative filter paper). The catalyst cake was washed with ethanol/water (50 mL, 5/1), and acetone (50 mL). Celite (1 g) was added to the filtrate to capture residual Pd/C particles. After filtration, the product solution was poured into deionized water (500 mL). The white precipitate that formed was collected on a Buchner funnel and washed with water (25 mL) and an ice-cold mixture of methanol/water (50 mL, 1/1). The product was further dried by standing in air. Further purification was performed by recrystallization from methanol by slow addition of water to initiate crystallization.

Catalyst recovery: The above catalyst cake was removed from the fluted filter paper with a stream of deionized water (ca. 200 mL). The slurry was filtered on a Buchner funnel (Whatman #1, qualitative filter paper) and washed with water (50 mL) and acetone (25 mL) and initially dried by suction then by standing in air.

Semi-Continuous Reactor and Procedure

Reactor. A schematic of the reactor is given in Figure 1. A catalyst charge (12 – 18 g, 0.1-0.2 mol %) of Pd/C 1/8" extrudates was supported between fine mesh, #316, stainless steel screens in a polished 6" x ¼" ss tube reactor. An 1/8" thermocouple was inserted to the midpoint of the catalyst bed for temperature measurement. Metal connections were made using Swagelok fittings and connections between other components made with Teflon tubing. A centrifugal pump was used for circulation at a rate between 200 – 400 mL/min. The exit from the catalyst bed led to an in-line, three-way stopcock which allowed for sampling of the reaction mixture during circulation. The reactor reservoir consisted of a three-necked, 1 L round-bottomed flask, fitted with a water-cooled condenser, thermocouple and an inlet for the pump. The top of the condenser was fitted with a nitrogen gas adapter to maintain and inert atmosphere during the reaction. A magnetic stirrer/heater was used to control the temperature of the reaction mixture in the reservoir.

Procedure. To a mixture of absolute ethanol/water (5/1, 345 mL) was sequentially dissolved potassium carbonate (26.4 g, 0.191 mol) and phenylboronic acid (11.4 g, 0.0934). Next, 4-bromobenzotrifluoride (15.2 g, 0.0676 mol) was added and the solution was transferred to the reservoir. Residual solution was transferred via 75 mL of EtOH/H$_2$O to give a total solvent volume of 420 mL. The reactor system was purged with three vacuum/N$_2$ cycles. The centrifugal pump was turned on to initiate circulation and circulation was allowed to proceed for ~5 min at room temperature to homogenize the solution. A sample was then drawn through the sampling stopcock and this was taken as the initial composition at zero time. Heating of the three-neck reservoir and catalyst bed was started and samples were withdrawn periodically for HPLC analysis to obtain the composition profile of the reaction with time.

ACKNOWLEDGMENTS

We thank Xiaodong Bu, Xiujuan Xie and Jane Wu for performing palladium analyses. JRS thanks Merck financial support and for sponsoring a sabbatical leave for this project.

REFERENCES

1. (a) R. A. Augustine, *Heterogeneous Catalysis for the Synthetic Chemist*, Marcel Dekker: New York, 1996. (b) H.-U. Blaser, A. Indolese, A. Schnyder, H. Steiner, M. Studer, *J. Mol. Catal. A*, **173**, 3-18 (2001).
2. Reviews on Suzuki coupling reactions: (a) J. Hassan, M. Sévignon, C. Gozzi, E. Schulz, M. Lemaire, *Chem. Rev.*, **102**, 1359-1470 (2002). (b) A. Suzuki, *J. Organometal. Chem.* **576**, 147-168 (1999). (c) N. Miyaura, A. Suzuki, *Chem. Rev.*, **95**, 2457-2483 (1995).
3. (a) G. Marck, A. Villiger, R. Buchecker, *Tetrahedron Lett.* **35**, 3277-3280 (1994). (b) D. Gala, A. Stamford, J. Jenkins, M. Kugelman, *Org. Proc. Res. Dev.* **1**, 163-164 (1997). (c) S. Sengupta, S. Bhattacharyya, *J. Org. Chem.* **62**, 3405-3406 (1997). (d) V. Bykov, N. Bumagin, N. *Rus. Chem. Bull.* **46**, 1344-1345, (1997). (e) D. S. Ennis, J. McManus, W. Wood-Kaczmar, J. Richardson, G. E. Smith, A. Carstairs, *Org. Proc. Res. Dev.* **3**, 248-252 (1999).
4. B. H. Lipshutz, J. Sclafani, P. A. Blomgren, *Tetrahedron*, **56**, 2139-2144 (2000).
5. Although Ni/C has been demonstrated for Suzuki coupling, up to 4 equiv of triphenyl-phosphine per equiv of Ni is required (ref 4). Because of its solubility, triphenylphosphine could complicate the product isolation step. In addition, transmetallation is known to occur between triphenylphosphine and aryl boronic acids which could contribute to product contamination see: (a) D. O'Keefe, M. Dannock, S. Marcuccio, *Tetrahedron Lett.*, **33**, 6679-6680 (1992). (b) F. E. Goodson, T. I. Wallow, B. M. Novak, *J. Am. Chem. Soc.*, **119**, 12441-12453 (1997).
6. C. R. LeBlond, A. T. Andrews, Y.-K. Sun, J. R. Sowa, Jr., *Org. Lett.*, **3**, 1555-1557 (2001).
7. H. Sakurai; T. Tsukuda; T. Hirao *J. Org. Chem.*, **67**, 2721-2722 (2002).
8. Currently, we have not been able to account for all of the lost palladium in the fixed-bed catalyst. For example the palladium concentration in the reaction solution is < 2 ppm which accounts for < 1 % of the lost palladium. The issue of Pd mass balance is under further investigation.

57

New Chiral Modifier for Enantioselective Heterogeneous Catalytic Hydrogenation

É. Sípos, A. Tungler, and I. Bitter
Budapest University of Technology and Economics, Budapest, Hungary

ABSTRACT

The use of *(S)-α,α*-diphenyl-2-pyrrolidinmethanol as a chiral modifier in asymmetric heterogeneous catalytic hydrogenation of ethyl pyruvate and isophorone is reported. Various solvents have been screened. The effect of modifier's concentration is described. The changes in optical yield of the saturated ketone as functions of the water content of the solvent has also been investigated.

INTRODUCTION

It is well known that enantioselectivity can be induced in heterogeneous catalytic reaction by adding chiral modifier to the reaction mixture. For producing any significant enantioselectivity the chiral modifier is required to have two functional part: one (a) for anchoring the modifier to the surface of the catalyst and another (b) which interacts with the substrate itself [2]. While the requirements are simple there are only a few known effective chiral compounds. The most studied reaction is the enantioselective hydrogenation of ethyl pyruvate to ethyl lactate on platinum catalysts modified with cinchona alkaloids (originally reported by Orito and co-workers [1]). Later a number of simple chiral amino-alcohol were synthesised possessing the crucial structural elements mentioned above and tested in the hydrogenation of ethyl pyruvate [4, 5, 6, 7]. The (R)-2-(1-pyrrolidinyl)-1-(1-naphtyl) ethanol was found to be the most effective among them. Another studied system is the hydrogenation of the C=C bond of isophorone [3], with the (-)-dihydroapovincaminic acid ethyl ester acting as the modifier.

In order to discover new modifiers in the hydrogenation of both C=O and C=C bonds we have tested (S)-α,α-diphenil-2-pyrrolidinmethanol (DPPM) (Scheme 1), which was used as a ligand in the transition metal complex catalyzed enantioselective reduction of prochiral ketons, like acetophenone and pinacolone [8, 9, 10].

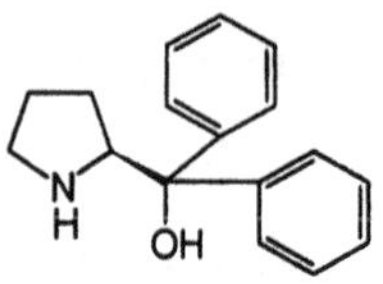

Scheme 1 (S)-α,α-diphenil-2-pyrrolidinmethanol (DPPM)

This chiral molecule is based on the (S)-proline, which in itself proved to be a good chiral auxiliary and synthon in asymmetric heterogeneous catalytic hydrogenation [11].

Our aim was to broaden the knowledge about the necessary structural parts of a potential chiral modifier for asymmetric induction. (S)-α,α-diphenil-2-pyrrolidinmethanol (DPPM) was tested in the hydrogenation of ethyl pyruvate and isophorone. Parameters optimisation was carried out - the influence of solvents, modifier concentration - was studied.

EXPERIMENTAL

Materials

The used Pt catalyst was commercial product: 5% Pt/Al_2O_3 Janssen. Pd black catalyst was prepared according to the following procedure: 18 mmol (6.0 g) K_2PdCl_4 was dissolved in 100 ml water and reduced at boiling point with 36 mmol HCOONa dissolved in 20 ml water. The pH of the solution during the preparation was basic, and the whole amount of the reducing agent (HCOONa) was added at the beginning of the reaction.

Isophorone was supplied by Merck, ethyl pyruvate by Fluka. They were distilled in vacuum before use.

(S)-α,α-diphenyl-2-pyrrolidinmethanol was prepared according to the procedure described in [12].

Hydrogenation

The hydrogenation of isophorone and ethyl pyruvate was carried out at 25 °C and under 50 bar hydrogen pressure in a stainless steel autoclave. (Technoclave).

Before the hydrogenation the reaction mixtures were stirred under nitrogen for 10 minutes in the reaction vessel.

Analysis

The reaction mixtures were analysed with a gas chromatograph equipped with a Supelco BETA DEX ^{TM} 120 Capillary Column (analysis temperature: dihydroisophorone at 110 °C, ethyl lactate at 90 °C) and FID. The chromatograms were recorded and peak areas were calculated with Chromatography Station for Windows V1.6 (DataApex Ltd., Prague). Enantiomeric excess was defined as:

$$ee\ (\%) = ([R]-[S])/([R]+[S])*100$$

RESULTS AND DISCUSSION

In the presence of (S)-α,α-diphenyl-2-pyrrolidinmethanol the hydrogenation of C=C bond of isophorone (Scheme 2) and C=O of ethyl pyruvate result in an excess of the (S) enantiomer. (Scheme 3).

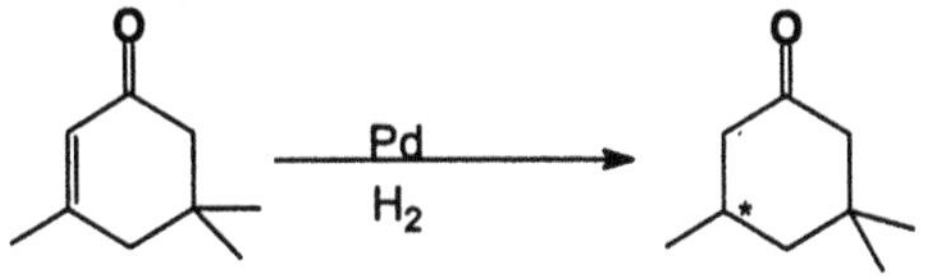

Scheme 2 Hydrogenation of isophorone

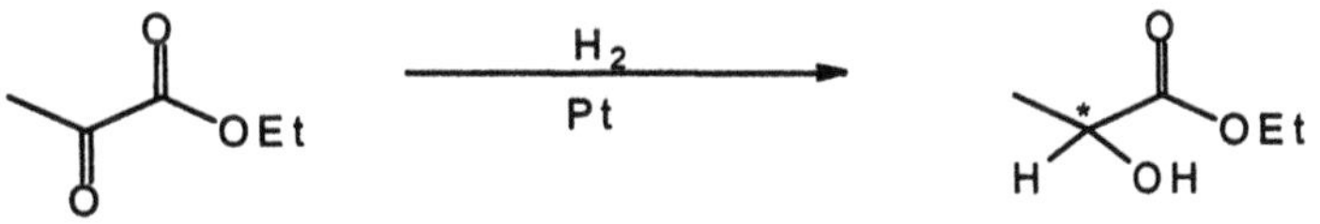

Scheme 3 Hydrogenation of ethyl pyruvate

The hydrogenations were carried out in different solvents and solvent mixtures in order to find out the best one. The enantioselectivities for ethyl pyruvate in different solvents are listed in Table 1.

The enantioselectivities for isophorone in different solvents are listed in the Table 2.

The apolar toluene and the acetic acid were found to be unfavourable both in the hydrogenation of ethyl pyruvate and in the hydrogenation of isophorone. It is interesting that adding water to methanol led to increase in the optical purity of dihydro-isophorone, but resulted in the complete loss of enantioselectivity in the case of ethyl pyruvate. The possible explanation may be that considerable hydration of ethyl pyruvate could be expected in water [13] and in that way the modifier could not interact with the substrate.

Table 1 The effect of solvents on the ee in the hydrogenation of ethyl pyruvate

Solvent	Conversion[a] (%)	ee (%)
MeOH	51	25
Toluene	52	5
AcOH	100	10
Acetonitrile	70	22
MeOH/H$_2$O 1/1	100	0

Reaction conditions: 0.01 mol Etpy, 0.14 mmol DPPM, 10 ml solvent, 0.05 g 5% Pt/Al$_2$O$_3$ Janssen catalyst, T=25°C, p=40 bar. [a]After 6 hours reaction time

Table 2 The effect of solvents on the ee in the hydrogenation of isophorone

Solvent	Conversion[a] (%)	ee (%)
Toluene	92	14
AcOH	97	5
MeOH	95	23
Acetonitrile	100	23
DMF	100	33
H$_2$O	100	15
MeOH/H$_2$O 1:1	100	40

Reaction conditions: 0.01 mol Isoph, 0.05 g Pd black, 0.035 g modifier (0.14 mmol), 10 ml solvent. [a]After 4 hours reaction time.

We have investigated the influence of the water concentration of solvent on the ee in the hydrogenation of isophorone (Figure 1).

We measured ee at three different substrate to modifier ratio, considering a wide range of water content of the solvent (10%-80%). The optimal was the equal mixture of water and methanol. At low water content even adding a small amount of water to the mixture was enough to increase the enantioselectivity considerably. Above the optimal water content the observed enantioselectivity decreased in smaller extent. It is obvious that water plays a crucial role in the solvatation, first of all of that of the modifier.

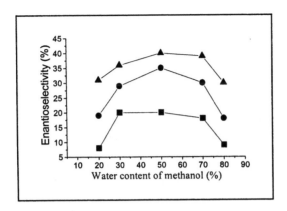

Figure 1 Influence of water content of methanol on ee in the hydrogenation of isophorone at 400/1 (■), 200/1 (●), 42/1 (▲) substrate/modifier molar ratios

In asymmetric reaction it is a basic question what is the necessary amount of the modifier. The influence of the concentration of DPPM on the ee is depicted in Figure 2.

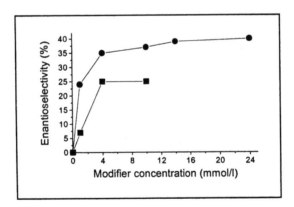

Figure 2 Influence of modifier concentration on ee in the hydrogenation of ethyl pyruvate (■) and isophorone (●).

Both hydrogenation reactions were carried out in the optimal solvent or solvent mixture. In both case the enantioselectivity increased abruptly with increasing amount of modifier and reached a plateau at 1.4 mol % with respect to the substrate isophorone, for ethyl pyruvate at 0.5 mol %. Increasing the modifier concentration above these values seems to have no significant influence on the optical purity of the products.

CONCLUSION

The assumption was correct that the *(S)-α, α*-diphenyl-2-pyrrolidinmethanol (DPPM) could be applicable as chiral modifier in heterogeneous hydrogenation. Although the afforded enantioselectivities are moderate (for ethyl pyruvate 25%, for isophorone 40%), this compound increases the narrow choice of chiral molecules that could be used as modifiers.

Important parts of the structure of DPPM are the basic amine function in a rigid chiral environment and an aromatic ring system. It can be assumed that the basic, secondary N atom in the less flexible pyrrolidine ring of DPPM is responsible for interaction with the reactant. The aromatic ring system is responsible for interaction with the catalyst surface.

This new findings improve our knowledge about the structural parts needed for asymmetric induction, how to design appropriate chiral modifiers for heterogeneous hydrogenation.

ACKNOWLEDGEMENTS

The authors acknowledge the financial support of the Hungarian OTKA Foundation under the contract number T 029557, of the Ministry of Education, FKFP 0017/1999 and 0308/1997, as well as of the Varga József Foundation.

REFERENCES

1. Y. Orito, S. Imai. J Chem Soc Jpn 1118, 1979.
2. A. C. Testa, R. Augustine. Catalysis in Organic Reactions. New York: Marcel Dekker, 2001, pp 465-470.
3. T. Tarnai, A. Tungler, T. Máthé, J. Petró, R. A. Sheldon, G. Tóth. J Mol Catal 102:41, 1995.
4. G. Wang, T. Heinz, A. Pfaltz, B. Minder, T. Mallat, A. Baiker. J Chem Soc Chem Commun 2047, 1994.
5. T. Heinz, G. Wang, A. Pfaltz, B. Minder, M. Schürch, T. Mallat, A. Baiker. J Chem Soc Chem Commun 1421, 1995.
6. B. Minder, M. Schürch, T. Mallat, A. Baiker, T. Heinz, A. Pfaltz. J Catal 160:261, 1996.
7. A. Baiker. J Mol Catal A: Chemical 115:463, 1997.
8. B. Jiang, Y. Feng, J. Zheng. Tetrahedron Letters 41 (52):10281, 2000.
9. C. Schunicht, A. Biffis, G. Wulff. Tetrahedron 56 (12):1693, 2000.
10. Jockel, Holger. J. Chem. Soc. Perkin Trans 2 1:69, 2000.
11. A. Tungler, M. Kajtár, T. Máthé, G. Tóth, E. Fogassy, J. Petró. Catalysis Today 5:159, 1989.
12. D. J. Mathre, T. K. Jones, L. C. Xavier, T. J. Blacklock, R. A. Reaner, J. J. Mohan, E. T. T. Jones, K. Hoogsteen, M. W. Baum, E. J. J. Grabowski. J. Org. Chem. 56 (2):751, 1991.
13. J. Zabicky. The Chemistry of the Carbonyl Group. London: Interscience, 1970, 2.

58

Synthesis of Dimethyl Carbonate from Propylene Carbonate and Methanol over Mesoporous Solid Base Catalyst

Tong Wei, Mouhua Wang, Wei Wei, Yuhan Sun, and Bing Zhong
Institute of Coal Chemistry, Chinese Academy of Sciences, Taiyuan, P.R. China

ABSTRACT

Mesoporous solid base catalyst was prepared by loading KI, K_2CO_3 and KOH on mesoporous active carbon and was used to catalyze dimethyl carbonate synthesis from propylene carbonate and methanol. The effect of preparation method, base strength, catalyst content, reaction time and temperature were investigated.

INTRODUCTION

Much attention has been paid to mesoporous materials in recent years (1). But little has been done to investigate them as catalysts, especially as solid base catalysts (2). Since most of organic reactions proceed in liquid phase, mesoporous catalysts are especially preferable in order to reduce inner diffusion resistance. Dimetyl carbonate is an important precursor of polycarbonate resins as well as a useful carbonylation and methylation agent (3,4). Because of the negligible toxicity of DMC, it is promising as a substitute for phosgene, dimethyl sulfate, or methyl iodide as well as a potential petrol octane enhancer. It is known that DMC can be synthesized concomitantly with propylene or ethylene glycol by the following liquid phase ester exchange reaction between propylene or ethylene carbonate and methanol (5,6), as shown in Scheme 1. For this reaction various base catalyst and group IV homogeneous catalyst are reported to be effective (6, 7). In the present work, KI, K_2CO_3 and KOH impregnated mesoporous active carbon was studied as solid base catalysts for the synthesis of dimethyl carbonate by transesterification of propylene carbonate with methanol. The effect of preparation method, strength of base site, catalyst content and reaction temperature were examined.

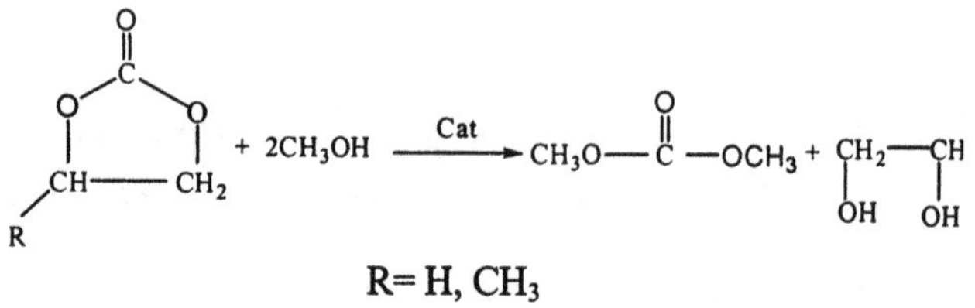

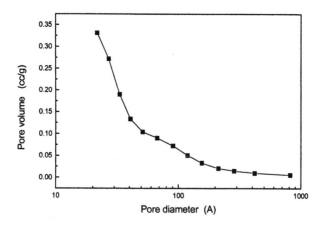

R= H, CH$_3$

Scheme 1

EXPERIMENTAL

Pore Size Distribution of Active Carbon

Coconut mesoporous active carbon was provided by 702 group of Shanxi Institute of Coal Chemistry, Chinese Academic of Sciences. The pore size distribution of this active carbon was determined by BET method as illustrated in Figure 1. It can be seen that diameter of most of pores is between 2~10 nm. The average pore diameter is 3.65 nm.

Figure 1 Pore size distribution of active carbon.

Preparation of Catalyst

Catalysts used in this paper were synthesized by two different methods. In first method (heat-treated method), mesoporous active carbon (20~40mesh) was heated from room temperature to 390K at a rate of 4 K/min and heated at 390K for 2 hr in N_2 atmosphere. After being cooled to room temperature, 5 g mesorporous active carbon was impregnated in 10 mL 0.3M K_2CO_3 aqueous by incipient wetness, then after being laid for 24 hr, the particles were dried at 373K for 24 hr in N_2 atmosphere, and subsequently heated from room temperature to 573 K at a rate of 4K min^{-1}, and treated for 10hr at this temperature in flowing nitrogen. In second one (vacuum-treated method), mesoporous active carbon was heated at 353K for 4 hr at 0.02MPa. After being cooled to room temperature, 5 g active carbon was impregnated in 10mL 0.3M KI, K_2CO_3 and KOH aqueous respectively by incipient wetness and was laid at room temperature for 24 hr in air, and then dried at 353K for 8 hr at 0.02MPa.

Catalytic Reaction

Ester exchange reactions were carried out in a 75mL autoclave. After catalyst, propylene carbonate (4.73g) and methanol (5.99g) was put into autoclave, the reactor was heated at the rate of 5 K/min. When expected temperature was reached, violent magnetic stirring was turned on. Products were analyzed on GC 920 Gas-Chromatograph with TCD using GDX-203 (3m) packed column after centrifugal separation of solid catalyst from liquid.

RESULT AND DISCUSSION

Effect of Preparation Method

In order to investigate the effect of preparation method, K_2CO_3 (0.6mmol/(g active carbon)) was loaded on mesoporous active carbon by two different methods (heat-treated method (HTM) and vacuum-treated method (VTM)). It can be seen from Table 1 that vacuum treated catalyst shows higher catalytic activity than that of heat-treated catalyst. In the process of evaporating water from the catalyst prepared by impregnation, the crystalline of active substance would grow with the increase of temperature. Therefore crystalline growth may be resisted in the process of vacuum desiccation because of the lower temperature, which would proba-

Table 1 The effect of preparation method on PC conversion and DMC yield[a]

Catalyst	PC conversion %	DMC yield %
K_2CO_3/C(HTM)	28.6	25.4
K_2CO_3/C(VTM)	30.2	27.7

[a] Reaction condition catalyst 1.82 (wt)%; temperature: 393K; reaction time: 2 hr.

bly result in the smaller K_2CO_3 particles and more active site on VTM catalyst than that on HTM catalyst.

Effect of Base Strength

The catalytic behavior of KI/C, K_2CO_3 /C, KOH/C with loading content 0.6 mmol/(g active carbon) is illustrated in Table 2. It can be seen that the sequence of PC conversion and DMC yield is KOH/C> K_2CO_3/C>> KI/C. As catalyzed by base, this transesterification reaction proceeds as shown in schem2 (6). As can be seen from the mechanism, the main function of base catalyst is to activate MeOH by hydrogen bound between hydrogen of MeOH and basic site on catalyst. While KI is a typical weak basic salt, K_2CO_3 and KOH is strong base with H_ being 18.4 and 15.0 respectively (8). Since the pKa of MeOH is 15.0 (9), both KOH/C and K_2CO_3/C can activate MeOH greatly and the activation capability goes up with the increase of base strength. Therefore the sequence of catalytic activity is KOH/C> K_2CO_3/C>> KI/C. Since potassium hydroxide has significant solubility in methanol (35 wt% at 301K), in order to make a clear view of whether the catalytic ability comes from supported KOH or from soluble KOH, the concentration of K on catalyst was determined by ICPAES. It seems that only a little KOH dissolved in reaction liquid and most KOH was loaded on active carbon by strong van der walls force, since the concentration of K was 2.32 (wt)% before the reaction and was 2.28(wt)% after the reaction.

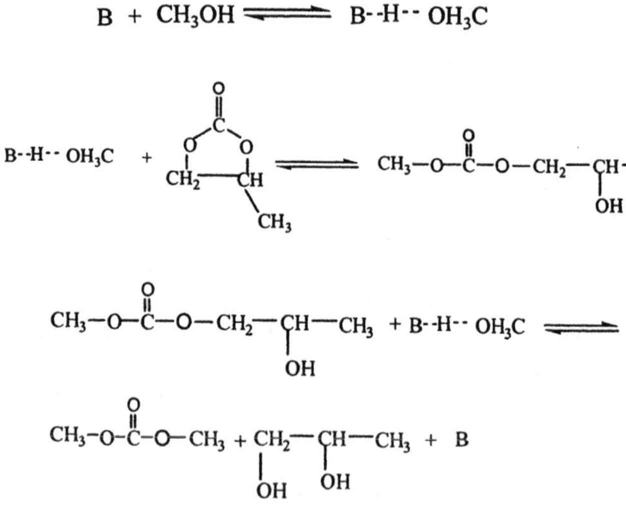

$$B + CH_3OH \rightleftharpoons B\text{-}H\text{-}\text{-}OH_3C$$

Scheme 2

Table 2 The effect of base strength on PC conversion and DMC yield [a]

Catalyst	PC conversion%	DMC yield%
KI/C	2.7	2.6
K_2CO_3/C	30.2	27.7
KOH/C	32.4	29.1

[a] reaction condition catalyst 1.82 (wt)%; temperature: 393K; reaction time: 2 hr.

The Effect of Catalyst Content and Reaction Time

As said above, because of its highest catalytic activity in this three solid base catalyst, KOH/C (0.6mmol/(g active carbon)) was used as a catalyst in order to investigate the effect of catalyst content and reaction time as illustrated in Fig. 2 and
Table 3. It can be seen that PC conversion and DMC yield increase greatly when
catalyst content increase from 0.91% to 1.82%, but increase only a little when
catalyst content increase from 1.82% to 4.70%. On the other hand, when reaction
time attains 2hr, PC conversion and DMC yield almost keep constant despite the
prolong of reaction time. Catalyst content mainly affects the activation of MeOH.
The amount of activated MeOH would increase greatly with the increase of catalyst concentration at first. But after catalyst concentration reaches certain value, it
would have only a little effect on activated MeOH concentration, and therefore the
reaction rate almost keep constant with the continual increase of catalyst concentration. As for the effect of reaction time, this reaction almost attains equilibrium
at 2hr 120□, so although the reaction time was extended, the reaction rate was so
slow that PC conversion and DMC yield almost keep constant.

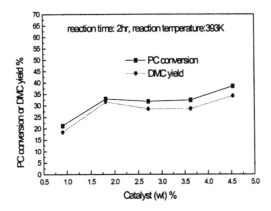

Figure 2 Effect of catalyst content on PC conversion and DMC yield

Table 3 The effect of reaction time on PC conversion and DMC yield [a]

Reaction time / hr	PC conversion%	DMC yield%
1	27.2	26.0
2	32.4	29.1
3	33.0	29.7

a: reaction condition catalyst: KOH /C; 1.82 (wt)%; temperature: 393K.

Effect of Reaction Temperature

Temperature is an important factor that could influence both reaction rate and reaction equilibrium. Figure 3 shows the effect of reaction temperature on PC conversion and DMC yield over KOH/C (0.6mmol/(g active carbon)). As can be seen from Figure 3, PC conversion and DMC yield increase greatly when temperature increase from 333K to 393K□but increase only a little when temperature continues going up. This reaction is exothermic with ΔrH=-7.092kJ/mol (10-13), which makes the reaction equilibrium being little affected by temperature. Therefore, The increase of temperature can elevate reaction rate greatly before the reach of reaction equilibrium. The more closely the reaction to equilibrium, the less the effect of temperature on reaction rate is. For reaction time is 2hr, the reaction is dynamically controlled when temperature is lower than393K and is thermally controlled when temperature is higher than 393K.

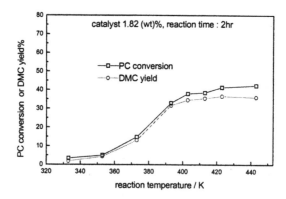

Figure 3 Effect of temperature on PC conversion and DMC yield

CONCLUSION

In a word, loading K_2CO_3 and KOH on mesoporous active carbon by vacuum impregnation can prepare highly basic mesoporous solid base catalyst, which is very active for the synthesis of dimethyl carbonate from propylene carbonate and methanol. As for this reaction, the higher the strength of base site of catalyst, the higher the reaction rate is. Moreover, because this reaction is a little exothermic, temperature has little effect on reaction equilibrium.

REFERENCES

1   T. Kyotani, Control of pore structure in carbon. Carbon. 38: 269-286 2000.
2   K R Kloetstra, H. V. Bekkum, Solid mesoporous base catalysts comprising of MCM-41 supported intraporous cesium oxide, Stud Surf Sci Cata 105: 431-438. 1997.
3   W B Kim, J S Lee. Gas phase transesterification of dimethyl carbonate and phenol over surpported titanium dioxide. J Catal 185: 307-313, 1999.
4   M A Pacheco, C.L. Marshall. Review of dimethyl carbonate (DMC) manufacture and its characteristics as fuel additive. Energy and Fuel 11: 2-29 1997.
5   J A Cella, S W Bacon. Preparation of dialkyl carbonates via phase-transfer-catalyzed alkylation of alkali metal carbonate and bicarbonate salts. J. Org. Chem 49: 1122-1125, 1984.
6   J F Knifton, R G Duranleau. Ethylene glycol-dimethyl carbonate cogeneration. J Molecular Catal 67: 389-399 1991.
7   Y Watanabe, T Tatsumi. Hydrotalcite-type materials as catalyst for the synthesis of dimethyl carbonate from ethylene carbonate and methanol. Microporous and mesoporous materials 22: 399-407, 1998.
8   J Zhu, Y Chun, Y Wang. Strong basic zeolite. Chinese Since Bulletin 44: 897-903,1999.
9   Organic Chemistry 2nd ed. Beijing: High Education press, 1992, pp88.
10 Y Yin, Z Xi, D Li. Physical Chemistry. 3rd ed. Beijing: Higher Education Press, 1994, pp569.
11 Z Dong . Synthesis of propylene carbonate by transesterification. Hangzhou Chemical Engineering, (3): 39-96,1979.
12 R C Reid, J M Prausnitz, B E Poling. The properties of Gases and Liquids, 4th ed. New York: McGraw-Hill Company, 1987, pp 683.
13 V Steele, R D Chirico, S E Knipmeyer, A Nguyen, N K Smith. Thermodynsmic properties and ideal-gas enthalpies of formation of dicyclohexyl sulfide, diethyl-enetriamine, di-n-octyl sulfide, dimethyl carbonate, piperazine, hexachloroprop-1-ene, tetrakis (dimethylamino)ethylene, N,N' –bis(2-hydroxyethyl)ethylenediamine, and 1,2,4-triazolo{ 1,5-a}pyrimidine. J Chem Eng Data 42: 1031-1052, 1997.

59

Esterification of Amino Acids by Using Ionic Liquid as a Green Catalyst

Hua Zhao and Sanjay V. Malhotra
Department of Chemistry and Environmental Science, New Jersey Institute of Technology, Newark, New Jersey, U.S.A.

ABSTRACT

Ionic liquids are emerging as viable media, in improving various organic reactions as green solvents or catalysts, and have shown promising applications in many chemical processes. This study, for the first time, shows the use of ionic liquid [EtPy][CF_3COO] as an excellent catalyst for the esterification of amino acids. Esterification of several amino acids was achieved with very good yields.

INTRODUCTION

Amino acid esters are very important intermediates for chemical and pharmaceutical industry (1). For instance, various methyl and ethyl esters of amino acids were resolved enzymatically in order to obtain chiral amino acids (2). D-phenylglycine and D-*p*-hydroxy-phenylglycine esters have been investigated in the enzymatic process to semi-synthetic penicillins and cephalosporins (3). Phenylalanine methyl ester was used in the Holland Sweetener Company (DSM-Tosoh joint venture) for enzymatic synthesis of the artifical sweetener aspartame (4).

The earliest method of esterifying amino acids was developed by Fischer (5). By continuously passing hydrogen chloride gas into ethanolic suspensions of amino acids, the amino acids dissolve and the solution becomes homogeneous. After adding base to the reaction mixture, the resulting ester hydrochlorides can be isolated and extracted with ether. In principle, this is not a very efficient method since it is an equilibrium reaction. However, it can be made useful by manipulation of the equilibrium. For example, using an excess of alcohol and removal of water as it is formed may force the reaction to the direction of the esters. Another disadvantage of this method is the difficulty of handling toxic hydrogen chloride gas.

Another chemical method (6) of synthesizing amino acid esters is to form acyl halides by reacting amino acids with thionyl chloride first. Then the appropriate alcohol is added to produce ester. A facile synthetic method under mild conditions was studied via cesium salts to yield amino acid esters (7). This process not only involves expensive cesium salts, but it also produces amino acid esters with one enantiomer in rich.

Lipases have been used for esterification of acids for a long time (8), but attempts to esterify amino acids have failed (9). A commercially available enzyme, *papain* was extensively investigated on the esterification of N-protection amino acids (10, 11). The average yield usually ranges from 60 to 80%. However, the resulting esters were enriched one of the enantiomers, which is undesirable for some applications.

Recently (12), the acid form of ultrastable zeolite Y (H-USY) has been studied as a solid catalyst in the reaction of α-amino acids with methanol at 100-130 °C (15-20 bar). The yields of the amino acids were DL-homophenylalanine, 68%, D-phenylglycine, 86%, L-phenylalanine, 77%, and D-*p*-hydroxy-phenylglycine, 14%.

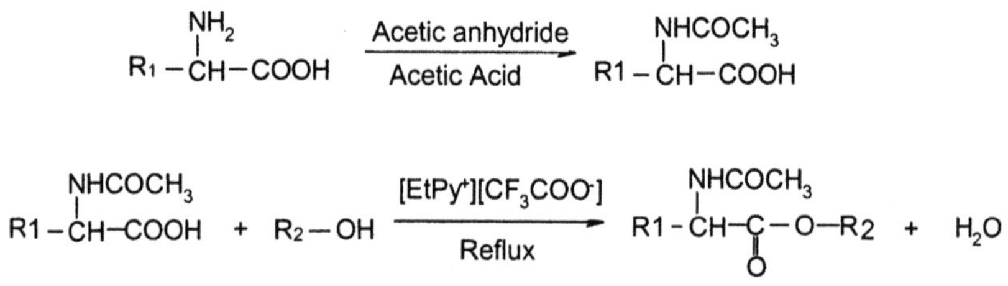

Figure 1 Scheme of acetylation and esterification of amino acid.

With the rapid use in the applications of ionic liquids as novel solvents and catalysts in organic reactions (13), it would be fascinating to try ionic liquids as catalysts for the synthesis of organic esters. The potential advantages of applying ionic liquids in this process are: (1) mild condition for the esterification reaction; (2) clean esters could be produced by using ionic liquids as environmentally friendly reaction medium; (3) most of the resulting esters could easily be separated due to their immiscibility with ionic liquids; (4) room temperature ionic liquids are usually moisture-stable and thermal-stable; therefore, they are suitable candidates as catalysts; (5) ionic liquids could be recovered and reused again. A representative esterification reaction of amino acid is shown in Figure 1.

EXPERIMENTAL SECTION

Materials

Amino acids, alcohols, acetic acid and acetic anhydride were purchased from Sigma-Aldrich. The preparation of ionic liquid [EtPy][CF$_3$COO] was based on the literature method (14).

Methods

(1) The preparation of N-acetyl amino acids followed a literature method (15): 5.0 g of amino acid was suspended and stirred in 80 mL of glacial acetic acid, followed by adding 1.2 molar equiv of acetic anhydride. The mixture was stirred at room temperature until the solid disappeared. If the amino acid does not appear to react to form a homogeneous phase, a gentle heat is supplied. The solvent is removed by rotary evaporation under vacuum, and the residue was taken into acetone and filtered. Rotary evaporation of the filtrate gave the N-acetyl amino acid.

(2) The general esterification process is described as follows: 5 g of N-acetyl amino acid was dissolved in 100 mL of anhydrous alcohol, followed by adding 1 mL of the ionic liquid. The solution was stirred and refluxed. And the reaction progress was monitoring by Varian GC CP-3800. After the reaction was complete, the solvent was evaporated under vacuum. The residue was dissolved in ethyl acetate and washed twice with distilled water. Rotary evaporation of the dried organic phase gave N-acetyl amino acid ester. The acetyl group was then reversed from the amino acid by refluxing in 3N HCl for ca. 3 hr.

RESULTS AND DISCUSSION

Reaction Strategy and Reaction Time

Esterification of amino acids is a difficult reaction because amino acids exist as zwitterions (dipolar ions) and carboxyl group is not a free group (but an anion). Since the esterification reaction is an equilibrium process, factors that shift the reaction toward the products will benefit the formation of amino acid ester.

First, the purpose of acetylation of amino acids before the esterification reaction are: (1) to make amino acid soluble in organic solvents like alcohol; (2) to free the carboxyl group by protecting the amino group; (3) to prevent possible polymerization and other side reactions caused by amino group. This step effectively increases the concentration of amino acid in the reaction phase which benefit to the esterification reaction. Other amino protecting groups may be used if they can promote the esterification progress. Second, the alcohol is used in excess in order to push the equilibrium to the products. Third, the reaction system was initially anhydrous, which helps to shift the equilibrium by reducing the water in the reaction medium. However, studies to understand the reaction mechanism of using ionic liquid as catalyst in the esterification process is underway.

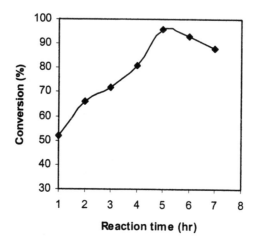

Figure 2 Time course of synthesizing homophenylalanine ethyl ester by ionic liquid [EtPy][CF$_3$COO]

Figure 2 shows the effect of refluxing time on the conversion of amino acid homophenylalanine. At first, increasing reaction times increases the reaction conversion as expected. However, extending the reaction time beyond 5 hours decreases the conversion. Possible side reactions may occur at this moment.

Esterification of Different Amino Acids

Syntheses were conducted with ethyl and isopropyl alcohol for each amino acid. Table 1 shows the reaction time and conversion for each ester. Interestingly, when the R$_1$ group (Figure 1) is small, e.g., in serine, the conversion is low. While in norleucine, with bulkier group R$_1$, the yield of corresponding ester increased. This could be due to (1) ionic liquid [EtPy][CF$_3$COO] shows better catalytic effect on amino acids with large side chains; (2) the bulky group increases the homogeneity of reaction system by increasing the solubility of amino acids and interaction with alcohol.

Comparing the results of 2-chloroglycine with serine, also 4-chlorophenylalanine with 2-phenylglycine and homophenylalanine, suggests that halide, such as chloride, improves the substrate activity and therefore increases the reaction conversion. By replacing the hydrogen by chloride, due to the strong electron withdrawing ability of chloride, the acidity of the amino acids is increased which benefits to the formation of esters (as shown in Figure 3).

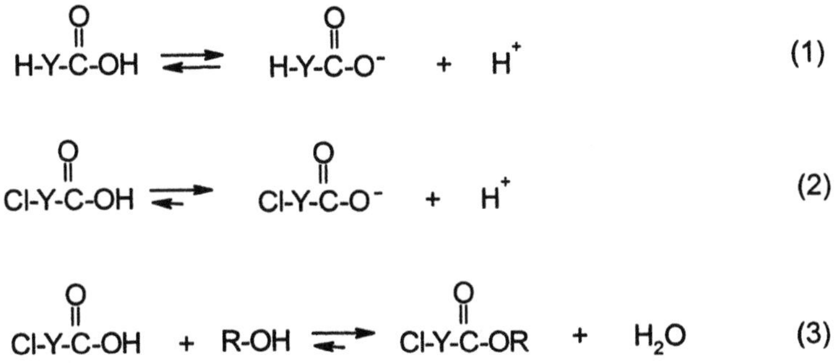

$$\text{H-Y-}\overset{\displaystyle O}{\underset{\displaystyle \|}{C}}\text{-OH} \rightleftharpoons \text{H-Y-}\overset{\displaystyle O}{\underset{\displaystyle \|}{C}}\text{-O}^- + \text{H}^+ \qquad (1)$$

$$\text{Cl-Y-}\overset{\displaystyle O}{\underset{\displaystyle \|}{C}}\text{-OH} \rightleftharpoons \text{Cl-Y-}\overset{\displaystyle O}{\underset{\displaystyle \|}{C}}\text{-O}^- + \text{H}^+ \qquad (2)$$

$$\text{Cl-Y-}\overset{\displaystyle O}{\underset{\displaystyle \|}{C}}\text{-OH} + \text{R-OH} \rightleftharpoons \text{Cl-Y-}\overset{\displaystyle O}{\underset{\displaystyle \|}{C}}\text{-OR} + \text{H}_2\text{O} \qquad (3)$$

Figure 3 Effect of halide substitute on the esterification of amino acid

Amino acids with side chains containing the benzyl and indoline groups usually show good performance in the esterification. In general, the higher yield is obtained for ethyl esters compared with isopropyl esters. One exception is indoline-2-carboxylic acid, where the yield of isopropyl ester is much higher that the ethyl ester. It is also believed that the N-acetyl ethyl esters are more stable than the N-unprotected ethyl esters and butyl esters (16), which implies that the N-acetyl ethyl esters studied are very valuable for further applications.

CONCLUSION

A green catalyst, ionic liquid [EtPy][CF$_3$COO], was first time used for the synthesis of amino acid esters including unnatural amino acid esters. The results show that under mild reaction conditions, satisfactory conversion can be achieved for the formation of amino acid esters.

REFERENCES

1. RA Sheldon. Chiraltechnology: Industrial Synthesis of Optically Active Compounds. New York: Marcel Dekker, 1993.
2. JY Houng, ML Wu, ST Chen. Kinetic resolution of amino acids esters catalyzed by lipases. Chirality 8: 418-422, 1996.
3. A Bruggink, EC Roos, E de Vroom. Org Process Res Dev 2: 128, 1998.
4. S Hanzawa. Encyclopedia of Bioprocess Technology: Fermentation, Biocatalysis and Bioseparation. New York: Wiley, New York, 1999.

5.   HD Jakubke, H Jeschkeit. Amino Acids, Peptides and Proteins: An Introduction. New York: John Wiley & Son, 1977.

6.   M Bodanszky, A Bodanszky. The Practice of Peptide Synthesis. Berlin: Springer-Verlag, 1984.

7.   SS Wang, BF Gisin, DP Winter, R Makofske, ID Kulesha, C Tzougraki, J Meienhofer. Facile synthesis of amino acid and peptide esters under mild conditions via cesium salts. J Org Chem 42(8): 1286-1290, 1977.

8.   G Langrand, J Batatti, G Buono, C Triantaphylides. Lipase catalyzed reactions and strategy for alcohol resolution. Tetrahedron Lett 27(1): 29-32, 1986.

9.   G Kirchner, M Scollar, A Klibanov. Resolution of racemic mixtures via lipase catalysis in organic solvents. J Am Chem Soc 107: 7072-7076, 1985.

10.   ST Chen, KT Wang, Papain catalysed esterification of N-protection amino acids. J Chem Soc, Chem Commun 327-328, 1988.

11.   D Cantacuzene, C Guerreiro. Optimization of the papain catalyzed esterification of amino acids by alcohol and diols. Tetrahedron 45(3): 741-748, 1989.

12.   MA Wegman, JM Elzinga, E Neeleman, F van Rantwijk, RA Sheldon. Salt-free esterification of alpha-amino acids catalysed by zeolite H-USY. Green Chemistry 3 (2): 61-64, 2001.

13.   T Welton. Room –temperature ionic liquids. solvents for synthesis and catalysis. Chem Rev 99: 2071-2083, 1999.

14.   JD Holbrey, KR Seddon. The phase behaviour of 1-alkyl-3-methylimidazolium tetrafluoroborates; ionic liquids and ionic liquid crystals," J Chem Soc, Dalton Trans 2133-2139, 1999.

15.   HK Chenault, J Dahmer, G Whitesides. Kinetic resolution of unnatural and rarely occurring amino acids: enantioselective hydrolysis of N-acyl amino acids catalyzed by acylase I. J Am Chem Soc. 111: 6354-6364, 1989.

16.   JP Greenstein, M Winitz. Chemistry of the Amino Acids, Vol. 2. New York: John Wiley & Sons, 1961, pp 1369.

60

Alkylation of Phenols by Caryophyllene on Acid Aluminosilicate Catalysts

V. V. Fomenko, D. V. Korchagina, N. F. Salakhutdinov, I. V. Sorokina, M. P. Dolgikh, T. G. Tolstikova, and V. A. Barkhash
Novosibirsk Institute of Organic Chemistry, Novosibirsk, Russia

The current worldwide tendency toward sustainable development calls for the improvement of ecological characteristics of synthetic processes. An important requirement is ecological purity of the process, implying the use of recyclable raw materials and safe technological methods, higher process selectivity and lower energy consumption per unit end product, as well as recuperation and safe utilization of the catalysts used and by-products. These requirements are met by the use of organized media (clays, zeolites) in various fields of organic synthesis and especially in the chemistry of terpenoids, which are accessible recyclable biologically active compounds possessing a unique structure. Using organized media improves the ecological characteristics of well-known processes involving natural compounds and leads to unusual transformations and hence new applications of recyclable raw materials to fine organic synthesis due to the unique properties of terpenoids such as the conformational mobility and multifunctionality. The rate ratio between the competing reactions in organized media may be changed by varying the statistical and concentration factors and/or due to the relative changes in the activation barriers of possible transformations compared to homogeneous media.

Using terpenoids in these studies involves a number of difficulties. This probably accounts for the scarcity of works dealing with the use of organized media as catalysts in terpenoid reactions. The available papers mainly concern isomerization of terpenoids and their interactions with lower alcohols.

Caryophyllene **(1)** is one of the most widespread and accessible sesquiterpene. Due to the presence of a strained e-substituted double bond in the 9-membered ring and trans-fusion of the cyclononane and cyclobutane fragments, this compound is one of the most interesting objects of study among medium-cycle polyenes. The rearrangements of caryophyllene are being actively discussed in the literature [1], [2]. Regretfully, few examples of additive reactions of caryophyllene have been reported in the literature because of its extremely high lability.

Previously, for reactions of the natural terpene camphene, we succeeded in selecting solvents promoting either phenol or ether formation [3]. It appeared that the tendencies found for these reactions may occasionally be extended to reactions of sesquiterpene **(1)** [4]. We have performed [4] alkylations of phenol **(2)**, o-cresol **(3)**, 2,6-dimethylphenol **(4)**, 3,5-dimethylphenol **(5)**, hydroquinone **(6)**, resorcinol **(7)**, pyrocatechol **(8)**, and eugenol **(9)** by caryophyllene **(1)**, forming terpenylphenols and terpenylphenyl ethers with predominantly caryolane structures.

Interest in these products may be motivated to some extent by the fact that analogous compounds isolated from Magnolia ovobata [5] affect neuron differentiation and chemotaxis and can act as neuron growth and choline acetyltransferase activity promoters. The compound with a caryolane framework exhibits the greatest activity. Thus most products obtained in this work are potential analogs of biologically active compounds and may be functionalized further. All reactions found in this study proceed at room temperature and in minimal amounts of solvent. The selectivity and the product yield may be increased and the reaction time may be reduced by varying the catalyst and reaction conditions. The synthetic procedures are therefore promising for further scaling and wide application in laboratory practice.

Of special interest among the products is compound **(10)** – the product of alkylation of resorcinol **(7)**, by caryophyllene **(1)**. It may be obtained with a high yield and selectivity on various solid acid catalysts [4]. The presence of two activating hydroxy substituents in the aromatic ring ensures further transformations of this compound. For biological tests (effects on the central nervous system), we have synthesized the disodium salt (11) (by adding of equimolar quantity of MeONa in absolute MeOH to (10), $LD_{50} > 1000$ mg/kg intraperitoneally) of phenol (10), possessing much better solubility in water compared to phenol (10).

Biological test on (11) was fulfilled on white mice weighing 20-22 g and held in standard vivarium conditions. The substance was injected intraperitoneally in a dose of 15 mg/kg 1hour before model reproduction. When introduced with a hypnotic chloral hydrate injected analogously in a dose of 300 mg/kg, salt (11) made an 100% rising of sleep time (Table 1).

Table 1 Interaction of (11) with Chloral Hydrate

Control (chloral hydrate)		(11)	
Fall-asleep time (min)	Sleep time (min)	Fall-asleep time (min)	Sleep time (min)
8.0±0.5	48.0±1.5	7.6±0.8	96.0±2.5

A test for exploratory activity (horizontal and vertical tests) showed that the compound (11) itself has no effect on animals but enhances the sedative effect on the tranquillizer seduxen injected intraperitoneally in a dose of 2 mg/kg (Table 2).

Table 2 Effect of (11) on the Exploratory Activity

Intact control		Control (seduxen)		(11)	
Horizontal test (No. of acts)	Vertical test (No. of acts)	Horizontal test (No. of acts)	Vertical test (No. of acts)	Horizontal test (No. of acts)	Vertical test (No. of acts)
12.0±0.8	18.0±1.5	4.2±0.2	8.1±1.5	0.7±0.2	1.8±0.5

At the same time, compound (11) was tested for myorelaxant activity. Experiments were performed on a rotating rod. The effect was estimated based on percent animals slipping off from the rotating rod within 2 min.

It was found that the compound **(11)** itself has no myorelaxant activity, but enhances the myorelaxant activity of seduxen injected intraperitoneally in a dose of 2 mg/kg 30 min after the injection of the target compound **(Table 3)**.

Table 3 Percent of Mice Slipping off from a Rotating Rod within 2 min.

Intact control	(11), 15 mg/kg	Seduxene, 2 mg/kg	(11), 15 mg/kg+seduxen
0	0	25.0	100

Thus one can assume that the disodium salt of **(10)** is capable of prolonging the sedative effect of tranquillizers. We continue to study the antioxidative activity of compound **(11)**. In case of positive results, salt **(11)** is expected to be useful as a perspective drug promoter.

REFERENCES

[1] A.V. Tkachev Chemistry of caryophyllene and related compounds. Chem.Nat.Compd.(Engl.Transl.) 23:393-412, 1987.
[2] I.G. Collado, J.R. Hanson, A.J. Macias-Sanchez Recent advances in the chemistry of caryophyllene Natur. Prod. Repts. 15:187-204, 1998.
[3] V.V. Fomenko, D.V. Korchagina, N.F. Salakhutdinov, I.Yu. Bagryanskaya, Yu.V. Gatilov, K.G. Ione and V.A. Barkhash Alkylation of phenol and some its derivatives on wide-pore β-zeolite. Zh. Org. Khim. 36:564-576, 2000.
[4] V.V. Fomenko, D.V. Korchagina, N.F. Salakhutdinov, V.A. Barkhash Alkylation of Phenols by Caryophyllene on Acid Aluminosilicate Catalysts Helvetica Chimica Acta. 84:3477-3487, 2001.
[5] Y. Fukuyama, Y. Otoshi, K. Miyoshi, K. Nakamura, M. Kodama, M.Megasawa, T. Hasegawa, H. Okazaki, M.Sugawara. Tetrahedron 48:377-392, 1992.

Index